POLICE ADMINISTRATION

An Introduction

Second Edition

POLICE ADMINISTRATION
An Introduction

Alfred R. Stone
Texas Department of Public Safety (Ret.)

Stuart M. DeLuca

Prentice Hall
Upper Saddle River, NJ 07458

Library of Congress Cataloging-in-Publication Data

Stone, Alfred R.
    Police administration : an introduction / by Alfred R. Stone and
Stuart M. DeLuca. -- 2nd ed.
        p.   cm.
    Includes bibliographical references and index.
    ISBN 0-13-681602-9
    1. Police administration--United States.   I. DeLuca, Stuart M.
    II. Title.
HV7935.S76 1994
351.74--dc20   93-14128
                                                                                            CIP

Editorial/production supervision, interior design: *Lynne Breitfeller*
Electronic composition: *Lynne Breitfeller/Julie Boddorf*
Cover design: *Wanda España*
Director of production and manufacturing: *David Riccardi*
Production coordinator: *Ilene Sanford*
Managing editor: *Mary Carnis*
Acquisitions editor: *Robin Baliszewski*
Editorial assistant: *Rose Mary Florio*

© 1994, 1985 by Prentice-Hall, Inc.
A Pearson Education Company
Upper Saddle River, NJ 07458

Printed in the United States of America

10   9   8   7   6   5   4

ISBN 0-13-681602-9

Prentice-Hall International (UK) Limited,London
Prentice-Hall of Australia Pty. Limited, Sydney
Prentice-Hall Canada Inc., Toronto
Prentice-Hall Hispanoamericana, S.A., Mexico
Prentice-Hall of India Private Limited, New Delhi
Prentice-Hall of Japan, Inc., Tokyo
Pearson Education Asia Pte. Ltd., Singapore
Editora Prentice-Hall do Brasil, Ltda., Rio de Janeiro

CONTENTS

Part Four Management of Police Operations

Appendices

Appendices

PREFACE

American law enforcement has never faced greater challenges than it does today.

On the other hand, American law enforcement has *always* faced enormous challenges. Many of those challenges are inherent in the concept of policing in a democratic, individualistic, multicultural society.

In an ideal world, people would never set out to prey on their neighbors. No one would commit acts of desperation due to mental illness, addiction to alcohol or drugs, or unrelieved poverty. People would care enough about one another, and about others' opinion of themselves, that committing criminal acts would be unthinkable.

This is not an ideal world.

For many years after the revolutionary war, American citizens steadfastly refused to establish any sort of permanent law enforcement institution. They believed that a self-governing, freedom-loving people did not need policing. Eventually, this idealistic vision of America gave way to a more realistic approach. The ideals, however, did not entirely disappear.

Once a police force is established, it must be given the authority and power to carry out its assigned task of protecting the public from the predatory, the desperate, and the disreputable. Americans, because of bitter past experience and because of the ideal of a self-governing nation, grant both authority and power grudgingly, hedged with restrictions and limitations. As a people, we do not entirely trust our guardians. Policing becomes the profession that everybody needs but nobody wants.

It is in the context of those realities, ideals, and challenges that we offer this introduction to police administration.

The practice of managing a law enforcement agency is fraught with serious difficulties. One segment of the community demands more, and more effective, police services to take criminals off the streets. Another segment of the community insists on paring public expenditures to the bone regardless of the consequences. Yet another segment of the community calls for greater sensitivity and compassion on the part of the police. When the beleaguered administrator looks within the agency for support, he or she finds subordinates who are sometimes poorly motivated, undisciplined, and clamoring for more pay and benefits.

The administrator's job is to choose the best available men and women for police service, train them, equip them, and guide them to perform their essential functions in a manner consistent with the public's expectations and the demands of the law—and to do so with the utmost efficiency and effectiveness.

The administrator has two major tools to carry out this task: the allocation of resources and the determination of policies. We address the use of these tools throughout the text.

"Allocating resources" means deciding what will be spent and for what purposes. Assigning personnel to particular jobs and scheduling their work hours is one form of resource allocation, just as much as buying equipment and supplies. In a broad sense, resource allocation determines what is to be done and by whom.

"Determining policies" means deciding *how* things are to be done, as well as what and by whom. For reasons we will explain in the text, we caution administrators not to overuse this tool: Too many unnecessary or unenforceable policies can confuse everyone and compromise the agency's effectiveness. Nevertheless, a very large part of the administrator's job will be to decide questions of policy.

Throughout the text, we have tried to present a range of alternative policies that represent both conventional or traditional American police practices and the more unconventional or innovative approaches that have been developed. We have tried to present the merits and drawbacks of each alternative in as objective and balanced a manner as possible. However, where it has seemed to us that a particular policy alternative is clearly superior or preferable to others, we have not hesitated to say so, and we have tried to give the reasoning behind our opinion.

In examining alternative policies and in offering our own recommendations, we have applied a set of criteria that we consider reasonable and necessary. These criteria are

- Every policy must promote, and not endanger, the safety, property, and liberty of all citizens equally.
- Every policy must be consistent with the requirements of the Constitution of the United States and the laws of the jurisdiction in which the policy applies.
- Every policy must promote, and not endanger, the safety and welfare of police officers.
- Every policy must contribute to the effectiveness of the police in its primary task of enforcing the law.

- Every policy must encourage the community's confidence in the police officers as fair, effective, and efficient public servants.

These criteria are not listed in priority order; a completely satisfactory policy must meet *all* of the criteria. Unfortunately, not all of the policies we discuss, nor all that we may recommend, satisfy the criteria perfectly. There is much room for improvement in every aspect of policing.

We anticipate that this text will be used in undergraduate courses in both two-year and baccalaureate colleges, with both preservice students (those who have not yet begun a career in law enforcement) and inservice students (those who are seeking to expand their work skills and career opportunities), as well as in police training academies. To meet the needs of this broad range of readers, we have made a concerted effort to establish a balanced presentation of the theory of police administration and the actual practices of police administrators.

Likewise, we have sought to take into account the fact that most police agencies in the United States are very small (ten or fewer employees), and that the needs and problems of the police administrator in a small agency are different from those in a medium-sized or large agency.

The text is divided into four parts. Part One is an introduction to basic concepts and theories of management, derived largely from business administration practices and applied to the specific requirements of law enforcement. Part Two represents both theoretical and practical approaches to the management of buildings, equipment, and other material resources required by a police agency.

Part Three covers the theory and practice of managing police personnel including recruitment, selection, training, promotion, leadership, and discipline.

Part Four treats the basic principles of organization and administration of police patrol, traffic law enforcement, criminal investigation, and other essential law enforcement functions. The final chapter of Part Four presents a discussion of the relationship between the police and various segments of the community, a relationship that is too often prickly and rife with potential misunderstandings on all sides.

Each chapter concludes with a brief review quiz, designed to check the reader's recall and comprehension of the main points in the chapter. The answers to the review questions are contained in Appendix A in the back of the book.

Throughout the text, we have referred to a mythical police agency, the City Police Department, when we have needed an example to illustrate some point. The *Instructor's Manual* that accompanies this text contains a fairly elaborate description of this fictitious, hypothetical agency and the community it serves, along with some "in-basket" simulation exercises that may be used to stimulate students' thinking and discussion.

Our development of this text has been enormously facilitated by the help that we have gratefully received from the following:

Chief Elizabeth Watson and Public Information Officers Gail Phillips and Charles Peters, Austin (Texas) Police Department; Major Kenneth Maurer and Sgt. Robert Shoun, Tucson (Ariz.) Police Department; Chief Travis L. Lynch, Macon (Ga.) Police Department; Chief Ernie Gallaher, Pendleton (Ore.) Police

Department; Chief Chris Wiggins, Durango (Colo.) Police Department; Chief Bill R. Myers, Birmingham (Ala.) Police Department; Chief Guy Meeks and the Planning and Research Division, Mesa (Ariz.) Police Department; Chief Alan E. Gollihue and Capt. Bill Facenda, Portsmouth (Va.) Police Department; Chief Byron R. Rookstad, Cheyenne (Wyo.) Police Department; Sheriff Charles Sharpe, Cumberland County, Maine; Chief W. W. Perrett and Public Information Officer Wallace I. Mason, Tacoma (Wash.) Police Department; Chief C. E. Swindall and Sgt. Mike S. Ward, Montgomery (Ala.) Police Department; Sheriff John Moran, Director of Planning Karen Layne, and Lt. Frank Barker, Las Vegas Metropolitan Police Department; former Chief Daryl F. Gates, Los Angeles Police Department; Deputy Superintendent Dennis E. Nowicki, Chicago Police Department; Captain R. Hampton, Fort Smith (Ark.) Police Department; Chief Thomas Vannoy, Temple (Tex.) Police Department; and Trooper Ray Finstad, Texas Department of Public Safety.

Also, Marcia Werner and Marty Lewis of Quality Composing, Austin, created the typography for several of the in-basket exercise materials in the *Instructor's Manual*; and Sue Bryant of the Texas Department of Transportation provided several photographs and documents.

We are grateful also to those who read and commented on the manuscript for this book in various drafts. For the first edition, they were Andrew P. Dantschisch, Robert E. Snow, Charles D. Hale, Robert B. Tegarden, Howard Abadinsky, Erik Beckman, B. W. Adams, Jr., and Wilson E. Speir. For this revised edition, they were Bruce L. Berg, John J. Sloan, Lois A. Wims, Thomas Dempsey, Richard R. Becker, Ronald G. Iacovetta, and Thomas McAninch, all of whom offered many valuable suggestions and criticisms.

BRIEF HISTORY OF POLICE AGENCIES

Any organization, regardless of its purpose or size, must have a certain minimum amount of structure to carry out its task successfully. It does not matter whether the task is simple or complicated, or whether the organization consists of two people or twenty thousand, except that the more complicated the task and the more people who must cooperate to get the job done, the more elaborate the organization is likely to be.

Consider, for example, a simple, routine task that you can accomplish entirely by yourself: preparing a meal. First you must decide what to eat, and then you must gather the various foods you will need. Next you must decide how to prepare the food—whether to boil, fry, bake, or otherwise process it, or simply to eat it in its present state.

Your choice of the manner of preparation also may influence the resources you will need. If you decide to fry some meat, for example, you will need a frying pan, perhaps some cooking oil, and a stove or other source of concentrated heat. Then you must actually perform the work of preparing the meal. If several steps are involved, you must coordinate them so that all the foods are properly cooked and everything is ready at the same time.

In this simple example, you will have performed each of the major functions involved in any large, complex organization, such as a business corporation, government bureaucracy, or the kind of organization we will discuss in this book, a police agency. These major functions can be defined as follows:

1. *Goal definition and policy setting.* The choice of what foods to prepare and how to prepare them.

2. *Resource acquisition and allocation.* The gathering of the foods you have chosen and the decision to use such resources as cooking oil, the stove, and so on.
3. *Task assignment.* The decision to perform some steps on the stove and others on the table or counter, and the decisions made as to the order in which various steps must be taken.
4. *Task performance.* The actual work of preparing the meal.
5. *Evaluation and modification.* Checking to see that the various steps are being carried out properly and making adjustments to correct any errors or deficiencies.

Routine, everyday household tasks are so much a part of our commonplace experience that they are performed largely out of habit, without thinking in terms of a series of specialized functions. But when several people must work together to carry out a task that is too big or complicated for one person to do alone, such an analysis is necessary. It is the only way to ensure that each person knows what to do and how to do it, and that each person has the necessary resources to carry out his or her tasks.

When people work together in more or less permanent cooperative organizations, these same functions must be performed regularly. In theory, each member of an organization could perform every function independently of the others, but centuries of human experience have demonstrated that greater efficiency (accomplishing the most work with the smallest possible expenditure of resources including human effort) can be obtained if the various functions are divided among the people in the organization. Such specialization permits each individual to do what he or she does best, using native talents, training, and experience to best advantage.

However, there are many different ways that the five basic functions listed earlier can be divided among a group of people. The study of the different ways to divide and distribute these functions is the science of management.

For example, one person could act as a solitary leader, carrying out all of the functions except task performance. The rest of the group could be assigned to perform the various tasks but have no role in making decisions. This is not an unusual arrangement in small groups with limited, well-defined tasks. A basketball team consists of five players (who perform the "tasks" of playing the game), and a single coach or manager (who performs all other functions).

The most common arrangement in any reasonably large and complicated organization involving a number of people over a long period is the structure that has evolved in business corporations, which we call the *corporate model,* although actually it can be traced back to earlier military, governmental, and religious organizations. In the corporate model, the five basic functions are allocated as follows:

- Administration (sometimes called executive level or upper-level management).

1. Goal definition and policy setting.
2. Resource acquisition and allocation.

- Midlevel management.
    3. Task assignment (with some responsibility for evaluation and modification).
- Supervisors.
    4. Evaluation and modification (with some responsibility for task assignment).
- Operating personnel.
    5. Task performance.

This four-level structure is a greatly simplified representation of what actually occurs in any corporate organization. In some organizations, portions of the responsibility for a given function may be assigned to two or more different levels of the structure. Midlevel managers and supervisors also have the additional function of serving as communication channels. They transmit the goals and policies of the administrative level to the operating level so that the latter will know what to do and how to do it. Simultaneously, the managers and supervisors transmit information about the work that has been accomplished from the operational level to the administrators, so that they will know how well their goals are being achieved, how well policies are being applied, and what additional resources may be needed.

In very large, complex organizations, this communications function can so overwhelm the five basic functions that everyone is busy shuttling information back and forth, and little real work is ever done. However, if the communications function is neglected, the administrators will not know the results of their goals and policies, and the operational personnel will be confused and uncertain about what they are supposed to do.

In modern police agencies, we can see the corporate model at work. The administrative level is occupied by the chief of police or agency head, who may have several deputies or assistants to help carry out the administrative functions. The middle-management level is occupied by the heads of the various major elements of the agency, often called bureaus or divisions; these personnel usually hold the rank of major or captain.

The supervisory role is carried out by lieutenants and sergeants, each of whom has a specific responsibility to supervise a given type of work or group of operating personnel. Finally, the operational personnel are the patrol officers, sometimes sergeants (if they have no supervisory duties), and clerks, secretaries, laboratory technicians, jailers, and so on.

In Chapter 2, we discuss in more detail the different responsibilities and functions of each level of the corporate model, and we describe some variations that are found in police agencies. For now, our point is that this basic structure is nearly universal, not only in police agencies, but in many other permanent organizations: supermarkets, universities, executive departments of government, and multinational corporations, for example.

The primary value of this model is that it gives us a tool to analyze many types of organizations and to understand how an organization is functioning or, perhaps, why it is not functioning well. The corporate model was not discovered as a fact of nature, nor was it invented overnight. It has evolved over many centuries in response to changes in human needs and the social environment. Police agencies have developed their own distinctive versions of the corporate model only in the past century and, in many cases, the process of development has been painful. It is still going on, and probably will continue to do so for the foreseeable future.

Before we can begin to analyze the management of modern police agencies, we need to gain a better understanding of current practices by reviewing a little history.

EVOLUTION OF THE AMERICAN POLICE SERVICE

We do not intend to provide in this chapter a thorough study of the history of American law enforcement. Rather, this section will serve as a review for most students, and as an overview with an emphasis on the administration of police services.

Just as our laws are derived from British legal traditions, our concept of criminal law and of the police service is based on British history and culture.[1] In most of the world, there is no clear distinction between the police function and the military function. The police operate as merely a local or specialized branch of the military. But in medieval England, for a variety of historical reasons, the idea arose of a community peace-keeping function entirely separate from the military function.

After the Norman invasion of England in A.D. 1066, many villages had a *shire reeve*, or head man, appointed by the local noble to maintain order and settle disputes. The Normans, under their king, William the Conqueror, reinforced and broadened the shire reeve's authority. He was not only the local police officer but also something like a modern justice of the peace.

Later, as a growing population crowded into the cities, a *bailiff* was appointed to keep track of the citizens and to discourage outsiders from causing trouble, and a *sergeans* was appointed to guard the city's gates. In rural villages, the local noble's army enforced the laws under the direction of the shire reeve. Whenever the troops went out on patrol, the stable keeper, or *constable*, was left in charge.

Each of these ancient positions had limited law enforcement responsibilities that were gradually expanded and refined over the centuries. The duty of the shire reeve became the basis for the modern *sheriff*, a position that no longer exists in England but that was adopted in this country. The bailiff became attached to the local courts as a keeper of the peace, jailer, or *marshal*. The role of the sergeans was assumed by watchmen while the title was absorbed into the military in its modern form, sergeant. The constable continued to function as an important law enforcement officer in small villages.

When British colonies were established in North America, they were governed primarily by the military. After the American Revolution, bitter memories of military oppression left the public mistrustful of any sort of permanent law enforcement agency. This distrust is clearly reflected in the Bill of Rights, five of whose ten major sections (the Fourth, Fifth, Sixth, Eighth, and Tenth Amendments to the Constitution) were designed specifically to limit the police powers of the federal government. The Fourteenth Amendment, adopted in 1868, extended those limitations to the states by guaranteeing due process rights to all defendants in criminal proceedings.

For many years after the Revolution, the police function simply did not exist. When a crime was committed, the victim's family or neighbors would attempt to find the criminal and bring him or her to court. In some of the larger cities, merchants organized a *night watch*, either rotating guard duty among themselves or hiring night watchmen. After the War of 1812, the practice of hiring a night watch became quite common in the larger cities.[2] There was no central administration of the watchmen, who were often transients looking for a few days' (or nights') easy work. They had no police authority. On the other hand, if they shot or beat a presumed thief, no one objected.

Both day and night watches were established by the city governments in New York, Boston, Philadelphia, and a few other large cities about 1820. There still was no central administration. However, by the mid-1820s, true police agencies began to develop. In the southern cities—Baltimore, Richmond, Atlanta, and several others—troops of police were organized mostly to capture runaway slaves.[3]

In the northern cities, police troops were established to deal with the frequent riots that had begun to occur as a result of massive immigration and the rapid concentration of unskilled labor in the new factories.[4]

In either case, the police did not function as individual patrol officers. They were not uniformed, they carried weapons only while they were under the immediate supervision of their commanders, and their principal function was to maintain order by brute force. Apprehending violators of the law was simply not part of their job. A few cities, notably Philadelphia and Washington, D.C., established separate detective agencies in the 1840s to carry out the law enforcement function.[5]

By 1850, the urban police function had begun to undergo several crucial changes. An officer was required to wear a uniform and was allowed to carry weapons.[6] More important, in a few cities the police began to operate in patrols, taking over the function of the old day and night watches. For the first time, the police were given the duty of preventing crime by their mere physical presence wherever crimes were thought likely to occur.[7]

The use of police patrols necessarily meant the decentralization of the police. Previously, the police troops remained at a central headquarters until they were called out to chase a runaway slave or to suppress a riot. This kind of centralization was impractical if patrol officers were to cover an entire city. Instead, several subheadquarters were established, one in each district or *precinct* of the city.

Patrol officers reported to their precinct headquarters at the beginning and end of each day's work, or watch. In some cities, reserve troops were also stationed in each precinct headquarters. An officer's working day might include eight hours on patrol plus another six or eight hours at precinct headquarters, on reserve.

At the end of the Civil War, this decentralized structure was the standard pattern in all urban police agencies. In smaller towns, the police function might be carried out by a town marshal or constable (the former usually appointed, the latter usually elected).[8] In rural areas, law enforcement was the responsibility of the sheriff, an elected official. There were no state or federal police agencies, although the Treasury Department formed a Secret Service during the Civil War to suppress counterfeiting.

The urban police agencies usually were organized under a police chief, appointed by the city's elected officials, or under a police commission, a supervisory board elected by the public. In theory, the chief or the commission held authority over the entire agency. In fact, however, the precinct captains held all effective power. Orders and policies issued by the chief or commission could be, and often were, simply ignored by the precinct captains. The captains hired, promoted, and fired their own personnel, assigned them to their beats, and provided all supervision (see Figure 1.1).

Figure 1.1. By the end of the nineteenth century, the "cop on the beat" was an integral part of many American urban communities. (Courtesy of the Bettmann Archive.)

During the 1870s, police work was a very low-status occupation. Officers were poorly paid and given no training; there were no pensions, paid holidays, or other benefits. The work was tedious and sometimes dangerous, and the hours were incredibly long. The public generally held the police in low esteem. Consequently, the job was attractive only to men who could find no other work, and that usually meant large numbers of recent immigrants. For most of them, policing was a temporary occupation until they could find something better. For others, it was a relatively undemanding, but not very promising, way to earn a meager living.

The precinct captains themselves came under the influence of local politicians. Political parties in the larger cities were organized geographically into wards that had boundaries similar to those of the precincts. Each ward had its *ward boss*, a functionary whose main job was to ensure that the party's candidates won every local election.

One way to reward loyal party followers was to secure them, or their unemployable relatives, government jobs, and the most plentiful jobs were in the police force. Thus, the precinct captains relied on the ward bosses to supply them with manpower. If the party in power lost an election, the new administration typically fired the entire police force and installed its own followers.[9]

Even where such extremes of corruption were not the case, political control of the police and other public employees, the so-called *spoils system*, undermined public confidence in government at all levels. Several unsuccessful attempts were made in Congress during the late 1870s and early 1880s to implement some sort of reform. When the reforms finally came, they were instigated by an unlikely source: President Chester A. Arthur, who previously had been the administrator of the spoils system for the Republican party in New York. Arthur strongly supported the *Civil Service Reform Act*, also known as the Pendleton Act. Congress passed the act in January 1883 and Arthur immediately signed it.

The Pendleton Act served not only to reform the federal employment system but provided a model for similar efforts at the state level. The essential feature of the act was the requirement that all public jobs be filled through a competitive process, based on written examinations. In theory, this process ensured that the best-qualified person would be appointed to each position, and that all equally qualified persons had a fair opportunity to gain public employment.

During this same period, America underwent one of its periodic waves of moral crusading. Church leaders and idealistic reformers were determined to improve the social conditions in the cities, which were, to be sure, much in need of improvement. To achieve their objectives, the reformers pressured state lawmakers and local governments to prohibit what the reformers considered immoral social behavior. Hundreds of laws were passed to forbid such vices as gambling, prostitution, public drunkenness, conducting business on Sundays, vagrancy, loitering, selling liquor to minors, and so forth. The police were charged with the responsibility of enforcing all these new laws.

But the police were completely incompetent to handle such a major expansion of their duties. They had had little enough success in dealing with serious crimes. At best, the police were able to prevent some property crimes and public disorders merely by their presence on the streets. Major crimes against persons, such as murders, usually were solved only if the perpetrator's identity was already known. Such crimes as rape, assault, and kidnapping were simply regarded as "not police problems" but as private matters to be settled by the victims' families.

Before long, the police found a way to accommodate their new duties. Moral offenses were not a serious problem in the rural areas and upper-class neighborhoods, where social pressure was a far more effective force than the law (as it is to this day). In the lower-class and working-class neighborhoods where police patrols were concentrated, and where the ward bosses had their greatest power, the public was not enthusiastic about the moralistic laws anyway.[10] Their own life-style and ethnic cultures were much more tolerant of so-called vices than was the strict Protestant culture of the upper-middle and upper classes. Thus the new vice laws created an opportunity for the police. Instead of arresting prostitutes, gamblers, and other vice offenders, the police readily accepted bribes from them. In effect, the police wound up "licensing" crime rather than suppressing it. Offenders (as defined by the laws) were allowed to operate as long as they paid tribute to the police and avoided flagrant scandal. An offender who crossed the police, of course, could expect to be arrested at once.

This system fit well with the prevailing political structure. The ward bosses decided which criminals would be allowed to operate, depending on their party loyalties, and distributed bribes to the police, judges, and any other officials whose tolerance was required. By the mid-1870s, many police officers, from patrolmen to precinct captains, and even a few police chiefs, received two or three times their annual salaries in bribes.[11]

First Reform Effort

The church leaders and other reformers did not fail to notice the extent of police corruption that their moralistic laws had produced. Their response, beginning in the 1880s, was to demand improvements in the police force.

At first, they tried to institute modest reforms such as centralized selection of police recruits and promotions on the basis of competitive testing.[12] However, these measures were easily thwarted; the ward bosses and precinct captains simply gave favored candidates a copy of the test, or otherwise helped them to cheat. In some cities, patrolmen's jobs were obtained not by competitive examination but by paying a fixed fee to the precinct captain.[13]

The reform crusaders stepped up their efforts, calling on state legislatures to deal with the urban corruption. In New York, the legislature formed an investigative committee chaired by State Senator Clarence Lexow in 1894. The committee compiled massive evidence of police abuses, corruption, and crime. However, no specific remedies emerged from the legislature.

At about the same time, a handful of reform-minded individuals appeared within the ranks of police agencies. Allan Pinkerton, who is best remembered as the founder of the Pinkerton Detective Agency, served briefly as chief of the

Chicago Police Department and tried to establish the idea that policing should be an honest occupation for talented individuals of unblemished integrity—in short, that it should be a profession. Pinkerton's ideas were not very successful in Chicago, but his concept of professionalism was attractive to other reform-minded police administrators.[14]

Several hundred police administrators joined forces in the National Police Chiefs Union in 1893, which became the International Association of Chiefs of Police (IACP) in 1901. This organization served as the forum and training ground for the slow but steady growth of the idea of police professionalism.[15]

Although different reformers held different ideas about police professionalism, the most widely shared definition included the following elements:

1. Law enforcement is an honorable occupation that serves a vital need in society; without effective policing, a society is doomed to anarchy.
2. Police officers must be carefully selected persons of great integrity, talent, and trained ability; to attract and retain such people, salaries and other benefits must be sufficiently high, and applicants must be evaluated on an objective, competitive basis.
3. The primary responsibility of the police is to prevent crime through techniques such as patrol; when prevention fails, the police have an obligation to enforce the law by apprehending the offender, using any scientific or other aids or methods available.
4. Corruption or graft within the police ranks is intolerable.
5. All citizens must be treated with respect and due regard for their constitutional rights; there must be no hint of favoritism or discrimination by the police; and anyone who commits a crime, however minor, must be subject to the penalties provided by law.

This concept of professionalism was very appealing to the upper-middle- and upper-class citizens who supported the era of moral reform. New laws were passed, increasing salaries and benefits for the police, requiring competitive testing for new recruits and for promotions, and placing greater administrative powers in the hands of police chiefs and commissions.

The most difficult task, however, was to remove the police from the control of the ward bosses. One after another, the reform-minded police chiefs faced immovable obstacles, and either resigned in frustration or were fired for ineffectiveness.[16]

The power of the ward bosses could not be swept away merely by issuing orders. The precinct captains and their favored subordinates were not at all inclined to accept the reforms that could cost them their supplementary incomes and privileges, or even their jobs. For their own protection, they formed "associations" that were not quite unions, but that served most of the same functions as unions.

Among the reform-minded police chiefs or commissioners whose efforts met with all-out resistance and eventual failure were Theodore Roosevelt, who was briefly commissioner of the New York City Police Department, and August K. Vollmer, who served at various times as chief of the Los Angeles, Berkeley, and Oakland police departments before withdrawing to the sanctuary of a professorship at the University of California at Berkeley.

However, Vollmer did not abandon the cause. As professor of criminology, he taught a generation of young men and women, many of whom became leaders in the long struggle for police reform.[17]

Although police reform was unpopular among the rank-and-file police officers, it was spectacularly popular with the general public. Even among the immigrants, ethnic minorities, and working-class people who generally provided the base of support for the corrupt political machines, the idea of a well-trained, competent, nonpartisan police force was most appealing. Thus the reform campaign, despite its many setbacks, continued through the end of the nineteenth century and the beginning of the twentieth century.

Outside of the major cities, criminal law enforcement remained largely in the hands of county sheriffs, constables, and village marshals. These officials were generally elected by the public for two-year terms or were appointed by local officials, and thus served only as long as the appointing officials did. Their success in winning and holding office had little to do with police professionalism, and, in fact, they were not generally effective in carrying out their law enforcement duties.

At the federal level, the desire for more effective criminal law enforcement led to the establishment of the Bureau of Investigation in the Justice Department in 1908. At first the bureau was given no real police authority. Its job was merely to conduct investigations for the Justice Department's lawyers. Soon the agency was dominated by partisan politicians who subverted it into a corrupt secret police agency.

In 1924, President Calvin Coolidge appointed a young Washington lawyer, J. Edgar Hoover, to take over and reorganize the discredited bureau. Hoover drew heavily on the support of reform-minded, politically influential citizens in transforming the bureau into a crime-fighting agency that relied on the latest scientific aids to investigation and a carefully maintained reputation for integrity. The name also was changed to the Federal Bureau of Investigation (FBI).

In 1935, Hoover established the National Police Academy, to train FBI agents and to provide inservice training, seminars, and short courses for state and local police officers.[18] In effect, Hoover represented the FBI as an exemplary law enforcement agency. Unfortunately, getting local agencies to follow the FBI's example proved difficult.

SECOND GENERATION OF REFORM

At the end of the 1920s, the status of the urban police in America was not much better than it had been at the turn of the century. However, the reformers had accomplished more than they may have realized. They had trained a whole new generation of young men who became the reform leaders of the 1930s and beyond. Among them were Lewis J. Valentine, Fred Kohler, Orlando W. Wilson, Herbert T. Jenkins, Eliot Ness, Virgil W. Peterson, Quinn Tamm, Franklin M. Kreml, William H. Parker, Clarence M. Kelly, Patrick V. Murphy, and A. C. Germann, some of whom are still active in police administration or in teaching police science.[19]

The young reformers, mostly college graduates, rose through the police ranks quickly or were appointed directly to upper-level administrative positions. By 1935, many of them were in a position to bring about the reforms they and their teachers had advocated.

The agenda of the second-generation reformers was not very different from the concept of police professionalism that had originated before the turn of the century. The young reformers sought absolute independence and freedom from political interference, and they called for a drastic upgrading of the quality of police personnel through better salaries and benefits, objective and competitive testing, and more thorough training. They believed that the traditional precinct headquarters must be abolished. In its place they would have specialized crime-fighting units, operating from a central headquarters, using scientific methods to attack specific types of crime. They also wanted to free the police from the distracting burden of various non–law enforcement duties, such as issuing taxicab licenses or inspecting boilers, that had accumulated over the years.

Police and the Mass Media

There is a popular impression that the 1930s, the decade of the Great Depression and Prohibition, was an era of total lawlessness in which America's cities were turned into playgrounds for ruthless gangsters. In fact, the condition of the cities was never quite as desperate and dramatic as Hollywood movies and television programs have portrayed it.

The Hollywood movie studios and the New York–based radio networks (and, later, television networks) did, however, influence the development of American law enforcement in ways that have been overlooked by most observers.[20]

Crime and law enforcement have been common themes in popular entertainment for centuries, and it is hardly surprising that one of the earliest successful motion pictures was E. S. Porter's *The Great Train Robbery* (1903). However, movies about crime did not become dominant until the late 1920s and early 1930s.

During the depression and throughout World War II, Hollywood produced an enormous number of murder mysteries, gangster movies, and other films in which criminals and the police were central figures. The radio networks, too, relied heavily on crime dramas and series about detectives and police heroes to attract an audience.

When television broadcasting began, after World War II, the new medium quickly adopted all of the genres of both radio and the movies. In 1950, the first year for which television ratings are available, three of the twenty most popular television series were crime shows: "Martin Kane, Private Eye"; "Man Against Crime"; and "Big Town."[21]

In 1952, "Dragnet," created two years earlier as a radio series by Jack Webb, premiered on the NBC television network. By the end of the year, it was the fourth most popular series, rising to the second most popular in the 1953–54 season and remaining on the air until September 1959.[22]

Webb's series was especially important because it purported to show actual crime cases, drawn from the files of the Los Angeles Police Department, and realistic, detailed studies of the procedures used by the police to identify and capture suspects.[23] Of course, it was no accident that only *successful* investigations were depicted!

In fact, the series was not much less fictional than any other television show. The heroes had unlimited time to concentrate on a single case, witnesses were almost always cooperative (if not always very clear-minded), and perpetrators were almost always disreputable characters. Each episode ended with a "mug shot" of the accused while an announcer informed the audience that the accused had been convicted and sentenced.

What made "Dragnet" important was, first, the extraordinary level of cooperation that Webb achieved with the Los Angeles police; second, the impact that the series had in creating a favorable image of, and high expectations for, the police.

When the series began, most police chiefs were understandably wary of any involvement with the entertainment media. Over the years, Hollywood movies conventionally portrayed the police as corrupt, brutal, and generally incompetent. Even the "good cops," the heroes of some movies, were portrayed as mavericks who must battle "the system" as much as the "bad guys."

Webb, however, persuaded the Los Angeles police chief at the time, William H. Parker, that his approach would be different. Parker ordered his staff to cooperate and to make files available to Webb and his writers, provided that reasonable precautions were taken to protect confidentiality. As the announcer reminded viewers at the end of each program, "Only the names were changed to protect the innocent." In return, Parker and his department got, in effect, a weekly half-hour advertisement, distributed nationwide over the most popular entertainment medium in the country.

What viewers saw in "Dragnet," aside from the entertaining "who-dun-it" stories, was a portrayal of the police as diligent, rational, soft-spoken, incorruptible, and unswervingly dedicated men (and, though rarely, women) who, despite the casual cynicism learned through hard experience, maintained a fierce loyalty to the public good and the American Way of Life.

The popularity of Webb's vision of American law enforcement spawned any number of imitators including his own later series, "Adam-12," which concentrated on patrol officers rather than detectives. Such recent series as "Hill Street Blues" and "Law & Order" are clearly descendants of "Dragnet," which itself is being recreated in a series for a national cable network.

By making superheroes of police officers, and by depicting police work as highly professionalized, Webb and his imitators have created a public expectation that *all* police agencies will be equally competent, dedicated, and successful. The public cannot be expected to know exactly where the dividing line falls between the factual and the fictitious in these "realistic" shows.

Police Reforms before World War II

The combination of Prohibition (1919–33) and the Great Depression (1929–39) gave organized crime a golden opportunity to reach new pinnacles of wealth and corruption. At the same time, public attention was focused on crime as never before.

One result was a rapid increase in public support for more effective law enforcement. The reformers capitalized on this shift in public attitudes, forcing politicians to adopt large parts of the reform agenda and insulating the police from political influence. The reforms might have proceeded even more rapidly, but economic conditions limited the resources available; as the specter of war in Europe loomed after 1935, the public grew distracted.

The reformers were fairly successful in upgrading the quality of police personnel through better recruitment, selection, and training methods. Unfortunately, these measures had an unanticipated consequence. The "objective" standards and tests inadvertently favored the dominant ethnic group at the expense of African Americans, Southern Europeans, and Hispanics. The descendants of earlier immigrant families, especially the Irish and Germans, gained a virtual monopoly on policing and were able to maintain their grip through such devices as civil service procedures and pension benefits that strongly favored those who remained in law enforcement for twenty or thirty years.

Ultimately, the idea of police professionalism was co-opted by the union-like police associations, such as the Patrolmen's Benevolent Association and the Fraternal Order of Police, and was used to protect the police from truly objective outside scrutiny.

The use of specialized crime-fighting units to centralize police authority also backfired. The first such specialized units were vice squads, answerable only to the police chief or commissioner and given citywide authority to combat vice. In many cities, the vice squads adopted the same crime-licensing tactics of the old precincts. Several reformist leaders were badly embarrassed, and hounded out of office, when their highly touted vice squads proved to be hopelessly corrupt.[24]

Nevertheless, the use of specialized squads continued and became a central feature of police agency organization. Nearly every category of crime was represented by a special unit: homicide, sexual assault, burglary, robbery, theft, fraud, and so on. The larger the agency, the more specialized the units became.

As specific crime-fighting duties were taken away from the general patrol force, few new responsibilities were added. By far the most important exception is the responsibility for traffic law enforcement, which was given to the patrol force in the 1920s and became an overwhelming burden in the 1930s. Other than traffic enforcement, the patrol force had little to do except to patrol their beats, presumably to prevent crime.

By 1940, the increasingly widespread use of motorized patrols and two-way radios, both of which were supposed to make patrol officers more responsive and effective, actually reduced their effectiveness. Previously, patrol officers on foot were expected to be continually alert to any suspicious activities or signs of crime and were supposed to be familiar with the people in the neighborhoods they patrolled. But the patrol officers cruising in their squad cars were assigned a much larger territory, had limited contact with the people in their patrol area, and had little opportunity to observe potentially suspicious activities.

Instead, motorized patrol officers were reduced to receiving complaints relayed by radio, interviewing complainants, and passing along reports to the detectives. In general, patrol officers were neither expected nor encouraged to do anything more, such as investigate the crimes that occurred in their patrol areas.

Another of the reformers' goals, the elimination of the precinct stations, was rarely achieved. There is a persistent public notion that the local police station contributes to a neighborhood's safety. The precinct captains and their supporters were able to muster public sentiment in favor of keeping most of the precinct stations open. However, the decline of the ward bosses' power and the increasing availability of modern communications such as the telephone gave the central administrators considerably better control over the precinct captains.

Another goal, the elimination of non–law enforcement duties, met with only marginal success. In many cities, the police had been assigned dozens of non–law enforcement tasks, and some of the earlier reformers had brought these burdens on themselves by arguing that the police had a larger role in society than merely catching criminals.

Some of the reformers had sought to act directly on social problems that are associated with crime, such as unemployment, poor housing, alcoholism, and broken families. A few police agencies went so far as to provide temporary housing for unemployed workers, to distribute food and clothing to the poor, and even to operate job-placement agencies.

In the 1930s and afterward, the second-generation reformers argued that these activities interfered with the real function of the police. Unfortunately, for every non–law enforcement duty the police were able to turn over to some other agency, or drop altogether, they seemed to gain at least two more.

Traffic control, which is only peripherally a matter of criminal law enforcement, has become firmly fixed in the minds of the public and of the police themselves as a major police responsibility. Juvenile delinquency prevention programs, the organization of marching bands and parade drill units, licensing of pets and bicycles, licensing and inspection of taxicabs and their drivers, and dozens of other miscellaneous duties are still borne by many police agencies.

Another aim of the reformers, to instill a sense of military pride and discipline in the patrol force, proved to be futile. Even when military-style ranks were adopted and recruits were trained in close-order drill and other militaristic rituals, the essentially nonmilitary nature of the American police undermined the reformers' efforts. To this day, many police authorities regard police service as paramilitary, but in fact there is only the slightest similarity between modern police service and the military.

Postwar Developments

When World War II ended, many thousands of young men came home to find that there were not enough civilian jobs to go around. However, their military experience was welcomed by the police agencies, which were undergoing rapid expansion to catch up to the population growth.

The influx of new recruits, used to military-style discipline and imbued with a strong sense of patriotism, greatly strengthened the power of the reform-minded administrators. The recruits, after all, bore little loyalty for the old political machines and their corruption. The young officers often responded enthusiastically to the call for professionalism and dedication to public service.

Thus the reform movement gained momentum during the postwar years. The concept of police professionalism gradually trickled down from the colleges and urban police departments to the smaller agencies. Journal articles and books by several of the reform leaders set out the doctrines of police professionalism, and their ideas were adopted as standards for police practice.

There were still glaring setbacks. The police in Chicago, New York City, Detroit, San Francisco, Houston, New Orleans, Pittsburgh, and Los Angeles all suffered the embarrassment of scandal from time to time.

Still, by the early 1960s most Americans regarded their local police departments more favorably and held their police officers in higher esteem than ever before. Graft, corruption, and the abuse of power were considered exceptions to the rule, the "rotten apple in the barrel," as Los Angeles Chief William Parker liked to put it.[25]

For the young men who entered police service just after World War II, the 1950s represented a golden age of policing, a period of domestic tranquillity during which it was easy to be proud of one's badge and uniform.

But that peace was soon to be shattered. In the southern states, the African American civil rights movement strained the existing social order to the bursting point. The police were caught between the demand for order (which was commonly interpreted to mean keeping African Americans in their segregated place) and the demand for law, as defined by government functionaries and judges in far-off Washington.

The cause of professional law enforcement was not helped by news reports, sometimes televised, of police officers using cattle prods, attack dogs, and fire hoses against demonstrators.

By the mid-1960s, demonstrations were widespread in the South and the North, East and West, on college campuses and in ghetto streets, for civil rights and against the war in Viet Nam. Most of the demonstrations were peaceful from beginning to end, and the police in most cases performed admirably under trying circumstances. But the exceptions, the demonstrations that turned into riots and the police who responded violently to provocation, were shocking and bewildering to the general public.

The bloody decade brought into question the entire idea of police professionalism. To some observers it seemed that the police had gained status by aligning themselves with an oppressive, elite "establishment" whose sole purpose was to protect its privileges at the expense of African Americans and young people. At a time when the cry for social reform was ringing in the air, the police found themselves in the position of guardians of the existing social order, whether or not they personally supported it (which, in fact, they most often did).

From the police viewpoint, the civil rights and antiwar protestors were radical troublemakers intent on subverting social order. The protestors' deliberate use of provocative rhetoric and occasional terrorist violence reinforced the fears of the police that they were, indeed, the last line of defense before Armageddon.

Presidents Johnson and Nixon appointed one high-level commission after another to examine the root causes of the riots and disorders. One after another, the commissions concluded that the police were more a part of the problem than a part of the solution. Rank-and-file officers felt betrayed, defensive, and increasingly embittered. The society they had sworn to protect seemed to be turning against them.

The leaders of the police reform movement were bewildered. The demand for community control of the police, heard in most of the larger cities, sounded dangerously like a reversion to the days of corruption and local political interference. The demand for safeguards against police brutality and abuses of police power—demands that were echoed by the Supreme Court in a series of decisions, most notably the *Gideon*, *Escobedo*, and *Miranda* decisions—seemed to hamstring the police, preventing them from discharging their most vital function, the war against crime. The demand for relaxed entry standards, to enable more African Americans and other minorities to enter the police service, seemed to undermine the century-long struggle to upgrade the quality of police personnel.

The most crucial gains made in some eight decades of the struggle for reform seemed to be swept aside. Even the champions of the second-generation reformers, among them Patrick Murphy, O. W. Wilson, Clarence Kelly, J. Edgar Hoover, William Parker, and others, were ultimately ignored, discredited, or hustled into the limbo of retirement.

At the same time, the public realization that the supercops of "Dragnet" and "Adam-12" were, at best, exceptions and not the rule, that there were serious and chronic problems in American law enforcement, created a new opportunity for reform.

President Nixon called for a resurgence of "law and order," a new war against "crime in the streets." In 1968, Congress passed the Omnibus Crime Control and Safe Streets Act, the first major federal effort to support local law enforcement.[26] Among other things, the act created the Law Enforcement Assistance Administration (LEAA) and allotted millions of dollars to bolster local police budgets and to carry forward the drive toward professionalism. Nixon's call for law and order took on ironic overtones as his own administration was torn apart by scandal; he finally resigned in disgrace in 1974, but the renewed public interest in improving law enforcement was not diminished.

America seemed to pause to catch its breath after the turbulent 1960s and the Watergate scandals. During the second half of the 1970s and throughout the 1980s, the police experienced no further cataclysmic changes either in the society they attempted to serve or in their own ranks.

The more radical demands for community control of the police, and other measures, quickly subsided. The police learned to live with the Supreme Court decisions and new laws that, at first, seemed to devastate their ability to fight crime. The LEAA doled out its grants, enabling the police to modernize equipment and try out new ideas.

Not all of the new ideas were successful, but some of the innovations of the 1970s have survived and continued to evolve into conventional police practices today.

For example, several communities experimented with nonmilitary uniforms that featured sports jackets and pocket patches in place of the traditional military-style jacket and metal badge. Usually neither the public nor the police themselves approved, and the style was quickly abandoned.[27]

Other new ideas were essentially refinements of traditional practices, such as the formation of specially trained tactical squads and Special Weapons and Tactics (SWAT) units to respond to crises. A few new ideas, such as patrol-oriented investigation and team policing, required extensive changes in police organization and techniques. These ideas were adopted very slowly, if at all.

In later sections of this text, we will discuss some of the innovations that were adopted and that seem to have found a permanent place in American police practice.

During the late 1970s, a floundering national economy forced cutbacks in federal expenditures for social services, and President Jimmy Carter decided that law enforcement was no longer a federal concern; the LEAA was abolished.

ACCREDITATION

One of the most controversial issues in police administration today is the nationwide accreditation program.

Accreditation is a process by which an institution is examined periodically by an impartial agency to determine whether the institution meets certain established standards. Public and private schools, colleges, hospitals and other medical facilities, and various other institutions are subject to accreditation either by private voluntary organizations or by governmental agencies.

In some cases, accreditation is required by law, or there may be financial penalties if an institution fails to maintain its accreditation. In other cases, accreditation has no legal force but is important to an institution's reputation. Being accredited means that the institution has won a "seal of approval" from its peers.

The idea of accrediting law enforcement agencies has been tossed around for some years. Finally, in 1979, four major law enforcement organizations decided to do something about it. The IACP, the National Organization of Black Law Enforcement Executives, the National Sheriffs Association, and the Police Executive Research Forum established committees to develop a set of standards that would define an "acceptable" level of performance for a modern police agency. The standards were first published in 1983 and have since been revised twice.

The accreditation procedure is managed by the Commission on Accreditation for Law Enforcement Agencies, Inc. The four police executives' organizations continue to be involved, but the development of the standards and the accreditation process is now managed by a professional staff under the supervision of the twenty-one-member commission.

The current standards manual lists nearly one thousand specific standards, covering eighty-four different areas of police operations, from defining an agency's law enforcement mission and jurisdiction to prisoner transportation, traffic law enforcement, crime prevention, and evidence handling. Some of the standards are mandatory; others are optional (or, as the manual puts it, "non-mandatory"). Some standards apply to all agencies; others apply only to agencies of a certain size. Most of the standards are phrased in terms of requiring a written policy, regulation, or order. For example,

> Standard 1.3.12. A written directive requires that only weapons and ammunition meeting agency-authorized specifications be used in the performance of duty.[28]

Accreditation is purely voluntary. No state requires its police agencies to be accredited, although it is possible that this will change in the future.

Any public law enforcement agency may apply for accreditation by submitting an application to the commission. The agency then will receive a copy of the standards manual and a handbook describing the accreditation process, which essentially involves two steps: *self-assessment* and *on-site examination*.

Self-assessment means that the agency compares its own management, operating procedures, and policies with those described in the accreditation standards. This is an exhaustive and time-consuming process that ultimately involves virtually every member of the agency's staff. Usually it requires the appointment of at least one commissioned officer, preferably with the rank of lieutenant or above, and in larger agencies a staff of assistants, to manage the self-assessment process.

The purpose of the self-assessment is to identify the areas in which the agency is deficient according to the standards for accreditation. The agency then must decide what, if anything, to do to correct the deficiencies. If the deficiencies found during self-assessment are substantial, a detailed plan for correction may be required.

Once an agency has decided that it meets the accreditation standards, the second step is taken: on-site examination. A committee of examiners from the commission visits the agency, examines its files, interviews its personnel and other people in the community, including officials of the parent government, and observes the agency's personnel in action.

The examining committee then prepares a report on its findings: either that the agency is substantially in compliance with the accreditation standards, or that there are substantial deficiencies that must be corrected before accreditation can be granted. Even if the agency meets most of the standards, the examining committee may recommend improvements to correct relatively minor deficiencies.

If an agency fails to meet the standards, it is given every opportunity to correct its problems and try again. The eventual hope is that every agency that seeks accreditation will obtain it. After all, the idea is to encourage police agencies to meet the standards.[29]

But critics are not so sure that voluntary accreditation really accomplishes anything worthwhile. The idea of accreditation for law enforcement agencies has drawn some criticism and skepticism.

The opponents claim that the accreditation standards are questionable and that the process is much too costly for the limited value that is produced. Some critics point out that every police agency is already subject to various sets of standards: those imposed by state law, those imposed by the local parent government, those imposed by limited resources, and those imposed by the public's expectations.[30]

Proponents answer that the standards are always subject to improvement, and that they represent a consensus among interested and qualified authorities as to what should be considered the minimum acceptable standards for a modern police agency.

It is certainly true that the accreditation process is time-consuming and expensive because of the personnel time required for the self-assessment, but the expense should be viewed as an investment in the improvement of the agency. Intensive self-examination should be a basic tool of management anyway, the proponents say.

As for the eventual rewards of accreditation, they are to be found in the increased prestige of the agency, the reassurance that accreditation gives to the public and elected officials that their police department is first rate, and the guidance that an agency receives toward achieving a high standard of performance.[31]

Because accreditation is entirely voluntary, and probably will remain so for many years, relatively few law enforcement agencies have undergone the process. Some found that the process was so expensive that they could not complete it. Those that have completed the process generally consider it worthwhile. However, as you might expect, agencies that volunteer for accreditation are most likely to obtain it with few major problems.

Perhaps the most interesting and valuable part of the entire accreditation system is the establishment of a set of standards that presumably defines what is to be regarded as a minimally acceptable level of performance. Because these standards are continuously under review and subject to revision, it would be pointless even to try to summarize them in this text. However, any police administrator or student of police administration who wants to know what is generally considered proper police practice should study the commission's standards manual.

ROAD TO PROFESSIONALISM

In retrospect, the American concept of the police function can be seen as an outgrowth of the English traditions of criminal law enforcement, but adapted to the unique circumstances of our own history.

The American police service began in the early nineteenth century as a crime prevention service, to which various law enforcement and non–law enforcement functions were added over the years. The system broke down in the late nineteenth century because of political interference and corruption. Since then, the police service has undergone a century-long struggle to achieve respectability, competence, independence, and effectiveness. As historian Samuel Walker remarks,

> The history of the police…becomes intelligible when we think of it in terms of its halting progress along the scale of professionalism.[32]

The concept of professionalism has been the organizing principle throughout the successive periods of police reform. Yet, as Robert Fogelson points out,

> The [professional] model was so vague, its meaning susceptible to so many different interpretations, that it united law enforcement figures who were divided on just about everything else.[33]

Thus, professionalism has been the rallying cry, but it could not serve as a yardstick for reform efforts because it is a yardstick of unknown length. There never has been a clear sense of what a truly professional police force would look like, or of what the police must do to become fully professional. When examined too closely, the whole idea of professionalism seems to disappear; it is a thin tissue of generalized goals and contradictory notions.

One of the basic principles of the corporate model that we mentioned earlier is that an organization's goals must be completely clear. Otherwise, neither the administrator nor the operating personnel can know whether the goals are being achieved. Therefore, before we proceed any further to discuss how a police agency should be managed, we should stop to consider just what it is supposed to do.

POLICE ROLE TODAY

If you were asked to define the function of a law enforcement agency, you probably would answer that its function is to enforce the laws. But it is not that simple. Two more questions would quickly follow: Which laws? What do you mean by "enforce"?

Concept of Law Enforcement

A crime is a violation of a law. Any legislative body—the Congress, state legislature, city council, or, in some states, county commission—may enact any law, so long as it does not conflict with the Constitution. When a law is enacted, the legislature may designate any violation of it as a crime. There are perhaps twenty thousand or thirty thousand different laws at the federal, state, and local levels in the United States that define crimes.

Any sort of human behavior can be designated as a crime. The deliberate taking of a human life (murder) is a crime everywhere in this country, except under certain circumstances defined in the law. The taking of property without the owner's permission may or may not be a crime, depending on the nature of the property, the circumstances under which it was taken, and the relationship, if any, between the taker and the owner. At one time, it was a crime in Boston to bathe without a doctor's prescription.[34]

There are also many laws that do not concern crimes. They involve relationships between private individuals, or the internal workings of government. In general, the police have no responsibility nor authority regarding noncriminal laws.

Therefore, our first answer (not necessarily our final answer) to the question *Which laws?* might be, "All criminal laws that have been enacted by the appropriate legislature and that are valid under the Constitution."

The enforcement of the laws, including criminal laws, is the duty of every citizen, not just of the police. The concept of democratic self-government is based on the belief that people who participate in making the laws, directly or indirectly, will be more inclined to obey them and to report violations to the proper authorities.

Unfortunately, it is difficult and sometimes dangerous for the average citizen to play an active part in law enforcement. Therefore, the duty of enforcing certain laws is partly assigned to a particular group of people, the police, and they are given the authority to act on behalf of the entire citizenry. However, it is not always very clear exactly what authority the police may have and what methods they are allowed to use to discharge that authority.

"Enforcing the law" means, at a minimum, that all citizens are required to obey all laws, and that anyone who fails to obey a law is subject to some form of punishment.

Our society takes rather elaborate precautions to ensure that punishment is given only to those who in fact have violated a law. Thus, a person accused of a violation is entitled to answer the accusation in a public forum, and the burden of proving the accusation is placed on the accuser. In all criminal cases, the state (that is, the government) acts as the accuser.

The first step in enforcing the law is to identify those persons who are thought to have committed violations. Next, evidence must be gathered to prove that the accused person is guilty. The accused and the evidence must be brought to the public forum, or court, and the accused must have every opportunity to answer the accusation. Either a judge trained in the law or a jury of citizens must decide that the accused is actually guilty. Then and only then may punishment be assessed.

Only some of the steps in this process have been assigned to the police: the identification of violators, the gathering of evidence, and the presentation of the accused in court. The rest of the process is left to other parts of the government: an attorney, representing the state, and the court. If the violator is found guilty and punishment is set, another part of government, the corrections system, is responsible for seeing that the punishment is carried out.

Thus, our initial answer to the question *What do you mean by enforce?* is, to identify and apprehend accused violators, to gather evidence of the violation, and to present both the accused and the evidence to a court.

However, as our review of police history suggests, this is not the only function of the police. In fact, "enforcing the law" might not be the most important function of a law enforcement agency.

Concept of Police Service

Law enforcement, as we have just defined it, is a complex task that requires the expenditure of a good deal of time and energy that might be saved if people could be prevented from committing crimes in the first place. If we knew in advance which individuals were going to commit crimes, or where crimes were going to be committed, it would be a simple matter to restrain the prospective violators.

The first duty assigned to the police and their predecessors, the watchmen, was to prevent violators from committing offenses by physically restraining them.

Unfortunately, in today's very complex society, it is nearly impossible to identify prospective criminals in advance, and it is not practical to station guards at every place where crimes might occur. Prospective criminals cannot be identified in advance because they are not clearly different from other citizens; any citizen might commit a crime.

Only a tiny percentage of the total population (about 1 or 2 percent) commit serious crimes, and even these individuals devote only a small part of their lives to criminal behavior.[35] It simply makes no sense to impose severe restraints on all citizens merely to keep a handful from committing crimes.

Those who do commit crimes may go to some effort to keep from being identified as violators. The presence of a guard at a particular place may prevent a crime from occurring there but does not necessarily prevent a crime from occurring somewhere else. Thus, the most that can be accomplished by stationing guards is to protect the places that would be most tempting to prospective criminals, such as places where valuable goods are stored.

As you can see, then, perfect crime prevention is not possible. Prospective violators cannot be distinguished from ordinary citizens, and all places where crimes might occur cannot be guarded.

Logically, if the police were perfectly successful in preventing crimes, there would be no need for law enforcement. Even if the police were imperfectly successful but managed to prevent almost all crimes, the enforcement function could be reduced to a minimal level.

The fact is that police crime-prevention efforts have had only a marginal effect on the incidence of crime. Greater and greater resources have had to be devoted to law enforcement. In this sense, enforcement always represents the failure of prevention.

This does not mean that the police are always at fault, although they usually get the blame. Not only is perfect prevention inherently impossible, but the police do not have complete control over the limited resources that they are able to apply to this task.

It is important to remember that law enforcement and crime prevention are two separate, distinct functions. They are often confused in the minds of the public, and even in the minds of the police. This confusion is compounded when people argue that law enforcement contributes to prevention by discouraging potential violators. The argument may be correct, but it is somewhat irrelevant. Preventing crime is preventing crime, and enforcing the law is enforcing the law. These are two separate functions that must be discharged in very different ways.

Law enforcement and crime prevention are not the only functions assigned to the police. A third function is the source of considerable argument: the maintenance of social order.

Remember that a legislature may define any sort of behavior as unlawful. Legislatures enact laws to prohibit behavior that is regarded as harmful to social order. Murder is a crime not just because it is harmful to the immediate victim, but because it disrupts the peaceful relationships that must exist in society if people are to live in reasonable comfort and safety. As far as the criminal law is concerned, the victim is not the individual who has been harmed by a violator but, instead, society as a whole.

But legislators cannot be expected to imagine, and therefore to prohibit in advance, every sort of human behavior that might be socially disruptive. "Social order" itself is a vaguely defined state. Whether it is disrupted depends not only on the actions of individuals, but on the reactions of other people.

Some behavior might be disruptive under certain circumstances but not under other circumstances. For example, if a person turns on a radio and plays rock music at high volume, is that socially disruptive? It certainly would be if the person is in a church where a religious service is going on. It might not be if the person is on a public beach.

To write a law that will prohibit social disruption, legislators must try to imagine and define exactly what circumstances will be regarded as a violation. If a law is written too broadly or is too vague, it is likely to be ignored by many people and will be enforced only at great expense in terms of the time and effort of the police and the judiciary.

During the nineteenth century, many social-order laws were written in very broad terms, and the police were expected to exercise discretion in deciding whether specific behavior was disruptive. The results were not very satisfactory. In many cases, the police took advantage of their discretionary authority to overlook violations committed by their friends and abused their authority to harass and intimidate anyone they happened to dislike.

Over the years, such vague, general laws have either been stricken from the books or overturned by the courts. According to the courts, the Constitution forbids vague laws on the grounds that every citizen has a right to know in advance whether certain behavior is illegal. If a law is so vague that a reasonable person cannot tell whether the behavior is permitted or prohibited, the law is not valid.

Furthermore, the police have no authority to interfere with the freedom of anyone unless the person has violated a specific law. Thus, the police have no authority to "maintain social order" except when a criminal act has occurred. The use of police power to intimidate or discourage people whose behavior may be obnoxious or objectionable to others, but not specifically illegal, has been drastically reduced.[36]

Unfortunately, the public tends to take a broader view of what the police should be able to do. Several studies have shown that most calls for police assistance do not concern crimes at all.[37] People call the police whenever there is any sort of emergency or any disruption, even a very minor one, in their lives. Clearly, the public expects the police to "preserve social order" whether they have the legal authority to do so or not.[38]

Furthermore, most police agencies encourage their officers to fulfill these expectations as far as possible. Consequently, when a police officer responds to a citizen's complaint, the officer is inclined to do whatever he or she can do to resolve the problem, regardless of whether it involves a violation of a law.

If it appears that no crime has occurred or that proving a violation would be unreasonably difficult, the officer does whatever can be done to resolve the problem to everyone's satisfaction. Often this means giving advice to the complainant; sometimes it means physically separating people who are arguing. Occasionally it means using verbal intimidation techniques to discourage the person causing the disturbance. In short, police officers often find themselves acting as negotiators between citizens whose disputes may have very little or nothing at all to do with criminal laws.[39]

These non–law enforcement services occupy a large part of police officers' time and energy, and they can produce enormous problems for individual officers and for their agencies. Strictly speaking, a police officer has no legal right to interfere in a private dispute between citizens where no violation of the criminal law has occurred. Yet the citizens clearly expect the police to do so.

If an officer responds to this expectation and the situation gets out of control, especially if someone is hurt, it is very likely that the officer and the agency will find themselves in serious trouble. Police officers are individually liable for any criminal act they commit, just as any other citizen would be, and there are some criminal laws that apply specifically to police officers: abuse of authority, false arrest, malicious prosecution, and others. Further, the officer and the agency can be liable for civil damages if there is misuse of police authority.[40]

Thus, the proper functions of the police are limited to only two.

- Preventing crimes by observing prospective violators when they can be identified and by guarding places where crimes are most likely to occur.
- Enforcing the criminal laws by identifying and apprehending accused violators, and by gathering evidence concerning violations.

The call for police professionalism is essentially a call for the police to discharge these two functions as effectively as possible, and to use an appropriate degree of discretion to provide non–law enforcement services, as the public expects, without going beyond their proper authority.[41]

CONCEPT OF POLICE PROFESSIONALISM

Sociologists have concluded that a true profession has certain characteristics that distinguish it from other occupations. Those characteristics are as follows.[42]

Social Grant of Authority.

The members of a profession have a special privilege or authority, sometimes granted to them exclusively, to practice their occupation. Although society has not granted the police an exclusive privilege to enforce the laws—in general, all citizens have the right to arrest a suspected offender, to register a complaint, or to gather evidence of a violation—only the police have an obligation to perform these tasks on behalf of the whole society. Furthermore, the police do have special privileges to carry sidearms, to use certain other weapons, and to use certain other "tools of the trade" such as nightsticks, handcuffs, sirens, and so on.

Autonomy of Practice.

The members of a profession exercise a high degree of discretion. What they do and how they do it are not subject to the criticism of people outside of their own occupational group. One of the major goals of the police reform movement has been to secure autonomy for the police agency but to reduce the individual police officer's discretion. The idea is that the police agency must be free of political interference and must be the sole judge, subject to the law and court rulings, of proper law enforcement techniques and procedures. However, an individual officer is supposed to have limited discretion in following the procedures, policies, and regulations of the agency.

In fact, most police officers work alone or in the company of no more than one or two of their peers. The nature of police work unavoidably requires individual officers to exercise a great deal of discretion.

An officer must decide whether to issue a ticket to a speeding driver, or merely to warn the driver, or to ignore the violation altogether. Even when the offense is more serious, officers must decide whether to believe the testimony of a victim or witness, what kinds of evidence to seek, when to arrest a suspected offender, and what charges to file.

Each of these decisions can have profound consequences for the people involved, but ordinarily an officer's decisions are not reviewed and criticized by anyone outside of the police agency until an arrest has been made and charges are filed. Even then, the only thorough review is performed by the state's prosecuting attorney, who usually tends to be sympathetic to the police.

Systematic Body of Knowledge.

The members of a profession are expected to have special competence because they have acquired a systematic body of knowledge including theories that explain the functioning of their work and rare skills that can be acquired only by special training.

The existence of books like this one and college classes in police science imply that there is a body of knowledge about police matters. There are also theories concerning the causes and effects of crime. However, these theories do not seem to be as systematic and conclusive as, for instance, theories about the causes and effects of infectious diseases.

As for special training, most police agencies in the United States require new officers to have no more than a high school diploma and two or three weeks of basic training. Nevertheless, the number and complexity of the special skills a police officer is expected to possess have risen in recent years, and so have the amount and quality of the training given to police officers.

Self-regulation.

The members of a profession are expected to govern themselves by controlling the entry of newcomers into their occupation and by enforcing a self-imposed code of ethics or discipline.

Police agencies do, in fact, regulate the entry of new members through selective recruitment, screening, and testing procedures. Most agencies also have some sort of internal inspection system to investigate complaints of improper conduct, and most agencies administer discipline when an officer is found to have behaved improperly.

Unfortunately, among police officers themselves, self-regulation does not appear to be very effective. Several studies have found that individual police officers are reluctant to criticize one another even when there are blatant improprieties. Instead, the tendency among police officers is to protect one another.[43] This tendency undermines the self-regulatory system and sustains a widespread belief among the public that the police are unable and unwilling to discipline themselves.

Often there are demands for civilian review boards or comparable outside agencies to exercise disciplinary authority over the police. In the few cases in which such a mechanism has been established, it has been sabotaged by the relentless hostility and uncooperative attitude of the police, or the "independent" agency has been captured by partisan political groups.[44]

Service Orientation.

Although the members of a profession are expected to be paid for their work like anyone else, they are also expected to be concerned primarily with performing an essential service to society. When necessary, they should willingly set aside their personal interests to provide that service.

From the very beginning of the police reform movement, policing has been represented as a noble endeavor and a vital service to the public. Unfortunately, studies of police behavior and attitudes suggest that this ideal is not universally accepted by the police themselves.

Lower-ranking officers frequently regard policing as "just a job, a way of earning a living," and not a very attractive job at that.[45] Some researchers have found that police officers regard the public with suspicion and hostility. Fogelson has described the common attitude among police officers as "occupational paranoia."[46]

According to some sociologists, there is no clear-cut distinction between "professional" and "nonprofessional" occupations but, instead, a scale or continuum between "fully professional" and "completely nonprofessional."

Overall, the police appear to rank somewhere in the middle, about the same as public school teachers or accountants.

This may be discouraging to the advocates of professionalism, but actually it represents a great deal of progress over the past century. Furthermore, there is no inherent reason that dedicated police officers, given adequate leadership, cannot raise their occupation a few more notches up the scale of professionalism.

REVIEW

1. During the 1850s, the urban police began to take over the crime-prevention function of the

   (a) Day and night watches.

   (b) Military.

   (c) Sheriffs.

   (d) Detectives.

   (e) State militia.

2. True or False: The Bill of Rights reflects the English tradition of relying on the military for law enforcement.

3. Accreditation is

   (a) The federal law requiring peace officers to receive various types of training.

   (b) A voluntary program in which a law enforcement agency's operations and policies are compared with a set of standards.

   (c) The requirement that each local police department operate according to established constitutional and legal principles.

   (d) A systematic means of determining the need for law enforcement services.

   (e) One of the most important benefits of professionalism.

4. The history of the police in America can be seen as a history of

   (a) Incompetence and corruption.

   (b) The long struggle for professionalism.

   (c) Patriotic service and courage.

   (d) Constant conflict between the police and the public.

   (e) Dismal failure in controlling crime.

5. Which of the following is *not* characteristic of a profession?

   (a) Social grant of authority.

   (b) Autonomy of practice.

   (c) Systematic body of knowledge.

   (d) High level of compensation.

   (e) Self-regulation.

   (f) Service orientation.

NOTES

[1]James F. Richardson, *Urban Police in the United States*. Port Washington, N.Y.: Kennikat Press, 1974, pp. 3–4.

[2]Richardson, *Urban Police*, pp. 19–22.

[3]Samuel Walker, *A Critical History of Police Reform*. Lexington, Mass.: Lexington Books, 1977, p. 4.

[4]Walker, *Police Reform*, p. 4.

[5]Robert Fogelson, *Big City Police*. Cambridge, Mass.: Harvard University Press, 1977, p. 15.

[6]Walker, *Police Reform*, pp. 12–13.

[7]Richardson, *Urban Police*, pp. 23–24.

[8]Richardson, *Urban Police*, p. 31.

[9]Fogelson, *Big City Police*, pp. 18–19.

[10]Richardson, *Urban Police*, p. 29.

[11]Fogelson, *Big City Police*, p. 21.

[12]Richardson, *Urban Police*, pp. 62–68.

[13]Fogelson, *Big City Police*, p. 26.

[14]Richardson, *Urban Police*, pp. 69–72.

[15]Walker, *Police Reform*, p. 48.

[16]Richardson, *Urban Police*, pp. 70–76.

[17]Charles B. Saunders, Jr., *Upgrading the American Police*. Washington, D.C.: The Brookings Institution, 1970, p. 16.

[18]Ginny Field, "The FBI Academy," in *FBI Law Enforcement Bulletin*, vol. 61, no. 1, January 1992, pp. 16–17.

[19]Fogelson, *Big City Police*, p. 141.

[20]Geoffrey P. Alpert and Roger C. Dunham, *Policing Urban America*, 2nd ed. Prospect Heights, Ill.: Waveland Press, 1992, pp. 1–3.

[21]Craig T. and Peter G. Norback, eds., *TV Guide Almanac*. New York: Ballantine, 1980, p. 546.

[22]Ibid., pp. 548–51.

[23]Les Brown, ed., *New York Times Encyclopedia of Television*. New York: Times Books, 1977, p. 465.

[24]Richardson, *Urban Police*, pp. 86–91.

[25]Not everyone agreed with the "rotten apple" idea. In 1972, an investigative commission appointed by New York City Mayor John V. Lindsay to seek the causes of corruption in that city's police force reported, "The 'rotten-apple' doctrine has in many ways been a basic obstacle to meaningful reform." Knapp Commission (Whitman Knapp, chairman), *Report on Police Corruption*. New York: George Braziller, Inc., 1972, p. 7.

[26]Saunders, *Upgrading the American Police*, pp. 2–3.

[27]D. F. Gunderson, "Credibility and the Police Uniform," in *Journal of Police Science and Administration*, vol. 15, no. 3, September 1987, p. 193.

[28]*Standards for Law Enforcement Agencies*. Fairfax, Va.: Commission on Accreditation, 1988, pp. 1–3.

[29]Raymond E. Arthur, Jr., "Accreditation: A Small Department's Experience," in *FBI Law Enforcement Bulletin*, vol. 59, no. 8, August 1990, pp. 1–5.

[30]W. E. Eastman, "National Accreditation: A Costly, Unneeded Make-Work Scheme," in James F. Fyfe, ed., *Police Management Today*. Washington, D.C.: International City Management Association, 1989, pp. 49–54.

[31]Jack Pearson, "National Accreditation: A Valuable Management Tool," in Fyfe, *Police Management Today*, pp. 45–48.

[32]Walker, *Police Reform*, p. x.

[33]Fogelson, *Big City Police*, p. 157.

[34]Barbara Seulig, *You Can't Eat Peanuts in Church and Other Little-Known Laws*. Garden City, N.Y.: Doubleday, 1975, p. 13.

[35]Anthony V. Bouza, *The Police Mystique*. New York: Plenum Press, 1990, pp. 100–108.

[36]David H. Bayley, "The Limits of Police Reform," in Bayley, ed., *Police and Society*. Beverly Hills, Calif.: Sage Publications, 1977, pp. 223–25.

[37]Lee W. Potts, *Responsible Police Administration: Issues and Approaches*. University, Ala.: University of Alabama Press, 1983, p. 2.

[38]Richard E. Sykes, "A Regulatory Theory of Policing," in Bayley, *Police and Society*, pp. 237–42.

[39]Potts, *Responsible Police Administration*, pp. 8–10.

[40]Potts, *Responsible Police Administration*, pp. 16–17, 40–63.

[41]Mary Jeanette Hageman, *Police-Community Relations*. Beverly Hills, Calif.: Sage Publications, 1985, pp. 22–24.

[42]Fredric D. Wolinsky, *The Sociology of Health*. Boston: Little, Brown, 1980, pp. 255–68.

[43]Harlan Hahn, "A Profile of Urban Police," in Jack Goldsmith and Sharon S. Goldsmith, eds., *The Police Community*. Pacific Palisades, Calif.: Palisades Publishers, 1974, pp. 19–21.

[44]Fogelson, *Big City Police*, p. 285.

[45]Alex Nicholas, *New York Cops Talk Back*. New York: Wiley, 1976; William Ker Muir, Jr., *Police: Streetcorner Politicians*. Chicago: University of Chicago Press, 1977, p. 45.

[46]Fogelson, *Big City Police*, pp. 238–42.

CONCEPTS OF MANAGEMENT

Before the last decades of the nineteenth century, management had never been considered as a particular discipline or body of skills and knowledge. In fact, until the industrial revolution, there were very few enterprises of any sort that involved more than a handful of people. The exceptions—government, the military, and the clergy—had each developed its own unique system of management and organization.

Industrial capitalism, which began in the seventeenth century and became the dominant economic system in the Western nations in the nineteenth century, required ever larger and more complex forms of organization and thus more skilled and knowledgeable management. At the same time, the triumph of science and technology contributed to a widespread belief that every sort of problem or human need could be resolved through the rational application of scientific principles.

Around 1880 a young factory worker, Frederick W. Taylor, observed that some of his colleagues in the machine shop produced more work with less effort than did others. This observation certainly was not original, but Taylor resolved to study the matter more closely.

He soon concluded that the more productive workers did not necessarily work harder. In fact, some of them did not seem to be working as hard as their less productive coworkers. The more productive workers had arranged their work in such a way that they lost as little time as possible in wasted movement, reaching for materials or tools, placing pieces on a machine, and so forth. Taylor found that he, too, could increase his productivity just by planning and organizing his work a little better.[1]

Based on this limited observation and experience, Taylor concluded that every job might be planned and arranged in one particular way that would be more productive than any other possible arrangement. When he explained his ideas to his superiors, he was encouraged to try them out.

First, Taylor analyzed each job, dividing it into the smallest possible steps. Then he tried various arrangements of the individual steps, placing the necessary tools and materials where they could be reached with the least possible effort exactly when they were needed. He also timed each step, using a stopwatch, and, when necessary, rearranged the work to reduce the time it would take.

Not every experiment succeeded on the first attempt, but before long Taylor was able to double and triple the productivity of every worker in the factory merely by defining and prescribing the "one best way" to perform each job.

At first, many of the workers were resentful of Taylor's methods. They were used to working in whatever way they pleased and at whatever pace they found comfortable. However, Taylor's methods proved to make their work more productive with less effort. Because many factory workers were paid on a piece rate, a set amount for each piece of work they accomplished, their increased productivity paid off immediately in larger paychecks.

Taylor began to publish his theories of *scientific management* about the turn of the century and attracted a great deal of attention. His techniques of job analysis and planning were significant in themselves, but even more important was the idea that the management of a business could be improved by the application of science and theory.[2] Taylor was the first of many management theorists.

THEORIES OF MANAGEMENT

Scientific Management Theories

Taylor's theories attracted several followers, one of whom was Frank Gilbreth. Gilbreth adopted Taylor's system of *time and motion studies* and the theory that there was one best way to perform every job. He also developed a method of depicting all the individual steps in a particular job through the use of a *workflow process chart*, a device that later proved to be indispensable in computer system design and programming.[3]

Perhaps Gilbreth's most important contribution, however, was to take Taylor's theories out of the factory and apply them to all kinds of work. Gilbreth trained a staff of efficiency experts and sent them with their stopwatches and clipboards into file rooms, offices, department stores, laboratories, and every type of work environment. He advised architects and industrial engineers on the design of production lines, office furniture and equipment, and even private homes. The modern kitchen, with its arrangement of appliances, cabinets, and countertops for the efficient storage and preparation of food, is a direct product of Gilbreth's theories.[4]

Meanwhile, other theorists tackled a whole range of management problems. C. B. Thompson developed new methods of accounting that gave managers a clearer picture of their businesses' financial status. Harrington Emerson applied Taylor's and Gilbreth's job analysis techniques to giant, complex industrial systems such as railroads. Henry L. Gantt created more sophisticated methods of recording complex work processes. By the 1920s the concept of scientific management and efficiency was firmly established in American business and had begun to have a profound influence on other kinds of enterprises including government.[5]

The early police reformers were not directly affected by the scientific management movement until after World War I, but even before then police administrators had begun to seek ways to make policing more efficient. Efficiency became a primary goal of the second-generation reformers.

The theories of scientific management were not universally admired. Critics complained that scientific management ignored the human factor in work, and that the methods advocated by Taylor and Gilbreth treated workers as if they were just another factory machine. Elton Mayo, a professor of business administration at Harvard University, was one of those critics.

Human Relations Management Theories

Mayo set out to learn whether Taylor's and Gilbreth's methods really resulted in increased productivity and whether there really was only one best way to perform any given job.

In the 1920s, Mayo was invited to assist in a series of experiments at the Hawthorne plant of the Western Electric Company in Chicago. The experiments were designed to apply the principles of scientific management by varying different aspects of the work environment, such as the lighting level and the number and duration of rest periods, to determine which measures would increase productivity the most.

To the astonishment of the experimenters, every change in the work environment produced at least a short-term increase in productivity. When light levels were increased, productivity rose; when light levels were decreased, productivity rose again! In each case, productivity soon dropped back to its previous level.

In another series of experiments, women assembly-line workers were given rest periods of varying duration and frequency. Each change in the schedule resulted in an increase in productivity. Even restoring the original schedule brought increased productivity. The experimenters were baffled.

Mayo solved the riddle of Hawthorne by interviewing the workers at great length. He discovered that they were responding as a group to the mere fact that the experiments were being conducted. The experimenters had asked the workers to cooperate, and so they did. This discovery that almost any change in the work environment, especially if it is introduced as an experiment, will produce at least a temporary increase in productivity, has since become known as the *Hawthorne effect*.[6]

According to Mayo and others who have studied this phenomenon, the explanation for the Hawthorne effect is not that the workers feel pressure to work harder because they are being watched, but instead that they feel encouraged to work harder because they are being given special attention and recognition.

Furthermore, Mayo found that the increased productivity was not the result of individual decisions made by the workers, but instead was a product of group interaction. The workers encouraged one another to accomplish more work, and, as far as possible, they helped one another.

This discovery led Mayo to examine more closely the conventional working environment in factories and offices. Once again, he found that workers as a group set their own production standards, deciding how much work they would accomplish in a given period. Slower workers were encouraged and assisted to keep up with the rest of the group, while faster workers were pressured into slowing down so that they would not show up their coworkers.

Mayo's discoveries led him to a generalization: In any working situation, the interpersonal relationships among the workers are at least as significant in determining productivity as the physical environment, work methods, or procedures dictated by scientific management. This generalization became the foundation of the *human relations theory of management*.

Mayo went on to make many other discoveries about human relations in the workplace. Taylor, Gilbreth, and the other scientific management theorists had assumed that people work only to earn money and that a larger paycheck would be a sufficient incentive to motivate workers to work harder. Mayo found that human motivation was much more complex. He showed that many workers paid on a piece rate could have earned more money by working just a little harder but did not. In part, workers voluntarily submitted to the productivity standards set by their working group. Besides that, most workers seemed to think that it was more important to function at a pace they found comfortable rather than exhausting themselves just to earn another dollar or two. The paycheck, according to Mayo, was an important motivator but not the only one.

Mayo also concluded that the worker's perception of the importance of his or her job was a significant motivating factor. Workers who believed that their jobs were unimportant, no matter how well they were paid, tended to perform poorly. People who believed that their jobs were important either to the economic success of their employer or to society as a whole tended to perform well no matter how poorly they were paid.

Furthermore, Mayo found that most people judged the value of their work according to the value put on it by their coworkers, friends, family members, and neighbors—that is, by the various social groups of which they were members.[7]

Elton Mayo's discoveries and his explanations for them had a great impact on the business world. Unfortunately, that impact was blunted by the difficulty of applying his theories. Taylor, Gilbreth, and their followers had offered concrete rules, specific procedures, and visible evidence of their methods' success. Mayo offered insights into human relationships and motivations that were not easily translated into specific actions to be taken by managers.

The human relations theories had even less impact in the world of law enforcement. The reform-minded police administrators were intent on introducing discipline, efficiency, and uniformity. If anything, they were more interested in breaking up group relationships such as the informal arrangements among police officers, precinct captains, and ward bosses that had allowed widespread corruption and incompetence to flourish.

There was an effort to convince police officers that their work was socially significant, as part of the campaign for professionalism. However, this effort to boost morale was intended mainly to reinforce the need for tight discipline and strict integrity, not to give the individual officer a greater sense of personal satisfaction.

Overall, Mayo's theories and those of his followers came to be treated as a sort of footnote to management theory. Managers sought to increase productivity by introducing more efficient work methods, offering better pay, and enforcing stricter supervision—and, oh, yes, paying some attention to human relations. Meanwhile, another group of theorists had come forward, pushing the human relations theories even further into the background.

Organizational Theories

This third group of management theorists concentrated not on the individual workers or the working environment, but on the relationships among all the elements in an organization.

Max Weber, a German sociologist, provided the most comprehensive theory of organizational management. He believed that the most efficient form of organization was the *bureaucracy*, in which the overall work of an organization is divided into sets of specific tasks that are assigned to the individual units of the organization. This theory applies to every sort of enterprise, whether private business or government. Briefly, Weber's theory of bureaucratic organization includes the following elements:

1. Each unit of an organization should be assigned one particular type of work, or set of closely related tasks, to be performed more or less continuously.
2. The units of the organization should be arranged into a hierarchy of offices, each having responsibility for a specific type of work and for the work of all its subordinate offices, and each office should have exactly the authority it needs to carry out these responsibilities.
3. Each person in an organization should be assigned to the office he or she is most competent to fill, and competence should be determined by objective means such as standardized tests or competition among applicants.
4. All of the relationships among the organizational units, and thus among the people who fill the various offices, should be regulated by a formal, uniform, and consistent set of rules and policies; each office should have authority to set the rules and policies for its subordinate offices, subject to the authority of higher offices.

5. The exact procedures to be followed for the completion of every task should be determined in advance and established by the appropriate authority.

6. All work should be reported by means of written documents, reviewed by supervising authorities as necessary, and kept in permanent files for future reference.[8]

Stated in these stark terms, Weber's theory seems harsh and authoritarian, but it was not meant to be. Weber himself thought that his theory, properly applied, would produce "precision, speed, unambiguity, knowledge of the files, continuity, discretion, unity, strict subordination, reduction of friction and of material and personal costs"[9]—in short, the kind of efficiency that a business manager can appreciate.

Weber's bureaucratic theory was made to order for the police reformers, and it was quickly adopted by the young police administrators of the 1930s. Above all, the theory promised an end to individual discretion and nonconformity. It seemed to show how to achieve a high degree of regularity in even the most complex endeavors.

Many other organization theorists followed in Weber's footsteps, especially during the 1930s and 1940s. They drew freely from military traditions, the hierarchical patterns of religious organizations and, in particular, the basic structure we defined in Chapter 1 as the corporate model. They blended these time-tested ideas with those of Weber along with the concepts of efficiency developed by Taylor, Gilbreth, and their followers, and produced what has become the standard model for nearly all governmental organizations and businesses.

In recent years these principles of organization have come under serious challenge. Weber-style bureaucratic organization has been criticized as excessively rigid, authoritarian, and unresponsive to changing human needs.[10] To some extent these criticisms are valid, but the fact remains that, for its time, Weber's theory of organization was a major step forward.

One of the most significant additions to organizational theory is the *systems approach*, developed in the 1950s and 1960s as an outgrowth of the design of computer systems. Briefly, the systems approach treats an organization not as a collection of separate elements, but as a single entity through which resources flow to produce something. This approach emphasizes the complex relationships and interactions among the various parts of an organization.

For example, a decision to manufacture a new product might be made at the highest level of administration. However, the decision affects many different elements of the organization. The purchasing department must obtain the raw materials to make the product, the production department must assign the necessary workers and machinery (or perhaps even build a new factory), the marketing department must arrange for suitable packaging and advertising, and the sales department must train the sales staff to sell the product. If all these effects are not considered when the decision is first made, the results might be disastrous.

The systems approach has been popular not only in the computer industry but also in many other industries. It has been adapted to several federal agencies, especially those that provide social welfare services where extremely complex social interactions must be considered. However, efforts to apply the systems theory to police administration have been limited and less successful.

Behavioral Theories of Management

The long-neglected ideas of Elton Mayo and his followers in the human relations school of management were revived during the 1960s and gained new influence in the 1970s as the basis for another set of management theories. These theories largely take for granted the kinds of organizational structures and work procedures advocated by the scientific management and organizational theorists of the past. Instead of seeking ways to make work more routine and efficient, the new theorists look for ways to make workers more productive by tapping their unused potential.

In 1943 psychologist Abraham Maslow published a theory of human motivation that attempted to explain virtually all behavior in terms of a relatively small set of universal human needs. According to Maslow, every person has essentially the same needs, although they may be expressed in different ways. All people must have food, drink, shelter from a harsh environment, and other basic physical elements of life. All people need to be free of pain or discomfort, and all people feel a need for physical security.

In addition to these physical needs, Maslow theorized that people also have a universal set of psychological needs: a need for friendship, affection, affiliation (the sense of belonging to a worthwhile group), self-respect, self-confidence, and the fulfillment of one's personal potential, or self-actualization.

Furthermore, according to Maslow, these needs exist in a hierarchical structure. Basic physical needs must be satisfied first because, if they are not, the person's very existence is threatened. Once these needs have been satisfied, a person seeks to achieve a degree of physical security; after that, social-emotional needs (affection, affiliation, and the like) must be satisfied. Next on the hierarchy are the needs for self-esteem and self-confidence; highest of all is the need for self-actualization.[11]

The practical significance of Maslow's work is that it seems to explain why people react differently to certain incentives.[12] A person who is hungry will be highly motivated by the prospect of earning some food, but once the hunger has been satisfied, an offer of more food would not motivate the person to perform any further work. Of course, ordinarily people are paid in money, not food, and one of the great advantages of money is that it can be used to buy many different things. Indirectly, money can even be used to buy psychologically gratifying experiences such as works of art, entertainment, or travel.

People also differ in the way their needs are expressed and in the degree to which they require satisfaction. Some people are perfectly content to accept a quantity of material goods that other people would consider completely inadequate. Many artists, for example, voluntarily maintain a rather low material standard of living to continue practicing their art, as a form of self-expression, without concern for financial rewards. In effect, they have suppressed some of their physical and social needs to satisfy their psychological needs.

Another psychologist, Douglas McGregor, built on Maslow's work in the late 1950s. McGregor pointed out that there is a fundamental conflict between the assumptions held by conventional management theorists and those held by Maslow and other behavioral theorists.

The conventional theorists have always assumed that most people do not want to work and would not work if they could avoid it. According to this assumption, which McGregor called *Theory X*, people must be forced to work and must be closely supervised to make sure that they keep working. Also, if Theory X is true, most people have no desire to hold and exercise responsibility, even for their own welfare, but would much prefer to enjoy the security of being cared for by someone else and being told what to do.

McGregor believed that there was ample evidence to support just the opposite view, which he called *Theory Y*: that work is natural to all people, that most people want to accept responsibility for themselves and for their own work once they have clear goals in mind, and that most people enjoy facing a challenge if they have a reasonable chance to meet it and if there is some reward for doing so.[13] According to McGregor, managers who rely on Theory Y should place much less emphasis on authority and accountability, the heart of Weber's bureaucratic model, and much more emphasis on helping workers to find personal satisfaction in their work.[14]

Several other behavioral theorists have contributed useful ideas to management science in the past two decades. The general theme in all cases has been to apply the knowledge gained from the behavioral sciences, especially psychology and sociology, to improve the productivity of business or government enterprises and to increase the satisfaction felt by workers.[15]

The late Robert R. Blake and Jane S. Mouton tried with some success to bring together all these theories into a single, systematic approach to management, which they called *grid organizational development*.[16] Briefly, their idea was that every organization has two dominant concerns: the concern for production (or, more broadly, for meeting the organization's overall goals) and the concern for the welfare (both physical and psychological) of the organization's members including employees.

However, organizations vary in the amount of attention that is paid to each of these concerns. Some managers are so obsessed with production that they neglect the welfare of the people in the organization. Other managers are so determined to "keep everyone happy" that production suffers.

Blake and Mouton believed that managers should use specific strategies to balance these two concerns so that they actually reinforce each other: Increased production brings greater rewards, both material and psychological, while more satisfied workers are naturally more productive.[17]

Management by Objectives

Another set of management theories begins with the idea that people are both more productive and more content when they know precisely what is expected of them and when they are working toward specific, challenging goals, especially if they have helped to set those goals. These theories, which have been developed by several theorists, are known collectively as *management by objectives* (MBO).[18]

According to the MBO theory, the most important task of management is to decide what the organization's goals will be and to define those goals in specific terms that can be understood by every worker.[19] As far as possible, the workers themselves should participate in setting goals, because people are likely to feel a stronger commitment to achieving a goal that they have set for themselves.

Once objectives have been set, everyone in the organization should be informed of their progress toward meeting them. There should be periodic assessments of progress at which time the goals might be amended to take into consideration factors that have changed since the original goal setting. Most MBO theorists agree that these feedback provisions and regular evaluations are just as important as the goal-setting process itself.[20]

Management for Quality

In the mid-1980s, a new challenge to American management theories came from an unexpected quarter. The challenge was to transform industrial organizations so that the quality of their products or services would be built in from the beginning, not merely evaluated after the fact. The source of the challenge was a country whose industrial products were once a target of American consumers' ridicule: Japan.

After World War II, Japan's economic and industrial base lay in ruins. With financial and technical help from the United States and other countries, the Japanese slowly began to rebuild. Because the country is severely lacking in most natural resources, the Japanese were forced to concentrate on developing consumer products that could be exported, thereby earning money to satisfy domestic needs.

During the early 1950s, an American management theorist, W. Edwards Deming, offered his services as a consultant to several large Japanese companies. Deming's specialty at the time was *statistical quality control*, a systematic method of sampling finished products to detect flaws.[21]

Deming's methods were adopted enthusiastically by the Japanese. Soon his seminars and lectures were attended by chief executives and managers from throughout the island nation.

Deming began to realize that simply detecting flaws in a sample of finished products was not an adequate method of maintaining, much less improving, product quality. In too many cases, he found, defects were built into the products from the very beginning. He began to revise his theories and to teach a new approach, one that began with the assumption that every element of an organization should be arranged to produce goods and services of the highest possible quality, and to improve their quality continuously. Management, Deming saw, played a crucial role in this process.[22]

But what exactly is "quality"? According to management theorist J. M. Juran, the practical definition of quality is "whatever the customer wants," or, more specifically,

- The presence of features in a product that satisfy or exceed the customer's expectations.
- The absence of defects or deficiencies.[23]

The theories of Deming, Juran, and other management-for-quality consultants are not the only explanation for the phenomenal success of Japanese industry in the 1970s and 1980s. Nevertheless, the fact is that Japanese companies accepted and adopted these ideas at a time when American companies continued to rely on the outmoded statistical methods of the 1940s. By the end of the 1970s, Japanese-made goods had driven American-made goods out of the marketplace for consumer entertainment products, and were making steady inroads on the American automobile market.

Startled American managers began in the early 1980s to take a closer look at the theories of Deming and his followers. Slowly and at first reluctantly, they studied Deming's famous "Fourteen Points," which summarizes his management philosophy and which includes such maxims as the following:

- Buy raw materials and components on the basis of quality, not just lowest price.
- Strive for constant improvement in small increments, rather than occasional large-scale improvements.
- Devote whatever resources are necessary to train and retrain workers continuously.
- Supervisors should be leaders.
- Employees should feel secure, not fearful for their jobs.
- Concrete actions are more important than slogans in motivating workers.[24]

According to Deming disciple Mary Walton, there are five essential assumptions that underlie managing for quality.

1. Decisions must be based on facts, not prejudices or habits.
2. People who do the work know the most about it, and their knowledge and ideas should be respected.
3. Teams of people who share responsibility for their work are more successful than individuals working alone.
4. Members of a work team must be trained in techniques of quality improvement—including how to work as a team.
5. Information presented in graphic form is more useful and easier to understand than written reports or tables of numbers.[25]

The management-for-quality theories of Deming and his followers were originally developed for goods-producing industries, but they have also been adapted to organizations that produce services.[26] More recently, efforts have been made to adapt Deming's ideas to public agencies.[27]

Does management-for-quality have a place in law enforcement? Yes, of course, although translating the industrial management model to the environment of a police agency is not easy. The first obstacle is the very definition of "quality." If "quality" is a matter of meeting the "customers'" expectations, who are the "customers" and what do they expect of their police service? Should the police administrator be concerned primarily with satisfying the expectations of, for example, victims of crime, taxpayers, influential citizens, or elected officials? To what extent are their interests and expectations the same, and in what important ways do they differ? Just finding and verifying answers to these questions is a major undertaking.

Nevertheless, as the concern grows for improved quality in industrial products and commercial services, there will be pressure on police administrators to improve the quality of the services they render. In some sense, of course, that has always been the chief task of administrators and managers. The advocates of Deming-style quality management are simply demanding that the question of quality be given the highest priority.

Contingency Theory of Management

Every one of the different management theories we have discussed has been put into practice by business and government managers. They have all proved successful in some cases and unsuccessful in others.

For the practical manager faced with immediate problems, this creates a dilemma: Which theory should he or she adopt?

According to advocates of the *contingency theory*, the only possible answer is that it depends.[28]

The contingency theory is based on the idea that there is no one universal best way to organize and operate a complex enterprise. Instead, managers must be responsive to the particular situation that faces them. They must choose management techniques that are appropriate to the needs of the organization itself, the people in the organization, and the outside world that the organization intends to serve.[29] The primary task of the manager is not to choose and apply slavishly any one system or theory but, instead, to integrate the needs of the individual with the needs of the organization by selecting and using any system that meets those needs.

Roy M. Roberg distinguishes between "mechanistic" and "organic" organizations. A mechanistic organization is one that is managed according to a rigid, highly authoritarian system, such as Weber's bureaucratic model. There is little room for personal preferences and the individual characteristics of the people in the organization. This organization can be successful when the organization's principal task is to carry out routine, standardized work; for example, a factory that produces products on a continuous, conveyor-belt basis.

However, organizations like police agencies whose goals are diverse, with constant interaction between the organization's members and the outside world, and with continual changes in the demands being made on the organization, are less likely to succeed with a mechanistic system. Instead, Roberg advocates an organic system that "grows out of the situation."

An organic system has as little structure as possible, it allows for and rewards individual initiative and discretion within limits that are set as far apart as possible, and it actively seeks opportunities for change and improvement in response to changing internal and external conditions.[30] Roberg believes that police agencies, if they are to be successful, must develop a more organic approach to their organizational structure and management than they have had in the past.

The challenges of contingency theory and organic management are very great. An organic system requires managers to be sensitive to the demands of the outside world and to the needs of the people in the organization, without losing sight of the organization's own ultimate goals. These managers must know a lot about many different theories and techniques of management, and about which techniques are appropriate to which circumstances.

If individual initiative and discretion are encouraged, mistakes inevitably will be made by lower-level personnel. Somehow the manager must make sure that when mistakes are made, the consequences will not be disastrous for some innocent citizen and that the person who made the mistake learns from it.

Most important, the successful manager must have a keen, clear understanding of his or her own role in the organization, and of the responsibility he or she must bear.

ROLE OF THE MANAGER

Duties of Management

In 1937, one of the scientific management theorists, Luther Gulick, attempted to define and list all the duties of a manager. He summarized them in the acronym, POSDCORB, which stands for *planning, organizing, staffing, directing, coordinating, reporting, and budgeting*.[31] Since then, many theorists have attempted to improve on Gulick's list by making minor changes or by redefining various terms. Some of the lists are so long and elaborate that few managers could ever hope to get around to all the duties they are supposed to discharge.

We will try to avoid this pitfall by simplifying the matter. Instead of listing every conceivable task a manager might undertake, we describe three general categories of managerial responsibilities.

Prospective Duties.

The first responsibility of the manager is to anticipate the future: to decide what is to be done, how it may best be done, and what resources will be needed to do it. In this category are the recognized duties of planning and budgeting, which we discuss in much greater detail in Chapters 4 and 5. Organizing is also partly a prospective duty, because it involves deciding which people and skills will be required to perform each task.

Perhaps the most important prospective duty of the manager is policy making. A manager makes policy when he or she decides how a given task or job is to be performed. Policies usually are expressed in the form of orders, rules, regulations, and directives.

Some police agencies have developed elaborate schemes for expressing policies as a hierarchy of orders, some of which (general orders) are supposed to serve as universal rules for all personnel at all times, while others (special orders) are supposed to apply only to particular situations or cases. These elaborate schemes usually break down. Some people in the organization never receive some orders, and many orders are simply ignored by most personnel.

Generally, managers are well advised to avoid making policies, in whatever form, whenever possible. Once a rule is made, it must be enforced. Otherwise, respect for all rules will decline, inconsistencies and injustices will occur, and personnel will become increasingly confused.

No manager can hope to anticipate every situation in which the operating personnel will find themselves. Thus, policies are likely to be riddled with exceptions and contradictions. The operating personnel will be forced to decide whether a particular rule made months or years ago applies to the immediate situation.

Some policies must be made to set realistic standards of performance, to give order to such routine events as the work schedule, and to ensure that all personnel abide by legal requirements in matters such as arrests, property searches, and the use of force. However, it is much better for the manager to make policies that set broad limits on the discretion of personnel rather than trying to make rules to restrict personnel to the one correct way of performing their jobs.

Current Duties.

Once the manager has decided what is to be done, how it is to be done, and what resources will be needed, it is the manager's job to acquire and organize those resources including people.

Most of these current duties are routine in nature: setting up periodic work schedules, transferring personnel from one unit to another, administering the payroll and purchasing systems, and so on. Because these duties are routine, they demand very little creativity on the part of the manager. Once the routine has been established, doing the work is relatively easy, though it may be tedious and time-consuming.

There is a danger that the manager will occupy more and more of his or her time with these routine, nondemanding chores at the expense of the other duties that are actually more important to the organization's success. Whenever possible, the manager should shift the burden of routine duties to subordinates: clerks, assistants, and so on.

Retrospective Duties.

The third part of the manager's job is to find out whether the tasks that were planned in advance have actually been performed and with what results. These "backward-looking" chores are perhaps the most often neglected part of the manager's duties.

Too many managers assume that once they have given an order, the job is done. Eventually they may discover that the order has not been followed, but by that time it is too late. The manager must constantly supervise, not in the sense of "cracking the whip" over subordinates but in the proper sense of "looking over" the work being performed to make sure that it is being done properly.

The goals and plans made previously serve as the standard of current performance. If the standards are not being met, the manager must determine whether the fault lies in the personnel (inability or unwillingness to do the work), the plans themselves (unrealistic expectations or inadequate resources), or some change in the environment that has made the plans unworkable. Adjustments and corrections then must be made, and new goals and plans must be set.

Thus the manager's retrospective duties provide the basis for the continual renewal of the prospective duties: There is a complete cycle, and ideally the organization moves steadily forward toward its ultimate objectives.

This summary of the manager's duties is not intended to define every task that a manager performs.[32] We will describe many tasks that are ordinarily assigned to management personnel in a police agency. However, in any police agency, there may be duties assigned to managerial and supervisory personnel in addition to those we will describe.

Lower-level supervisors almost always have some operational duties. For example, a sergeant may be expected to serve as a backup patrol officer in addition to supervising a patrol squad, or may have clerical duties at the police station.

Because most police agencies are chronically short of personnel, there is a natural tendency to distribute routine chores to anyone whose time does not appear to be fully occupied. Even executive-level administrators should be willing to perform some operational-level duties, if only to ensure that they have continuing, personal experience with the day-to-day reality of police work.

However, remember that the most essential duties of the manager are prospective and retrospective, planning for the future and evaluating past performance, and that it is dangerously easy to let these crucial tasks be overwhelmed with routine and trivial chores that could be performed just as well by someone else.

Leadership Function

It is not too difficult to compile lists of the various jobs that managers must perform, but it has proved to be very difficult to describe how managers should perform their work. Many management theorists agree that managers must exercise something called leadership, but there is little agreement on what leadership is, what it looks like, or how managers should exercise it.[33]

Leadership has been defined as "the ability to cause other people to do things."[34] Presumably, the "things" done are those desired by the leader but not necessarily by the people who are caused to do them. Otherwise, leadership would require nothing more than getting out of the way.

If we adopt McGregor's Theory X view of management, the definition makes good sense: a leader's (manager's) job is to influence people to do things that they otherwise would not want to do.

This does not mean that a leader must operate only by physical force and intimidation. Anthropologists George Maclay and Humphry Knipe have found that not only all human societies, but many nonhuman groups, exhibit patterns of *dominance* in which certain individuals are accepted as leaders within their own group. According to Maclay and Knipe, physical force is rarely used to achieve or to maintain dominance, because if a leader must resort to violence to maintain authority, there is constant turmoil in the group and valuable members are needlessly injured.[35]

A somewhat more subtle idea of leadership is that it involves "the ability to cause people to want to do things that the leader wants to have done." This implies that the leader psychologically manipulates the group's members, persuading them to change their own desires until the leader's desires are satisfied. People may be made to think that they are acting voluntarily. Most theories of motivation start with the premise that ordinarily people would not want to do certain things, but somehow the leader must persuade them to want what he or she wants.[36]

McGregor's Theory Y suggests that neither coercion nor manipulation may be necessary. If people actually want to do the things that need to be done, the leader's only task is to make sure that the subordinates know what is needed.

Of course, there is a little more to it than that. There may be confusion or uncertainty about what tasks are necessary or about the best way to perform the tasks. There may also be conflicts among the subordinates as to what they really want. For example, two patrol officers patrolling the same beat might have very different ideas about how they should relate to the people in the neighborhood, how they should respond to suspicious activities or complaints of crimes, and so on. The leader, therefore, must work to remove confusion and uncertainty wherever it exists and to resolve the subordinates' conflicting needs and desires.[37]

Bennis and Nanus list four requirements of an effective leader.

1. The leader must create a "compelling vision of the future," an image of the organization's ultimate goals and objectives.
2. The leader must arrange the organization to achieve that vision.
3. The leader must establish the organization's position in its environment (that is, its relationship with the rest of the world including competing organizations) to achieve the vision.
4. The leader must acquire whatever information or knowledge is needed to achieve the vision.[38]

Some theorists have taken a different approach to the subject of leadership. Instead of trying to define its nature, they have tried to discover the common characteristics and personality traits of individuals who are generally regarded as successful leaders. Unfortunately, different theorists have compiled lists of traits that, when combined, include hundreds of different "traits of leaders."[39]

More recent theorists have tried to simplify these lists by determining which traits were truly universal among leaders. According to Roberg, "In 1971, [Edwin] Ghiselli...concluded that the most important trait for successful managers is supervisory ability, followed closely by the desire for occupational achievement."[40] But if leadership is supposed to be a prerequisite of supervisory ability, and supervisory ability is a characteristic of leadership, we know nothing more than we did when we began except the assertion that successful managers want to succeed.

Some behavioral scientists have suggested that there may be no universal leadership traits, but that leaders may operate successfully in any of several different styles.[41] Robert Tannenbaum and Warren H. Schmidt have described a continuum of leadership styles, from highly authoritarian to highly democratic, according to the amount of freedom and discretion allowed to subordinates. A highly authoritarian leader simply makes decisions and issues orders. At the other extreme, a highly democratic leader makes very few decisions, but allows subordinates to do their work however they see fit. Between these extremes are several different styles.

According to Tannenbaum and Schmidt, no one style is universally effective. The most successful leaders consider several variables including their own personalities, the abilities and personalities of their subordinates, and the nature of the demands made on their organizations.[42]

Thus, a leader might function in a highly authoritarian style at one time but in a more democratic style at another time. For example, the chief of a small police department might allow the patrol officers to set their own duty schedule, requiring only that a minimum number of officers be on duty at any time. However, the same chief might decide and announce detailed regulations to govern the procedure for a drug raid, without consulting subordinates and without giving them any opportunity for discussion or debate.

Behavioral scientists also have considered the possibility that the nature of leadership and the traits or style of the leader may represent only half the picture, and perhaps not the more important half. The other half may be the nature, traits, or style of followers. The whole subject of followership has only begun to be explored.[43]

Some sociologists believe that every group chooses its own leaders with only minimal regard for the official position a person happens to hold, and that these decisions, usually made informally or even unconsciously, are based on the group's perceptions of its needs, not so much on the abilities of the leader.[44]

For example, it is not unusual for a police captain, who is supposed to hold authority over an entire bureau, to find that his or her orders are frequently ignored unless they are transmitted to the lower-ranking personnel by a respected lieutenant. In effect, the group has chosen the lieutenant as its leader and rejected the leadership of the captain.

It is also common for a lieutenant to find that he or she has almost no influence on subordinates unless every order is endorsed and confirmed by the captain. In this case, the group has rejected the lieutenant's leadership. The personal qualities and abilities of the captain and the lieutenant might be essentially the same. The difference lies in the group's perception of their superiors and of their needs.

However, we know very little about how groups choose their leaders or whether the choices generally are good or bad.

REVIEW

1. Match the following management theorists with the type of theory they helped to develop:

(a) Elton Mayo.	(i) Behavioral theory.
(b) Frank Gilbreth.	(ii) Organizational theory.
(c) Jane S. Mouton.	(iii) Scientific management theory.
(d) Douglas McGregor.	(iv) Human relations theory.
(e) Frederick W. Taylor.	
(f) Max Weber.	

2. What is the *Hawthorne effect*?

(a) The use of time and motion studies to increase efficiency.

(b) The tendency of workers to allow their social relationships to interfere with their job performance.

(c) A temporary increase in productivity that usually occurs whenever there is any change in the work environment.

(d) The decrease in productivity that usually occurs when there are unexpected changes in the work environment.

(e) The loss of motivation that occurs once a worker's material needs are satisfied.

3. True or False: According to the *contingency theory*, every manager must choose the one management system or technique that is likely to be successful no matter what changes may occur in the situation.

4. A *bureaucracy* is

(a) An organization that is divided into several subunits, arranged hierarchically, each of which is responsible for performing specific tasks.

(b) An organization that is incapable of responding to changes in external demands.

(c) A manager who is more concerned with productivity than with the personal needs and desires of subordinates.

(d) Any governmental agency, as distinguished from a business or other private organization.

(e) A German theory of human motivation.

5. According to some management theorists, the quality of goods or services should be measured by

(a) Standards established by law.

(b) The competitive forces of supply and demand.

(c) What customers expect.

(d) The lowest common denominator.

(e) Objective facts determined by scientific research.

NOTES

[1] Edwin B. Flippo and Gary M. Munsinger, *Management*, 5th ed. Boston: Allyn and Bacon, 1982, p. 10.

[2] Ernest Dale, *Management: Theory and Practice*, 3rd ed. New York: McGraw-Hill, 1973, pp. 114–17.

[3] Flippo and Munsinger, *Management*, p. 11.

[4] Dale, *Management: Theory and Practice*, pp. 117–20.

[5] Flippo and Munsinger, *Management*, p. 12.

[6] Flippo and Munsinger, *Management*, pp. 12–14.

[7] Dale, *Management: Theory and Practice*, pp. 136–43.

[8] Alan O. Bates and Joseph Julian, *Sociology: Understanding Social Behavior*. Boston: Houghton Mifflin, 1975, pp. 223–24.

[9] Max Weber, *Theory of Social and Economic Organization*, trans. A. M. Henderson and Talcott Parsons. New York: Oxford University Press, 1947, p. 333.

[10] Charles T. Goodsell, *The Case for Bureaucracy*. Chatham, N.J.: Chatham House Publishers, 1983, pp. 1–12; James Q. Wilson, *Bureaucracy*. New York: Basic Books, 1989, pp. 316–32.

[11] Abraham H. Maslow, *Motivation and Personality*. New York: Harper and Row, 1954.

[12] Flippo and Munsinger, *Management*, pp. 268–70.

[13] Douglas M. McGregor, *The Human Side of Enterprise*. New York: McGraw-Hill, 1960.

[14] Flippo and Munsinger, *Management*, pp. 270–72.

[15] Dale, *Management: Theory and Practice*, pp. 179–83.

[16] Flippo and Munsinger, *Management*, pp. 347–48.

[17] Robert R. Blake and Jane Srygley Mouton, *The New Managerial Grid*, 2nd ed. Houston: Gulf Publishing, 1978.

[18] Flippo and Munsinger, *Management*, pp. 91–99.

[19] Wilson, *Bureaucracy*, pp. 25–36.

[20] Roy R. Roberg, *Police Management and Organizational Behavior*. St. Paul: West Publishing, 1979, pp. 91–92, 98, 114.

[21] Mary Walton, *Deming Management at Work*. New York: G. P. Putnam's Sons, 1990, p. 12.

[22] Walton, *Deming Management*, p. 13.

[23] Joseph M. Juran, *Juran on Planning for Quality*. New York: Free Press, 1988, pp. 4–5.

[24] Walton, *Deming Management*, pp. 17–19.

[25] Walton, *Deming Management*, p. 21.

[26] D. Keith Denton, *Quality Service*. Houston: Gulf Publishing, 1989.

[27] Jeffrey W. Vincoli, "Total Quality Management and the Safety and Health Professional," in *Professional Safety*, vol. 36, no. 6, June 1991, pp. 27–32.

[28]Flippo and Munsinger, *Management*, pp. 21–34.

[29]Roberg, *Police Management*, pp. 43–47.

[30]Roberg, *Police Management*, pp. 208–11.

[31]Luther Gulick, "Notes on the Theory of Organization," in Luther Gulick and Lyndall Urwick, eds., *Papers on the Science of Administration*. New York: Institute of Public Administration, 1937.

[32]For a somewhat different list, see Flippo and Munsinger, *Management*, pp. 8–10.

[33]Warren G. Bennis and Burt Nanus, *Leaders: The Strategies for Taking Charge*. New York: Harper and Row, 1985, pp. 4–5.

[34]John P. Kotter, *The Leadership Factor*. New York: Free Press, 1988, p. 16.

[35]George Maclay and Humphry Knipe, *The Dominant Man*. New York: Delacorte, 1972.

[36]Dale, *Management: Theory and Practice*, pp. 459–65.

[37]Roberg, *Police Management*, pp. 160–61.

[38]Bennis and Nanus, *Leaders*, pp. 21–30.

[39]For example, Joe D. Batten, *Tough-Minded Leadership*. New York: American Management Association, 1989, p. 14; Alan Bryman, *Leadership and Organizations*. Boston: Routledge and Kegan Paul, 1986, pp. 18–35.

[40]Roberg, *Police Management*, pp. 159–60.

[41]Paul Hersey, *The Situational Leader*. New York: Warner Books, 1984, pp. 29–37, 57–70; Bryman, *Leadership and Organization*, pp. 36–125.

[42]Roberg, *Police Management*, pp. 165—69.

[43]Flippo and Munsinger, *Management*, pp. 334–37.

[44]Allan R. Cohen and David L. Bradford, *Influence Without Authority*. New York: John Wiley & Sons, 1990, pp. 1–4; Flippo and Munsinger, *Management*, pp. 221–22.

ORGANIZATION OF POLICE AGENCIES

Any activity that involves more than two or three people over any relatively long period demands some sort of organization.

Even if the people involved are determined to avoid a formal structure, such as a communal group dedicated to a belief in perfect democracy, there seems to be an inescapable tendency for one person to become dominant. Either that person's superior skill and knowledge causes the person to assume leadership, or the other members of the group prefer to rely on the dominant person's leadership.

Once there is a leader, the group has a structure. At the very least, it consists of the leader at the top and the rest of the members at the bottom.

If the group is so large or its activities are so complicated that the leader is unable to supervise everyone at once, subleaders may be given the authority and responsibility for portions of the group's overall task.[1]

It is important to understand that this tendency toward hierarchical structure appears to be universal in human societies. It is not a feature of any one ideology or culture. In fact, anthropologists say that the same kind of hierarchical structure appears in groups of social animals other than humans. Even some flocks of birds and schools of fish exhibit approximately the same structure.[2]

However, this does not mean that an organizational structure, to be successful, must be so rigidly authoritarian that the people at lower levels feel oppressed and exploited. Instead, the principles of hierarchical structure must be recognized and observed, but applied in such a way that every individual in the organization feels that his or her contribution is needed and valued.

THEORY OF HIERARCHICAL ORGANIZATION

We introduced the basic principles of organization in Chapter 2, so they do not need to be repeated here. However, some points do require further clarification.

Authority and Responsibility

The word *authority* means that a person has a right or privilege to perform some action, to require others to perform some action, or to make certain kinds of decisions. In principle, all legal authority originally rests in society as a whole. The rights and privileges held by any particular individual or group of individuals have been granted by the rest of society, and this grant must serve some need or purpose of the whole society.

Thus, the authority of government derives from the "consent of the governed," or the desire of the people in general to have a government. By the same token, the people have an inherent right to limit the authority they grant, by setting restrictions or conditions on it.

Every grant of authority carries with it an equal *responsibility*: an obligation for the person to perform the authorized duty, to observe the limits or restrictions set by the people, and ultimately to serve the public need or purpose for which authority has been granted.[3]

In accord with this principle, both the authority and the responsibility to make laws have been granted by our society to the various legislatures: the Congress, state legislatures, city councils, and, in some states, county commissions or boards of supervisors. The authority and responsibility to enforce the laws have been granted to various executive bodies: the president, the governors, the mayors or city managers, and, again, the county commissions or boards of supervisors.

These grants of authority and responsibility, and the terms and conditions that limit these grants, are contained in the federal and state constitutions, city charters, and the state laws establishing local governments.

A grant of authority with its accompanying responsibility can be *delegated* to another person, just as society as a whole granted the authority in the first place. However, when authority is delegated, the person or agency to whom it was originally granted remains responsible for its proper exercise and has the right to revoke the delegation at any time.

There can be a series of such delegations as different parts of a task are passed along and divided into smaller and smaller segments. No matter how long the chain of delegations may be, the ultimate authority always rests with the person or agency at the beginning of the chain. Usually this person is represented as being at the top of a triangle or pyramid. Authority is said to flow downward from the top of the pyramid toward the bottom, while responsibility is said to flow upward from the bottom to the top.

Management theorists often point out that it is unfair and ineffective to delegate responsibility for a particular task without also delegating the authority to carry it out. Actually, it is not only unfair and unwise; it is impossible. Authority and responsibility are bound together like the two sides of a coin.

Unfortunately, it sometimes happens that a person in authority will try to shift the blame to subordinates when something goes wrong by claiming that the subordinates had responsibility for taking some action. However, if it is clear that the subordinates never had the authority to act, then they could not have had the responsibility to act.

When responsibility and authority are granted or delegated, the person who receives the grant or delegation also must be accountable to the person (or agency, or society as a whole) who made the grant or delegation. *Accountability* simply means that each person in the organization may be called on at any time to show that he or she has exercised the authority properly.[4]

Chain of Command

The fact that authority and responsibility are delegated to the various parts of an organization creates a *chain of command*. The chain of command is the route or channel along which authority and responsibility have been subdivided and parceled out among the elements of the organization (see Figure 3.1).

For example, the Chief of Police has both authority and responsibility for everything done by the entire agency. Obviously the chief is not expected to personally perform every task.

The chief's authority and responsibility for law enforcement operations might be delegated to a Deputy Chief of Operations, who in turn might delegate the authority and responsibility for criminal investigations to the captain in charge of the Detective Bureau. The captain, in turn, delegates the authority and responsibility for investigating certain crimes to the lieutenant in charge of, say, the Crimes against Persons Section, and the lieutenant delegates the authority and responsibility for investigating specific cases to the several detective sergeants in that section.

The chain of command, then, begins with the Chief and passes through the Deputy Chief, captain, and lieutenant to the sergeants.

Unity of Command

Traditional organization theorists have maintained that there must be *unity of command*, that each person in an organization must be accountable to one and only one higher official. The reason for this rule is that if a person is placed under the authority of two different officials, they might give conflicting orders that would confuse and immobilize the subordinate.[5]

As a general rule, the principle of unity of command is valid and ought to be observed. However, it is sometimes the source of as much confusion as it prevents. For example, in some very large metropolitan police agencies, detectives are assigned to the various precinct headquarters. To whom are the detectives accountable: to the Precinct Commanders or to the Chief of Detectives for the whole department?

Most agencies, following the principle of unity of command, place the precinct detectives under the authority of the departmental Chief of Detectives.

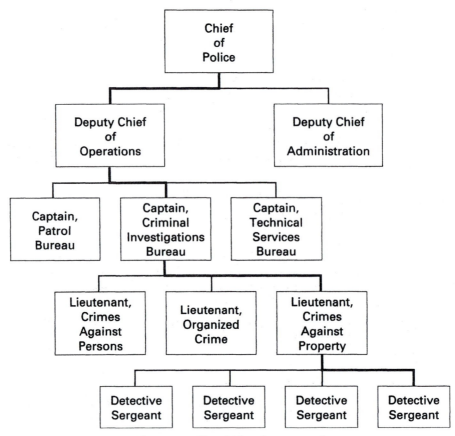

Figure 3.1. The chain of command.

But this puts the detectives in the position of resident aliens in the precinct stations. Because they are not accountable to the Precinct Commanders and must carry out assignments from the departmental headquarters, they are not available to assist in precinct-level operations and they are not necessarily bound by the Precinct Commanders' rules and policies.

Conversely, making the detectives answerable only to the Precinct Commanders diminishes the role of the Chief of Detectives to that of a technical adviser, and breeds inconsistency in the investigative policies and procedures followed from one precinct to the next.

The traditional answer to this dilemma is that the Precinct Commander should go through channels and ask the Chief of Detectives for the assistance of the precinct detectives when it is needed, or for their compliance with precinct rules and policies. That way, the authority of the Chief of Detectives is not diminished, but in a sense is delegated to the Precinct Commander, and an acceptable degree of cooperation and coordination can be achieved.

In practice, however, the procedure of going through channels to obtain coordination often fails, and it is frequently more time-consuming and troublesome than it is worth.

An alternative, and perhaps more practical solution, is to place the detectives under *dual jurisdiction*, stating explicitly the areas of policy and procedure for which each commanding officer has authority. The Chief of Detectives might retain authority for developing investigative procedures and policies, for transferring personnel from one precinct or station to another, for recommending promotions, and so on. Meanwhile, the Precinct Commander might have authority to assign work schedules, to assign cases for investigation within the precinct, and to enforce precinct-level rules and policies that apply to all personnel.

Inevitably, there will be occasions when the Chief of Detectives and the Precinct Commander issue conflicting or ambiguous orders. However, these two officials should be able to resolve these conflicts between themselves. If they cannot, they should refer the matter to the next higher authority in the chain of command.

A more serious problem develops when personnel in different chains of command are required to work together. Often there is no time or opportunity for questions of authority to be resolved by rational compromise.

For example, if a regular patrol officer and a traffic division officer arrive simultaneously at the scene of a serious accident, their immediate life-saving duties do not allow time for them to debate over which of them should take charge. The responsibility in such a situation should be established in advance, by agency policy. If it is not, the two officers should be able to work together cooperatively without concern for which of them has the right to take charge. Such matters can be settled after the immediate task is completed.

If we seem to be belaboring a point that ought to be a matter of common sense, it is only because there have been far too many instances of the sort described by Roy Roberg.

> In one incident, [a student of Roberg's] was the first officer to arrive on the scene of a traffic accident in which one of the cars had overturned and pinned the driver inside. The officer could not free the accident victim from the car. Not knowing the extent of the injuries, and fearing that the car might catch fire and explode, the officer immediately radioed for an ambulance and a fire truck to help free the victim. The firemen, after some effort, removed the victim from the car and the ambulance took the victim to the hospital. Instead of receiving praise for his efforts, the officer was later reprimanded by the department for not going through channels and receiving permission from his immediate superior (sergeant) to call for the ambulance and fire truck.[6]

What is most discouraging about this incident is that the reprimand was an official one, from the department itself, which in effect means from some higher level of authority, and not just from a misguided sergeant. No doubt the reprimanded officer's experience became common knowledge among the rest of the patrol force. One wonders how many future accident victims might die while a confused officer tries desperately to locate a superior officer to authorize what obviously must be done at once.

Coordination and Span of Control

The principles of the chain of command and of unity of command define the manner in which people are arranged into a hierarchical order of authority and responsibility. If everyone performed his or her tasks independently, these principles probably would be sufficient to describe an ideal organization.

In some simple enterprises, a very simple organizational structure is quite adequate. For example, in a factory that produces a single product more or less continuously, the workers on the assembly line each have a particular job to do, each line foreman supervises a group of workers doing essentially the same job, and each supervisor is responsible for a group of foremen whose workers perform a clearly defined portion of the overall work. As long as everyone does his or her job properly, the entire operation should proceed smoothly and efficiently.

Police work, however, is vastly more complicated. The "workers" at the end of the chain of command have not one task but many different tasks to perform at different times. Some tasks can be performed by individual workers, while other tasks require the joint efforts of several people, not all of whom are necessarily accountable to the same supervisor. The tasks themselves are not always routine. Many of them require a variety of skills and the exercise of mature judgment and discretion.

In the single-product factory, the task of *coordination*, bringing various parts into the proper order or relationship, is easily performed by a shop supervisor who assigns the workers to their different tasks and sees that raw materials are delivered to each work station at the proper time.

Coordination is not so easy to achieve in a police agency. In theory, the responsibility for coordinating any two elements in an organization rests with the person who has authority over both of them.

For example, if the Chief of Detectives has authority over both the Crimes against Persons Section and the Property Crimes Section, it is the chief's responsibility to coordinate the activities of the two sections so that every new case is assigned to one section or the other. Each section should know what procedures to employ in each type of case. Cases that involve both crimes against persons and property crimes, such as an armed robbery and assault, are assigned to be investigated by the most competent person.

Similarly, any operation that involves both detectives and patrol officers should be coordinated by the official who has authority over both: usually the Deputy Chief of Operations or some equivalent official.

The difficulty is that in any organization large enough to have a Deputy Chief of Operations, that individual could not hope to take charge personally of every case that might require the cooperative efforts of both patrol officers and detectives.

Even in a comparatively small agency, one that has, say, eight detectives divided equally between two sections, it would be unrealistic to expect the Chief of Detectives (usually a captain) to supervise every case that requires some degree of coordination. Instead, the usual practice is to try to establish coordination by rules and policies set in advance.

The danger in relying on formal rules, as we have said earlier, is that no set of rules can anticipate every circumstance that may arise, without creating a mass of rules so complicated and elaborate that no one can understand or remember them.

When a situation occurs that demands coordination between two units, the people most immediately involved usually try to decide which rules apply. If the rules are unclear or the individuals are uncertain as to which rules apply, they appeal for help to their respective supervisors, who may have to appeal to a higher level of authority to resolve the impasse.

Most of the time, this cumbersome procedure is accepted as a necessary product of the principle of unity of command. Most officials are relieved of the need to make nearly all routine decisions, because these are resolved in formal rules and policies, and the period of confusion and lost production while unusual situations are resolved ordinarily has no serious consequences. However, police work does involve life-and-death crises sometimes, and there may be no opportunity for the work to wait while someone resolves questions of coordination.

A better solution, once again, is to encourage all personnel to coordinate their own work, to cooperate with one another to the maximum extent possible, and to avoid referring jurisdictional questions to their superiors.[7] Some formal rules are necessary, of course, but their intent should be to establish broad guidelines and general principles for coordination. The emphasis always should be on getting the job done first, and afterward straightening out any questions of authority, jurisdiction, or procedure.

If mistakes are made, they should be brought to the attention of supervisors, and the personnel involved should be corrected, not to punish them for the mistake or for exercising discretion and initiative, but to help them avoid similar mistakes in the future. Such after-the-fact evaluation and correction is an important duty of every supervisor. It is a far more potent method of coordinating the work of subordinates than any volume of rules and policies.

One of the traditional concerns of organizational theorists is the extent of a supervisor's or manager's *span of control*, which simply means the number of people or organizational elements that are accountable to one person.[8] For the supervisor or manager to be aware of the subordinates' work, to coordinate their activities, to evaluate and correct their performance, and so on, the span of control must be reasonably small.

For example, the Chief of Police of a small agency may have no great difficulty in supervising the department's twenty-five employees organized into two main sections. However, as the agency grows and is divided into more and more sections, a point may be reached at which the chief is attempting to control the activities of two hundred people in twenty major sections. There would not be enough time in the day for the chief to learn what has been done in all the sections, let alone to make important decisions and resolve conflicts. The chief's span of control has gotten too large.

The solution is to insert an intermediate level of management, such as three deputy chiefs, each of whom assumes responsibility for several of the sections. Now the chief merely has to keep track of the activities of the three deputy chiefs, evaluating and correcting their work, receiving reports from them on the work of the various sections, and resolving disputes or conflicts (see Figure 3.2).

As this example illustrates, any reduction in the span of control for an individual manager is obtained at the expense of inserting a new level of management into the organization. The chain of command is thereby made longer, further removing the workers at the bottom of the pyramid from the ultimate source of authority and further removing the person at the top of the pyramid from intimate, day-to-day knowledge of activities at the operational level.

Management theorists have debated for years which is worse: a flat structure in which the manager's span of control is extended too far, or a tall structure in which the chain of command is too long. Although there can be no absolute, universal formula, most theorists have concluded that the best structure is as flat as possible, provided no one manager has too great a span of control (see Figure 3.3).

How great is "too great"? At one time, management theorists believed that there was a simple answer: No manager should be responsible for more than six subordinates (or six organizational elements). However, cases were found where a manager easily supervised twice or three times that many people, and other cases were found where a manager was overtaxed with only two or three subordinates. Today, most theorists reject any magic number but rely instead on the consideration of three factors.

- *Nature of the work.* Routine work performed more or less continuously will require few decisions on the part of the manager, and therefore a manager can control a greater number of subordinates. Work that varies in content, that requires the exercise of discretion and judgment, or that is relatively complex will require closer supervision by the manager, and therefore the span of control must be reduced.

- *Character of the workers.* If employees are highly skilled, intelligent, well trained, and highly motivated, they will require little supervision regardless of the nature of their work, and therefore the manager's span of control can be enlarged. If workers are poorly skilled and unenthusiastic about their work, close supervision will be required, and the span of control must be reduced.

- *Character and personality of the manager.* Some managers seem to be able to be everywhere, to keep track of thousands of details about their subordinates' jobs and performances, to make decisions quickly, and through it all to remain alert and relaxed. Other managers agonize over decisions, or feel compelled to immerse themselves in the details of each individual job, or hesitate to make any demands on their subordinates. Certainly, the first manager would be able to handle a much larger span of control than the second.

Again, there is no firm rule that dictates the ideal span of control. The nature of police work is such that, even given exceptionally well-qualified personnel and unusually competent managers, it is unlikely that anyone would be able to provide adequate direct supervision of more than about ten subordinates

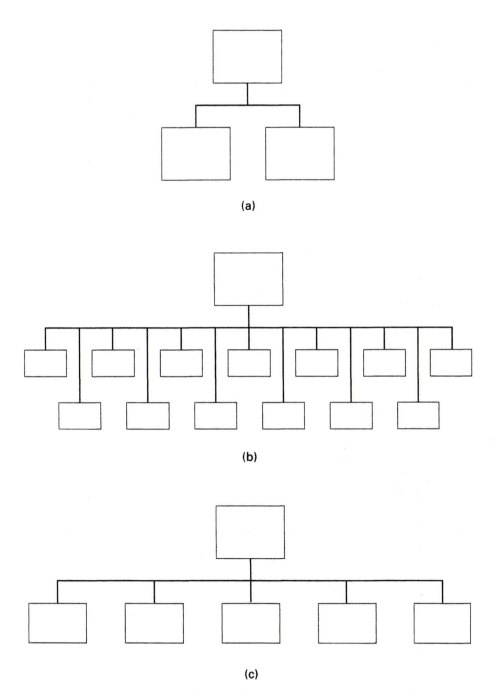

Figure 3.2. The span of control. (a) Very narrow. (b) Very broad.
(c) Moderately broad.

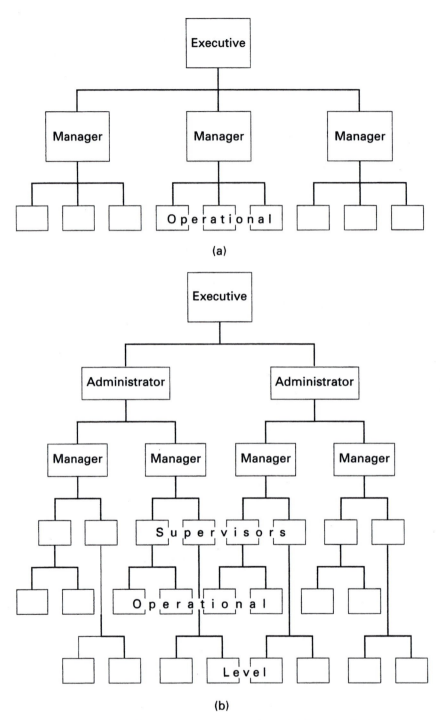

Figure 3.3. "Flat" and "tall" structures. (a) Flat. (b) Tall.

at the operational level or more than five at the administrative level. Conversely, any manager who supervises fewer than five subordinates, except perhaps at the highest administrative level, probably is not being overworked, and the organizational structure probably is taller than necessary.[9]

Unfortunately, the effort to organize an agency so that the span of control is appropriate for every supervisor often conflicts with the tendency of many police administrators to overspecialize.

Specialization

The fundamental purpose of any organization is to divide the tasks that make up the total enterprise among the people who are available to perform them, so that each person is assigned to the job for which he or she is best qualified.

A basketball team could not function effectively if it were made up of five centers or five guards or five forwards. But one center, two guards, and two forwards, if they work together harmoniously and if each is skillful at his or her task, should be able to win any given game. Some basketball teams specialize even further. One forward may be especially skilled at driving under the basket for a lay-up shot while another may be better at long jump shots.

Specialization is often carried to a considerable degree in other sports. A certain football player may be not just a defensive tackle but specifically a defensive left tackle who only plays when the other team is likely to pass.

As we discussed in Chapter 1, there was little if any specialization in the earliest police agencies. Police officers were either troops or commanders. After the Civil War, more advanced ideas of crime-prevention patrols and criminal investigation brought about a degree of specialization: An officer might be assigned to patrol, or to investigate crimes, or to supervise other officers. Intermediate supervisory ranks such as patrol lieutenant and precinct captain were established.

Later still, the first wave of police reform around the turn of the century involved greater degrees of specialization. Separate units were established to deal with particular types of crime or to provide specific services. Since then, the tendency toward specialization has continued and if anything has grown stronger over the years.

The advantages of specialization have been cited frequently by management theorists and authorities on police administration. Specialization is supposed to enable a person to make the best use of individual skills and interests, to learn a great deal about a specialty through repeated experience, and to develop a high level of motivation through pride in one's special expertise.[10] Also, specialization is supposed to enable administrators to assign responsibility more clearly, because every task becomes the duty of one or another special unit, and to organize training more effectively, because specialized individuals need to be given only the training required for their particular specialty.[11]

The disadvantages of specialization have become more widely recognized in recent years. The most significant disadvantage is that specialization narrows the range of competencies of every individual in the agency. A detective may know everything there is to know about, say, auto thefts but almost nothing about homicides. If the detective is assigned to a case that involves both crimes, the homicide may be improperly investigated, or the detective must ask for help from a homicide specialist.

Even a more limited degree of specialization has the same disadvantage. For example, in many traditional police agencies, the patrol officers are neither expected nor encouraged to know anything about criminalistics techniques. Their responsibility at a crime scene is merely to obtain a preliminary report by interviewing the complainant or victim, then to notify the detectives, protect the crime scene from contamination, and assist the evidence technicians when they arrive. These policies mean that much of the patrol officers' time is wasted. At least one additional person (a detective) and often three or four (evidence technicians, photographers, etc.) must be involved in every routine investigation. Almost always, valuable evidence is lost or overlooked.[12] At the very least, time is lost in which the perpetrator may escape beyond the agency's reach.

Excessive specialization complicates the already difficult task of coordinating the activities of police personnel, and it leads to inefficient allocations of personnel. One section of the agency may have little or no work to do while another section is overburdened, but it may be difficult to shift people from the first section to the second because the people have not been trained in the appropriate specialty.

Some degree of specialization is obviously necessary. For example, you would not expect a criminalistics technician to issue parking tickets.

In general, specialization is most practical in the administrative and technical sections of a police agency, where the work often requires particular skills or training, and where the tasks are continuous or repetitive in nature, such as checking and filing reports or keeping financial accounts. Even here, it is desirable to see that employees who perform similar or related work are *cross-trained* so they can assist one another or fill in if there is a shift in the workload or someone is absent. Cross-training also promotes better coordination and cooperation because employees have a better understanding of one another's jobs.

Among operational personnel, those engaged directly in crime prevention and law enforcement duties, specialization should be avoided where possible and kept to a minimum where it is necessary. There should be no rigid lines of demarcation between regular patrol officers, traffic patrol officers, investigators, and other officers.

Any police officer who is called on or who sees an opportunity to perform any police service for the public, whether it involves investigating a crime, arresting an offender, or directing traffic, should do so without hesitation, to the limits of his or her abilities.[13] All personnel should be trained in advanced techniques and specialized areas of law enforcement such as traffic control or criminal investigation, and should be discouraged from committing themselves and their careers exclusively to one specialty.

Certainly, differences in talent and personal preferences ought to be accommodated, and some officers will be better at certain tasks than at others. Effective, ongoing training programs should be designed to expand every officer's repertoire of skills over the whole spectrum of law enforcement responsibilities rather than narrowing each officer's range of competencies. As far as possible, every position in a police agency should be given responsibilities and tasks that challenge employees and encourage them to develop their abilities, not simply to perform the same limited range of tasks repeatedly.[14]

From an organizational viewpoint, the tendency to specialize should be resisted. Officers performing related work should be grouped into units of sufficient size to permit flexibility in work assignments and to maintain an optimal span of control for supervisors.

It makes little sense for a medium-sized agency to divide its Detective Bureau into six or seven specialized units, each containing only two or three officers supervised by a lieutenant. A better arrangement would be to consolidate the units into at most two, cross-train the officers in the various investigative specialties, and rotate case assignments to give each investigator experience in different kinds of cases. Those officers who have developed special expertise in a particular area should assist their supervisors in cross-training their colleagues and should readily give technical advice when it is needed.

If nothing else, this arrangement ensures that the agency's overall abilities are not impaired if an officer with a particular specialty leaves the department or is temporarily absent.

There is still a place for individuals with unique specialties within a more generalized operational structure. For example, most smaller agencies do not need a full-time polygraph specialist. However, a patrol officer or detective could be given advanced training in polygraph techniques and then could be assigned to conduct all polygraph interviews and interrogations. The rest of the officer's time would be spent on ordinary patrol or investigative duties. Such a part-time specialist also should train other members of the department in his or her specialty, acting as a technical adviser and resource person to the agency.

ORGANIZING A POLICE AGENCY

Few individuals will ever have the opportunity to establish and organize an entire police agency from scratch. Such opportunities usually arise only when a town is newly incorporated and hires its first police chief, who may be the only salaried employee in the agency.

However, small agencies generally grow if the community they serve grows. Medium-sized and even larger agencies periodically must be reorganized to accommodate changes in the community's size, population, political system, or its needs and desires for police services.

An agency may have internal needs that require reorganization too. Rapid growth in the number of personnel, changes in the availability of special skills, or changing ideas about the police service and the agency's responsibilities may make reorganization desirable.[15] Fortunately, the procedures that should be followed in organizing a new agency or in reorganizing an existing one are essentially the same.

Terminology: Ranks, Functions, and Elements

Before we discuss how to organize a police agency, we should establish a standard vocabulary for the various parts of the organization. There is often a great deal of confusion when two different police agencies are compared because there is no standardized terminology in use among all law enforcement agencies. What one agency may call a unit may be equivalent to another agency's section, while the first agency's section is more or less the same as someone else's bureau.

Adding to the confusion is the traditional use of military-style ranks. As traditions go, this is not a very old one. It extends back to about the post–Civil War period, around the same time that military-style uniforms were first introduced to law enforcement, and for much the same reason: to instill pride and discipline by drawing on military traditions. It is by no means clear that military-style ranks are necessary or appropriate in the police service at all.

Nevertheless, it does not appear likely that the use of military-style ranks will disappear from the police service anytime soon. Because rank designations vary from one agency to another, we list in Table 3.1 the terms we use in this book, the more or less equivalent terms used by some agencies, and a rough description of what the title means.

Table 3.1 **POLICE RANK DESIGNATION**

Our Terminology	Equivalent Terms	Description
Chief of police	Director of public safety[a] Commissioner[a]	Executive head of police agency, appointed by city government (or, in a few cases, elected)
Deputy chief	Assistant chief Major	Second-highest administrator, head of a major element of the agency
Captain	Commander	Head of a major element, if there are no deputy chiefs; head of a precinct, district, or bureau
Lieutenant	Same	Head of an element with a specific area of responsibility: a unit, section, watch, or shift
Sergeant	Same	Head of a group of personnel (unit, watch, or shift); in some units, such as detectives, the lowest-level operating personnel
Patrol officer	Officer; patrolman, patrolwoman	Operational personnel; the lowest-ranking commissioned officer
(Not used)	Corporal	Intermediate rank between officer and sergeant
(Not used)	Cadet	Probationary patrol officer; temporary rank given to new recruits during or just after training, while assigned to on-the-job training

[a]These titles are sometimes used to designate officials who are responsible for overseeing the police department and other public safety agencies such as the fire department and ambulance service. Often these officials are elected and do not take an active, day-to-day part in law enforcement operations.

In addition to the ranks listed in Table 3.1, some agencies use the ranks of senior patrol officer and senior sergeant. These ranks may be intermediate between patrol officer and sergeant, or sergeant and lieutenant, or they may merely indicate that the individual has remained at the same rank long enough to deserve a higher salary.

In some agencies, intermediate ranks are awarded to individuals who show exceptional promise and initiative, and who are expected to exert leadership among their colleagues. Other agencies use the "senior" designation to indicate that the individual, although competent in his or her present position, is not likely to be promoted to any higher rank but is being rewarded for faithful service.

There is often confusion concerning the fact that most police officers have two titles: their rank and their functional, or office, title. Rank indicates nothing more than a person's salary level and general range of authority and responsibility. Functional titles indicate what a person does for the agency and approximately where he or she fits into the agency's hierarchy. There is virtually an infinite list of possible functional titles, because there is a title for every position in an agency, and agencies are free to make up their own titles. The list in Table 3.2 is by no means exhaustive; it merely includes some of the more common functional titles that we use here.

In addition to the titles in Table 3.2, there may be other functional titles, such as Juvenile Officer, Training Officer, and Field Training Supervisor, whose meaning is more or less evident from the title itself. The important point to remember is that these functional titles do not necessarily reflect a person's rank or vice versa.

In this and the preceding chapters, we have referred to the various parts of a police agency by various terms. Again, there is no uniform terminology used by all police agencies. In Table 3.3 we compare the terms we use with some of the roughly equivalent terms used by many agencies.

Function Analysis

Once the various parts of an agency have been identified, the police administrator is ready to consider the best way to arrange them into an effective organizational structure. The different categories of elements listed in Table 3.3 are the "boxes" into which the functions and ranks listed in Tables 3.2 and 3.1 must fit.

One more step must be performed before any decisions about structure are made: the administrator must know what tasks are to be performed by the agency.[16]

Every police agency's list of functions is unique to that agency because it is derived from the needs of the community and from the laws and ordinances that grant authority to it. However, there is enough similarity among all American police agencies to permit some generalizations.

In Table 3.4, we have tried to include every specific function that is common to most police agencies including municipal police departments, sheriff's departments, and state police. Some agencies might have additional functions not listed here, and others may omit some of these functions. We have not

Table 3.2 FUNCTIONAL TITLES

Our Terminology	Equivalent Terms	Description
Patrol officer[a]	Uniformed officer; patrolman, patrolwoman	Any low-ranking police officer assigned to general police duty such as patrol
Traffic officer	Traffic patrolman	Patrol officer assigned exclusively to traffic law enforcement
Detective	Investigator, detective sergeant, plainclothesman, inspector (rarely)	Any officer assigned to criminal investigation duties; usually does not wear a uniform
Evidence technician	Criminalist	Any person, either a police officer or civilian, who assists in locating, collecting, and evaluating evidence
Supervisor	Same	Sergeant, lieutenant, or (sometimes) captain who directly supervises operational personnel
Chief of —	Head of — Section (or unit) commander	Supervisor of a functional element such as a bureau, section, or unit
Commander	Chief of — Head of — Commanding officer	Manager (usually ranked captain or higher) of a major functional element such as a precinct, bureau, or division
Ranking officer	Commander	Officer, regardless of rank or title, who has the authority and responsibility for a given task; usually the highest-ranking officer present in a situation
Manager, administrator	Supervisor	Every officer above the operational level
Executive	Same	Chief of police; highest-ranking person within the agency

[a]We capitalize Patrol Officer when we mean a given individual, and lowercase patrol officer when we mean the function.

attempted to include non–law enforcement functions such as ambulance service, funeral escorts, and so forth.

A list of functions similar to that in Table 3.4 can be developed for any police agency. Once the list has been compiled, the next step is to match the list of functions to the personnel available. To do this, the administrator must know three things.

- How many people are available? What is the total number of employees, both full-time and part-time? Are some of them available only at certain times of the day or only on certain days of the week?
- What competencies do the personnel have? How many are certified police officers (if there is a state certification program)? How many have completed college preparation programs or advanced training in such areas as investigation, evidence collection, traffic law enforcement, crisis intervention, and so forth? How many have had actual working experience in one of these special areas, or in supervision or management?

Table 3.3. **ELEMENTS OF A POLICE AGENCY**

Our Terminology	Equivalent Terms	Description
Division	Bureau, section	One of the two to five major parts of the agency, headed by a deputy chief, major, or captain
Bureau	Division, section, office	One of the parts of a division including several smaller units, headed by a captain or lieutenant
Section	Unit, office, watch, detail, division	One of the parts of a bureau; may be divided into smaller parts; headed by a lieutenant or sergeant
Unit	Section, office, watch, detail, shift, squad	One of the parts of a section; not further divided by function; may be divided by time of day; headed by a lieutenant or sergeant
Precinct	District, sector	Geographical section of a city with its own police subheadquarters (station)
District	Precinct, area	Geographical section of a city, used to assign patrol and other personnel, but with no permanent station
Sector	District, area	Part of a district or precinct containing two or more adjacent patrol areas
Area, beat	Sector, section, route	Part of a district or sector to which a patrol officer (or squad of patrol officers) is assigned
Shift, watch	Same	Unit of time: a part of the working day, which is not necessarily divided into equal parts
Squad, team	Patrol, unit	Several officers who work together in a coordinated activity

- What changes can be made in the number of personnel and their competencies? Are funds available to hire more people? If so, what competencies are most needed, and can they be obtained by hiring the right people? Can present personnel be given additional training to expand or change their competencies, and how long will the additional training take?

For a newly organized department, the answers to these questions will depend almost entirely on the size of the budget, which in turn is dictated by the overall availability of tax revenues and other funds to the government. A new police chief establishing a department in a newly incorporated city of, say, 10,000 people could not expect to hire 150 certified police officers and 75 civilians; a total staff of 10 to 15 would be much more reasonable.

For an existing agency, the number of people available and their competencies usually are determined by the agency's previous budgets, organizational structure, and personnel practices. Unless the agency has been chronically underfunded and the parent government indicates a willingness to make more money available, the administrator cannot count on being able to add many new people.

Except in unusual circumstances, it is not possible to fire the existing personnel to replace them with others who have the desired competencies. To some extent, *attrition*, the normal loss of personnel through death, retirement, and resignation, will provide some opportunities to bring in new people with new com-

Table 3.4 FUNCTIONS OF A POLICE AGENCY

Function	Purpose	Description of Tasks
I. Law Enforcement Functions		
1. Receipt and disposition of citizens' complaints	1. To ensure that complaints of crime, prospective crime, emergency, etc., receive prompt attention	1. Receive complaint, record on standard form; assign police officer to respond as needed
2. Response to crime in progress and other emergencies	2. To protect lives and property; to apprehend offender if possible	2. Proceed rapidly to scene of incident; determine nature of incident; obtain additional help if needed; act to protect lives and property and to arrest any suspected offenders found at the scene
3. Preliminary investigation	3. To determine the nature of the incident and identity of offenders if known	3. Interview complainant, victim, or witnesses; record condition of the crime scene; gather and preserve evidence found at scene
4. Continuing investigation	4. To determine the identity and whereabouts of, and to apprehend suspected offender	4. Interview witnesses, interrogate suspects, gather and preserve evidence away from the crime scene; apprehend and arrest suspected offender
5. Analysis and development of evidence	5. To identify, characterize, and rationalize physical evidence	5. Apply criminalistics techniques to physical evidence, as needed; prepare reports; prepare exhibits for presentation in court
6. Polygraphic interrogation	6. To test the veracity of witness, suspect	6. Apply polygraphic techniques; interpret and report findings; prepare court presentations
7. Investigation of organized crimes	7. To determine the identity of persons involved in organized crime; to gather evidence against organized criminals	7. Observe sites and persons believed to be involved in organized crime; penetrate criminal organizations by use of undercover techniques, by locating cooperative and knowledgeable informants, and by legally acceptable surveillance techniques
8. Preparation of case reports	8. To organize and present evidence for use by prosecutors	8. Prepare a written statement of the nature of the crime and of the evidence; deliver the report to prosecutors; confer with prosecutors on any improvements needed in the report.
9. Presentation of evidence in court	9. To present evidence against accused offenders	9. Testify in court proceedings, such as hearings and trials, on the evidence produced by an investigation
II. Traffic Law Enforcement Functions		
10. Direction of traffic	10. To ensure the safe movement of vehicles and pedestrians	10. Direct traffic in and around congested areas; coordinate the traffic flow with traffic control devices and other officers

Table 3.4 CONTINUED

Function	Purpose	Description of Tasks
11. Detection and apprehension of traffic violators	11. To discourage violations; to protect other citizens from unsafe and illegal acts of violators	11. Observe motorists and pedestrians, using appropriate devices such as radar speed detectors; apprehend violators; identify violators and their vehicles; issue verbal or written warnings or citations as appropriate; arrest violators who present an imminent danger to others (such as reckless or intoxicated driver)
12. Investigation of accidents	12. To protect the lives or property of citizens; to relieve traffic congestion; to create legal record of incident	12. Proceed rapidly to scene of reported accident; observe need for and obtain assistance if hazard exists; establish traffic control around accident site; interview drivers and witnesses; record conditions at scene; arrange for and supervise removal of disabled vehicles
13. Enforcement of parking regulations	13. To ensure safe traffic movement; to protect revenue from meters; to provide citizens with fair access to convenient parking areas	13. Observe legal parking areas adjacent to roads; place citations on illegally parked vehicles; arrange for and supervise removal of vehicles parked in a hazardous or obstructive location; instruct motorists to move away from improper parking locations
III. Crime Prevention Functions		
14. Preventive patrol	14. To observe criminal acts, behavior, or conditions that might lead to criminal acts or that create a hazard to lives or property; to discourage persons from committing crimes; to identify and apprehend offenders	14. Observe the activities of suspicious persons and places where criminal behavior frequently occurs; observe hazardous conditions and remove the hazard or report it to appropriate authorities
15. Traffic patrol	15. To discourage motorists from violating traffic laws	15. Conspicuously patrol streets, highways, intersections where traffic law violations and accidents occur most often
16. Public education	16. To encourage all citizens to obey all criminal and traffic laws, and to protect themselves from offenders	16. Advise public, individually or in groups, of the requirements of the laws, reasons for the laws, consequences of violating the law, and methods of protecting themselves and property
IV. Family Protection Services		
17. Juvenile crime prevention	17. To discourage juveniles from engaging in crime	17. Observe juvenile groups and individuals; advise them as to the requirements of the laws and consequences of committing crimes; encourage constructive youth activities

Table 3.4 CONTINUED

Function	Purpose	Description of Tasks
18. Apprehension of juvenile offenders	18. To identify and apprehend juvenile offenders; to promote rapid, effective reformative treatment	18. Interview witnesses who have knowledge of juveniles engaged in criminal activities; gather evidence concerning juvenile crimes; apprehend suspected juvenile offenders and present them, with evidence, to appropriate court or other officials
19. Truancy law enforcement	19. To ensure that juveniles and parents obey school attendance laws	19. Obtain names and likely whereabouts of truant students from school officials; locate and apprehend truant students; notify and counsel parents; refer chronic offenders to juvenile court or other appropriate officials
20. Child protective law enforcement	20. To protect the lives and welfare of children	20. Respond promptly to all complaints or reports of possible child abuse, neglect, mistreatment, or exploitation; gather evidence of offenses; assist child welfare authorities in taking legal steps against abusive adults; when necessary, arrest abusive adults and assume custody of abused children
21. Family crisis intervention	21. To protect the lives and welfare of all persons; to reduce disturbances of the peace; to assist family members in obtaining needed assistance	21. Intervene in family disturbances to the extent necessary to protect the lives of all persons present; counsel and advise citizens to obtain appropriate help in resolving their problems constructively; if necessary, restrain or arrest belligerent family members
V. Operational Supervision Functions		
22. Patrol supervision	22. To ensure that patrol functions are discharged properly	22. Assign officers to patrol shifts and beats; create and maintain work schedules; observe patrol officers in the field; receive, review, and correct all routine field reports; inform patrol officers of applicable policies, regulations, crime and hazard conditions on their beats; instruct officers in policing techniques; assist in conducting potentially dangerous activities
23. Investigative supervision	23. To ensure that investigative functions are performed properly	23. Assign officers to investigative duties; create and maintain a work schedule; receive, review, and correct all investigative reports; ensure that officers observe all applicable policies and regulations; inform officers of current crime trends and conditions; assist officers in applying appropriate investigative techniques, in preparing case reports and other documents, and in preparing court testimony

Table 3.4 **CONTINUED**

Function	Purpose	Description of Tasks
VI. Technical Support Functions		
24. Communications	24. To maintain continuous communication with all field units and to provide all other communications services needed	24. Operate telephone, radio, and other communications systems
25. Document entry	25. To ensure that all reports and documents are complete and accurate; to place all documents correctly in files	25. Receive incoming reports and documents; review them for completeness and accuracy; return incomplete or inaccurate documents to originator for correction; create cross-index entries for each new document; place new documents in correct file position[a]
26. Document retrieval	26. To provide agency personnel with needed documents	26. On proper request, retrieve any needed document from files, record its removal, and transmit to requesting officer; assist personnel in identifying and locating needed documents[a]
27. Public records service	27. To provide needed documents to citizens	27. Receive requests from citizens; determine whether release of document is authorized by law and agency policy; if so, retrieve document from files, record its removal, and transmit it (or a copy) to citizen[a]
28. Vehicle servicing and maintenance	28. To ensure that all police vehicles are available for use when needed	28. Service all patrol vehicles at end of each shift and all other vehicles on regular basis; perform all needed repairs on vehicles; provide field repair and towing for disabled vehicles; maintain records of all maintenance and repairs; maintain and repair vehicle-related equipment (siren, radio, etc.)
29. Evidence and property service	29. To maintain legal custody of all personal property in agency's possession; to maintain legal chain of custody of evidence	29. Provide secure storage for all nonevidence personal property (found property, recovered stolen property, prisoners' property); maintain records of all personal property in agency's possession; provide legally secure storage for all physical evidence
30. Jail	30. To maintain custody of prisoners according to law and agency regulations; to maintain records of persons in custody	30. Receive prisoners; record prisoners' identity, physical condition, and personal property; detain prisoners in a secure place; provide prisoners with food, clothing, medical care, recreation, access to visitors, communication with legal counsel, and other services required by law and agency policy; maintain records of current location of each prisoner until final release is authorized by appropriate official

Table 3.4 **CONTINUED**

Function	Purpose	Description of Tasks
31. Crime analysis	31. To assist agency personnel in detecting and understanding crime trends and patterns	31. Receive and review all crime reports; analyze crime reports to determine trends or patterns in criminal behavior; report analyses to administrative and operational personnel
VII. Administrative Support Functions		
32. Personnel recruitment and selection	32. To provide the agency with needed personnel	32. Inform the public of the availability and benefits of employment in police service; conduct recruitment campaigns aimed at specific population segments; select qualified applicants in a manner consistent with laws and Civil Service regulations; provide administrators with lists of eligible recruits; advise and assist recruits during entry process
33. Personnel training	33. To ensure that all agency personnel have needed skills and knowledge	33. Establish and conduct training programs for recruits, inservice personnel, and personnel eligible for promotion; maintain records of training; assist personnel in taking advantage of external training opportunities
34. Personnel records	34. To maintain accurate records of employment of all personnel	34. Create a permanent file on entry of each new employee; enter information concerning training completed, rank and duty assignments, promotions, awards, current salary, and all other matters as agency policy requires; locate and retrieve files on proper request of administrators
35. Accounting	35. To create and maintain complete and accurate records of agency's finances	35. Receive and record all receipts; prepare and issue all disbursements (payroll and purchase vouchers or checks); transfer funds to and from proper depositories (banks); keep a current record of the budgeted and expended funds in each of the agency's accounts
36. Purchasing	36. To ensure that the agency acquires needed supplies and equipment at best price	36. Prepare descriptions and specifications of needed items; issue requests for bids; receive and review bids; issue purchase order to lowest qualified bidder; receive purchased items, determine that they meet specifications, and deliver them to ordering unit or store for future use
37. Inventory	37. To ensure that all agency property is accounted for	37. Periodically perform a count of supplies and equipment; determine the location and condition of equipment having a useful life of one year or more; report discrepancies (equipment not located) to administrators

Table 3.4 **CONTINUED**

Function	Purpose	Description of Tasks
38. Building maintenance	38. To ensure that the agency's buildings and equipment are properly maintained	38. Clean all buildings on regular basis; repair facilities and equipment as needed
39. Public information	39. To inform the news media of significant events and activities; to answer inquiries from news media or the public	39. Inform the news media of all events, agency actions, and related matters that are likely to be considered newsworthy; answer inquiries or relay them to appropriate personnel; prepare and disseminate information about agency activities and matters of continuing public interest; assist crime prevention personnel in publicizing information about crime prevention
40. Planning	40. To assist administrators in developing policies; to determine future needs for personnel, other resources	40. Review all aspects of agency's operations and performance of all elements of agency; identify needs for change in policies, resource allocations, or methods of operation; prepare program plans and research reports for administrators' use
41. Budget preparation	41. To prepare a fiscal budget for future use	41. Solicit and receive budget requests from each element of the agency; develop a consolidated budget for approval by administrators and submission to parent government
42. Budget supervision and execution	42. To ensure that budgeted funds are spent as planned	42. Periodically review current accounts of all agency elements; advise administrators of any deviations from approved current budget; prepare and execute budget amendments authorized by appropriate officials

[a]These tasks may be somewhat different if the agency has a computerized records system; see Chapter 8.

petencies. However, the attrition rate in most police agencies is something less than 10 percent per year.

Furthermore, there is a strong tradition in policing that upper-level vacancies are filled by the promotion of people already in the agency. For the most part, new employees are hired only at the lowest level. If they have any special competencies that may be desired at higher levels, it will be several years before they work their way up to those levels through promotions.

In short, the administrator's options are somewhat limited. Personnel can be reassigned to new functions if their present competencies are compatible with their new duties, or they can be retrained to equip them for new functions. Usually reassigned personnel are less productive for a while than they would be otherwise, because they must learn the details of their new jobs.

There also may be a period of reduced morale, because people often resist being reassigned unless their new job is perceived as a promotion. Sometimes a temporary loss of morale and productivity is an acceptable price to pay if the anticipated gains from reorganization are sufficiently high. However, the administrator must realize that these costs will be borne and that the benefits may not appear for months or even years.

With all these factors in mind, the administrator must complete the task of organizing, or reorganizing, the agency. Every function must be assigned to someone, and everyone must have a specific function.

Assigning functions to specific organizational elements is a matter for the administrator's judgment and, to some extent, for trial-and-error experimentation. If an organizational structure clearly is not working, there may be no good reason not to change it. The procedure for assigning functions to organizational elements is fairly simple.

1. First, list on a sheet of ruled notebook paper or graph paper all functions of the agency, as in Table 3.4, but adapt them to the needs of the actual agency.

2. Across the top of the sheet, establish column headings for each of the kinds of personnel that are available to the agency, showing the number of each kind.

3. Assign each function to one or more of the kinds of personnel (see Figure 3.4).

This functional assignment matrix should be regarded as only tentative, because assignments may be shuffled around somewhat. However, every function must be assigned to someone, and every person in the organization must be assigned to one or more specific functions.

Next, the administrator must group related functions into organizational elements. Personnel who are to perform the same function usually should be assigned to the same element. These groupings are obvious from the functional assignment matrix. Again, judgment must be exercised in grouping related functions into specific elements. Figure 3.5 shows how these groupings can be developed.

Finally, the elements are arranged into a hierarchical structure to provide for authority, responsibility, and accountability. At this point it may be necessary to add specific elements for supervision and management. The most common way to represent the organizational structure is to draw an organizational chart, such as the one illustrated in Figure 3.6.

It is important to realize that there is no one right way to organize a police agency. The organizational chart often will resemble the charts of many other police agencies, and the administrator will wonder whether the effort of going through all the steps we have outlined has been wasted if the results are so predictable. The real value of this procedure lies in the series of rational analyses and decisions that the administrator must make.[17]

Figure 3.4 — A functional assignment matrix.

Group	Function	Asst. Chief (2)	Captain (5)	Lieutenant (12)	Sr. Sgt. (6)	Det. Sgt. (18)	Sergeant (11)	Sr. Patrol Off. (21)	Patrol Off. (49)	Secretary (3)	Clerk (3)	Bookkpr. (4)	Criminalistics (4)	Evidence Tech. (2)	Comm. Svc. Aide (8)	Parking Officer (6)	Jail Attendant (18)	Animal Control Off. (3)	Dispatchers (12)	Janitors (6)	#
Patrol	Receive complaints	A	A						42										X		1
Patrol	Complaint response	A	A					X	42												2
Patrol	Prelim. invest'g'n.	A	A					X	42												3
Inv.	Cont'g invest'g'n.	A	B			10		X													4
Inv.	Evidence collection	A	B			X		X							X						5
Inv.	Evidence analysis	A	B										X	X							6
Inv.	Polygraph operator	A	B	1																	7
Inv.	Case review	A	B	2																	8
Traffic	Direct traffic	A	A						7												9
Traffic	Violator appreh'n.	A	A					X	X												10
Traffic	Accident invest'g'n.	A	A						7												11
Traffic	Parking Enforcem't.	A	A						X							X					12
Patrol	Prev. Patrol: car	A	A					18	30												13
Patrol	Prev. Patrol: cycle	A	A					2	6												14
Patrol	Prev. Patrol: foot	A	A						6												15
Community Svcs.	Crime prev/educ'n.	B	C			1	1								1						16
Community Svcs.	Juvenile offenses	B	C			2									2						17
Community Svcs.	Family offenses	B	C		1	1									3						18
Community Svcs.	Org. crime/spec. cr.	B	C	1		3															19
Sup.	Patrol supervision	A	A	2	1		3														20
Sup.	Inv. supervision	A	B	3	1																21
Technical Svcs.	Communications	B	D	2			1												X		22
Technical Svcs.	Records & ID	B	D	1							6										23
Technical Svcs.	Vehicle maint.	B	E				1														24
Technical Svcs.	Evidence/Property	B	D				1								1						25
Technical Svcs.	Jail	B	E				1										X				26
Technical Svcs.	Planning & analysis	B	D			2															27
Admin. Svcs.	Personnel	B	E	1																	28
Admin. Svcs.	Training	B	E	(1)																	29
Admin. Svcs.	Accounting	B	E									3									30
Admin. Svcs.	Purchasing/Stores	B	E									1									31
Admin. Svcs.	Bldg maint.	B	E																	X	32
Admin. Svcs.	Public information	B	D	1											2						33
Misc.	Internal affairs	Chf	–	1	1																34
Misc.	Animal control	B	D															X			35
Misc.	License issuance	B	D								2										36

Figure 3.4. A functional assignment matrix.

EXAMPLES OF POLICE AGENCY ORGANIZATION

In Figures 3.7 to 3.15, organizational charts for several typical police agencies are shown. The charts represent agencies of different sizes and with somewhat different functions. Studying these charts will show how different administrators,

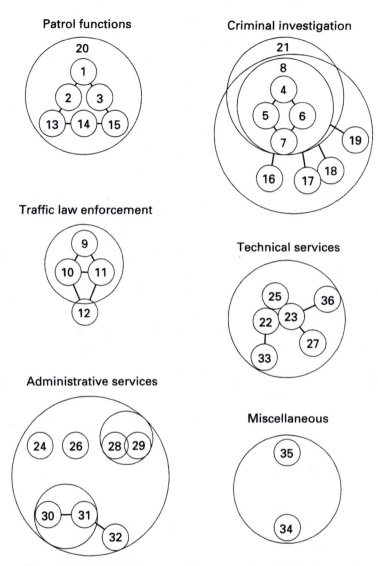

Figure 3.5. Functions grouped into organizational elements.

given similar situations, have arrived at somewhat different solutions to the problem of organizing a police agency. However, it is also clear that most police agencies are organized along similar lines, largely because of the strong and pervasive traditions of American policing.

UNCONVENTIONAL ORGANIZATIONAL PLANS

In recent years, several police agencies have tested organizational structures that are significantly different from those we have just discussed. In some cases, the

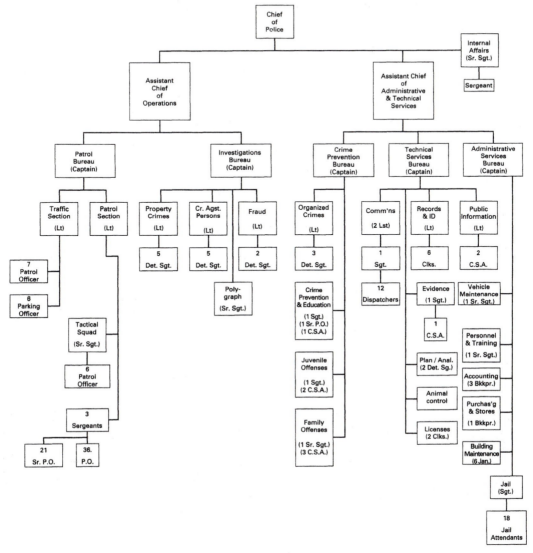

Figure 3.6. An organization chart for a hypothetical police agency.

experiments have been quite successful; in other cases, they have not. However, the mere fact that experiments were made indicates that police administrators are more willing to try variations on traditional themes than they have been in the past. The unconventional plans that we discuss here are only examples of the directions some agencies have taken.

Patrol-Oriented Investigation

In a conventional police agency structure, the job of investigating crimes is divided between the patrol officers and the detectives. Typically, patrol officers

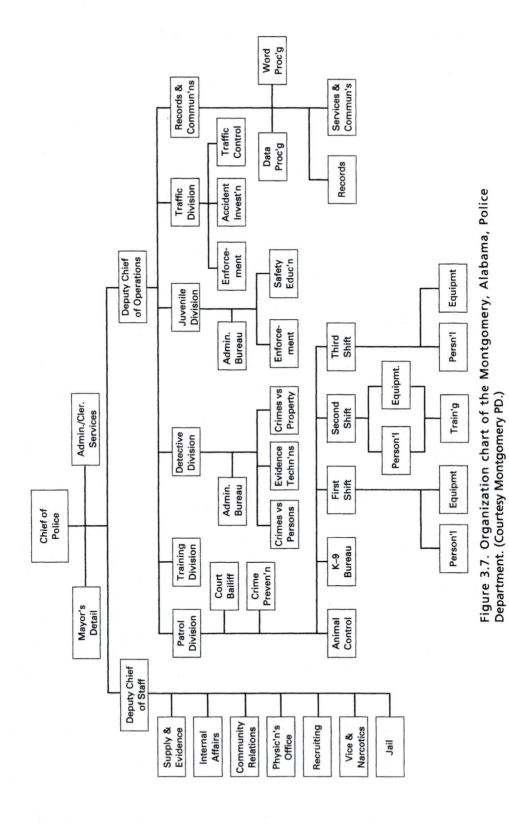

Figure 3.7. Organization chart of the Montgomery, Alabama, Police Department. (Courtesy Montgomery PD.)

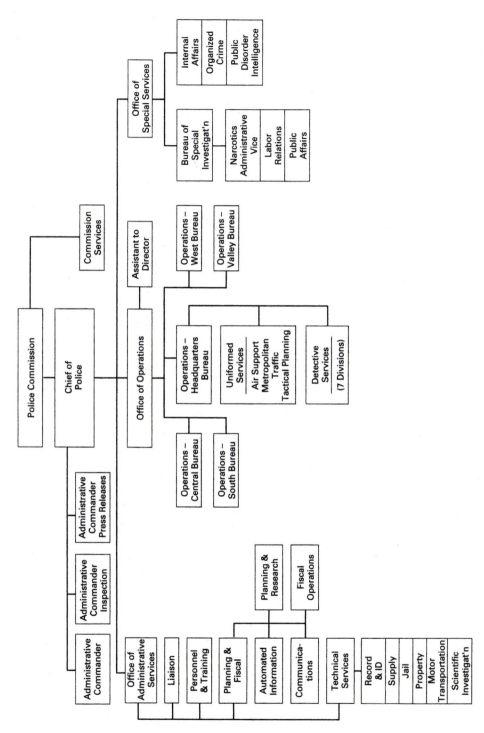

Figure 3.8. Organization chart of the Los Angeles Police Department (not current). (Courtesy LAPD.)

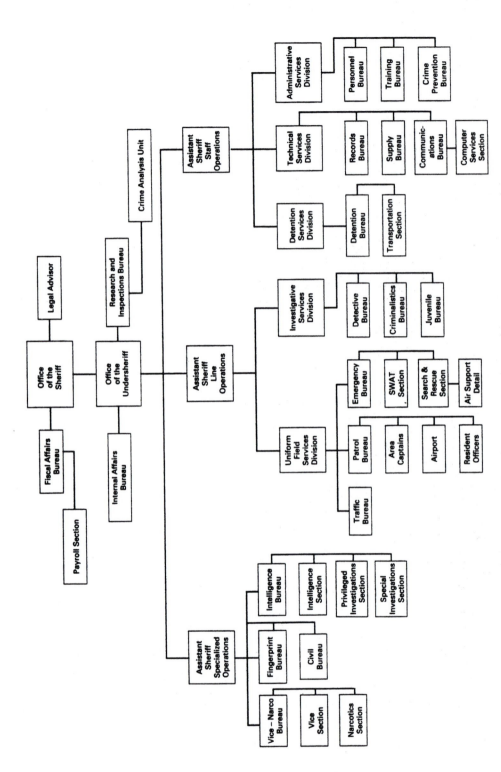

Figure 3.9. Organization chart of the Las Vegas Metropolitan Police Department. (Courtesy Las Vegas MPD.)

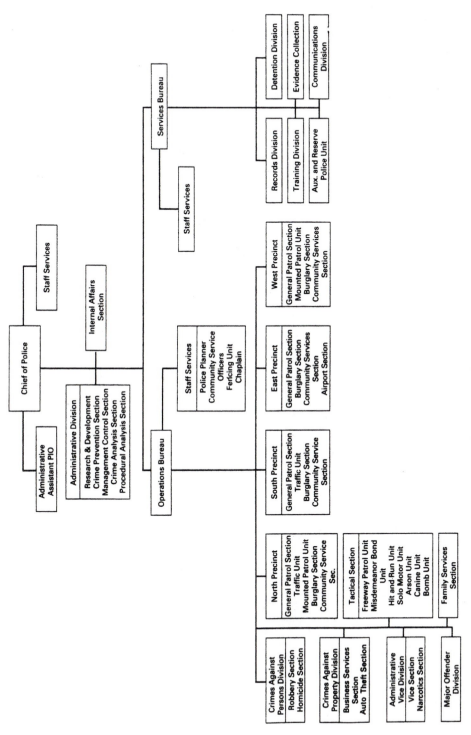

Figure 3.10. Organization chart of the Birmingham, Alabama, Police Department. (Courtesy Birmingham PD.)

79

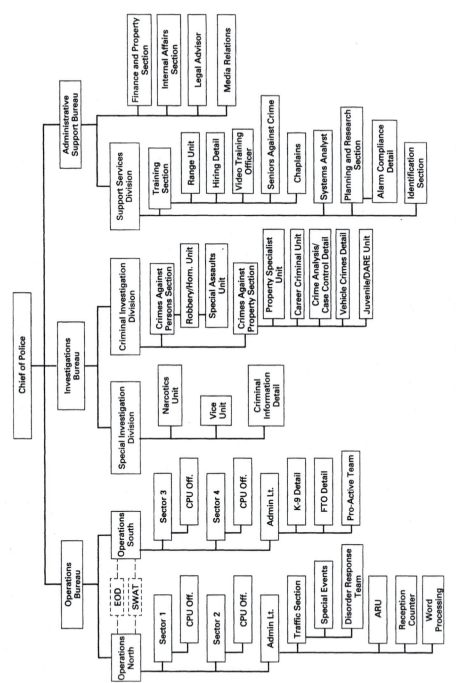

Figure 3.11. Organization chart of the Tacoma, Washington, Police Department. (Courtesy Tacoma PD.)

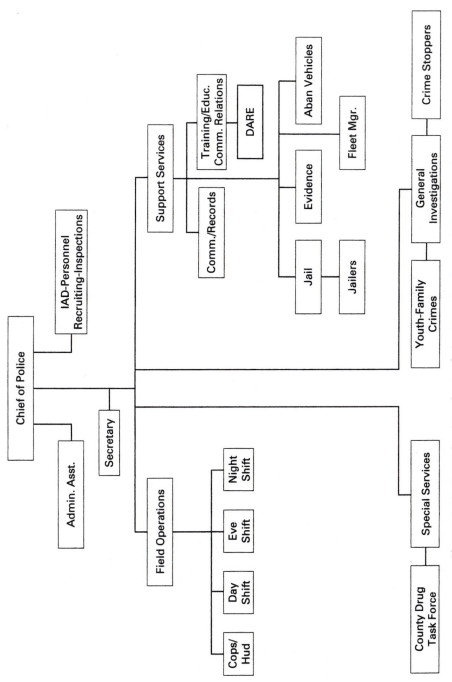

Figure 3.12. Organization chart of the Temple, Texas, Police Department. (Courtesy Temple PD.)

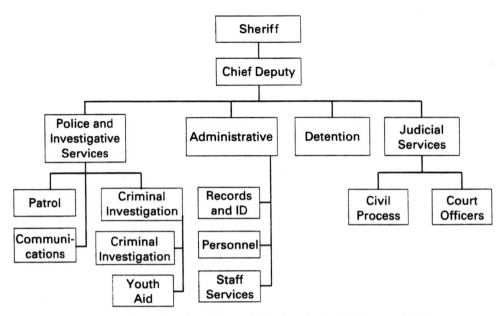

Figure 3.13. Organization chart of the Cumberland County, Maine, Sheriff's Office. (Courtesy Cumberland County Sheriff.)

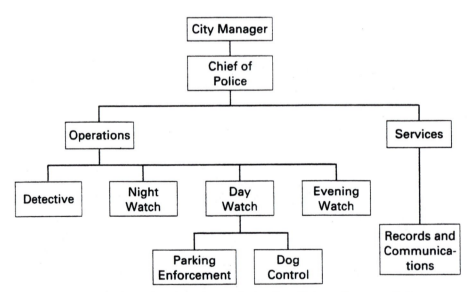

Figure 3.14. Organization chart of the Pendleton, Oregon, Police Department. (Courtesy Pendleton PD.)

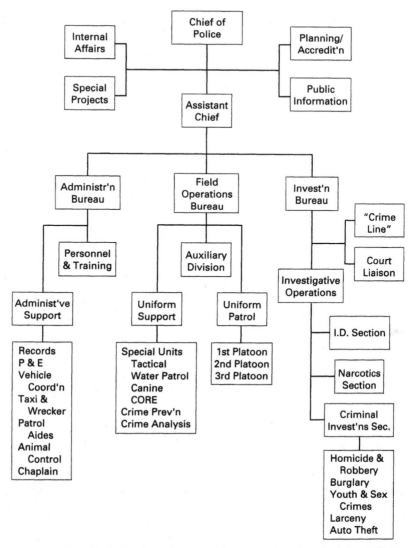

Figure 3.15. Organization chart of the Portsmouth, Virginia, Police Department. (Courtesy Portsmouth PD.)

respond to the initial report of a crime, take a preliminary report by interviewing the victim or witnesses, and, if there appears to be any physical evidence at the crime scene, call evidence technicians and detectives to make a more thorough investigation. Once these tasks have been completed, the investigation is turned over to the detectives and the patrol officer has no further role in the case.

Critics have claimed that this common practice wastes the talents of personnel. They say that specialist detectives are not significantly more effective in solving crimes than are properly trained patrol officers.[18]

Giving patrol officers a larger role in criminal investigation requires some changes in organizational structure. A thorough investigation may require an hour or more at the crime scene, with additional time spent on the follow-up investigation, case reporting, and so on. To give patrol officers the time to spend on investigations, they must be freed from routine patrol duties.

One approach is simply to increase the number of patrol officers assigned to each sector. Most officers are assigned to a specific beat, in the conventional manner. However, there are also patrol officers who have no specific beat assignment.

When one officer becomes involved in an investigation, one of the unassigned officers takes over the beat. Or the extra patrol officers can be designated as patrol officer investigators, who are dispatched to respond to any complaint that is likely to require extensive investigation. When they are not involved in an investigation, the patrol officer investigators perform routine patrol duties.

The increased number of patrol officers usually is obtained by sharply decreasing the number of specialist detectives. Instead of three or more sections of detectives, there may be only one or at most two. Instead of dividing each section into several highly specialized units, the sections are consolidated into general-purpose units. The detectives serve as resource advisers to the patrol officers, offering guidance and assistance in complicated investigations and taking charge of only the most difficult or complicated cases.

An organizational chart for an agency that has adopted a patrol-oriented investigation structure is shown in Figure 3.16.

Team Policing

One result of the urban disorders of the late 1960s and early 1970s was a public demand for more effective law enforcement service, and especially for police services that would be more responsive to the needs and interests of the community.

Team policing was proposed as a way to make better use of the patrol force and to improve law enforcement service by maintaining a closer relationship between police personnel and the community they serve.

The original concept was to assign a team of officers, usually consisting of one or more patrol officers, a detective, and sometimes other specialists, to assume responsibility for all crimes that occur in a given area. Another approach to team policing involved forming several teams, each consisting of a patrol officer, a detective, and sometimes another specialist for each area. Whenever a complaint was received, the entire team responded as a unit to perform whatever investigation was needed. This second approach, in practice, usually amounted to putting detectives in patrol cars.

The relative merits and disadvantages of various forms of team policing have been the subject of vigorous debate in law enforcement.[19] Figure 3.17 shows the organizational charts for two different types of team policing structures.

Community Policing

During the 1980s, team policing began to decline in popularity as a broader concept of police service became popular: community policing, community-oriented policing, or neighborhood policing.[20] It is difficult to know exactly what any of

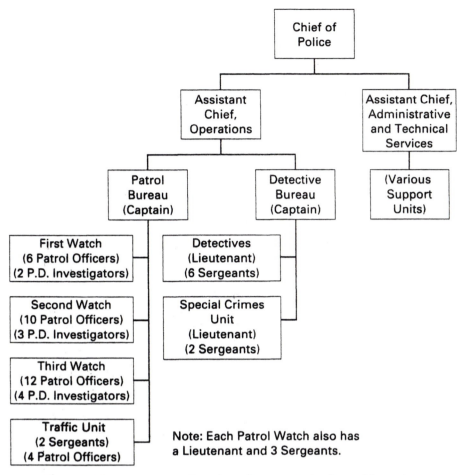

Figure 3.16. An organization chart for a patrol-oriented investigative structure.

these terms means, since various administrators and theorists defined them in different ways.[21] However, all have one common theme: the desire to bring policing, especially at the patrol level, closer to the citizens whose lives and property are supposed to be protected.[22]

All of these organizational plans have involved making changes in the way patrol officers are deployed. Some community policing programs have included increasing the number of officers on the street, eliminating two-officer patrol cars in favor of one-officer cars, and requiring patrol officers to spend at least part of each day out of their cars, getting acquainted with the people in their assigned territories.[23] Many departments have reintroduced foot patrols that had been abandoned in the 1960s. A few departments have begun to use horse patrols, bicycle patrols, and other specialized forms of patrol. A few agencies have gone to the extreme of asking officers, as volunteers, to live in high-crime neighborhoods as a way of increasing the law enforcement presence.[24]

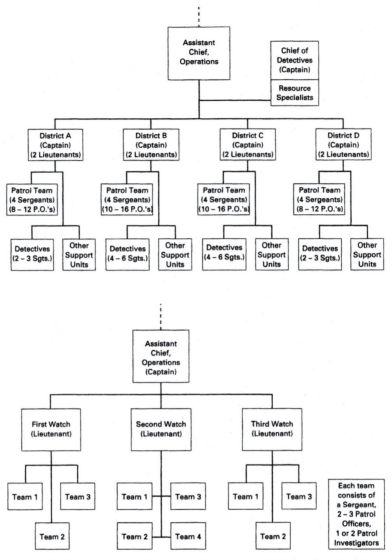

Figure 3.17. Two types of team-policing organizational structures.

Several agencies, among them the Detroit and Houston police departments, established networks of small "storefront" police stations at which citizens could receive a variety of police services. Usually these "storefront" operations are oriented toward crime prevention rather than law enforcement: "storefront" officers give workshops on self-protection, conduct security surveys, and organize neighborhood crime watches, for example (see Figure 3.18).

These innovations obviously have a significant effect on an agency's organizational structure. In Detroit, for example, the "storefront" crime-prevention stations were completely separate from the preexisting precinct stations. As a result, there was no communication between the crime-prevention officers and

Figure 3.18. New technologies and new ideas about the police role in the community have led to a revival of the foot patrol.

the patrol officers and investigators assigned to the precincts, until the crime-prevention officers developed their own informal methods of breaking down the organizational barriers such as inviting the patrol officers to stop by the "store-front" station for a cup of coffee and a doughnut on winter mornings.[25]

Community-oriented police organizations are difficult to evaluate, partly because there is no standard method of evaluating police agencies' performance, and partly because conducting a full-scale evaluation would be expensive.[26] There are substantial obstacles to exporting even the most successful innovations from one agency to another. However, the favorable publicity that has been given to some of these innovations will encourage administrators in many agencies to see what they can do along the same lines. The result may be a wealth of new ideas in police administration, and consequently in the organization of police agencies, as a new century begins.[27] We have much more to say about community policing and its effects on police administration in later chapters of this text.

REVIEW

1. The responsibility for coordinating any two elements in an organization rests with

   (a) The chief of police.

   (b) The executive.

   (c) The person who has authority over both of them.

   (d) The subordinate.

   (e) None of the above.

2. True or False: Subordinates should be encouraged to avoid referring questions of coordination to their superiors.

3. *Cross-training* is one method of

(a) Ensuring the accountability of subordinates.

(b) Increasing specialization.

(c) Reducing the length of the chain of command.

(d) Reducing the cost of training programs.

(e) Avoiding some of the undesirable effects of specialization.

4. Once a list of an agency's functions has been compiled, the administrator must match the list to the personnel available. To do this, the administrator must know

(a) The competencies of the available personnel.

(b) What changes can be made in the number of personnel and their competencies.

(c) The number of available personnel.

(d) All of the above.

(e) None of the above.

5. In organizing a police agency, the most important principle is that

(a) There can be only one chain of command.

(b) Every function must be performed by someone, everyone must have at least one specific function, and there must be a reason for every assignment.

(c) Every member of the agency must have one, and only one, function to perform.

(d) All ranks and functional titles must be standardized to avoid confusion.

(e) Administrators must not be permitted to delegate their responsibilities.

NOTES

[1] Arnold S. Tannenbaum, Bogdan Kavcic, Menachem Rosner, Mino Vianello, and Georg Wieser, *Hierarchy in Organizations*. San Francisco: Jossey-Bass Publishers, 1974, p. 2.

[2] George Maclay and Humphry Knipe, *The Dominant Man*. New York: Delacorte, 1972.

[3] Eugene J. Benge, *Elements of Modern Management*. New York: American Management Association, 1976, pp. 129–31.

[4] James E. Morgan, Jr., *Administrative and Supervisory Management*, 2nd ed. Englewood Cliffs, N.J.: Prentice Hall, 1982, pp. 62–63.

[5] Morgan, *Administrative and Supervisory Management*, p. 62.

[6] Roy R. Roberg, *Police Management and Organizational Behavior*. St. Paul, MN: West Publishing, 1979, p. 199.

[7] Walter R. Mahler, *Structure, Power, and Results*. Homewood, Ill.: Dow Jones-Irwin, 1975, pp. 42–43.

[8] Morgan, *Administrative and Supervisory Management*, p. 63.

[9] Anthony V. Bouza, *The Police Mystique*. New York: Plenum Press, 1990, p. 122.

[10] Mahler, *Structure, Power, and Results*, p. 40.

[11] O. W. Wilson and Roy C. McLaren, *Police Administration*, 3rd ed. New York: McGraw-Hill, 1972, pp. 80–81.

[12]Peter W. Greenwood, Jan M. Chaiken, and Joan Petersilia, *The Criminal Investigation Process*. Lexington, Mass.: D. C. Heath, 1977, pp. 115–20.

[13]Bouza, *The Police Mystique*, p. 24, points out that less than 20 percent of calls to the police from citizens involve emergencies or reports of criminal violations; the rest are requests for non–law enforcement services.

[14]Mahler, *Structure, Power, and Results*, pp. 137–39.

[15]David N. Ammons, *Municipal Productivity*. New York: Praeger, 1984, p. 35; Mahler, *Structure, Power, and Results*, pp. 20–23.

[16]Mahler, *Structure, Power, and Results*, pp. 31–33.

[17]Bouza, *The Police Mystique*, pp. 26–29, has some pithy comments on the traditional police organizational structure.

[18]Jerome H. Skolnick and David H. Bayley, *The New Blue Line*. New York: Free Press, 1986, pp. 4–6; Greenwood et al., *Criminal Investigation*.

[19]Ammons, *Municipal Productivity*, p. 34, but also note the difficulties in evaluating any type of police organization, pp. 64–66.

[20]Skolnick and Bayley, *The New Blue Line*, describe several examples. See also Ammons, *Municipal Productivity*.

[21]Lisa M. Reichers and Roy R. Roberg, "Community Policing: A Critical Review of Underlying Assumptions," in *Journal of Police Science and Administration*, vol. 17, no. 2, June 1990, pp. 105–14.

[22]Bouza, *The Police Mystique*, pp. 30–31; Skolnick and Bayley, *The New Blue Line*.

[23]Gary Enos, "Cities Take Diverse Paths to Community Policing," in *City and State*, vol. 9, no. 14, July 27, 1992, pp. 16–17.

[24]Clay Pearson and Charles Gruber, "Their Neighborhoods Are Their Beats," in *American City and County*, vol. 107, no. 5, April 1992.

[25]Skolnick and Bayley, *The New Blue Line*, p. 73.

[26]Ammons, *Municipal Productivity*, pp. 88–90.

[27]For one overview of community policing, see Jack R. Greene and Stephen D. Mastrofski, eds., *Community Policing: Rhetoric or Reality*. New York: Praeger, 1988.

RESEARCH AND PLANNING

One of the most frequent criticisms of police agencies is that they fail to plan for the future. Too often, according to the critics, police administrators are content to operate on a day-to-day basis without regard for the long-range needs and direction of their agencies.[1] As a result, they are often faced with crises that demand immediate attention but that could have been foreseen and perhaps avoided, if appropriate steps had been planned in advance.

Planning is an indispensable human activity; all people plan, at least for themselves. Setting an alarm clock at night is an act of planning. Buying groceries once a week requires planning. Putting money into a savings account or signing a loan agreement to buy a car represent forms of planning.[2]

In its broadest sense, planning is "the design of a desired future and of effective ways of bringing it about."[3] It is a continuous process: The plans we have made in the past are subject to revision, and new plans are made for the future because of new information about our current status or new ideas of what the future may hold for us.[4]

Left to itself, the future will not turn out the way we might like. Present conditions may continue indefinitely, or they may get worse. A supreme optimist might refuse to plan anything on the assumption, or hope, that future conditions will be better than those of the present. However, an administrator who is responsible for the operations of an agency that protects the public's lives and property cannot afford such blind optimism. Planning is an essential management function that is "directed toward producing one or more future states that are desired and that are not expected to occur unless something is done."[5]

THEORY OF PLANNING

Role of Research

As management theorist Arthur B. Troan, Jr., points out, "One of the most interesting and disquieting things about plans is that they rarely work out quite as anticipated."[6] There are a number of reasons for this to be true.

- Planning requires predictions to be made about the future, and these predictions are likely to be less than perfectly accurate.
- Planning is based on perceptions about current conditions, and these may be based on inaccurate or incomplete information.
- Planning is based on current objectives, or notions of what is desirable, and these can change over time so that before a plan is fully accomplished its objective may have been discarded.
- The actions chosen to carry out a plan may have results different from those that were expected.

It should be apparent from this list that all plans must be based on knowledge about current conditions, about the future, and about the effects of current actions, and that the accuracy and completeness of this knowledge will have a considerable bearing on the success of plans. This knowledge must be carefully assembled from a variety of sources. It does not simply come into existence by itself.[7]

Research is the systematic gathering of knowledge. To most people, research means one of two things: either conducting scientific experiments in a laboratory filled with exotic equipment or poring over dusty volumes in the dark recesses of a library. In fact, these stereotypes represent only two forms of research; there are several others. Briefly, these are the forms of research that are most often useful to police administrators:

Observation.

Events, people, or natural phenomena can be observed, described, and counted, and the results of observations can be compared.

Experimentation.

Processes, techniques, or actions can be carried out under controlled conditions and the results can be observed to determine whether the researcher's ideas about what will happen are correct.

Review of Previous Research.

Previous researchers' observations, the results of past experiments, and other researchers' theories or generalizations can be studied and compared to determine what is already known about a problem or subject.

Forecasting.

Predictions can be made about future conditions on the basis of observations of current conditions and apparent trends or patterns of change.

Of these four forms of research, the one most often used by police agencies is observation. Most police agencies amass a tremendous volume of records and documents concerning current activities and conditions.

These records and documents, although they are not often considered in this way, actually represent observations of events, people, and to a lesser extent natural phenomena. A detective who looks through a file of criminal *modus operandi* (method of operation) records is conducting research, comparing observations of past criminal behavior to current cases.

Ordinarily, a police agency's records are considered important only because they are needed to prosecute offenders and to demonstrate that the agency has acted in a lawful and responsible manner. However, these same records, if they have been assembled and maintained properly, contain valuable information that can contribute to the agency's planning for the future.

Experimentation rarely has been used as a method of research by police agencies. Police administrators in the past believed that they could not afford to divert their already limited personnel and other resources into experimental activities that might not succeed. An experiment that failed might attract public or political criticism or, worse yet, might result in damage to some citizens.

However, in recent years administrators have begun to recognize that without experimentation, new ideas about police techniques and procedures would never be implemented. A reasonable degree of caution certainly is necessary. An administrator who recklessly endangers public lives and property by trying every new idea that comes along could only be regarded as irresponsible. But it may be just as irresponsible to ignore promising new ideas, if they can be tested safely, only because "we've never done that before."[8]

Police administrators also have not taken advantage of previous research as well as they might. Information about current police practices, crime conditions in almost every community in the country, and the results of research on police techniques can be found in dozens of journals, magazines, and books. Not all of this information is useful. Nevertheless, a responsible administrator should be aware of the information available and should be willing to devote time and energy to finding information that may shed new light on current problems.

A forecast is a statement of expectations for the future based on historical experience, and it can be one of the most valuable forms of research for the police administrator. Forecasting future trends is an art and science in itself, practiced primarily by mathematicians and economists. However, relatively simple forecasts can be made by anyone who has learned the proper techniques.

For example, if records of past experience indicate that the volume of automotive traffic in a city has grown by an average of 2.5 percent per year over the past five years, it is reasonable to assume that this trend will continue over the next five years. This would mean that the volume of traffic five years from now will be 113 percent of the current volume. Such a simple forecast could be valuable to the administrator in planning for the number of personnel who will be needed in the future or changes in traffic law enforcement practices, for example.

Most large cities have their own forecasters in the city planning department whose job is to predict the city's future population, economic conditions, patterns of growth, and so forth. Economic and demographic (population measurement) forecasters also can be found in various college departments, especially in schools of business administration and economics; in private research bureaus; and in state and federal agencies. Detailed economic and demographic forecasts are published by these agencies and may be obtained directly from the source or through a public or university library. The U.S. Bureau of the Census, a part of the Department of Commerce, and the Bureau of Labor Statistics in the Department of Labor publish thousands of detailed economic and demographic studies for individual cities and counties. Police administrators should obtain this information and use it in making their own plans.[9]

Of course, even the best forecast is no guarantee that conditions or events at some future date will occur exactly as predicted. That is one reason plans must be revised and updated frequently. By and large, forecasts for the near future are more likely to be accurate than those for the distant future. Unfortunately, the long-range forecasts are more likely to be important to the administrator because they are necessary for long-range planning. Short-range plans are more often based on the assumption that current conditions will continue without any major changes. This is not a dilemma that can be solved; it is merely one of the uncertainties of life that every administrator must accept.

All four of the forms of research we have discussed can contribute to the planning process. What matters most is not the type of research, but that administrators base their plans on accurate, complete knowledge, obtained in a systematic way.

Responsibility for Planning

One of the responsibilities of the executive—the Chief of Police or equivalent position—is to see that planning is carried out. In smaller agencies, nearly all planning is actually done by the executive, because everyone else is occupied with other vital functions. All too often, the executive is also "too busy" and somehow never gets around to planning at all.

Larger police agencies sometimes go to the opposite extreme. They set up elaborate planning departments staffed with computer programmers, information specialists, and other experts. The Planning Department analyzes data, projects future trends, and prepares bewildering charts and proposals. Eventually the planners prepare a master plan that calls for the entire department to be reorganized over the next five or ten years. This plan is copied and distributed to the agency's upper-level administrators, who point out the impracticalities and inconsistencies in the master plan, which is then filed away and never seen again.[10]

Practical planning is not only possible, it is absolutely necessary. However, the responsibility for planning cannot be delegated to a separate planning department and then forgotten. The planning process must be carried out by the people who will be responsible for putting the plans into action: the working administrators of the agency. If the agency is large enough to afford a staff of planning experts and specialists, they can be immensely helpful in carrying out detailed research, in preparing forecasts, and in assisting the agency's other personnel during the planning process.[11] However, every police officer from the patrol force to the chief should have some part in that process.[12]

Types of Plans

A plan can be anything that represents a conscious decision to influence some future action or condition, such as setting an alarm clock. More formal kinds of plans, such as those we consider in this chapter, vary according to the

- Scope and nature of the action that is to be taken according to the plan.
- Period encompassed in the plan.
- Number of times the plan, or minor variations of it, will be repeated.
- Definiteness of the plan.

Some management theorists consider all policies, standard operating procedures, and rules to be "standing plans," in the sense that they represent conscious decisions that are intended to influence future acts.[13] For example, a policy or set of regulations that describes the proper procedures for issuing a traffic citation could be regarded as a "standing plan for the issuance of traffic citations." However, for our purposes we will set aside the "standing plan" as a special case.

The kinds of plans that will concern us are designed not just to standardize employees' work performance, but to deal with unusual future events or to bring about desirable changes in an agency.[14] We discuss four types of plans: operational plans, contingency plans, programmatic plans, and strategic plans.

Operational Plans.

An operational plan is one that concerns a limited scope of actions, encompasses a relatively short period and may be repeated with some variations. Its purpose is to provide for the coordination of several persons, activities, or processes to achieve a particular desired outcome.

The most common operational plans are schedules of people, things, and processes. For example, a work schedule or duty roster lists the various jobs that are to be performed or the work that is to be done at different times and shows which personnel are assigned to each job or work period. The desired outcome, of course, is that all the work is to be completed.

Similarly, a schedule for the processing of the agency's payroll is an operational plan. Various supervisors and managers must report to the accounting office the number of hours each employee has worked; the personnel office must report the correct salary rate for each employee; the salaries, deductions for benefits and taxes, compensation for overtime, and so forth must be calculated; and, finally, the checks must be prepared and distributed. Without a schedule, there is no assurance that everyone would be paid on time.

Another type of operational plan involves the coordination of personnel in carrying out complex actions to achieve a specific, limited purpose such as a felony arrest, a raid on a suspected crime site, or a surveillance action. This plan does not have to be recorded in any elaborate form, but it should be outlined on paper in advance and a record should be kept for future reference. The purpose of the plan is to ensure that everyone involved in the operation knows what to do and when to do it, and that as far as possible everything that might occur during the operation has been taken into account.

Contingency Plans.

A police agency's responsibilities are much broader than those of most governmental units. To some degree, the police must be prepared for almost anything that might happen to endanger citizens' lives and property. Most police agencies are expected to assume responsibility for both security and rescue operations in all sorts of emergencies, from violent weather to riots to plane crashes.

Such events are so infrequent that it is impractical to establish standardized policies and operational plans to deal with them. But once an event occurs, it is too late for thoughtful, deliberate planning; the police must respond immediately in an efficiently coordinated manner. Thus, plans must be made well in advance and kept ready in case they are needed. They will go into effect only if certain events occur; that is, the plans are contingent on the event or condition and thus are known as contingency plans.

It is impossible to give detailed recommendations as to the number and variety of contingency plans a department should maintain. But at the very least, even a small police department should have contingency plans to deal with the following kinds of events:

I. Natural disasters.
   A. A storm (hurricane, flood, tornado, earthquake) that causes widespread property damage, personal injuries, and disruption of communications.
   B. A major fire or explosion in a commercial district.
   C. A major fire or explosion in a residential district.
II. Transportation disasters.
   A. A commercial aircraft crash.
      1. At the local airport.
      2. Away from the airport, in a commercial, residential, or undeveloped area.
   B. A multivehicle collision on a major highway or arterial street.
   C. A railroad derailment or accident.
      1. Involving passengers.
      2. Involving bulk freight (coal, grain, etc.).
      3. Involving hazardous cargo (chemicals, radioactive material, gases).
   D. A pipeline or storage tank explosion or major leakage.

III. Public events and civil disorders.
    A. A protest demonstration.
        1. In a confined area.
        2. Involving a large area or parade.
    B. A riotous attack on, or invasion of, a public building.
    C. An actual or threatened bombing of a public building.
    D. A potentially riotous disturbance at a public event.
    E. A parade or other public appearance of a celebrity or controversial person requiring special security.
IV. Miscellaneous events.
    A. An aircraft hijacking.
    B. Kidnapping or hostage taking of a public figure or group of individuals.
    C. A barricaded sniper.

Each contingency plan should describe in as much detail as possible the kind of event it is intended to cover and all the circumstances that can be anticipated. A single plan might cover several variations. For example, the police procedures outlined in a plan for a disturbance at a public event (III-D in the preceding list) would vary depending on whether the event is a dance, rock concert, street party, or sports event, and also would vary depending on whether the event is occurring inside a dance hall, in a stadium, or in a city park.

Each contingency plan should specify

1. Who will decide that the plan is to be implemented.
2. Who will be in charge of carrying out the plan (the commanding officer).
3. What departmental units and personnel will be assigned to the operation, in what capacities, and how they are to be informed.
4. Where personnel are to assemble for instructions, distribution of special equipment, and deployment.
5. What actions may be taken at the discretion of individual personnel and what actions are to be taken only on the instructions of the commanding officer.
6. Who will serve as subcommanders or intermediate officers.
7. When and by whom the operation will be terminated, and how personnel will be informed.

In preparing contingency plans, care must be taken not to include resources that may be unavailable. For example, it would be foolish to plan to call up off-duty officers by telephone in the event of a major disaster, because the telephone system is likely to be disrupted by the disaster itself. The plans themselves may need to address the availability of specific resources, the competencies of personnel whose participation will be required, and so forth. Sometimes the process of developing contingency plans will reveal deficiencies in personnel that need to be corrected, either through selective recruiting or new training programs.

A great deal of time will be required to develop a complete set of contingency plans if none exists. However, once the plans have been developed, reviewed by all the administrators who are likely to be involved in carrying them out, and approved, they can be retained on file indefinitely.

At least once a year, the file should be reviewed. Any plans that are clearly out of date should be pulled and revised, and new plans should be drafted to cover situations that were not previously anticipated.

Programmatic Plans.

The most important plans for any organization, including a police agency, are those that specify a course of action that, if successfully accomplished, will contribute in an important way to carrying out the organization's basic purpose or function.[15] These plans represent the agency's program for fulfilling its function. They are crucial to the success of the whole organization. Yet this is the kind of planning that is often not done at all by police agencies.

Programmatic plans involve a series of complex actions to be undertaken by several people over a relatively long period, usually no fewer than six months and sometimes as much as two to five years.

The most common type of programmatic plan is the budget.[16] In its simplest form, a budget is merely a statement of the financial resources one expects to have over a given period and how those resources are to be used. Virtually all police agencies are required to prepare a budget on an annual basis to obtain operating funds from the parent government. Because budgeting is such an essential process, we devote a separate chapter to it; for now, we merely wish to point out that a budget is a plan.

A programmatic plan can be designed to do nothing more than to maintain the status quo, the current level of accomplishment in fulfilling the agency's basic functions. It is easy for administrators to persuade themselves that the status quo is entirely satisfactory or that improving current conditions would be either impossible or not worth the extra effort.

At the opposite extreme, administrators sometimes develop plans for an ideal world in which their resources are limitless, and all contemplated activities are successful. In most cases, a better approach avoids both extremes. An administrator should plan for specific changes that bring the organization closer to ideal conditions by adopting a realistic estimate of resources and courses of action that are likely to be successful in a reasonable period.[17]

Planning for constructive change is one of the most challenging, and potentially most rewarding, tasks a manager can undertake. Most managers feel a great sense of pride and satisfaction when they see their plans successfully put into effect, bringing about measurable progress toward their agency's fundamental purpose. It is not unlike the feeling that athletes get when they have won an important victory: It is the payoff for the creativity and hard work they have devoted to their efforts.[18]

Programmatic plans can take one of two basic forms: the project and the program.

A project involves the introduction of a single, specific change in an organization over a limited period.[19] Once the change has been introduced and the project ends, a new situation may exist that will continue indefinitely thereafter.

For example, a project might involve installing a computer-assisted dispatching system in the police communications network. First the system must be designed, proper equipment must be chosen and purchased, and space must be found to house it. While the equipment is being installed and tested, the existing system will continue to be used but the communications personnel will be trained on the new system. Perhaps for a while both systems will be operated in parallel, or perhaps the new system will be introduced in stages. Eventually the old system will be discontinued and the new equipment will take over. At that point, which may be six months or a year after the first steps, the project would end.

A project also can be used to experiment with a new idea, technique, or type of equipment. At the end of a predetermined period, the project is evaluated and a decision is made to discontinue the innovation, to install it permanently, or to continue the testing period, perhaps with some modification.

For example, a police department might decide to experiment with the use of two-wheel motorcycles for traffic patrol. Initially perhaps only a few motorcycles would be purchased and assigned to one section of the city. After six months or a year, the experiment would be evaluated. Either more cycles would be bought and assigned throughout the city, or the motorcycles would be assigned to another part of town for further testing.

A project can be used when the need for a particular kind of activity is likely to be temporary. A police department might conduct an anti–shoplifting campaign just before the Christmas shopping season, for instance.

Finally, a project can be used when there is an ongoing need for a certain kind of activity, but the agency does not have adequate resources to carry it out continuously. For example, a police department might conduct a short-term campaign to educate the public about burglary prevention, the dangers of driving while intoxicated, or any number of other matters. Few agencies could afford to devote officers full-time on a permanent basis to such projects, but personnel may be able to spend a few weeks on each campaign. Often a series of short-term projects of this sort can be linked together to form a continuous effort.

A program is also a type of programmatic plan that is designed to meet a specific need through a specific course of action, with the desired results stated in advance, and a program should be designed for a specific period.

The difference between a project and a program is that the latter is intended from the beginning to be a permanent part of the organization.[20] Changes may be introduced from time to time, but ordinarily a program continues from year to year until the need for it no longer exists.

Furthermore, a program is designed to fulfill one of the basic, permanent functions of the agency. Conversely, each of the agency's basic functions or purposes should be represented by a particular program. Each program should be designed to bring about desired changes that will enable the agency to fulfill its purposes more completely and effectively in the future, considering reasonable expectations of future resources and the kinds of activities that will be feasible.

A program can be composed of a single set of actions that are carried out continuously, or it can be made up of a series of separate projects. For example, a department's crime prevention program might consist entirely of the patrol bureau's routine patrol activity, or this activity could be supplemented with a series of short-term projects, each aimed at reducing the incidence of a particular type of crime.

Program planning and budgeting are so closely related that we will have a great deal more to say about programs and how they are developed in the next chapter.

Strategic Plans.

Every police agency—indeed, every organization of any sort—undergoes change more or less continuously. The programmatic plans we have just discussed are intended to bring about desired changes, to improve an agency's ability to serve its community.

However, it is not always easy to see how several individual programs and projects fit together into an overall plan of progress. What is needed is a strategic plan, a master plan for the entire organization, showing how all the bits and pieces of programmatic planning are interrelated and coordinated to achieve, over a substantial period (perhaps five to ten years), large-scale improvement in the whole organization.[21]

A strategic plan, because it covers such a long period, is most vulnerable to unpredictable future changes. By the same token, it is most dependent on accurate knowledge of current conditions and accurate forecasts of future conditions. One way to avoid the possibility that a strategic plan will become obsolete is to take into consideration several different possible futures.

For example, if the agency's present personnel strength is one commissioned officer for every 1250 citizens in the community, the strategic plan might call for increasing overall strength to a ratio of 1:1000. However, the actual number of officers would vary according to the rate of population growth in the community. The strategic plan should consider these variations. Thus, the plan might show the personnel levels for three different growth rates.

Present Personnel: 100
Present Population: 125,000
Personnel/Population Ratio: 1:1250

If Growth Rate Is	Population in 5 Years Will Be	Personnel in 5 Years[a]
1% per year	131,400	131
2.5% per year	141,400	141
5% per year	159,500	160

[a]To achieve personnel/population ratio of 1:1000.

The strategic plan should contain several projections of the same sort showing how personnel needs, the community's need for police services, and so forth are likely to change over the period encompassed by the plan and how the agency proposes to respond to these changes. The value of a strategic, or master, plan is that it can serve as the overall guide to the development of specific program and project plans, and each year's budget can be compared with the predictions given in the strategic plan. Of course, like any other long-range plan, the strategic plan itself should be reviewed and updated periodically, at least every two years. Otherwise it is likely to become obsolete and useless except as a historical curiosity.

PLANNING PROCESS

Perhaps one reason many police administrators do not plan well, or do not plan at all, is that they are intimidated by what seems to be a hopelessly complicated and mysterious planning process. Actually, planning need not be complicated or mysterious if the administrator follows an orderly, logical procedure. However, planning is undeniably time-consuming, and the better it is done, the more time it takes. It can and should be done on a group basis, with various parts of the process carried out by different individuals, so that no one must invest a disproportionate amount of time in the whole process.

Briefly, there are six steps in the planning process:

1. Needs assessment.
2. Goal and objective setting.
3. Definition of alternative methods.
4. Cost-benefit assessment.
5. Selection of methodology.
6. Definition of evaluation method.

Needs Assessment

Before any plan can be designed to bring about some action or condition in the future, first it must be determined that the action or condition does not already exist and that it is desirable. In other words, needs assessment requires the administrator to ask, "What should we do (or have) that we do not already do (or have)?"[22]

Management scholars Newman, Warren, and Schnee compare this step with the diagnostic process in medicine. They warn, "Accurate diagnosis is the essential first phase of sound decision-making. Unless the diagnosis is correct, subsequent planning will be misdirected and wasteful."[23] They suggest several ways to make an accurate diagnosis.

- Compare existing conditions with some basic value, such as a statement of the organization's purpose.
- Compare existing conditions with conditions in another, comparable organization.
- Compare existing conditions with long-term goals.

Each of these comparisons may reveal a gap between the existing condition and the desired condition. For example, suppose one of the basic purposes of the City Police Department is to promote traffic safety and reduce, to the minimum attainable, deaths and injuries resulting from traffic accidents. How well is City Police Department serving this purpose?

According to the agency's records, last year there were 36 traffic deaths, or 2.73 per 10,000 population, and 624 traffic injuries, or 47.27 per 10,000 population. Is that acceptable or not?

The city of Springdale has about the same population as City. Last year, there were 21 traffic deaths and 476 injuries in Springdale; clearly, then, City is not doing as well as Springdale.

How well should City be doing if it is to serve its ultimate purpose to the best of its ability? One way to measure progress is to set a realistic but challenging goal. In this case, City Police Department might decide that traffic deaths should be reduced to not more than 1 per 10,000 population per year, and injuries from traffic accidents to not more than 25 per 10,000 population per year.

These comparisons can be quite dramatic:

	Traffic Deaths		Traffic Injuries	
	Number	Ratio[a]	Number	Ratio[a]
City, current[b]	36	2.73	624	47.27
Springdale	21	1.62	476	36.62
City, long-term goal	13	1.00	330	25.00

[a]Number per 10,000 population.
[b]Based on most recent year's records.

There are several things that you should notice about this simple example. First, information has been collected from various sources; in this case, from the two police departments. The accuracy of a diagnosis based on this information will be no greater than the accuracy of the information itself. Most police departments routinely keep all sorts of records. Unless the records are correct and are compiled in a consistent manner, important administrative decisions are likely to be wrong.

Also, it is tempting to look for needs only where the information is readily available. If City Police Department did not keep records of traffic deaths and injuries or if the records were not readily accessible, the administrator might never think of this as a possible area of need, unless a deliberate effort is made to evaluate the entire agency's performance, in comparison with basic values and long-term goals, from time to time. When this is done, deficiencies in the information system are likely to be revealed when the administrator discovers that necessary information is not available.

Notice too that we have not indicated how the agency's basic values and long-term goals were produced, or by whom, or where they came from. In fact, they can come from many sources: the parent government, statewide or national law enforcement advisory groups, planning agencies, public demands, or the administrator's own personal values and beliefs.

Wherever the values and long-term goals come from, they are useless as a basis for comparison unless they are expressed in specific terms. In our example, the value statement is fairly vague: "to promote traffic safety and to reduce, to the minimum attainable, deaths and injuries due to traffic accidents." In the absence of any comparison, one might argue that the current death and injury rates are the "minimum attainable" and that City Police Department is satisfying this goal. But the long-term goal is stated in precise terms and there is no room for argument: The current conditions are far short of the goal.

Finally, notice that in this example we have not tried to suggest why there is a gap between current and desired conditions. To use Newman, Warren, and Schnee's medical analogy, we have identified only the "symptoms" but not the underlying cause of the "illness."[24] Sometimes that is as much as can be done, and many problems can be treated "symptomatically." If you relieve or eliminate the symptoms, it is reasonable to assume that the underlying illness is controlled if not "cured." If City Police Department took any action that resulted in a reduced traffic death and injury rate, no one would care whether the real cause of the problem had been corrected.

In fact, often the real cause of a problem cannot be discovered or is not accessible for "treatment." Many social scientists believe that a large part of the crime problem in our society is the result of fundamental social conditions such as poverty and ethnic discrimination. Although this theory may or may not be correct, it hardly matters from the police administrator's viewpoint because the police have little opportunity or authority to change these social conditions. The police can only treat the symptoms, the crimes themselves.

Not all problems can be treated symptomatically. For example, a police administrator might find that the agency is experiencing an abnormally high rate of personnel turnover. Instead of an acceptable rate of 8 to 10 percent per year, the rate has climbed to 15 percent. Before any action can be taken to correct the problem, the causes of the problem must be discovered.

If young officers with less than five years' experience are quitting to go into some other kind of work, that would suggest problems with the agency's recruitment and selection process. If young officers are leaving to join another police agency that pays better, that would suggest an urgent need to adjust the salary schedule.

But if the increased turnover rate can be attributed to an unusual number of older, experienced officers who happen to be reaching the normal retirement age at the same time, the administrator's response would be entirely different. Perhaps it would be necessary to bring in experienced personnel from another agency, possibly by offering higher salaries, to fill the middle-management ranks, or perhaps less-experienced younger officers can be quickly trained and promoted. In any case, the turnover problem should not be treated symptomatically, such as by rushing to the City Council for an immediate across-the-board pay raise, until the root cause can be determined.

It is also important, especially at a time when the general public is demanding more involvement in decision making about their government, for police administrators to consider the needs and interests of their "clientele": the people who every law enforcement agency is intended to serve.

In the past, the police too often have ignored the voices of "outsiders" and "mere civilians," sometimes using the excuse that professional law enforcement should not be "politicized."

The fact remains that the people of a community pay the taxes that allow the police department to exist, and they either benefit or suffer from the agency's daily operations.

Unfortunately, determining the public's needs and wishes is no easy matter. There is no single, monolithic public; there are many separate publics, even in a small community, and each segment of the community may have its own unique needs, interests, values, and desires for police service. If a law enforcement agency tries to accommodate every segment of the community, it could wind up with a mass of inconsistent policies and procedures, or it could be justifiably accused of paying heed only to the loudest and most insistent voices.

One solution is to conduct a periodic, systematic survey of citizens' expectations, needs, and concerns for police service.[25] Few law enforcement agencies are likely to have the expertise, or for that matter the time, to conduct such a survey, but perhaps the local college has a class in sociology or in market research that would be interested in undertaking the survey as a class project. Another alternative might be to find a commercial market research firm willing to do the survey free of charge, for community good will.

Once a need has been identified—that is, the administrator has determined that some current condition falls short of what is desirable—the next step is to decide the extent to which the undesirable condition should be corrected within a given period.

Goal and Objective Setting

All of an organization's goals and objectives ultimately are derived from its basic values and purposes. As we saw earlier, basic values usually are stated in such general terms that, by themselves, they cannot be applied directly to immediate problems; first they must be restated in specific terms.[26]

A specific statement of an agency's basic values or purposes is an expression of a goal toward which the agency is working. In the example of City's traffic death and injury rates, the goal is to reduce deaths to a ratio of 1 per 10,000 population and to reduce injuries to a ratio of 25 per 10,000 population. The obvious advantage in stating a goal in such specific terms is that progress toward meeting it can be measured, and particular activities can be designed to achieve it. Such a goal is sometimes called an *operational definition* of an agency's values: It defines the value in measurable terms.

Once values and goals have been operationally defined, it is possible to assess needs in comparable terms and to determine the root causes for those needs. For example, careful study of the pattern of traffic accidents in City might point to three major causes of the unacceptable death and injury rates.

1. More than 25 percent of deaths and more than 30 percent of injuries occur in accidents on the Central Expressway, and the majority of these accidents occur outside of the Loop Expressway.

2. More than 65 percent of the accidents resulting in a death and more than 45 percent of accidents resulting in injuries involve a person who is driving while intoxicated (DWI). About 35 percent of the DWI drivers are males under 30 years of age who are either military personnel assigned to Wingtip Air Force Base or students at City University.

3. Of the accidents resulting in death or injury other than those on the expressway, and other than those involving drunken drivers, the majority occur at street intersections where there are traffic controls. Nearly all such accidents involve a driver who fails to observe the signal and yield the right of way.

This information suggests the directions that City Police Department might take to achieve its goals. Apparently there are several interrelated causes of the present unacceptable condition. The next step for the planner or administrator is to translate these findings into objectives that can be attained within a reasonable period. Setting objectives is an exercise in judgment based on the planner's own past experience, the experience of others, and creative imagination. Objectives must be stated in measurable terms, and they should be high enough to challenge the organization but not so high that success will be impossible.[27] In our example, the following goals might be set:

1. The number of serious accidents on the Central Expressway outside of the Loop Expressway will be reduced by 15 percent in one year and 25 percent in two years.

2. The number of accidents involving DWI drivers who are military personnel or students will be reduced by 25 percent each year during the next two years.

3. The number of accidents at major intersections controlled by traffic signals will be reduced by at least 10 percent each year during the next two years (see Figure 4.1).

The administrator can check these specific objectives against the long-term goals by simply computing how well the goals will be achieved if the objectives are achieved. In this case, if all three objectives are met, the result after two years would be 12 fatalities (0.91 per 10,000 population) and 201 injuries (15.23 per 10,000), which is better than the long-term goal. However, these calculations assume that there is no overlap among the three causes of traffic accidents. In fact, it is very likely that some of the expressway and intersection accidents involve drunken drivers. Still, even if there is some overlap, the attainment of the three specific objectives listed above would mean that City Police Department's ultimate goals would be met.

The next question that must be answered is how best to achieve these objectives.

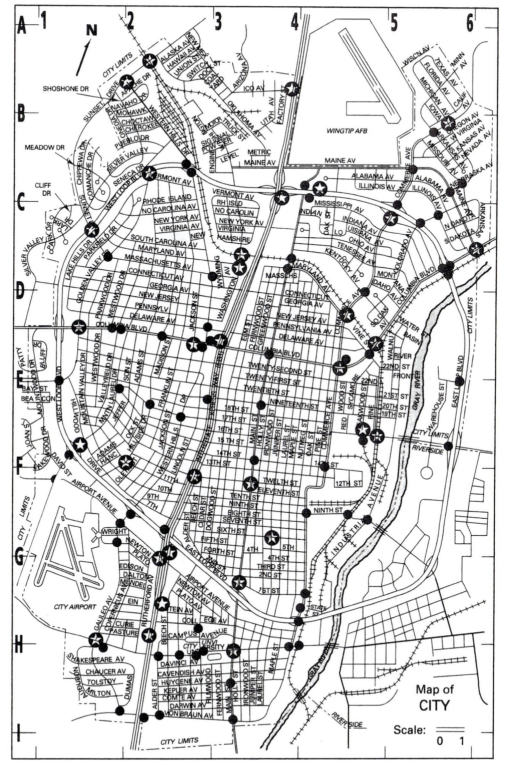

Figure 4.1. Map of City showing traffic accident incidence.

DEFINITION OF ALTERNATIVE METHODS

Almost any objective might be reached by several different methods, in the same way that you could get from where you are to where you want to be by any of several different routes. The search for the best method is one of the most creative parts of the planning process because it relies largely on the ingenuity, experience, and knowledge of the planner.[28]

There are three sources of alternative methods to reach an objective.

1. What has worked in the past.
2. What other, similar organizations have done.
3. What the planner can imagine.

The search for alternative methods should not be limited. The idea is to produce the maximum possible number of alternatives, no matter how impractical or unrealistic they might appear at first. Often a ridiculous idea can be modified or adapted until it is made practical, or it may suggest another approach that is practical.

A separate list of alternative solutions should be made for each objective, even though some methods might contribute to two or more related objectives. In our example, the planner might list the following methods of attaining the second objective:

A. Intensively patrol taverns and other drinking places frequented by military personnel and students, especially during high-accident-frequency periods.

B. Equip all patrol units with field breath analyzer kits, and train all officers in field sobriety testing and DWI arrest techniques.

C. Conduct a public education campaign on the dangers of DWI and on City Police Department's strict enforcement policy.

D. Ask the City Council for an ordinance requiring tavern operators to limit the sale of alcohol to one drink per hour per patron during the evening.

E. Ask tavern operators to inform the police whenever a patron who may be intoxicated leaves the tavern.

F. Ask City University officials to change their current policy, which forbids alcoholic beverages on campus and at school-related events; if students drink on campus, they are less likely to drink away from campus and drive home.

G. Ask Wingtip AFB officials to cooperate by imposing stiff penalties on any personnel who return to the base intoxicated, especially if they are driving.

These examples certainly do not exhaust the possibilities. You should be able to add several more alternative solutions to this list. The more alternatives one can develop, the greater the likelihood that the most effective, successful method will be chosen to meet the plan's objectives.

In looking over the list of alternative methods, notice that some would be much more expensive or difficult to implement than others. Some methods also might be more likely to produce beneficial results than others. Of course, an expensive solution that works may be preferable to an inexpensive solution that does not work. The next step in the planning process involves just such considerations.

Cost-Benefit Assessment

Every activity an agency undertakes involves some cost and, if the activity is worthwhile, produces some benefit. The administrator's hope is to produce the maximum possible benefit for the minimum possible cost. Cost-benefit analysis or assessment simply means comparing the costs and benefits of alternative activities.

In carrying out this analysis, it is important to include all costs, even though some of the resources needed to carry out an activity might already be available. If you assign patrol officers to a new activity, they will no longer be available to do what they are now doing. Either that activity must be dropped, at the expense of whatever benefit it may have produced, or the officers must be replaced. However, some indirect costs such as administrators' salaries and general overhead can be omitted from the analysis because these costs generally do not change regardless of the agency's activities.

Let us see what a cost-benefit analysis for our seven proposed activities might show:

OBJECTIVE 2. REDUCE NUMBER OF ACCIDENTS INVOLVING DWIS BY 25 PERCENT PER YEAR FOR EACH OF TWO YEARS

Activity	Estimated Costs	Estimated Benefits (%)[a]
A	Three patrol officers × 6 hours a day × 313 days per year at $30 an hour per officer = $169,020	60
B	Breath analysis kits, 20 at $250 + training program, 40 officers × 24 hours of training at $30 an hour per officer = $33,800	40
C	Campaign development, 2 officers × 80 hours at $30 an hour per officer + literature and materials, $5000 + implementation, 2 officers × 500 hours at $30 an hour per officer = $39,800	20
D	Administrative time to develop ordinance and obtain council action, 40 hours at $75 an hour = $3000	5
E	Development of literature, 1 officer × 8 hours at $30 an hour + distribution of literature, 2 officers × 10 hours at $30 an hour = $840	5
F	Administrative time to consult with City University officials, 4 hours at $75 = $300	5
G	Administrative time to consult with Wingtip AFB officials, 4 hours at $75 = $300	2.5

[a]The benefits are shown as a percentage of the total objective that each activity is likely to produce. These estimates are based on past experience and intuition, and may later prove to be incorrect. In this example, if Activity A is carried out, 60 percent of the objective should be achieved.

The analysis in this example can be converted into a dollar cost for each percentage of the objective that each activity is expected to produce.

Activity A	$169,020/60%	=	$2817
Activity B	33,800/40	=	845
Activity C	39,800/20	=	1990
Activity D	3000/5	=	600
Activity E	840/5	=	168
Activity F	300/5	=	60
Activity G	300/2.5	=	120

In this example, no single activity is expected to attain the entire objective, but two or more activities could be combined. If Activities A and B were combined, the predicted benefit would be 100 percent of the objective at a total cost of $202,820, or $2028.20 per percent. Or you could combine Activity A with all of the others except B. This would cost $213,260 and would meet 97.5 percent of the objective at a cost of $2187.28 per percent. No other combination of proposed activities would be expected to achieve more than 90 percent of the objective.

In calculating costs, it may be possible to consider alternative resources. For example, in Activities C and E, it was assumed that commissioned officers would be involved in the development of literature and related materials. However, at least some of the development work might be undertaken by civilian employees, probably at much less cost, or even by volunteers at no direct cost to the agency.

Some of the activities might be funded through a special grant from a state or federal source, in which case there would be no added cost to the agency, or perhaps a greatly reduced cost if the agency is required to match part of the grant. We discuss possible funding sources in Chapter 5.

Once the costs and benefits for each proposed activity have been calculated, the choice of activities to be implemented may seem obvious. Sometimes, in fact, cost-benefit analysis points to one alternative as the obvious choice. More often, as in this case, other considerations may affect the planners' decisions.

Selection of Methodology

The estimates of cost are based on factors that should be easily known and accurately estimated such as the number of people needed to carry out an activity, their salaries and related expenses such as the cost of operating their vehicles, and the number of hours they are likely to spend on the activity.

The estimate of benefits is not always so clear-cut. Unless the same activity has been carried out in the past, there may be very little experience on which to base an estimate. However, the estimate always should be based on an assumption that the activity will be successful.

For instance, in the cost-benefit analysis for Activity G, we would assume that the Wingtip AFB officials will agree to cooperate. Even so, the activity will have relatively little effect because the military personnel will soon find ways to avoid getting in trouble with their superiors, but probably would continue drinking away from the base.

Before actually choosing an activity, the administrator must consider whether there actually is a reasonably high probability that an activity will be successful. If the Wingtip AFB officials are not likely to cooperate, the expense of Activity G, though relatively small, would be completely wasted. Similarly, Activity D is not very expensive, but is it likely that the City Council would pass an ordinance limiting the sale of liquor? How vigorously would tavern and restaurant owners object? If the ordinance were passed, how would it be enforced? Would City Police Department then have to increase patrol activities to enforce the new law?

Activity E involves soliciting the tavern owners' cooperation. If completely successful, this activity would produce about the same benefit as Activity D but at much lower cost. The planner must consider whether, in fact, the tavern owners probably would be responsive to this idea.

The administrator also must consider the effects of each activity on the other operations of the agency. The tavern patrols proposed in Activity A would require three patrol officers for six hours each day. If regular patrol officers are diverted from their beats for this purpose, other types of criminal activity might increase. If the taverns are concentrated on one or two patrol beats, the officers assigned to those beats might spend nearly all their time on tavern patrols and the rest of their duties might be neglected. Or additional officers may have to be hired just to perform the tavern patrols, and such specialization usually is not desirable.

Ultimately, the administrator's choice of methodology must be based on experience, knowledge of other agencies' successful practices, understanding of the community, intuition, and judgment. Any program adopted by a law enforcement agency must meet three essential criteria.

- The program must be effective; it must accomplish some part of the agency's central purpose.
- The program must be legally and constitutionally acceptable.
- The program must conform to the community's sense of what is appropriate and proper for its law enforcement agency.

Whatever decision is made, it should never be regarded as final and immutable. If a particular activity does not work the way the administrator thought it would, it should be discarded and replaced.

First, however, the administrator must have some way of knowing whether the activity works.

Definition of Evaluation Method

Whenever a plan is first developed, it should contain an explicit statement of the methods that will be used to evaluate its success. Usually the method of evaluation is implied by the objectives set for the program. In the example we have been using, the objective was to reduce the number of accidents involving drunken drivers by 25 percent each year for two years. The method of evaluating this program, then, would be to study the accident records at the end of each year to determine whether DWI-related accidents have been reduced by 25 percent.

If a program involves several related activities, it may be desirable to evaluate whether each activity is contributing as much benefit as was predicted. Again, the method of evaluation should be specified in advance, when the activity first begins. However, it is sometimes impossible to tell which activity has caused a certain kind of event not to happen. In our example, if DWI-related accidents were reduced by 25 percent, it would be difficult to know which of several related activities were responsible.

Not only should all new projects and programs be evaluated, but the agency's existing programs and use of resources, especially personnel, also must be evaluated periodically. In many cases, an administrator will discover that officers, civilian employees, facilities, or equipment can be diverted from programs that are ineffective or no longer needed, and can be applied to new activities that will increase the agency's effectiveness at essentially no additional cost.

Almost always, careful evaluation will show that some programs are successful, perhaps even more so than was expected, whereas others are not entirely successful. At that point, the evaluation becomes, in effect, a needs assessment for the next round of planning and the process has come full circle.

REVIEW

1. Plans often do not work out as anticipated because

(a) Employees are unable or unwilling to do their jobs properly.

(b) Most police agencies cannot afford the highly specialized experts needed to make successful plans.

(c) Plans usually are based on less than perfectly accurate information and predictions.

(d) Administrators are not competent to develop good plans.

(e) Most planners are unrealistic about the resources that will be available to carry out their plans.

2. _____ is a form of research in which processes, techniques, or actions are carried out under controlled conditions to determine whether the researcher's ideas about what will happen are correct.

3. A programmatic plan that involves the introduction of a single, specific change in an organization during a limited period is

(a) A project.

(b) A budget.

(c) A program.

(d) A forecast.

(e) An experiment.

4. A _____ plan is intended to show how all the separate programs and projects are interrelated and coordinated to achieve large-scale improvements in an organization during a substantial period.

5. An operational definition

(a) Serves as a comparison of current conditions with a desired future condition.

(b) Expresses an agency's values in measurable terms.

(c) States a specific objective to be achieved in a specific period.

(d) Is a type of operational plan.

(e) Defines the cost-benefit ratio for an activity.

NOTES

[1] For example, see Jerome H. Skolnick and David H. Bayley, *The New Blue Line*. New York: Free Press, 1986, pp. 225–26.

[2] William R. Osgood, *Basics of Successful Business Planning*. New York: American Management Association, 1980, pp. 1–2.

[3] Russell L. Ackoff, *A Concept of Corporate Planning*. New York: Wiley-Interscience, 1970, p. 1.

[4] Merritt L. Kastens, *Long-Range Planning for Your Business*. New York: American Management Association, 1976, p. 151.

[5] Ackoff, *Corporate Planning*, p. 3.

[6] Arthur B. Troan, Jr., *Using Information to Manage*. New York: Ronald Press, 1968, p. 25.

[7] Jack L. Kuykendall and Peter C. Unsinger, *Community Police Administration*. Chicago: Nelson-Hall, 1975, p. 111.

[8] Skolnick and Bayley, *The New Blue Line*, p. 127, point out a curious variation on this theme: police officers, and administrators, who dismiss innovative ideas by saying either, "We're already doing that" (when in fact they are not), or "We tried that and it didn't work" (when in fact they have not tried anything remotely similar).

[9] For more information about forecasting, see John C. Chambers, Satinder K. Mullick, and Donald D. Smith, *An Executive's Guide to Forecasting*. New York: Wiley, 1974; William Ascher, *Forecasting: Appraisal for Policy-makers and Planners*. Baltimore: Johns Hopkins University Press, 1978.

[10] Louis A. Allen, *Making Managerial Planning More Effective*. New York: McGraw-Hill, 1982, pp. 5–8.

[11] Kastens, *Long-Range Planning*, p. 4.

[12] Allen, *Managerial Planning*, p. 277.

[13] William H. Newman, E. Kirby Warren, and Jerome E. Schnee, *The Process of Management*, 5th ed. Englewood Cliffs, N.J.: Prentice Hall, 1982, pp. 56–58.

[14] Osgood, *Successful Business Planning*, pp. 7–10.

[15] Allen, *Managerial Planning*, pp. 251–67.

[16] Ibid., pp. 8–9.

[17] Ackoff, *Corporate Planning*, pp. 8–20.

[18] Allen, *Managerial Planning*, p. 272.

[19] Newman, Warren, and Schnee, *The Process of Management*, p. 77.

[20] Ibid., pp. 73–80.

[21] Osgood, *Successful Business Planning*, pp. 31–32.

[22] Newman, Warren, and Schnee, *The Process of Management*, pp. 111–12.

[23] Ibid., p. 110.

[24] Ibid., pp. 112–14.

[25] R. M. Patterson, Jr., and Nancy K. Grant, "Community Mapping: Rationale and Considerations for Implementation," in *Journal of Police Science and Administration*, vol. 16, no. 2, June 1988, pp. 136–43.

[26] James Q. Wilson, *Bureaucracy*. New York: Basic Books, 1989, pp. 32–36.

[27] Newman, Warren, and Schnee, *The Process of Management*, pp. 41–42.

[28] Ibid., pp. 119–21.

BUDGETING

There is one kind of plan that every police agency must develop periodically, usually on an annual basis: the agency budget.

For some administrators, budgeting is a distasteful, frustrating chore that is undertaken only because the agency's parent government requires it. In most states and municipalities, a governmental agency cannot receive funds from the public treasury until it has submitted a budget and the budget has been approved by the legislature, city council, county commission, or other governing body. Sometimes this process is marked by strident public controversy and hot competition among department heads or between the agency administrators and the legislature.[1]

Budgeting can, and should, be an orderly, rational process whereby an agency's programmatic plans and goals are weighed against competing claims for resources, and the public's will is expressed in terms of an allocation of those resources. Much depends on whether the budget itself is based on solid facts and considered judgments of what is necessary and desirable. Developing such a budget is the subject of this chapter.

PURPOSE OF BUDGETING

Even if it were not necessary to submit a budget to obtain funds from the parent government, police administrators would find the budgeting process an indispensable tool of management. Newman, Warren, and Schnee point out, "The greatest strength of budgeting is probably its use of a single common denominator—dollars—for many diverse activities and things."[2] By reducing all of an agency's resource requirements to a single, easily compared quantity, conflicting claims can be evaluated, and reasonable decisions can be made as to which activities and what things are truly needed.[3]

Once a budget has been adopted, it serves as the standard for the control of expenditures. Funds can be spent only for the purposes stated in the budget. If the budget is executed properly, expenses cannot exceed resources and all resources are used for their intended purpose. Without a budget, there is no standard of accountability, no way to determine whether funds are being used properly or whether the resources one expects to receive in the future will be adequate.[4]

Finally, a budget serves as a standard for evaluating the efficiency of many different kinds of activities. In the previous chapter, we showed how cost-benefit analysis can be used to compare different activities that have been proposed. The same kind of analysis can be applied to current activities by comparing the budgeted cost with the actual benefit being produced. This kind of analysis often plays an important role in the needs assessment stage of the planning process.

TYPES OF BUDGETS

Any budget consists of a statement of expected or required income (the funds an agency expects to receive in the future or will need to operate) and a statement of expected or proposed expenses, how the organization plans to use its funds.[5] Ordinarily a police agency's budget should show that its income and its expenses will be precisely equal over the budget period, which we will assume to be one year.

The expense portion of a budget usually is divided into two parts: the capital budget and the operating budget. The *capital budget* lists expenditures for items such as buildings; durable equipment that will last several years, such as vehicles; and anything else that has a useful life beyond the immediate budget period. These costs can be prorated over the entire useful life of the object to be bought, so that the true annual cost can be compared meaningfully with other current costs. However, there are different ways to treat capital expenditures in a budget, as we will see shortly.

The *operating budget* includes the expenses that occur regularly and continuously such as personnel costs, utilities, and other routine costs, and any items that are not expected to last beyond the immediate budget period such as office supplies or nondurable equipment.

Some government agencies also have *special fund budgets* in addition to their operating and capital budgets. A special fund budget is used when certain revenues are required to be spent only for a certain purpose. For example, federal taxes on gasoline, tires, and other automotive products go into a special Federal Highway Trust Fund. Money from this fund can be used only for highway construction and maintenance and for the support of public transportation.

Special funds are not always a concern to police administrators, but there may be instances when funds from a state grant or other source are dedicated to a particular use such as traffic law enforcement or crime prevention.

Capital Budget Formats

Capital expenditures are separated from operating expenditures simply because the objects bought on a capital basis are expected to last beyond the budget period. If the whole cost of a capital expenditure were shown in one year, a comparison between the two types of expenditures would be misleading.[6] Furthermore, many capital expenditures are funded by borrowing money, usually through the sale of bonds, so that the cost of the object can be spread out over its expected useful life, and the government is not burdened with the entire cost in a single budget period.

Nevertheless, the parent government and the public should know the total cost of everything that is to be bought. The capital budget, then, includes a list of things that are to be purchased during the budget period and their expected useful life. If the expense is to be funded out of bond revenues, that fact should be shown in the budget.

The total cost is not included in the operating budget. Instead, the operating budget should include only the cost of repaying the debt during the immediate budget period.

For example, suppose the police agency needs a new headquarters building. The building will cost $5 million and is expected to be used by the police for 25 years. If bonds are sold to pay for this building, the interest on the bonds will be 10 percent per year. The average annual payment of principal and interest will be $468,000 per year (the total cost being $11,700,000). This figure should be shown in the operating budget as the current year's *debt service* for the building.

Not all capital expenditures, even large ones such as the cost of a new building, are funded through debt. Some states, including Texas, and many cities do not permit their government to borrow money except for certain limited purposes. They are required by law to spend in any one budget period only as much as they expect to receive in taxes and other revenues. Also, a city or county government might be unable to sell its bonds at an acceptable interest rate, or may be unable to persuade its citizens to approve the issuance of new bonds. Or a city might have a windfall of revenue, perhaps from the sale of some city-owned land that is no longer needed, or funds from the state or federal government might be available to finance a needed capital item.[7]

Whenever the total cost of a capital item will be paid in the current budget period, the capital budget should show both the total cost and the *amortized cost*, the total divided by the expected useful life. In our example, the amortized cost of a $5 million building that will last 25 years would be $200,000. However, this figure is not added to the operating budget, because no current funds will be needed each year.

Vehicles, furniture, and equipment of every sort may be included in the capital budget, if their useful life is expected to be greater than the budget period. These items are often bought out of current funds, but the expenditure should be amortized over the useful life of each item. Both the total cost and the amortized cost should be shown in the capital budget, but neither amount should be added to the operating budget.

Usually a capital budget also shows the amortized cost of previous years' capital expenses, throughout the useful life of each item.

There is still another way that capital expenditures can be treated. When an agency buys, say, a patrol car for $16,000, the administrator expects the vehicle to last for perhaps 4 years. At the end of that time, the patrol car will have to be replaced because it will no longer be serviceable. Even though the original cost of the patrol car has already been paid, it would be a good idea to save up the money for its replacement.

One way to do this is to show the current expense for the vehicle in the capital budget, but, instead of amortizing the vehicle's cost, the cost can be *depreciated* by adding to the operating budget a portion of the replacement cost.

In this case, the depreciation would be shown as $4000 per year. Even though depreciation is shown as an expense on the operating budget, the money would not actually be spent; it would be set aside to be used when a new vehicle is needed.

Most agencies, if they have been in existence for several years and they are reasonably stable in size, reach a point at which their annual capital expenditures for equipment, vehicles, furniture, and so on are about the same from year to year. The amounts set aside in the operating budget for depreciation then should be sufficient to cover the actual capital expenditures for each year.

Operating Budget Formats

Operating budgets are generally more complex than capital budgets because there are so many different kinds of things, from people to paper clips, that must be included. Experts in accounting and management have devised a number of ways to list these expenditures so that reasonable comparisons can be made among the costs of different kinds of activities. Robert D. Lee, Jr., and Ronald W. Johnson have listed three basic types of operating budget formats.[8]

Object-of-Expenditure Budgets.

This type of budget (see Figure 5.1) is also sometimes called a *line-item budget*. Only the things or types of things to be purchased are stated, along with their cost. For example, the budget might list, in part, the following:

Personnel	
Salaries, 130 commissioned	$3,068,826
58 noncommissioned	905,000
Benefits at 15%	596,074
Subtotal, personnel	$4,569,900
Utilities	$ 540,000
Vehicles	
Depreciation, 42 autos	$ 189,000
Depreciation, 4 trucks	26,000
Operating costs, maintenance	322,000
Subtotal, vehicles	$ 537,000

ANNUAL BUDGET – CITY OF PORTSMOUTH, VIRGINIA

DEPT.: PUBLIC SAFETY – POLICE FUND: GENERAL FUND CODE: 09010
FUNCTION: POLICING, INVESTIGATION AND TRAFFIC CONTROL

DEPARTMENTAL DETAIL

CATE-GORY-SUB-GROUP	CLASSIFICATION	PERSONNEL CUR-RENT	PERSONNEL PROPS'D DEPT.	PERSONNEL RCMD. MGR.	ACTUAL EXPENDITURE 78-79	ESTIMATED EXPENDITURE 78-79	PROPOSED BY DEPARTMENT 80-81	RECOMMENDED BY MANAGER 80-81	FINAL BUDGET 80-81
110	**Personal Services**								
111-01	Police chief	1	1	1		31,610	33,473	35,542	35,542
111-02	Assistant chief	2	2	2		52,157	54,414	58,892	58,892
111-07	Field operations commander	1	1	1		21,701	22,974	24,386	24,386
111-09	Captain	8	8	8		163,137	168,399	178,882	178,882
111-13	Lieutenant	11	11	11		206,840	213,098	225,938	225,938
111-15	Sergeant	28	28	28		453,979	474,622	500,733	500,733
111-20	Police officer	156	156	156		2,049,798	2,154,819	2,287,607	2,287,607
111-21	Police planning analyst	1	1	1		16,198	17,186	18,775	18,775
111-23	Property and evidence clerk	1	2	2		11,523	18,886	20,074	20,074
111-28	Administrative assistant	1	1	1		15,485	16,150	17,176	17,176
111-30	Secretary	1	1	1		11,437	12,418	13,195	13,195
111-31	Senior clerk	1	1	1		9,576	10,152	10,795	10,795
111-32	Supervising clerk	1	1	1		10,716	11,299	12,021	12,021
111-33	Clerk stenographer	2	2	2		15,988	16,410	17,461	17,461
111-34	Senior account clerk	1	1	1		9,465	9,719	10,334	10,334
111-35	Fingerprint technician	1	1	1		9,288	9,848	10,469	10,469
111-36	Intermediate clerk typist	12	15	13		87,255	113,120	106,476	106,476
111-39	Clerk typist	1	1			6,843			
111-47	Court pay					50,000	55,000	55,000	55,000
111-48	Field training officer allowance					5,040	10,920	10,920	10,920
111-51	Janitor	1	2	1		8,104	15,091	9,058	9,058
111-52	Special assignment					71,550	90,000	85,000	85,000
	Subtotal	231	235	232	3,099,335	3,317,690	3,527,998	3,708,734	3,708,734
	CETA Personnel								
181-37	Intermediate clerk typist CETA	2	3	3		12,773	15,214	15,214	15,214
181-40	Clerk typist	1	1	1			6,924	6,924	6,924
191-37	Intermediate clerk typist CETA	4	2	2		25,294	12,835	12,835	12,835
191-40	Clerk typist CETA	3				25,294			
191-43	Janitor – CETA	1				6,412			
	Total	241	241	238	3,099,335	3,387,463	3,562,971	3,743,707	3,743,707

(Continued)

Figure 5.1. An object-of-expenditure budget. Only the items to be purchased (including personnel salaries) are listed, without showing their purpose. (Courtesy Portsmouth, Virginia, PD. The budget shown is not current.)

| DEPT.: | PUBLIC SAFETY – POLICE | | | FUND: GENERAL FUND | | CODE: 09010 |

FUNCTION: POLICING, INVESTIGATION AND TRAFFIC CONTROL

DEPARTMENTAL DETAIL

CATE-GORY-SUB-GROUP	CLASSIFICATION	PERSONNEL			ACTUAL EXPENDITURE 78-79	ESTIMATED EXPENDITURE 78-79	PROPOSED BY DEPARTMENT 80-81	RECOMMENDED BY MANAGER 80-81	FINAL BUDGET 80-81
		CUR-RENT	PROPS'D DEPT.	RCMD. MGR.					
210	**Contractual Services**								
211-00	Telephone service					32,246	32,246	32,246	32,246
213-00	Travel and expenses					5,000	6,500	5,200	5,200
221-00	Repairs office equipment					2,000	2,000	1,500	1,500
222-00	Repairs other equipment (video)						2,000	1,500	1,500
225-00	Training expense					10,000	14,908	12,000	12,000
231-01	Contractual services – other					18,000	21,178	19,000	19,000
231-02	Police community relations					32,760	32,550	32,550	32,550
231-03	Vehicle equipment rental					8,450	7,200	7,200	7,200
232-00	Medical costs					7,986	7,986	7,986	7,986
235-00	Data processing charges					101,712	121,365	131,631	131,631
	Total				241,269	218,154	247,933	250,813	250,813
310	**Fixed and Sundry Charges**								
311-00	Rental equipment garage					552,555	630,520	598,634	598,634
321-00	Clothing allowance detective					37,380	42,720	42,720	42,720
325-00	Parking rental					3,840	5,760	5,760	5,760
327-00	Membership dues					350	350	350	350
333-00	Maintenance K-9 corps					15,786	18,370	16,000	16,000
340-09	Fees, other					1,800	1,800	1,800	1,800
346-00	Coroner fees					5,500	5,500	5,500	5,500
	Total				581,785	617,211	705,020	670,764	670,764
410	**Materials and Supplies**								
411-00	Office and printing					29,700	32,670	30,000	30,000
411-03	Micro film supplies						2,800	1,500	1,500
416-00	Laboratory supplies					19,702	25,242	23,000	23,000
429-04	Uniform cost					35,376	41,542	37,500	37,500
429-12	Weapons and ammunition					13,731	23,531	16,000	16,000
	Total				94,995	98,509	125,785	108,000	108,000
520	**Equipment**								
521-01	Equipment other					4,155	437	437	437
522-01	Office machines						1,600		
	Total				187	4,155	2,037	437	437
	Total Budget.				4,017,571	4,325,492	4,643,746	4,773,721	4,773,721

Figure 5.1. Continued.

This is the simplest type of operating budget, but it also provides the least information. There is nothing to show how many of the personnel will be captains, how many lieutenants, and so forth, nor does it show what these people will be doing. It is not clear from the line, "operating costs, maintenance," what the average annual mileage per vehicle is likely to be, how much the administrator expects fuel to cost, and how much is likely to be spent on repairs and routine upkeep. Nor does the budget show what the vehicles will be used to do, or by whom. Of course, much of this information could be included in a separate document, but that would reduce the primary value of the budget: to allow expenditures for different kinds of activities to be compared.[9]

Performance Budgets.

In addition to the items to be purchased, the purpose of each item can be shown in terms of the function it is to perform.[10] For example, instead of merely showing the total number of personnel to be employed and their total salaries, the figures might be broken down as follows:

PERSONNEL	
Chief of police	$ 42,554
Executive head of the agency; responsible for management of all agency functions	
Division chiefs (captains) (4)	$ 121,032
Administrators of principal agency functions: patrol, investigation, technical support, and administrative services	
Bureau chiefs (lieutenants) (15)	$ 412,680
Managers of agency's operational units; supervisors of agency personnel and activities at operating level	
Sergeants (38)	$ 904,788
Supervisors of operating personnel in patrol bureau (14); perform operational functions in other bureaus (24)	
Patrol officers (46)	$ 958,868
Operating personnel in patrol bureau and traffic section; provide 24-hour-per-day patrol, complaint response, and routine law enforcement services on 12 patrol beats	
Civilian personnel (20)	$ 331,000
Radio communications operators (8) provide 24-hour-per-day communications services; jail staff (6) assist in providing security and custodial services to prisoners; clerical staff (6) perform essential secretarial and bookkeeping services	
Subtotal, salaries	$2,770,922
Personnel benefits	$ 415,638
Employer contribution to Social Security (6.75%), pension plan (5%), group insurance plan (3.25%); total, 15%	

This example does not exhaust the possibilities of what can be shown in a performance budget. For instance, the item for patrol officers might include the number of police services rendered per year, based on an average of previous years, plus an increase if more officers are to be hired. However, the example does indicate the greater amount of detail that a performance budget offers as compared with an object-of-expenditure budget. The added detail makes it possible for decision makers, both within the agency and in the parent government, to choose between, say, adding two more sergeants to the Investigation Bureau or four more patrol officers to the Patrol Bureau.

A performance budget, although it offers more information than an object-of-expenditure budget, still concentrates on the things and people to be bought without showing how those things and people will be used to provide services or benefits to the public.[11]

Program Budgets.

In a program budget (see Figure 5.2), expenditures are grouped according to the agency's major programs and subprograms. It is then possible to see not only how much money is to be spent, but how activities are planned and coordinated to achieve specific goals.[12]

A program budget, if it is done properly, is much more detailed than either an object-of-expenditure or performance budget. Naturally, this means that a program budget is also more complicated and time-consuming to prepare. The time and effort are well spent, though, if the result enables the agency administrators and the parent governing body to make better decisions about the allocation of scarce resources.[13]

In practice, the type of budget and its format usually are dictated by the parent government. The police administrator often must follow detailed instructions from the city's or county's budget officer or the state legislature's budget committee. However, there is nothing to prevent the police agency from preparing its budget in the format it finds most useful for internal planning purposes, and then adapting the figures to the format required for submission. In a small city or county, where the police department often is the largest single agency, the administrator may be able to persuade the parent government to adopt the budget format he or she prefers.

PREPARATION OF BUDGET

The way a budget is prepared and presented reflects the amount of thoughtful planning that has gone into it. A budget is a plan; it is either a well-conceived, organized, logical plan, or it is a hastily thrown-together one.

Budget Cycle

The preparation of the budget is the first step in the *budget cycle*, the series of events that occur during the entire budget period. The second step in the cycle is the adoption of the budget; the third step, implementation or execution of the budget; and fourth and last, review and evaluation of the previous budget. These cycles necessarily overlap to some extent because the current year's budget is being executed while the previous year's budget is reviewed and the succeeding year's budget is being prepared and adopted. Thus, the budget cycles for three successive years might look like the following (on page 122):[14]

City of Tacoma, Washington
Program Summary

Functional Category	Program Category		Low Org.
Security of Persons & Property	Law Enforcement		2111
Program Title	**Fund Name**	**Fund No.**	**Bars No.**
Patrol Division	General	0010	52122

Program Description

The Patrol Division provides uniformed police service 24 hours a day, 7 days a week. Specific duties are crime suppression, citizen education and empowerment, and order maintenance. This includes investigating crimes, arresting criminals, assisting citizens and maintaining "peace" to serve and protect.

Program Objectives

1. To improve police delivery of enhanced community education and involvement through the utilization of current and projected crime prevention education specialists.

2. To proactively address violent crime trends by utilizing community oriented policing techniques and innovative technology.

Required Resources	Biennium		Biennium		Biennium	
	1989 Actual	1990 Actual	1991 Actual	1992 Appropriation	1993 Budget	1994 Budget
Personal Services	$6,147,535	$7,812,022	$8,599,060	$9,998,203	$10,919,564	$11,518,997
Variable Maintenance & Operations	$19,707	$50,039	$57,342	$21,073	$44,522	$45,587
Fixed Maintenance & Operations	$8,339	$16,379	$19,508	$23,668	$10,266	$10,266
Capital Outlay	$21,752	$30,612	$7,410	$109,114		
Other	($89)					
Total	$6,197,244	$7,909,052	$8,683,320	$10,152,058	$10,974,352	$11,574,850
Personnel (Budgeted)	145	180	182	182	197	197

Key Performance Measures.	Biennium		Biennium		Biennium	
	1989 Actual	1990 Actual	1991 Actual	1992 Estimate	1993 Estimate	1994 Estimate
1. Initiate 40 specific police/community educational and/or informational contacts by March 31, 1994.				N/A	20	20
2. Equip and train the Operations South (Patrol) Proactive Team by March 31, 1994.				N/A		100%

Figure 5.2. A portion of a performance budget. Each proposed expenditure is explained in terms of the purpose to be served. (Courtesy Tacoma, Washington, PD. The budget shown is not current.)

City of Tacoma, Washington
Program Summary

Functional Category Security of Persons & Property	Program Category Law Enforcement		Low Org. 2157
Program Title SET	Fund Name General	Fund No. 0010	Bars No. 52122

Program Description

The Special Operations Team (S.E.T.) Program allows the police department to address the escalating problem of street level drug and gang criminal activity without sacrificing police services currently provided by existing personnel. S.E.T. officers, working voluntary overtime, are assigned to target locations which have been identified as drug, gang or disorder nuisances.

Program Objectives

1. To dispatch approximately 3,600 hours of designated S.E.T. patrol time in 1993.
2. To dispatch approximately 3,600 hours of designated S.E.T. patrol time in 1994.

	Biennium			Biennium			Biennium	
	1989 Actual	1990 Actual	1991 Actual	1992 Appropriation		1993 Budget	1994 Budget
Required Resources							
Personal Services	$392,386	$513,966	$204,003	($7,789)		$100,107	$100,107
Variable Maintenance & Operations	$83	$202		$4,000			
Fixed Maintenance & Operations							
Capital Outlay	$42						
Other							
Total	$392,511	$514,168	$204,003	($3,789)		$100,107	$100,107

Personnel (Budgeted)

	Biennium			Biennium		Biennium	
	1989 Actual	1990 Actual	1991 Actual	1992 Estimate		1993 Estimate	1994 Estimate
Key Performance Measures							
Dispatch S.E.T. patrol hours							

Figure 5.2. Continued.

121

Month	Last Year	This Year	Next Year
September	I	A	—
October	I	A	—
November	I	A	—
December	I	A	—
January	E	I	P
February	E	I	P
March	E	I	P
April	—	I	P
May	—	I	P
June	—	I	P
July	—	I	P
August	—	I	P
September	—	I	A
October	—	I	A
November	—	I	A
December	—	I	A
January	—	E	I
February	—	E	I
March	—	E	I

Note: P = preparation; A = adoption; I = implementation; and E = evaluation.

Each of the four stages of the budget cycle is important in itself, and each can contribute to the ability of an administrator to manage the agency and to plan for the future. As always, the key to success is good planning, which in this case means careful, thorough preparation of the budget.

Methods of Budget Preparation

The most common procedure for preparing a budget is the *incremental method*.[15] An increment is simply the amount of an increase or decrease. When a budget is prepared according to this method, the current level of expenditure for each item is used as a base that needs no justification. However, any increase or decrease from that base must be explained. For example, consider the following:

Personnel	Current	Proposed
Salaries, 124	$3,068,826	$3,068,826
Four patrol officers (Traffic Section)		81,952
Two dispatchers (night shift)		37,000
Three Clerk-typists (Records Section)		43,500
Subtotal, salaries	$3,068,826	$3,321,278
Benefits, 11.75%	444,980	468,535
Increase in group insurance, 0.25%		166,059
Total personnel cost	$3,513,806	$3,955,772

The principal advantage of the incremental method is that it focuses attention on the changes from the previous budgets that an agency is requesting. Much more often than not, these changes are an increase over the previous level of expenditure, and increases are understandably the major concern of the parent government.

The danger in this procedure is that past practices and activities are likely to continue indefinitely, whether or not they are still appropriate, because it is not necessary to justify them. What is even worse, personnel and equipment previously budgeted for one purpose may be applied to some entirely different purpose or activity without this significant change being reflected in the budget.

The incremental method is most often used in preparing an object-of-expenditure budget, although it can be used in developing a performance or even a program budget. However, there are much better methods of preparation for the more detailed budgets.[16]

One alternative is the zero-base budget method (see Figure 5.3). This method can be used to prepare either a performance or program budget. Each category of expense in a performance budget, or each type of activity in a pro-

LVMPD Budget Unit _____ FY 1993-94 Current Level
\# _____ Schedule MB1

BUDGET SUMMARY

	BUDGET UNIT REQUEST	APPROVED REQUEST
TOTAL SALARY		
TOTAL BENEFITS		
TOTAL SUPPLIES/SERVICES		
TOTAL CAPITAL		
TOTAL		

For Budget Office Use Only	CURRENT	E1	E2	E3	TOTAL
SALARY					
BENEFITS					
SUPPLIES					
CAPITAL					
TOTAL					

Figure 5.3. A few of the forms used to prepare an annual budget. (Courtesy Las Vegas Metropolitan PD.)

```
┌─────────────────────────────────────┐
│        PERSONNEL COST SUMMARY        │
└─────────────────────────────────────┘
```

	BUDGET UNIT REQUEST	APPROVED REQUEST
5010-Permanent Salaries		
5013-Shift Differential		
5014-Acting Higher Capacity		
5015-Call Back Pay		
5016-Holiday Pay		
5017-Assignment Pay		
5018-Court Pay		
5020-Temporary		
5030-Overtime		
5050-Longevity		
5060-Leave Sellback		
5061-Separation		
5071-Uniform/Clothing Allowance		
TOTAL SALARY		
6100-Retirement		
6200-Group Insurance		
6300-Industrial Insurance		
6330-Social Security		
6340-Medicare		
6350-Unemployment Insurance		
TOTAL BENEFITS		

Figure 5.3. Continued.

gram budget, is presented as if it were being proposed for the very first time—that is, as if it were at "base zero." Thus, an explanation or justification must be given for every expenditure, usually in the form of a statement of the expected benefit or level of performance.

The usual procedure is to show three or more possible levels of activity, with the anticipated costs and benefits of each. In theory, this allows the parent government to decide exactly how much should be spent to buy a given level of service.[17] For example, taking just one item from our sample budget, presented in the following performance format (on page 124):

SUPPLIES AND SERVICES SUMMARY

CLASS	ACCOUNT NAME	BUDGET UNIT REQUEST	APPROVED REQUEST
7010	Office Supplies		
7020	Groceries		
7030	Operating Supplies		
7060	Small Equipment		
7110	Maint. of Vehicles		
7120	Rental-Equipment/Space		
7130	Data Processing		
7140	Telephone		
7150	Utilities		
7160	Liability Insurance		
7210	Professional Services		
7212	Attorney		
7214	Psychiatrist		
7220	Medical Services/Supplies		
7230	Maint. of Facilities		
7235	Maint. of Grounds		
7240	Maint. of Equipment		
7250	Postage		
7260	Conventions/Seminars		
7265	Education/Training		
7270	Travel/Transportation		
7280	Printing/Reproduction		
7310	Dues/Subscriptions		
7330	Fees/Licenses		
7350	Laundry		
TOTAL SERVICES/SUPPLIES			

Figure 5.3. Continued.

Patrol Officers

Level I. Provide crime-prevention patrols and routine law enforcement services on 12 patrol beats for 18 hours a day (rely on county sheriff for service at other times); 24 officers plus 4 relief officers, total 28; no patrol officers assigned to Traffic Section $576,948

Level II. Provide crime-prevention patrols and routine law enforcement services on 10 patrol beats for 24 hours a day; 30 officers plus 6 relief officers, total 36; no patrol officers assigned to Traffic Section $742,494

Level III. Provide crime-prevention and law enforcement services on 12 patrol beats, 24 hours a day; 36 officers plus 8 relief officers, total 44; plus 2 patrol officers assigned to Traffic Section, total 46 (current level of service) $950,658

Level IV. Twelve patrol beats (36 officers) plus 8 relief officers; plus 6 patrol officers assigned to Traffic Section for increased coverage during rush hours and high-accident-rate periods; total 50 $1,032,610

In a similar manner, proposed expenditures for personnel benefits, vehicles, and all other types of equipment and supplies would be shown at each of several possible levels of activity.

In the preceding example, two levels have been shown below the current level of activity. Level I represents the administrator's concept of the absolute minimal level of operation that could be tolerated. Level II represents a reduction from the current level, but not so drastic as Level I. Level III is the current level, and Level IV represents an increase in activity.

Additional levels could be proposed as well, up to the point at which further expenditures would be wasteful. In preparing a zero-base budget, most administrators show the current level as Level II, with one lower level and one higher level.[18]

The brief preceding example, showing four possible levels of activity, indicates the considerable complexity that a zero-base budget involves. Often no real purpose is served by treating all of an agency's activities as if they did not exist. In theory, the parent government might decide not to fund any patrol officers at all (level zero), but in fact that is quite unlikely. Furthermore, presenting this information in a performance budget format can lead to great confusion.

For example, the parent government might decide to fund the patrol officers at Level III but the vehicles at Level II, which would mean that some of the patrol officers would have no vehicles in which to patrol! This difficulty is reduced by using a program format budget, but even so, the presentation of several levels of activity can cause confusion.[19]

Another problem in the zero-base method is that inappropriate measures of performance often appear. In the examples we have been using, there are currently eight civilians employed as radio dispatchers, and they provide 24-hour-a-day service. Presumably, then, there are three 8-hour shifts with two people on duty on each shift and two who serve as a relief shift for people on their day off.

In our example of an incremental budget, the agency is requesting two more dispatchers for the night shift. It is unlikely that the night shift is getting twice as many calls as the other shifts, so the likely explanation for this request is that the new employees will be assigned to the night shift, and the present night-shift dispatchers will be reassigned to one of the day shifts. If we were to present the budget for the Communications Section in a zero-base format, it might look like the following:

Level I.	Two dispatchers, providing telephone response and radio operation services for 43,800 calls a year	$37,000
Level II.	Eight dispatchers, providing telephone response and radio operation services for 175,200 calls a year	$148,000
Level III.	Ten dispatchers, providing telephone response and radio operation services for 219,000 calls a year	$185,000

Here, the level of activity has been shown for each proposed level of expenditure, and the administrator could reasonably explain that in past years the agency has handled an average of about 10 calls per hour. Thus, the number of dispatchers, multiplied by the number of hours they are on duty, multiplied by 10 calls an hour, gives a projected number of calls handled.

However, the dispatchers are going to be paid the same no matter how many calls they handle. Furthermore, the police department has no control over how many calls it receives. Reducing the dispatching staff to two persons will not cause the number of calls for police service to be reduced from 175,200 to 43,800, nor will adding two dispatchers cause the number of calls to increase. What the police agency is buying is a certain number of hours of the dispatchers' services, and that is the appropriate measure of benefit.[20]

The zero-base budget method has been popular in recent years among many state governments and federal agencies. However, it is one of the most difficult of all types of budgets to prepare well. If it is not done well, the results can be confusing and misleading. The various levels of activity can be manipulated by administrators to show whatever they want to show. The current level can be shown as "Level I" (the minimum reasonable level), with all other levels showing a higher expenditure. The lower levels can be portrayed as so unappealing that the parent government would be extremely reluctant to reduce the current level of expenditures.

The parent government also may "play games" with the zero-base method. An arbitrary decision may be made by the parent government that Level I must represent at least a 10 percent reduction from the current level, and Level II may not show more than a 5 percent increase from current expenses. These restrictions may discourage administrators from trying to manipulate the levels, but they also may prevent reasonable proposals from being presented.[21]

A third alternative method of preparing a budget is the *planning-programming-budgeting system* (PPBS).

The original PPBS was developed by the RAND Corporation and first used by the U.S. Department of Defense in the mid-1960s. Later, other federal agencies adopted variations of the PPBS, which President Johnson especially favored, and it has been adopted by some state and local governments.[22]

As the name suggests, a PPBS is particularly appropriate for a program budget format. In essence, a PPBS simply carries the planning process, as we described it in Chapter 4, a step further to include the budget preparation process. Once a program has been planned, the analysis of the program's cost forms the basis for the program's budget. The budget for an entire agency consists of the separate budgets for all of its programs. The justification or explanation of each program also serves as the justification for the expenditure.[23]

Ideally, every cost should be assigned to one program or another. However, this leads to a difficulty, because some costs are shared among several programs. There are three common solutions to this problem.[24]

1. Every cost can be allocated to one or more programs; shared costs are allocated according to the proportion that each program is responsible for the cost. For example, if the Chief of the Technical Services Division manages five separate bureaus, his or her salary can be divided evenly and each bureau's program budget can include one-fifth of the chief's salary; or the salary might be divided according to the amount of time the chief devotes to each bureau. The same principle can be applied to the cost of utilities, office supplies, and anything else that is shared among several programs.

2. Costs can be divided into two types: direct costs, which are attributable to one program, and indirect costs, which are shared among several programs. Each program's budget includes all of its own direct costs plus a portion of the total indirect costs. The proportion may be based on the number of personnel in each program, the proportion of each program's direct costs to the agency's total budget, or any other reasonable formula. In this case, the administrators' salaries, utilities, office supplies, and all other shared costs would be lumped together as indirect costs.

3. Instead of charging each program for its share of the agency's indirect costs, these costs can be presented separately as the agency's administrative program. This is the simplest solution; the only danger is that program managers may try to reduce their own budgets by shifting more and more items to the administrative program, until eventually the administrative program contains most of the agency's expenditures.

The PPBS almost requires every manager and supervisor in the agency to contribute to the budget preparation process. Each supervisor can be asked to prepare a proposed budget for his or her program. These program budgets are then reviewed by section or bureau managers who make adjustments or corrections. Each section's or bureau's program budgets are then submitted to divisional administrators, who review them for accuracy and, if necessary, make further adjustments or corrections. The divisional budgets are then consolidated into the agency budget, which is submitted to the parent government.

A program budget, by its very nature, includes the most important feature of a performance budget: the statement of the activity, function, or benefit that is to be obtained for each expenditure. A program budget also should show a comparison of proposed expenditures with current costs, either in an incremental format or in a zero-base format.

Thus the program budget incorporates the most useful features of each type of budget while adding an important new feature: the presentation of all costs in relation to the goals that each program is expected to achieve.[25]

Once again, we can see what a small portion of a program budget might look like for our hypothetical agency.

32. Traffic Law Enforcement Section

Program Purpose

1. To reduce traffic deaths, injuries, and property damage from traffic accidents, to the lowest levels attainable.
2. To promote the safe and rapid flow of vehicular and pedestrian traffic.
3. To encourage motorists and pedestrians to observe traffic laws.
4. To identify and cite violators of traffic laws.

Program Objectives

1. To reduce traffic deaths by 10%, injuries from accidents by 20%, and property damage from accidents by 25%.
2. To improve traffic flow during rush hours, increasing average vehicle speed through the downtown area from the current 8 to 12 miles per hour.
3. To conduct a traffic safety campaign in local junior and senior high schools, providing 6 hours of instruction to a total of 200 students.
4. To reduce the incidence of speeding by 20% through the use of radar speed detectors and to reduce the incidence of DWI by 25%.

Program Methods

1. Staff will consult with City Streets and Bridges Department and with consultants from U.S. Department of Transportation to identify roadway and intersection features that contribute to accidents and to determine other causes of accidents. Staff will implement consultants' recommendations to the extent possible.
2. Four downtown intersections have been identified as the most congested during the evening rush hour (4:30 to 6:30 P.M.). Officers will be stationed at these intersections during weekday rush hours to direct traffic, thereby increasing traffic flow. Other officers will be assigned to monitor traffic movement on Central Expressway and major arterials to coordinate traffic flow. The average motorist's time in congested traffic will be reduced by 10 minutes, which corresponds to an increase from an average 8 miles per hour through the downtown area to 12 miles per hour.
3. Two officers will develop a traffic safety presentation for each ninth-grade social science class in City's high schools and each eighth-grade civic class in junior high schools. Total enrollment in these 6 classes is 200 students.

4. Two radar speed detectors will be purchased; Traffic Section officers will be trained in their use and assigned an average of 4 hours per day to locations selected because of a high accident rate, special safety problem (such as school zones), or records showing numerous speeding violations. Afternoon and evening shift patrol officers will concentrate on areas with large numbers of drinking establishments and locations where numerous drunken drivers have been observed. All suspected drunken drivers will be stopped and checked.

PROGRAM BUDGET

Capital Expenditures	Current	Proposed
Four patrol cars at $18,000	0	$72,000
Two speed radar detectors	0	2,500
Four hand-held radios	0	3,000
Total capital expenditures	0	$77,500

Note: No capital expenditures are budgeted for the current year. Four patrol cars are needed for additional officers. Two radar speed detectors are needed for program method 4. Four hand-held radios are needed for use by officers directing traffic at intersections (program method 2) to enable them to coordinate traffic flow with mobile officers.

Operating Expenses	Current	Proposed
Lieutenant (chief of section)	$26,186	$27,512
Sergeant	23,507	24,095
Two patrol officers	39,976	42,000
Four patrol officers	—	79,952
Subtotal, salaries	$89,669	$173,559
Personnel benefits, 14.5%	13,002	—
Personnel benefits, 15.0%	—	26,034
Total personnel costs	$102,671	$199,593

Vehicle Operation and Maintenance

	Current	Proposed
Three patrol cars, 18,000 miles at $0.28 per mile	$15,120	—
Seven patrol cars, 18,000 miles at $0.28 per mile	—	$35,280

Note: Mileage is based on Traffic Section's experience during past 5 years. Cost per mile figure is provided by Vehicle Maintenance Section and includes $0.21 per mile for fuel and $0.07 per mile for maintenance.

	Current	Proposed
Consumable materials and supplies	$5,200	$8,500
Literature, printing, etc.	500	2,500

Note: Includes materials for distribution as part of junior/senior high school traffic education program.

	Current	Proposed
Total operating expenses	$123,491	$245,873

In studying this example, notice that the program budget includes capital expenditures for the proposed budget period. These expenditures are not amortized, nor is debt service included in the operating expense budget. The only purpose in showing the capital expenditures is to relate them to the program. Debt service or depreciation might not be considered a direct program cost, but may be included in the agency's total administrative costs, which then might be divided proportionally among all programs and added as an indirect cost to each program's operating budget.

Also, in this example an explanation is given for each significant change from the current budget. Presumably the explanation for current items has been given in previous program budgets, and no real purpose would be served by repeating the explanations here. Anyone who wants to know exactly what functions are performed by the lieutenant, sergeant, and the two patrol officers who are already on the staff should get this information from previous budgets. However, any significant changes in these officers' duties would have to be shown and explained here, just as the duties of the additional four officers have been explained.

The sample may seem excessively detailed, especially when one considers that the Traffic Law Enforcement Program is probably one of forty or fifty separate programs in a medium-sized agency and a comparable budget presentation would be needed for each and every program. Eventually the mass of detail could reach a point at which the budget is almost incomprehensible, and that certainly is not desired.

The program budget statement should be just detailed enough that the administrator and the parent government can understand what the program intends to do, what those activities are expected to cost, and what choices can be made among different activities.[26] In this example, the City Council might decide not to fund the requested four additional patrol officers and thus would not need to buy the extra patrol cars and radios. Or the City Council could decide to fund only two additional officers, cars, and radios. Or they could agree to everything proposed here except the educational program, which would eliminate the extra $2000 for literature.

Finally, a program budget should contain a single, summary statement of all of the agency's programs and their current and proposed costs. In the summary statement, capital expenditures can be amortized if that is how the parent government prefers to handle them, or the debt service or depreciation figures can be added to the total budget if they have not been included in indirect costs.

The summary statement should include a listing of all programs, showing their current and proposed capital and operating expenses. It may also be desirable to rearrange the summary figures into an object-of-expenditure listing, to show the total amounts being spent on personnel, equipment, supplies, and so on.

Using the PPBS is no small task. However, it is a task that can be divided among all of the agency's managers and supervisors, which is often difficult to do with other budget formats. Once a program budget has been done the first time, it will be easier in succeeding years because each new budget will contain the programs that are being continued and that require no additional explanation, excluding those that have been discontinued, and adding in detail any new ones that are proposed for the coming year.[27]

Budget Adoption

Once the budget has been approved by the agency's chief executive, the preparation phase of the budget cycle ends, and the adoption phase begins.

The details of the adoption process vary enormously from place to place. Usually the budget must be submitted first to the parent government's chief executive, the city manager, the county commissioner, or the governor. It is then reviewed in detail by the administrative staff.[28]

The chief executive may order changes, such as a reduction in the total amount, or may suggest specific items that are unlikely to be approved and therefore should be omitted. Once these adjustments and corrections have been made, the police department's budget is consolidated with all the other city, county, or state agency budgets. In some cases, the capital expenditure budget and operating expense budget will become two completely separate documents, especially if all capital expenditures are financed through the sale of bonds or some other form of debt.

The consolidated budget is then presented to the legislative body. The legislature may have its own separate staff to review the budget. Usually there are public hearings during which the representatives of each agency must appear to defend their proposals.

Many police administrators resent the budget adoption process. Rarely will a budget go through the entire process without changes. Items or whole programs may be sharply reduced or eliminated altogether. The legislature may even insist on increasing an item or program that the police administrator considers unnecessary or undesirable. Even when the administrator is confident that the budget has been prepared with careful attention to detail, and that every item in it is justified, the prospect of having to defend the budget in public hearings or debates is enough to make anyone uncomfortable. Of course, a poorly prepared budget gives the administrator even more reason to dread the adoption process.

Police administrators often dismiss the budget adoption process and the decisions made by city, county, or state officials, as "purely political." Indeed they are!

Elected officials generally respond to what they perceive to be the wishes of the public. If their perceptions of the public's wishes are correct, and if they are honest in attempting to represent the wishes of the majority and not just the interests of some special-interest group, then they are fulfilling their proper duty in a democratic government.[29]

Nothing is accomplished by the administrator who ignores the budget adoption process, who merely submits the budget and hopes for the best. An intelligent, reasonable presentation of a well-prepared budget may or may not succeed, but at the least it will help to inform the public about the police department's needs and intentions. A careless presentation will merely persuade the public that the department is ineffective and poorly managed.

Some administrators routinely "pad" their budgets by 10 or 20 percent, on the theory that the legislature can then cut the excess without damaging the essential items in the budget. Unfortunately, legislators often detect this sort of trickery and lose respect for administrators who use it.

Eventually, a budget will be approved and adopted. Otherwise the police department would cease to exist. The adoption of the budget completes the second phase of the budget cycle, often just a few weeks and sometimes only a day or two before the new budget period begins.

Budget Execution

The implementation, or execution, of the budget is one of the executive's most important responsibilities. The chief administrator of a governmental agency can be held personally accountable for every penny of public money that the agency spends.

In most relatively small police agencies, the chief executive usually acts as the primary budget administrator. In larger agencies, the day-to-day work is delegated to a budget officer or finance office, but the ultimate responsibility is still the chief's.

The budget, as adopted by the parent governing body, determines exactly how much money can be spent and for what purposes. The items listed in the budget become budget accounts, and the money credited to an account can be spent only for the purposes stated in the budget. Once all the money has been spent, nothing more can be spent on that item unless funds are transferred to the account from some other account.

There are three basic ways that money can be paid out of a budget account, or *disbursed*.

Payroll Payments.

In most governmental agencies, the largest account and the largest part of the budget is the payroll, or payments to employees.[30] Payroll payments are made on a regular, scheduled basis. The budget or accounting office is informed by the agency's supervisors of the people who are to be paid, the personnel office (in most cases) confirms the amount to be paid, and the accounting office prepares and distributes the payroll checks. The accounting office also makes various deductions from each employee's pay check to pay federal and sometimes state income taxes, the employee's share of Social Security, and, depending on what benefits the agency offers its employees, things such as pension contributions, group insurance premiums, and union dues. These deductions are held in the payroll account, or transferred to their own separate accounts, until they are paid out.

Transfer Payments.

Funds can be transferred from one account to another. For example, as we just said, payroll deductions are transferred from the payroll account to any of several other accounts. This may be merely a bookkeeping action: The funds are subtracted from one account (a page in a ledger) and added to another. However, federal income taxes and Social Security payments, and some other deductions, must be transferred periodically to federally approved banks. Technically, this is still only a transfer because the bank accounts that hold these deposits are legally under the control of the employer.

Purchases.

Any object or service can be purchased if funds for that purpose are available in the budget. Some accounts are restricted to specific purchases. For example, a capital account may be used only to buy the items whose purchase has been approved.

Most accounts are established for general categories of things to be purchased such as office supplies, communications equipment, postage and shipping services, and utilities. We will discuss in Chapter 10 the procedures that are commonly used to make purchases. Briefly, the guiding principle is that no purchase may be made until it has been approved in advance by the administrator who has the authority to give this approval, and the accounting or budget office has confirmed that there are funds in the appropriate account to pay for the purchase.

For every purchase, there should be a permanent record showing who requested the item, who approved it, who confirmed that budgeted funds were available, who determined that the item should be bought from a certain vendor, when the purchase was made, when the vendor demanded payment, and when payment was made. The records also should show, of course, what was bought and how much was paid. These records not only ensure that purchases are made properly but also that administrators have the information they need to make purchases wisely and efficiently.

Most computerized accounting systems, including those designed for small computer systems, include budget-tracking features that can be immensely helpful to the police administrator. We have more to say about computers and software of interest to law enforcement agencies in Chapters 8 and 9.

Budget Review

At the end of the budget period, ideally every account should be empty, except depository accounts such as those that hold federal income taxes and Social Security contributions that may not be due until two or three months later.

In practice, not all of the budgeted funds will be spent in some accounts. Usually these unexpended funds are simply "given back" to the parent government. However, some cities and a few states allow their agencies to "carry over" unexpended funds into the next budget period or permit the money to be spent on capital items that were not previously budgeted.

Many administrators routinely check all their budget accounts about three months before the end of the budget period. If there appear to be any funds that are not going to be expended, the administrator transfers the money to some other account where it can be spent before the budget period ends. This practice may be limited, because some governments prohibit budget transfers of more than, say, 10 percent of the original budgeted amount, and some prohibit any transfers at all during the last three months of the budget year. Furthermore, spending a lot of money near the end of the year may attract unfavorable attention from the parent government; it is usually regarded as a sign of sloppy budget management.[31]

Once the budget period is over, a new budget period begins. The previous period is then subject to review and evaluation.

Internally, the agency's own accounting staff, or better yet a separate auditing staff if the agency is large enough to have a separate staff for this function, conducts a *sample audit*, a check of a random sample of purchase vouchers, payroll records, and other records of disbursements, to make sure that all disbursements were made and recorded properly. If any serious discrepancies are found, a complete audit of the entire year's records may be necessary.

A complete audit ought to be performed not less than once every five years in any case. Most agencies also should have an *external audit*, performed by accountants who are not employees of the agency, not less than once every two years. Some governments are required by law to have an external audit conducted every year.[32]

The purpose of the audit is not so much to detect cheating or stealing, although a properly conducted audit will do so. More important, the audit tests the agency's disbursement and record-keeping procedures for any weaknesses or errors. A small error, if it is repeated frequently during a long period, could cause an agency to lose a great deal of money. Too often small errors go undetected, even when purchases must be reviewed and approved several times before a disbursement is made, simply because everyone assumes that the erroneous information must be correct—or someone else would have caught it!

FUNDING

Throughout most of this chapter, we have discussed mainly the procedures for spending money: developing the capital and operating expense budgets, handling disbursements, and so on. We have paid little attention to the question of where the money will come from.

The reason is that virtually all police agencies are funded out of the general tax revenue of their parent governments and from no other source. City police departments obtain their funds from the city's local taxes, county governments fund the sheriff's office out of county taxes, and state police agencies are funded out of statewide general taxes. Usually all of these taxes are in the form of property and sales taxes. Some states and a few cities also use local income taxes to support law enforcement.

However, there are other sources of funds that could be used by police agencies. For example, federal community development funds, allocated to the states and cities, can be used for certain capital expenditures such as the construction of police buildings.

Obtaining funds from these sources is not easy. The funding agencies have strict rules and guidelines concerning what kinds of things can be funded and how application must be made. The police administrator should consult with the parent government's budget officer about these funds and how to apply for them.

Most states have established a separate fund, often under the administration of a separate agency, to promote the development of law enforcement and other public safety agencies. These funds and the agencies that administer them were originally established under the federal Law Enforcement Assistance Administration (LEAA), a program that was discontinued in the late 1970s. Nevertheless, many states continued the planning and funding agencies that had been established to handle LEAA grants, using state funds instead of federal funds.

Typically, a state law enforcement or criminal justice planning agency has a limited amount of money that it can disburse to local law enforcement agencies and other public safety or criminal justice agencies. The state agency develops a plan to improve the entire state's criminal justice system. The specific activities within the state plan are assigned priorities according to their relative importance. Grants can be awarded to local police agencies if their proposed programs or projects meet the criteria given in the state plan.

Applying for one of these grants is not as difficult as many police administrators expect it to be. The first step is to obtain a copy of the state plan. The administrator then compares the local agency's own programs with the state plan, to see whether any local programs meet the state criteria. If so, and if the kind of activity planned locally has been given a reasonably high priority, there is a good chance that a grant can be obtained.

A formal application must be submitted to the state planning agency, whose staff usually will assist the administrator in filling out the forms and gathering supporting documents. The application is then reviewed by the planning agency's staff to make sure it meets the technical requirements and that the proposed activity is eligible for funding. After this review, the application is forwarded to an advisory board, which makes the final decision on whether a grant should be awarded.

In some states, other agencies such as regional planning councils also must review and approve criminal justice grant applications. Again, the state planning agency's staff will assist the police administrator in getting the application through this procedure.

Most administrators who apply for funding from a state planning agency or other source do so only for programs or projects that cannot be funded locally, either because they are just too expensive for the local government to undertake or because there is not enough public support for the proposed activity. Ordinarily it is not a good idea to seek these external funds for activities of an ongoing nature or for anything that could be funded locally. There is never any guarantee that the external source will continue to provide funds from year to year. However, external funding is an excellent way to start new or experimental projects. If they are successful and attract public support, it will be that much easier to include them in the local budget after a year or two.

In addition to funds from state criminal justice planning agencies, external funds are sometimes available from various federal agencies. Earlier we mentioned federal community development funds, which generally are limited to capital improvements such as new buildings. However, there may be funds available from other federal sources for specific projects.

Some philanthropic foundations have funds that are available to police agencies. There are catalogs and guides, usually available at a public library, that describe the various kinds of federal grants and private charitable organizations, what kinds of activities they are interested in funding, and how application should be made. Generally, federal and foundation grants are more difficult to get than state grants, but if a worthwhile program cannot be funded in any other way, the administrator should explore this possibility.

REVIEW

1. A budget serves as a standard for

    (a) Evaluation of the efficiency of different kinds of activities.

    (b) Comparison of conflicting claims on scarce resources.

    (c) The control of expenditures.

    (d) All of the above.

    (e) None of the above.

2. Depreciation is

    (a) The reduced value of revenues caused by inflation.

    (b) An annual charge for the replacement of capital items.

    (c) A capital expense.

    (d) The transfer of funds from the capital budget to the operating budget.

    (e) The interest paid on borrowed funds.

3. An operating budget that shows only the items to be purchased and their cost is

    (a) An object-of-expenditure budget.

    (b) A capital budget.

    (c) An incremental budget.

    (d) A program budget.

    (e) A zero-base budget.

4. An indirect cost is

    (a) An expense that has been assigned to a particular program.

    (b) An expense that is not included in the current year's budget.

    (c) An expense that is shared among all of an agency's programs.

    (d) The current portion of a capital expense.

    (e) An operating expense that is transferred to a federally approved bank.

5. The three types of disbursements are

  (a) Accounts.   (e) Payroll payments.

  (b) Transfers.   (f) Purchases.

  (c) Depreciation.   (g) Deposits.

  (d) Amortization.   (h) Audits.

NOTES

[1]Robert D. Lee, Jr., and Ronald W. Johnson, *Public Budgeting Systems*, 3rd ed. Rockville, Md.: Aspen Publishers, 1983, pp. 5–9.

[2]William H. Newman, E. Kirby Warren, and Jerome E. Schnee, *The Process of Management*, 5th ed. Englewood Cliffs, N.J.: Prentice Hall, 1982, p. 493.

[3]Ronald G. Lynch, *The Police Manager*, 3rd ed. New York: Random House, 1986, pp. 133–34.

[4]Cole B. Graham, Jr., and Steven W. Hays, *Managing the Public Organization*. Washington, D.C.: Congressional Quarterly Press, 1986, pp. 195–96.

[5]Lee and Johnson, *Public Budgeting Systems*, p. 11.

[6]Jackson E. Ramsey, *Budgeting Basics*. New York: Franklin Watts, 1985, pp. 18–19.

[7]Richard W. Lindholm and Hartojo Wignjowijoto, *Financing and Managing State and Local Government*. Lexington, Mass.: Lexington Books, D. C. Heath, 1979, pp. 67–71.

[8]Lee and Johnson, *Public Budgeting Systems*, p. 12.

[9]Lynch, *The Police Manager*, pp. 135–37.

[10]Ibid., pp. 137–38.

[11]Lee and Johnson, *Public Budgeting Systems*, p. 72.

[12]Lynch, *The Police Manager*, pp. 138–39.

[13]James L. Mercer and Edwin H. Koester, *Public Management Systems*. New York: American Management Association, 1978, pp. 73–75.

[14]Lee and Johnson, *Public Budgeting Systems*, pp. 51–63.

[15]Lindholm and Wignjowijoto, *Financing*, p. 44.

[16]Graham and Hays, *Managing the Public Organization*, pp. 215–17.

[17]L. Allen Austin and Logan M. Cheek, *Zero-Base Budgeting*. New York: American Management Association, 1979, pp. 1–7.

[18]Graham and Hays, *Managing the Public Organization*, pp. 220–21.

[19]Lee and Johnson, *Public Budgeting Systems*, p. 73.

[20]Mercer and Koester, *Public Management Systems*, pp. 24–39.

[21]Lee and Johnson, *Public Budgeting Systems*, pp. 117–21.

[22]Ibid., pp. 81–95.

[23]Lindholm and Wignjowijoto, *Financing*, pp. 41–43, 45–53.

[24]Ramsey, *Budgeting Basics*, pp. 26–32.

[25]Michael Babunakis, *Budgets*. Westport, Conn.: Greenwood Press, 1976, pp. 11–25.

[26]Ibid., pp. 39–43.

[27]Ibid., pp. 6–8.

[28]Lee and Johnson, *Public Budgeting Systems*, pp. 138–39.

[29]Ibid., pp. 183–218.

[30]Ibid., p. 276.

[31]Ibid., pp. 222–28.

[32]Ibid., pp. 230–38.

POLICE BUILDINGS AND VEHICLES

Managing a police department efficiently, or any other enterprise for that matter, means making the best use of the available resources. Those resources include time, things, and people. This section of the text concerns the management of things.

The largest and often most visible thing a police agency uses is the building or buildings it occupies. Buildings serve a great many different purposes: They are places to store and protect equipment, supplies, records, and even people (such as prisoners). They are places where people meet to interact with one another and transact business. In the first part of this chapter, we will consider how police administrators can make the best use of their physical facilities.

If there is any one thing that distinguishes modern law enforcement from the police practices of the nineteenth century, it is certainly the use of the automobile. It would be difficult to imagine a police force today that did not rely on automotive patrol as the primary tool in combatting crime. Later in this chapter, we consider the different kinds of vehicles used by police agencies, and how administrators can make certain that their officers are equipped with the most appropriate vehicles.

POLICE BUILDINGS

Police administrators do not often have an opportunity to design and build a new police building. However, the planning process that is required to design an effective new building also can be applied to the remodeling of an existing building, and the same kinds of decision making can be used to evaluate existing buildings.[1] Later we discuss some ways to make existing facilities more efficient, but first we consider the process of developing a new police building.

Most police buildings, like other publicly owned structures, are designed to last as long as possible because this reduces the effective cost of the building.

Unfortunately, this sometimes means that the building outlasts its usefulness. A police building that was designed and built thirty or forty years ago may not meet today's needs; it may be too small for the number of people who must use it, or it may be in the wrong location, or it may be arranged inconveniently. But public officials are extremely reluctant to abandon and replace a building for any of those reasons if the structure is still sound. Instead, the building may be enlarged somewhat to make room for a larger staff, without resolving the other problems of an obsolete building. Or, more often, replacement is postponed until the obsolete building is no longer serviceable at all.[2]

There are ways that an obsolete building can be managed to make it more efficient in serving today's needs, and we discuss some of those methods later. Occasionally an entirely new building will be built, either to replace an existing one or to meet the needs of a new agency. In either case, the administrator should take full advantage of the opportunity to make a new building as functional and useful as possible; the opportunity may not come again for several decades.

The police administrator does not need to know much about architecture or structural engineering; those matters can be left to hired experts. However, it is unwise to assume that architects and engineers will know anything about police operations and the physical facilities needed by an efficient police force. Those matters require the police administrator's knowledge, experience, and attention. Together the administrator and the architectural experts can design a building that is both architecturally sound and operationally satisfactory.

A great deal of planning must go on before any thought is given to the shape, arrangement, or size of the building itself. If the planning procedures described in Chapters 4 and 5 have been followed, the administrator is already well prepared to begin the specific planning for a new building.

Building Design: Planning Process

Planning anything is largely a matter of making choices: what you want to have happen, when it should happen, what it should look like, and so forth.

One of the first choices to be made for a new building is the site where it is to be built. If the site is not large enough to allow a building of the necessary size on one floor, then a multistory building will be required. This means that extra space must be set aside for stairways and, if the building will be more than two or three stories high, elevator shafts. Similarly, the foundation and supporting structure must be sturdier.

Because choices are so interrelated, it is a good idea to postpone making any hard and fast decisions until fairly late in the planning process. Choices should be considered according to a set of planning criteria that have been established in advance and that define an ideal police building.

The planning process can be time-consuming. Six months may not be long enough to properly plan a new building.

Everyone who will occupy and use the building should have some role in the planning process.[3] The choices made in the design of the building will affect the way the building is used, and thus how the police department functions, for years to come. Nearly every member of the department should help to gather the information on which important decisions will be based. Field and operational personnel, including police officers, technicians, and clerks, should be given an opportunity to contribute their own suggestions and ideas.

First-line supervisors should provide the basic information about personnel patterns and physical requirements for each unit in the agency. Higher-level managers such as division and bureau chiefs should correlate the information from lower levels and add their own perceptions of what is needed for each unit to function most effectively. Finally, all of the agency's upper-level administrators should participate in making choices and design decisions. These decisions then should be reviewed by the chief executive, who may add his or her own suggestions and ideas.

Almost always, the design of a new building is subject to various factors imposed on the project from outside. For example, sometimes a city council will inform the police department, "We will set aside a certain amount of money for your new building, and not a penny more." In this case, the budget for the building would act as a constraint on the design.

Basic Planning Criteria

Planning criteria are the standards against which all decisions and design choices will be evaluated. Thus, the criteria serve to define what the agency regards as desirable in a building. Each planning criterion is important in itself. In addition, the criteria should be ranked according to their priority, so that conflicting design choices can be resolved on the basis of which criterion has the greater priority.

The planning criteria for a police building reflect the philosophy of the police agency. According to the National Clearinghouse for Criminal Justice Planning and Architecture,

> . . . a police station is one of the most important public buildings in a community. In its design and architectural character a police facility often has a significant impact on the image of law enforcement in the community. It can also have a direct influence on the morale and operational efficiency of police department employees.... Those who pass by or enter the building should be reminded of the community's commitment to modern law enforcement standards.[4]

Some of the criteria that should influence the choice of police building designs are examined.

Flexibility.

Every police agency experiences changes over time. When the period is the typical life span of a building, thirty or forty years or more, the extent of change can be substantial. Consequently, a building should be designed to adapt to those changes. It should be relatively easy to rearrange spaces in the building for new uses that were not foreseeable when the building was first designed.

For example, fifty years ago few police agencies had or needed well-equipped photographic laboratories. Today, photography is used routinely in all sorts of police operations. Fifty years from now, some other technology may replace photography. Although a modern police building might need a photo lab with its stainless steel sinks, floor drain, light-proof doors, and other specialized fittings, the designers should realize that all of these features might be obsolete and unnecessary at some future time.

The number of people who must be served by a building changes constantly. If the community is growing, it is likely that the police agency will expand, and allowance must be made for future expansion. If the community is declining, the building might contain more space than will be needed in the future, in which case it should be possible to adapt the space for some other use.

Flexibility, then, means that the building as a whole and all of its component parts should be adaptable to different uses, and consideration must be given to the building's future expansion if that is likely.

Public Accessibility.

A police building is a public facility, owned by the public through its government. It exists only to serve the public. Citizens may need to come into the police building for any number of reasons. Juveniles and their parents may need to meet with juvenile officers. Journalists may need to interview officers and administrators. Volunteers may need to meet at the police building to carry out their programs in support of law enforcement. Neighborhood groups may need to gather for a crime-prevention lecture by a police officer.

In short, the building should be designed so that members of the public have convenient access to police personnel and to those areas where public services are delivered. Convenient access also must be provided for citizens who are physically handicapped.

Nothing about the building should discourage public access. It should not be intimidating, remote, uncomfortable, impersonal, or frightening. The citizens coming into a police building may be emotionally upset, embarrassed, or reluctant. The building's design should contribute to a comforting, civil, and respectful atmosphere (see Figure 6.1).

Serviceability.

A police building is one of very few public buildings that is typically in use twenty-four hours a day, throughout the year, with heavy traffic loads at almost all hours of the day. Some of the people using the building may be violent, emotionally agitated, or physically ill. A few people may express hostility by attack-

Figure 6.1. The open design of this police headquarters lobby creates a nonintimidating atmosphere for citizens. (Photo by the authors, courtesy Austin PD.)

ing the building and its fixtures. Consequently, police buildings must be designed to withstand heavy use and abuse, while still retaining the attractiveness and comfort that provide an inviting atmosphere to the public.

Building surface materials (walls and floor coverings), fixtures (lighting, water fountains and other plumbing, handrails, and so on), and equipment (including furniture) should be selected for durability under hard use. The building itself should be designed to minimize maintenance requirements by reducing, or even eliminating, hard-to-clean devices and surfaces.

Serviceability also means that the building should be economical to operate. The energy needed for lighting, electrical equipment, heating, air conditioning, and hot water should be conserved as much as possible.

All of the building's working areas should be designed to promote efficient operations. Each working area should be planned for maximum efficiency and convenience in itself. In addition, the relationships among the working areas should allow the department's employees to interact with one another as conveniently and efficiently as possible.

Aesthetic Qualities.

In the past, police officials have not always paid attention to the attractiveness of their physical surroundings. This may be a serious mistake. Many psychological studies have demonstrated that human behavior reflects the environment in which people find themselves. Attractive surroundings encourage people to act toward one another in a pleasant, harmonious way.

Both public and working areas of a police building should be designed with bright, eye-appealing colors and textures. Areas where more serious, emotionally upsetting interactions are likely to occur should be designed with more subdued, relaxing, and inviting qualities. Even in areas where high security is required, an effort should be made to preserve an atmosphere of dignity and respect for the individual.

Police administrators should study contemporary designs for office buildings, schools, hospitals, banks, and other facilities whose operational needs are similar to theirs. This is also an area in which the advice of architectural experts should be sought early in the planning process.[5]

Security.

There are so many misconceptions about the security needs of police buildings that we discuss the subject in some detail.

In the past, security has often been given the highest priority as a design criterion. Too often, the result has been a monumental structure resembling a medieval fortress. Such a building often is inflexible, uninviting to the public, difficult to clean, expensive to operate, inefficient in its working areas, and ugly.[6]

We do not mean to suggest that security needs should be ignored or overlooked. However, they should be assessed realistically. The first question to be asked is, "Security from what?" What are the dangers to be avoided? What kinds of events might occur in a police building that would not occur anywhere else? How likely are these events? Who might cause them?

In the most general terms, the dangers are three: attacks against people, attacks against property, and attacks against the operations of the agency. There is also one danger that is peculiar to police buildings: the escape of persons who are being detained (that is, prisoners).

Attacks against people might mean either personal attack or the use of weapons against police officers, civilian employees, or innocent bystanders. The most likely source of these attacks would be a violent prisoner. Another possible source would be hostile or mentally deranged citizens who are not known to be criminal.

The use of weapons by prisoners can be avoided simply by maintaining a firm policy of searching every prisoner when he or she is brought into the building. Personal attacks by prisoners can be controlled by the standard techniques of personal restraint and by detaining prisoners in areas that are separated from other areas by architectural barriers such as locked doors.

Attacks by nonprisoners are more difficult to control, but they are also considerably less likely to occur. The danger can be controlled by limiting public access to those specific personnel and areas where interaction with the public must occur and by providing the personnel in those areas with protected spaces within easy reach in the event of threat.

Attacks against police property can take a great variety of forms. Buildings and vehicles can be vandalized; working areas can be attacked by fire, bomb, or weapons. Records are especially vulnerable to destruction. Both paper records and other forms, such as computer files, can be destroyed by fire, water, contamination, or physical attack. Records and other police property also can be stolen or damaged by tampering. Criminal evidence in police custody can be made worthless if there is even the mere suspicion of tampering.

The likelihood of any of these dangers is difficult to estimate. In this case, a better measure of the need for security is the intrinsic value of the property that might be attacked. The building itself and its ordinary furniture and equipment can be replaced if damage occurs. There is some expense and inconvenience, of course, but the cost might be considerably less than the expense of taking extreme precautions to avoid every conceivable danger.

Records, weapons, and certain other police equipment, and criminal evidence in police custody, represent a much higher value. These items, if lost or destroyed, often cannot be replaced. In the case of police weapons, any loss into criminal hands could represent a serious danger to the whole community. All such property must be protected by placing it out of the reach of anyone other than employees of the department and by using protective architectural devices, such as fireproof barriers, wherever possible.

A related security need is protection *from* some police property. Chemicals used in the criminalistics and photography labs are often toxic, flammable, or explosive. Ammunition stored for police weapons is, of course, explosive. These items must be protected from attack and at the same time the building and personnel must be protected from them. Limited access and architectural barriers are the most practical measures.

Attacks against police operations might mean efforts to disrupt police communications or records systems, physical interference with police activities, and harassment of police and civilian personnel. The people most likely to attack police operations are criminals, whether or not they are prisoners at the time, and hostile or emotionally unstable citizens.

Actually, attacks of this kind are not very common. They are most likely to occur as a response to a police activity, such as an arrest, rather than as a preconceived plan to disrupt the police agency. The best protective measure is to conduct every police operation in a manner that is least likely to provoke a physical confrontation or to be emotionally upsetting to the individuals involved. Creating a fortresslike atmosphere in a police building and maintaining attitudes of suspicion and hostility toward the public are more likely to provoke than to prevent attacks.

When security hazards are carefully considered, it becomes apparent that the building design should provide for different levels or degrees of security. The building might be divided into *security zones* that are given different architectural treatment (see Figure 6.2).

Zone 1.

Maximum public accessibility, minimum protection from attack. This zone includes the public lobby, service counters, meeting rooms, witness interview rooms, juvenile counseling rooms, and any other area in which unrestricted access between agency personnel and the public is necessary. Aside from the selection of materials and equipment that are capable of withstanding heavy use and possible abuse, no obvious security measures are necessary.

Police personnel should not be present in the public areas except to give information or directions to citizens, to receive visitors, or to carry out specific activities that involve the public. Personnel at service counters should be protect-

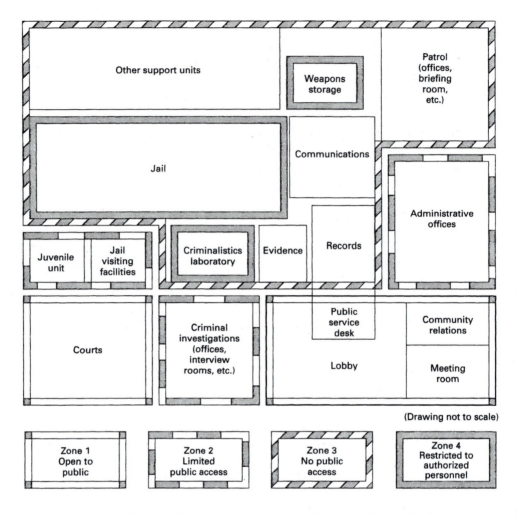

Figure 6.2. Security zones. Each unit is placed in the part of the building with appropriate security design. The drawing is not to any particular scale.

ed by bullet-resistant glass and protective barriers, but these devices need not be obvious. Architectural designers who are familiar with the design of banks and similar institutions should be able to offer suggestions.

Zone 2.

Limited public access. This zone includes all areas to which some members of the public require access on occasion, but the general public is excluded. Administrative offices, classrooms, interview rooms, prisoners' visiting rooms, and similar areas should be included in this zone. Security is maintained primarily by permitting access only to individuals who are properly identified and who have sufficient reason to enter the restricted zone. Anyone who might pose a threat to police personnel or property should be accompanied by a police officer at all times.

Zone 3.

No public access. This zone, often called the security core of the building, includes all records facilities, the communications center, evidence storage, and other sensitive facilities. With few exceptions, only agency personnel should have access to this zone, and any citizen who enters it must be accompanied by a police officer at all times. Architectural devices such as fireproof walls or bullet-resistant glass should be used to protect irreplaceable property and to prevent disruption of vital police operations.

Zone 4.

Maximum security area. This zone includes any police facility that is a likely target of attack or that is potentially dangerous in itself. The jail, criminalistics lab, armory or weapons storage area, and any area in which toxic or explosive chemicals are stored should be in this zone. With the exception of the jail, only police personnel should be permitted in this zone under any circumstances. In a large agency, there may be restrictions on which police personnel may enter this zone.

The placement of a particular facility or work area into one zone or another (for example, whether the communications center should be in Zone 3 or Zone 4) is a matter for the building planners to decide. There may be a range of architectural treatments that blurs the distinctions between zones; parts of Zone 2 may be essentially as secure as parts of Zone 3, for example.

Once the specific planning criteria have been defined and given appropriate priorities, the planners must consider the basic physical requirements for the new building.

Physical Requirements

There are two ways to approach the design of a new building. The more common way is to begin with the maximum overall building size that the budget and other constraints will allow, then divide the total space among the various components of the agency.

All too often, this method results in many mistakes. Some units are assigned to areas that are too small to accommodate their present personnel, much less future expansion; other units are given more space than they really need. Proper physical relationships among the various components are overlooked because of the pressure to squeeze every existing operation into the available space.

A far better method is to determine first the physical needs of each unit, then add them together to determine the optimum overall building size. If the total is greater than the maximum space that the budget will allow, which is likely, adjustments and compromises can be made. Nevertheless, it is much more likely that the building will be efficiently arranged if this method is used.

The first step is to survey the space requirements of each component of the agency. This can be done simply by sending a brief questionnaire to the supervisor or manager of each program and operating unit in the agency, with instructions to consult with their subordinate personnel. The questionnaire should contain questions concerning three points.

1. The number of personnel in the unit at present and the projected number of personnel at specific points in the future (say, in five and ten years).

2. The activities and operations carried out by the unit's personnel in the police building. The unit's working hours also should be determined; areas that must be accessible at all hours may need to be located in a different part of the building from areas that are in use only during the day, for instance.

3. The furniture and equipment used by the unit, especially if there are any special requirements such as high-voltage electricity, plumbing, and so on.

Once the questionnaires have been completed, they should be reviewed by management and administrative personnel to correct any errors or discrepancies and to ensure that the information is complete.

The questionnaires serve as the basis for two important planning steps: the allocation of space based on the total number of employees in each unit, with allowance for future planned expansion, plus the amount of working space required by each unit, and the arrangement of the individual spaces to promote efficient interactions. This is one area in which architects and design experts can be extremely helpful.

Building Alternatives

Until recently, nearly all police buildings in the United States were surprisingly similar in their concept and design. In large communities and small ones, the central police station contained every kind of program and facility the agency had. In major metropolitan areas, where the agency's jurisdiction was divided into several precincts or districts, each precinct or district station was a complete, independent replica of the central downtown headquarters: It contained essentially the same facilities and programs on a smaller scale.

Police facilities do not have to be designed with such "cookie-cutter" duplication, nor do all programs and facilities have to be crammed into one building whether they fit or not. Communities differ in the kinds of police programs they require, and thus police agencies differ in their physical needs. Small agencies generally need, and can afford, only one comprehensive police building. However, there are alternatives that should be considered by medium-sized and larger agencies. Small agencies should seriously consider the possibility of sharing some specialized facilities with neighboring agencies, thereby avoiding wasteful duplication and underutilization.

Independent Police Station

A conventional police building contains all of the offices and operational facilities that the police agency requires. We use the term *independent police station* for this kind of facility.[7]

An independent police station can be divided into several distinct areas.

Public areas	Space for the general public to enter, to interact with police personnel (obtaining records, being interviewed, etc.), and in some cases to meet (volunteer organizations, community crime prevention meetings).
Administrative	Offices for administrative police officials; supervisors' offices and work spaces; records storage; work space for functions such as accounting and personnel.
Operational	Work spaces for operating personnel including patrol, investigative, and special units (juvenile, organized crime, etc.).
Technical support	Work spaces for functions such as communications, records and identification, criminalistics, photography, fleet maintenance.
Prisoner custody	Jail or detention and related facilities.
Personnel training	Briefing, classroom training, physical training, vehicle training, firearms training, and other special training facilities.

The extent to which each of these areas must be developed depends (1) on the size of the agency and of the community it serves, and (2) on the kinds of programs it conducts. Every independent police station, even the very smallest, must devote some space to each of these six areas.

Specialized Building Network

One alternative for medium-sized and larger agencies is to distribute the various programs and functions among several buildings at different sites. At first, this may seem expensive and inconvenient. However, careful allocation of programs to different facilities can reduce the inconvenience to a minimum, and the expense can be more than outweighed by the efficiency of placing each type of program in the most appropriate facility.[8]

A specialized building network could be developed on a single site (what might be called a "campus plan"), thus avoiding some of the inconvenience of travel between sites. However, the campus plan negates the advantages of having certain facilities in the locations where they are most needed.

Any number of different arrangements of building networks might be practical, depending on local needs and preferences. A typical arrangement might divide the agency's facilities into seven types.

Administrative and technical services center	Houses all administrative offices plus some central operating programs (such as an organized crime unit) and most technical support units (such as communications, criminalistics).

Community police station	Houses subunits or branches of all operating programs (patrol, investigation, etc.) and some technical support services (such as duplicated records that are significant only for the area served); similar to a conventional precinct station but without duplicating administrative, technical support, and prisoner custody facilities (see Figure 6.3).
Neighborhood police station	Sometimes called a "store-front" or "walk-in" police station; intended primarily for community relations and crime prevention; limited operational use; no technical support or prisoner-holding facilities.
Maintenance and supply center	Warehouse for police supplies and stored property; maintenance shops for vehicles and equipment.
Training center	Often called a police academy; includes classrooms, specialized training facilities, materials production facilities (see Figure 6.4).
Adult detention center	Complete jail for adult prisoners.
Juvenile detention center	Complete facility for holding juveniles on a temporary basis (not intended for long-term detention after sentencing).

One great advantage of a specialized building network, aside from the fact that each type of facility can be placed where it is most needed and designed for its specific purpose, is that neighboring agencies often can share some of the facilities.

For example, several small police agencies could pool their funds to build one maintenance and supply center, one adult detention center, one juvenile detention center, and one training center rather than each agency attempting to build and maintain all of these special facilities.

In a metropolitan area, several suburban police departments could build a single administrative and technical services center, and scatter several community police stations throughout their areas of jurisdiction. It might seem peculiar to have the police chiefs from three or four different cities occupying offices in one building, but the advantages in lower building costs and in promoting interagency cooperation would be enormous.

Figure 6.3. A community police station. Located in a semisuburban residential area, the station has facilities for a patrol district, a magistrate court, and a meeting room available to the public. (Photo by the authors, courtesy Austin PD.)

Figure 6.4. A police training academy. Placed in a semirural area, the academy has ample room for such facilities as a pistol range and driver training course, as well as classrooms, conference rooms, and so forth. (Photo courtesy Austin PD.)

Multiagency Facilities.

Another way to reduce costs through facility sharing is to house several different governmental agencies in the same building.

It is quite common for small police agencies to occupy only a few rooms in the city hall or county courthouse. In slightly larger communities, the police and fire departments frequently are in the same building. The reason is primarily economic: It costs much less to consolidate several spaces into one building than it would to construct several small buildings.

However, the same principle applies even in large communities, and there are other significant advantages to the police agency in sharing a building with other agencies (see Figure 6.5).

No matter how successful a police agency's community relations program may be, there are still people who are extremely reluctant to walk into a police building even when they have good reason to seek help. Some people are intimidated by the authority of the law, while others may fear that they will be subjected to some embarrassment or discomfort. But if the building also houses other community service agencies, much of the intimidation is removed.

Furthermore, it is vital to a police agency's functions for it to act as an integrated part of the community. One of the best ways to demonstrate this principle is through a direct, physical association with other community service organizations.

It is perfectly feasible, and often very practical, for the municipal police, county sheriff, and state highway patrol or police to share facilities. Usually one government body owns the building and the others rent the space they need or pay a proportional share of costs. Again, considerable savings can be achieved through such sharing, by arranging facilities to serve common needs while avoiding wasteful duplication.

A few examples of multiagency facilities will illustrate some of the possibilities.

Public safety administrative center	Houses administrative offices, technical support, and some operational units for police, fire, and Emergency Medical Services (EMS).

Figure 6.5. A combined police-fire station. Shared facilities are common in small towns, but even in urban areas there are attractive economies in interdepartmental sharing. (Courtesy Tucson, Arizona, PD.)

Public safety services center	Houses police operational units plus fire station and perhaps EMS station.
Community services center	Houses police operational units plus various community service agencies such as welfare office, employment office, or public library.
Regional law enforcement center	Houses all units of municipal police and county sheriff, plus local office of state police or highway patrol; some technical support facilities shared among all agencies; could house two or more local police agencies.

These examples do not exhaust the possibilities but merely illustrate them.

Recycling of Existing Facilities

As we said earlier, police agencies do not often have the opportunity to build new facilities. Because public buildings are expensive, they are designed to last as long as possible, even if that means they outlast their usefulness. Many police agencies find themselves stuck in overcrowded, inefficient buildings, with little hope of getting the money to put up a new building.

The alternative is to remodel an existing structure or to put some other building to a new use. Often this is much less expensive than constructing an entirely new building, and the results can be surprisingly successful.

Remodeling of Existing Police Building

The most obvious reason to remodel an existing structure (usually an independent police station) is to relieve overcrowding. Other reasons include the deterioration of the building through heavy use or abuse, and sometimes inadequate maintenance, and the need to accommodate changes in the physical requirements of police programs.

Before any effort is made to plan a remodeling, a qualified structural engineer should be hired to evaluate the soundness of the building. There is little to be gained by trying to remodel a building that is likely to fall down in a few years. In some cases, the effort to remodel could hasten the building's demise.

If the building is structurally sound, the engineer also should be able to evaluate the kinds of remodeling that would be practical. It may seem unlikely that overcrowding can be relieved without adding to a building's overall size, but that is often the case. In many older buildings, valuable space is used for unnecessary circulation areas (corridors and lobbies), storage, and inefficient arrangements of work spaces.

With these principles in mind, the agency's planners and an architect should proceed exactly as if a new building were being designed. The same planning process should be used, except that there are more constraints to be considered.

Conversion of Other Facilities to Police Use

There is nothing about a police facility that makes it substantially different from any other kind of building, except perhaps the unusual security needs of a jail. Any structurally sound building can be converted to police use. If constructing a new building is not feasible and the existing building is not suitable for remodeling, police officials should give careful thought to the possibility of finding some other existing building that can be converted.

The present use of the building is not important. A warehouse, office building, grocery store, school, even a church, or a drive-in restaurant can be successfully converted into a practical, efficient police facility. Any building that is a candidate for conversion should meet three important criteria.

- The building must be structurally sound.

- It must contain at least as much space as the agency needs immediately and should contain ample room for future expansion.
- It must be in a suitable location for the kind of police operations that are to be housed.

Converting an existing building is often the most practical approach when a network of specialized facilities is needed. By placing the various functions of the police agency into different kinds of buildings, in different parts of town, a great deal of flexibility is obtained.

A maintenance and supply center can be built in an industrial district. An administrative center can be housed in an office building in the downtown business district. A community police station might occupy a former grocery store or large restaurant, a former school building, even a former church. A neighborhood police station could occupy a store-front, a drive-in restaurant, even a gas station.

Shopping centers and industrial office-warehouse complexes should not be overlooked as possible sites for police facilities. A shopping center might be an ideal location for a community police station; what better way to integrate the police department into the fabric of the community!

The point we want to emphasize here is that the physical facilities should serve the police program, not the other way around. If the administrator begins by considering what each of the agency's programs is intended to accomplish, and what kinds of physical facilities would best support each program's mission, some very unusual possibilities may develop.

For example, the Elgin, Illinois, police department considered various types of facilities for its community policing program, which was intended to attack crime problems at the neighborhood level and help to reduce citizens' fear of crime. One solution that the agency tried was to assign teams of police officers (generally a sergeant and four or five patrol officers) to live in apartment buildings or public housing projects in high-crime neighborhoods. The officers, who were volunteers for this project, became members of the community in every sense. Their mere presence discouraged some potential criminals, and gave residents a new sense of security.[9]

If it is true that a police facility reflects "the community's commitment to modern police service methods and procedures"[10] then it is an important part of the police administrator's responsibilities to see that the agency's facilities are adequate to its operational needs and responsive to the needs and interests of the community. Police officers and civilian employees who are crowded into outmoded, inefficient, and unattractive facilities cannot serve the public to the best of their ability, and the public cannot have pride and confidence in a police department housed in shabby quarters.[11]

POLICE VEHICLES

Automobiles were first introduced in American police agencies after World War I. Before then, police officers patrolled almost entirely on foot in urban areas or on horseback in rural and semirural areas. Bicycles were used by some municipal police departments after about 1895, but they were not common.

At first automobiles were not used for patrol, but merely to transport officers to their beats or to deliver squads of five or six officers to the scene of a disturbance; thus the obsolete term "squad car" applied to the basic police vehicle.

Apparently the idea of using the automobile as a moving platform from which an officer could continuously observe activities along a patrol beat was first developed in Detroit, which may have been the first city to employ motor vehicles for law enforcement.

Those early auto patrols were not very impressive (see Figure 6.6). The vehicles were balky, difficult to operate, and slow. There was no convenient method to communicate with a patrolling officer; he had to stop the vehicle every so often to call headquarters by telephone or by use of a police call box. The same lack of communication was a problem for foot patrols too, of course, but foot patrolmen covered comparatively smaller beats and generally were within reach of a call box.

The automotive patrol first became reasonably practical when it became possible to install a mobile radio in each vehicle. This, too, was a Detroit innovation, in the mid-1920s.[12] The combination of the automobile and the mobile radio seemed so powerful that it quickly dominated law enforcement practices and has continued to do so to the present day.

In recent years law enforcement authorities have begun to question the efficiency of this most basic of crime-prevention tools. Doubts have been raised

Figure 6.6. An early police patrol car. Notice that this 1936 Texas Highway Patrol car has no radio antenna. Mobile radios began to be introduced in urban areas in the early 1930s, but did not serve rural areas until a decade later. (Photo courtesy Archives Division, Texas State Library.)

about whether a police officer riding around the streets, presumably observing activities of the citizenry while listening to radio calls and keeping an eye on traffic, to avoid being involved in an accident, is in fact performing a worthwhile service. A careful study of standard patrol techniques in Kansas City, Missouri, reported in 1974, indicated that the answer might be no.[13] We discuss this study further in Chapter 14.

Despite the discouraging results of the Kansas City study, police departments have not abandoned the use of patrol cars. The fact remains that the patrol vehicle is nearly indispensable to law enforcement because of its many functions other than serving as a mobile platform for preventive observation (see Figure 6.7).

Functions of Police Vehicles

The term *police vehicle* immediately brings to mind the familiar marked patrol car. Actually, there are several kinds of vehicles used by most police agencies; each kind has its own distinctive functions and purposes.[14]

Patrol Vehicles

In the most basic sense, the function of the patrol vehicle is to transport one or two patrol officers from place to place within their assigned beat or district; to transport and protect the equipment they may need from time to time; and occasionally to transport a witness, prisoner, or other person.

A patrol car is most effective if it is highly conspicuous. Conventional patrol car markings include the use of contrasting colors of body paint, such as black and white, and the word "police" in large letters on each side of the vehicle and on the rear trunk lid. Patrol cars usually are fitted with flashing lights, a siren, and sometimes a public-address speaker (often combined with the siren) on the roof.

One additional marking is important: the vehicle's unique, assigned unit number. The unit number should be highly visible from every direction. It

Figure 6.7. A modern patrol car. (Photo by the authors, courtesy Austin PD.)

should appear on each side, the rear deck, and in large letters or numbers on the vehicle's roof where it can be seen from police aircraft.

A patrol car should be immediately recognizable from any direction, even at night or in inclement weather.

Whether a patrol car should carry one officer or two has been extremely controversial. We examine this subject in Chapter 14, but for now we simply point out that a one-person patrol vehicle may need to be equipped and arranged in a somewhat different fashion from a two-person vehicle. For example, in a one-officer vehicle the driver must have ready access to the controls for the siren, lights, radio, and any other equipment that may be needed while the vehicle is in motion. In a two-officer vehicle, the operation of this equipment might be divided between the driver and second officer.

A patrol car serves as, among other things, a mobile office for the patrol officer. There must be provisions for the storage of report forms, blank paper, writing instruments, maps, and other printed materials. Some patrol cars are outfitted with a small desk. A mobile computer terminal may eliminate the need for some of this material.

It is standard practice in most agencies to equip each patrol car with a shotgun or other weapon in addition to the sidearm each officer wears. Various devices have been developed to provide for the safe storage of long-barrel weapons including a sort of under-the-seat holster and the familiar locking rack attached to the dashboard. The latter, however, may get in the way of the extensive communications equipment now carried in most patrol cars. Whatever arrangement is used should ensure the safety of the officer and others in the event of a traffic accident, should prevent unauthorized persons from tampering with the weapon, and should provide the officer with rapid access to the weapon.

By far the single most important piece of equipment next to the vehicle itself is the police radio. Usually the main section of the radio, the combination transmitter and receiver, is stored in the trunk or underneath a seat. A small "control head," containing the operating controls for volume, channel selection, and so on, is attached to the dashboard or the hump between the seats, within easy reach of the driver and, in the case of a two-person car, of the second officer.

Often more than one radio is installed in a patrol car. There may be a scanner (a receiver that monitors several channels) to enable the patrol officer to hear calls for the fire department, ambulance service, or neighboring police departments. There may be a second police radio enabling the officer to both hear and respond to such calls.

There also may be a Citizens Band radio, a commercial broadcast receiver for AM and FM broadcasts, a Weather Service broadcast receiver, and sometimes other types of radios. Radio equipment manufacturers now provide some models that integrate two or three different types of radios into one unit, saving space.

A mobile computer terminal usually is attached to the dashboard or sits on the transmission hump between the front seats. Some mobile terminals include radio receiver-transmitters as well.

The trunk of the patrol car should contain a variety of other kinds of police equipment, depending on the particular law enforcement and emergency services that an agency's patrol officers are expected to provide. For example, there might be a fingerprint kit, evidence collection kit, and photographic equipment if patrol officers are expected to perform fairly extensive preliminary investigations at crime scenes.

Most patrol cars also should carry well-stocked first-aid kits and basic rescue equipment such as pry bars and collapsible stretchers. In communities where severe weather is common, there should be a small, flat shovel, a set of snow chains, and other portable devices for helping motorists out of thick mud or snow. All patrol cars should carry flares or reflectors, a fire extinguisher, a chain (capable of being used to tow a disabled vehicle), and at least 100 feet of strong rope. Portable barricades, a battery-powered high-intensity spotlight, blankets, and other emergency gear might be added to the list.

Obviously, a patrol car's trunk can become a very crowded place! In fact, few patrol cars carry everything we have listed. Decisions must be made as to which equipment will be carried; these decisions should be based on officers' actual experience as well as the agency's policies.

Command Vehicles

A command vehicle is a general-purpose vehicle, usually a standard sedan. It is used primarily to transport agency administrators, plainclothes officers (investigators), and nonuniformed supervisors, along with any specialized equipment they might need. Occasionally a command vehicle may be used to transport a witness, prisoner, or other person.

The standard practice among nearly all police agencies in the United States is to leave command vehicles unmarked. Usually a command vehicle is equipped with a siren mounted under the front hood and some kind of portable flashing light that can be set on the dashboard or attached to the roof by magnets when it is needed. The police radio usually is mounted out of sight, under the dashboard or in the glove compartment.

Leaving command vehicles unmarked may be unwise. A command vehicle does not have to be as readily identifiable as a patrol car, but there might be good reasons to make the command vehicle more recognizable. At the scene of a crime, major traffic accident, public disturbance, or other police business, a semi-marked command vehicle will be more easily identified both by the public and by other police officers.

The markings may be unobtrusive: black lettering, perhaps 2 inches high, on the side doors and rear deck; a small reproduction of a police shield on the side doors; a flashing light permanently mounted on the dashboard and on the shelf behind the rear seat; and unit numbers on the front or rear fenders. Plastic magnet-backed signs could be used on the doors, although these are fairly expensive and have a tendency to fall off while the vehicle is moving (see Figure 6.8).

If nothing else, these markings will increase the public's awareness of the presence of the police in the community and will reassure citizens that appropriate police attention is being given in emergencies.

Figure 6.8. A semimarked command vehicle. The vehicle is equipped with a portable flashing red light that can be placed on the dashboard or roof when appropriate. (Photo by the authors, courtesy Texas Dept. of Public Safety.)

Undercover Vehicles

Whether command vehicles should be unmarked or semimarked, there clearly is a need for most law enforcement agencies to maintain some vehicles that are not recognizable as police equipment. These vehicles are used for undercover investigations and covert surveillance. They ordinarily contain no police equipment other than a police radio that is fully concealed in the glove compartment or under the driver's seat.

An undercover vehicle, if it is to serve its purpose, must be completely inconspicuous. It should not be the same kind of automobile that is ordinarily used as a patrol or command vehicle. Older-model sedans, convertibles, sports cars, pickup trucks, vans, and station wagons may serve as undercover vehicles. They should not be equipped with sirens, flashing lights (portable or otherwise), tax-exempt license plates, police parking lot stickers, or anything else that can be recognized as police related. Any police equipment such as weapons or emergency gear should be kept in a locked trunk, preferably under a false floor or in a padlocked crate.

Police officers who use an undercover vehicle must be careful not to leave any law enforcement equipment or material such as report forms or stolen car lists in the car. An undercover vehicle loses most of its usefulness and may actually pose a danger to the police officers who use it, if it is identified by the criminal element in the community.

Many police agencies encourage their undercover officers to use their own private automobiles, rather than the agency's vehicles, and provide a generous mileage allowance for this purpose. Some officers have even been known to buy a car just for a particular undercover operation, then sell it when the operation has been concluded.

Single-Purpose Police Vehicles

Some kinds of police services require the use of special-purpose vehicles; in addition, it is sometimes desirable, in the interests of efficiency and economy, to use special-purpose vehicles for a particular type of police function even though a general-purpose vehicle such as a patrol car could be used.

For example, in many metropolitan areas, a *patrol wagon*, sometimes called a paddy wagon, is used for the single purpose of picking up drunks and transients in skid-row neighborhoods. The patrol wagon is simply a van that is outfitted with a metal screen or clear plastic barrier between the driver and the rear compartment, and screens or bars over all the windows. Benches are provided in the rear compartment for the passengers while they are transported to jail. The patrol wagon also may be used at the scene of a civil disturbance or a mass arrest, such as a vice raid, when large numbers of prisoners are to be transported.

Some police agencies use station wagons and vans with special equipment as mobile criminalistics laboratories, mobile radio command centers, and so on. These special-purpose vehicles are usually expensive to buy, equip, and operate. Too often, police agencies buy these vehicles with the best of intentions, but find that the vehicles simply sit in the parking lot day after day because they are not truly needed.

Conversely, an investment in a special-purpose vehicle is perfectly sensible, even if the vehicle is used only on rare occasions, if it serves some function that could not be handled in any other way (see Figure 6.9).

For example, a disaster response vehicle might be used only once in every few years when a major disaster occurs. If the vehicle is appropriately equipped, the expense might be justifiable. However, the equipment and supplies must be

Figure 6.9. A single-purpose police vehicle. Such vehicles may serve an important purpose even if they are used only infrequently, but a clear, well-conceived plan for their use (including personnel training as necessary) should exist before a single-purpose vehicle is acquired. (Photo by the authors, courtesy Austin PD.)

maintained so that the vehicle is ready for use at all times, and someone must be trained to operate it.

Police Motorcycles

Both two- and three-wheel motorcycles are used by many urban police agencies. In most cases, their use is limited to traffic patrol and crowd-control functions. A motorcycle is capable of carrying one person and a very limited quantity of equipment and materials.

These limitations may be outweighed by the advantages of the motorcycle. Two-wheel motorcycles are extremely maneuverable even in very dense traffic and they are capable of being operated at relatively high speeds. Three-wheel motorcycles are useful primarily for low-speed applications with frequent stops and starts such as parking law enforcement and crowd control. In general, motorcycles are considerably less expensive to operate than even the smallest, most economical automobiles (see Figure 6.10).

However, motorcycles, especially the two-wheeled ones, are extremely dangerous. Some police agencies have reported that their motorcycle officers have experienced eight times as many accidents resulting in death or injury as their automobile-patrol officers.[15]

Two-wheel motorcycles cannot be operated in rainy or otherwise inclement weather not only because of the discomfort to the driver but also because these conditions make the vehicle even more unsafe. Three-wheel motorcycles, especially if they are equipped with canopies, can be operated in poor weather to some extent. However, they too are involved in a disproportionately high number of serious accidents.

Figure 6.10. Police motorcycles. (Photo by the authors, courtesy Austin PD.)

The reason for the high rate of injuries and deaths, of course, is that a motorcycle offers no protection at all in the event of an accident, and the inherent instability of a two-wheel motorcycle contributes to a high accident rate even when the drivers are well trained and competent.

An unusual alternative has been adopted by several small towns in Georgia that have numerous unpaved roads in their jurisdictions. For low-speed and off-road patrol, they have found electric golf carts to be extremely effective.[16]

Nonmotorized Patrol Vehicles

Strictly speaking, a horse is not considered a "vehicle," but horses can and do serve some of the same functions as patrol cars. Horses were widely used in law enforcement as late as the 1960s in rural and suburban areas, and have continued to be used in some urban areas for specific purposes. In recent years, horse-mounted patrol has made something of a comeback (see Figure 6.11).

The chief advantages of a horse as a patrol "vehicle" are its size and mobility. An officer mounted on horseback usually is high enough to see over pedestrians and most motor traffic, and a well-trained horse can quickly and easily move through even a dense crowd. Horses are therefore ideal for use in areas where there is a great deal of pedestrian traffic: parks, amusement areas, and community events. Some officers also have reported another benefit: horses are widely admired, especially by children and adolescents, and therefore attract favorable attention to the police.

The major disadvantage of horses is their cost. The initial cost for a suitable, fully trained horse can be as much as, and sometimes more than, the cost of a motor vehicle. The "operating costs" include feed and veterinary care, plus maintenance of related gear such as saddles and bridles. Horses also require proper stabling facilities.

Figure 6.11. Horse patrol. The use of horses has become increasingly common in dense urban areas, parks, and recreational areas. (Photo by the authors, courtesy Austin PD.)

Before being put into police service, a horse must be trained not only for riding, but also for use under normal conditions such as dense crowds and under such difficult circumstances as a riot. Not all horses are suitable for police work. Large, sturdy horses are preferred, but a patrol horse also must be agile. A tranquil disposition also is essential.

Patrol horses often are owned by the officers who use them, which helps to ensure that the animal and its rider work smoothly as a team. However, this practice may make the horse useless if the officer is not available for some reason.

Some police agencies have had success in acquiring suitable horses by borrowing or leasing them from civic-minded owners. This practice removes the initial cost of acquiring the horse but may create problems if the owners decide to discontinue the arrangement.

Another type of nonmotorized patrol vehicle also has become popular in recent years: the bicycle. Again, bicycles are certainly no substitute for the standard patrol car but are used in special circumstances where automobiles are at a disadvantage (see Figure 6.12).

For example, bicycles often can move through heavy traffic in urban areas more quickly than a motor vehicle, a fact that messenger services in metropolitan areas have known for many years. Bicycles also can be extremely useful in narrow alleys, on park trails, on beaches, and in many other settings where an automobile cannot be operated normally. In these settings, it is even possible for bicycles to be used to pursue an automobile, as well as fugitives on foot.

Bicycles are relatively inexpensive and require very little routine maintenance. Officers assigned to bicycle patrol usually are given a patrol car as well, equipped with a bicycle rack on the roof or trunk.

Both horses and bicycles used in patrol work have the disadvantage that they are capable of carrying only a limited amount of equipment and supplies. A rifle or shotgun can be carried on horseback, but the risk of having it stolen is

Figure 6.12. Bicycle patrol. Notice the uniform adapted to this special assignment. (Photo by the authors, courtesy Austin PD.)

high. Some material, such as report forms and the like, can be carried in saddle bags or bicycle carriers (preferably enclosed, not wire baskets). Officers mounted on horseback or bicycle usually are restricted to hand-held radios, which may have limited range.

Despite these limitations, both horses and bicycles are extremely useful in certain circumstances, and their use is likely to continue for many years.

Police Aircraft

Both fixed-wing and rotary aircraft are used by many police agencies for a variety of purposes.

Fixed-wing aircraft (conventional airplanes, generally with one or two engines) were first used by state police in 1929, and continue to be used by state, county, and rural police agencies. Their most common use is for the transportation of personnel between distant locations. However, they also can be used for observation, rescue searches, traffic surveillance, and other patrol-type functions, especially in rural areas.[17]

The major advantages of fixed-wing aircraft are their speed and relatively large load capacity. Most small, single-engine airplanes have a cruising speed of 80 to 120 miles per hour, a very considerable advantage over the speed limit that applies to ground vehicles. Most automobiles are capable of being operated at higher speeds, but very few can outrace an aircraft over a sustained period. A small aircraft can be operated for three or four hours at a time, and sometimes longer if it is equipped with high-capacity fuel tanks. This operating limit can be a significant advantage in traffic patrol, rescue searches, searches for fugitives, and similar police activities.

Rotary aircraft (helicopters) have the unique advantage of being able to hover over one location or fly at very slow speeds. They are ideal for surveillance and observation from the air, especially in densely built urban areas. They also are invaluable in some rescue operations where the person to be rescued is inaccessible from the ground and there is no place for an aircraft to land. The helicopter can hover over the rescue site while a rope or basket is lowered, or a rescuer can be dropped down on a rope to complete the operation.

The principal disadvantages of the helicopter are its noisy operation, its high fuel consumption and therefore high operating expense, and its relatively high risk. If a fixed-wing aircraft has engine problems, usually it can glide to an emergency landing. When a helicopter malfunctions, it may plummet to the ground. Some helicopters also have much more extensive, and costly, maintenance requirements than most fixed-wing aircraft. Most helicopters are able to run for no more than an hour or two on a full fuel load, which limits their ability to patrol continuously.

Both fixed-wing and rotary aircraft require specially trained, licensed pilots. Some police agencies prefer to hire experienced pilots and train them as police officers. Other agencies invite experienced police officers to become trained as pilots voluntarily. Usually the officer-pilot is promoted to a higher rank or given other special compensation.

Aircraft require specialized facilities such as an airport, maintenance hangar, and licensed aircraft mechanics. These costly operations make sense only if the aircraft are used frequently for purposes that cannot be served in any other way with equal efficiency.

In fact, studies of helicopter use in several metropolitan areas have demonstrated that a single aircraft, when properly used, can perform about as much police work as ten ground vehicles. We discuss the use of aircraft in Chapters 14 and 15.

Police Marine Craft

Any community that has a substantial body of water—a river, a lake, an ocean beach, and so on—requires a police agency capable of responding to marine emergencies and crimes. This in turn means that the law enforcement agency must have some kind of marine craft.

The kind of craft needed depends mostly on the size of the body of water and the kinds of police functions that are to be served. A city that borders on a major river, such as the Mississippi or the Columbia, will be concerned mainly with shore patrols and rescues. A city in which there is a major port, such as Houston or Boston, requires far more extensive police marine activity. Some major port cities have established separate law enforcement agencies, or harbor patrols, specifically for this purpose.

In some cases, there is no need for marine patrols on a full-time, year-round basis. If the body of water is used primarily for recreation during the warm-weather months, daylight patrols during the recreational season should be sufficient. However, in any city or county where there is a substantial body of water, the police agency should have appropriate marine craft readily available, with trained personnel ready on short notice to handle rescues, searches for drowning victims, and routine police tasks.

Selection and Acquisition of Police Vehicles

Police aircraft and marine craft are so specialized in nature and the types of vehicles available for police use are so varied that there is not a great deal we can say to guide the police administrator in selecting and acquiring this kind of equipment, other than to follow the general rules for property acquisition that we discuss in Chapter 10.

Before any aircraft or marine craft is purchased, there must be a clear plan for its use: not only how it is to be operated, but who will operate it, what kinds of trained personnel will be required, how the equipment will be stored and maintained, and how operating expenses are to be paid.

Police automobiles are another matter. Their general purposes are so much a part of conventional police practice that, unfortunately, they usually are taken for granted. Too little thought is given to the operational requirements of police vehicles. Many agencies routinely buy the same kind of automobiles, often from the same manufacturer and local dealer, year after year without ever reexamining whether the vehicles being bought are the most appropriate and economical ones for the intended purpose.

Standard urban patrol techniques do not vary a great deal from one community to the next, and it is reasonable to suppose that the same kinds of automobiles are equally appropriate in most communities. For many years, the three major U.S. automobile manufacturers offered standardized vehicles with standardized "police packages."

These vehicles were simply fleet versions of the company's least-expensive sedan. Essentially the same cars were offered to taxicab companies, utility companies, and other fleet operators. The biggest difference between a taxicab and a police car is that the latter may have a larger engine, radiator, and air-conditioning system, and sturdier suspension system, plus some accessories that are intended to meet the special needs of police.

In recent years, even these "police packages" have grown very skimpy. At one time, police agencies could count on the major automakers to provide vehicles with the necessary features for high-speed pursuit driving and the hard day-in-day-out use to which police vehicles are exposed. These vehicles were sometimes called police cruisers in manufacturers' literature and in police circles. Today, a true police cruiser is as readily available as dinosaur eggs.

Several factors including government regulations concerning pollution and fuel conservation, the general economic decline of the American auto industry since the 1970s, and the rising public demand for smaller and more fuel-efficient cars served to eliminate the traditional police cruiser. In its place the manufacturers offer a limited version of the police package on their standard models, which are smaller, lighter, and less durable than equivalent cars of the past.[18]

Ideally, a police agency should be able to select vehicles that are perfectly suited to their intended use. The police administrator should be able to determine first how the vehicle will be used: the operational requirements. From the definition of operational requirements, the administrator should be able to prepare detailed specifications and standards that the vehicle must meet. Manufacturers and dealers then should be invited to offer vehicles that meet the specifications, at whatever price they consider appropriate. The police agency then should test each offered vehicle to make sure that it actually meets the specifications and standards; vehicles that fail these tests would be disqualified. From those vehicles that meet the tests, the police administrator then could choose on the basis of the best price offered.[19]

Unfortunately, few police administrators have the time or expertise to develop complete statements of operational requirements and then to develop comprehensive specifications and standards. Only a handful of very large police agencies could afford the extensive testing that would be ideal.

Most police agencies assume that the manufacturers have given some thought to the operational requirements of a police patrol vehicle. Therefore, the administrator typically copies the vehicle's specifications from manufacturers' brochures.[20] Bids may or may not be solicited from several manufacturers and dealers. If there is any bidding, the police agency usually buys whichever vehicle is offered at the lowest price, without making an independent decision on whether the vehicle meets established specifications.

Quite often, especially in smaller communities, only one dealer will bid on the police fleet because the police department already has indicated which model it favors by using that manufacturer's specifications. It also happens in some small communities that three or four dealers will "take turns" bidding on the police fleet, a practice that is clearly illegal.

There is one alternative that ought to be used by nearly all law enforcement agencies. The Technology Assessment Program of the National Institute of Justice, part of the U.S. Department of Justice, publishes an annual report of the vehicle testing program conducted by the Michigan State Police.

Because the Michigan State Police annually buys several hundred new vehicles and has ready access to both the manufacturers and the test-track facilities, the department decided in 1979 to begin its own comprehensive testing program and to share the results with other law enforcement agencies. The information produced by the Michigan State Police testing program is considerably more comprehensive than most municipal police agencies could acquire in any other way. Copies of recent test reports are available from the National Institute of Justice.[21]

The Michigan test results should not automatically determine which of the tested vehicles will be chosen by a particular police agency. Different agencies do have different needs. In some cases, economical operation may be more important than high-speed performance, while in other cases maneuverability in dense traffic might be most important.

Furthermore, the Michigan tests do not include every factor that might be significant to the police administrator. For example, no effort is made to test the durability of the vehicles or their maintenance costs, although comparative fuel economy is considered. Before a final decision on any fleet of new vehicles is made, as much information from as many different sources as possible ought to be gathered and analyzed.

At present, there is no comparable testing program for police motorcycles. However, most police agencies use relatively few motorcycles and buy them once every several years, whereas automobiles are used in large numbers and usually purchased new every year or two. Thus it is more practical for motorcycles to be selected and purchased on the basis of more carefully developed specifications and some kind of performance testing.

Vehicle Operation and Maintenance

Police administrators must assume responsibility for the proper maintenance of the agency's vehicles. Most police vehicles are in constant, hard use. The maintenance procedures that are appropriate to one's private automobile, which usually can be summarized as, "When something breaks, fix it," do not apply to police vehicles.[22]

The key concept is preventive maintenance. Every vehicle should be inspected and serviced on a regular, scheduled basis, whether there is anything apparently wrong with it or not. Often preventive maintenance inspections will result in the discovery of potential problems before they become serious enough to affect the vehicle's performance. Preventive maintenance also helps to ensure that all of a vehicle's complicated mechanical and electrical systems are operating at their optimum level of performance.

Most automobiles come with an owner's manual that specifies the proper interval for routine servicing. Some manufacturers provide much more extensive documentation, including shop manuals, when several vehicles are purchased at once. The manufacturers' service recommendations should be followed scrupulously, except where actual experience with a type of vehicle indicates that more frequent servicing is needed because of the hard use to which the vehicle is subjected.[23]

Proper fleet maintenance requires some facility where the vehicles can be maintained and repaired when necessary. Some small agencies are able to get by without their own garage, by contracting with a private garage for maintenance services. However, any agency that has ten or more vehicles should have its own maintenance and repair facility. It is almost certain to give closer attention, better care, and lower vehicle operating costs in the long run than a contract mechanic.

The officers assigned to drive a police vehicle also bear part of the responsibility for proper maintenance. Every officer should inspect his or her vehicle at the beginning of each shift. The inspection should include the appearance of the exterior and interior of the car; the operation of lights, wipers, and horn; the extent of wear of the tires; and any mechanical component (brakes, transmission, steering, suspension, and so on) that does not appear to be operating properly. Any defects found during the inspection should be corrected before the car goes out onto the street.

The automobile is so much a part of the American way of life that it is often not appreciated as fully as it might be. Motor vehicles have made modern law enforcement practices possible. There is no doubt that police agencies will continue to rely on motor vehicles for the foreseeable future. Aircraft, marine craft, and nonmotorized vehicles are important supplements, but it is difficult to imagine a replacement for the patrol and command vehicles as they are presently used in American law enforcement.

REVIEW

1. True or False: The planning and design of a new building should be entrusted to qualified experts such as architects and engineers. It is too complicated and too time-consuming for police administrators to get involved.

2. A police building might be divided into several zones for the following purpose:

    (a) To economize on utilities such as heat and electricity.

    (b) To prevent the public from interfering with police operations.

    (c) To provide appropriate levels of security for each police function.

    (d) To prevent prisoners from escaping.

    (e) To make routine cleaning and maintenance easier.

3. The automotive patrol first became practical when

   (a) High-speed patrol cruisers were developed.

   (b) Foot patrols were found to be inefficient.

   (c) Mobile radios were installed in automobiles.

   (d) Traffic signals were invented.

   (e) Telephone call boxes came into widespread use.

4. True or False: The patrol vehicle is the only means of providing a mobile platform for preventive observation.

5. Police aircraft and marine craft are alike in that

   (a) They are specially designed for police use.

   (b) They require specially trained personnel for effective operation and maintenance.

   (c) They are too expensive for most police agencies to have.

   (d) They are useful only for limited purposes such as transporting personnel over long distances.

   (e) Their use is resented by most citizens.

NOTES

[1]John Sturner, Sheldon F. Greenburg, and Deborah Y. Faulkner, "Equipment and Facilities," in William A. Geller, ed., *Local Government Police Management*, 3rd ed. Washington, D.C.: International City Management Association, 1991, p. 416.

[2]David A. Varrelman, "Facilities and Materiel," in Bernard L. Garmire, ed., *Local Government Police Management*, 2nd ed. Washington, D.C.: International City Management Association, 1982, p. 335.

[3]Ibid., pp. 336–37.

[4]National Clearinghouse for Criminal Justice Planning and Architecture, *Guidelines for the Planning and Design of Police Programs and Facilities*. Urbana, Ill.: The University of Illinois, 1973, p. D-3.1.

[5]Varrelman, "Facilities and Materiel," p. 338.

[6]Ibid., pp. 342–45.

[7]National Clearinghouse, *Guidelines*, p. D-3f.

[8]James D. Munger and Edward Spivey, "Facility Planning in Colorado: Form vs. Function," in *FBI Law Enforcement Bulletin*, vol. 59, no. 6, June 1990, pp. 10–14.

[9]Clay Pearson and Charles Gruber, "Their Neighborhoods *Are* Their Beats," in *American City and County*, vol. 107, no. 5, April 1992.

[10]Varrelman, "Facilities and Materiel," p. 335.

[11]Edwin J. Delattre, *Character and Cops*. Washington, D.C.: American Enterprise Institute, 1989, pp. 27–29.

[12]Alan Burton, *Police Telecommunications*. Springfield, Ill.: Charles C Thomas, 1973, p. 4.

[13]*The Kansas City Preventive Patrol Experiment*. Washington, D.C.: The Police Foundation, 1974.

[14]Varrelman, "Facilities and Materiel," pp. 354–355.

[15]Anthony Scotti, *Police Driving Techniques*. Englewood Cliffs, N.J.: Prentice Hall, 1988, p. 5.

[16]"Operation Golf Cart," in *Law and Order*, vol. 38, no. 11, November 1990, pp. 49–50.

[17]Sturner, Greenburg, and Faulkner, "Equipment and Facilities," in Geller, *Local Government Police Management*, 3rd ed., p. 412.

[18]Scotti, *Police Driving Techniques*, pp. 161–62.

[19]Varrelman, "Facilities and Materiel," pp. 355–57.

[20]Scotti, *Police Driving Techniques*, pp. 162–66.

[21]Patrol vehicle tests conducted by the Michigan State Police are published by the Technology Assessment Program, U. S. Department of Justice, National Institute of Justice, Washington, D.C. 20531; or contact the Technology Assessment Program Information Center, Box 6000, Rockville, MD 20850, or call (800) 248-2742.

[22]James N. Auten, *Law Enforcement Driving*. Springfield, Ill.: Charles C Thomas, 1989, pp. 6–7.

[23]Varrelman, "Facilities and Materiel," pp. 357–58.

PERSONAL EQUIPMENT

Law enforcement officers perform their vital duties on the streets, in public places. This is equally true for plainclothes investigators, undercover agents, and patrol officers. They must carry their working equipment, the tools of their trade, on their persons. In some cases, the very clothes they wear *are* tools of their trade.

Unfortunately, the police officers' personal equipment, like their vehicles, is too often taken for granted and overlooked as the proper subject of administrative attention, planning, and budgeting. The result is that officers are left to shift for themselves, to accumulate their own collection of equipment that may or may not be suitable to their duties, and certainly will not conform to a reasonable degree of standardization.

Standardization is important for several reasons. First, the equipment used by police officers, regardless of their specific assignment, should be the best available for the tasks they are to perform. Police work is difficult and sometimes dangerous, both to the officers and to the citizens with whom they interact. Inadequate or inappropriate equipment increases the danger.

Second, police officers must work cooperatively with one another; sometimes this means sharing equipment. Because police equipment can be used effectively only with proper training, an officer may be hindered more than helped if a borrowed piece of equipment is unfamiliar.

In some cases, the sharing of equipment may be impossible unless there is a high level of standardization. For example, if officers provide their own handguns, and there is no common type of weapon used by all officers, one officer's ammunition may not fit another officer's weapon.

Finally, standardization of police personal equipment helps the public to recognize and identify a police officer without hesitation. In some small police agencies, each officer supplies his or her own uniform, with no well-defined standards. As a result, the uniforms are not—in a word—uniform. Even though all officers may wear the same shoulder patch, badge, and general color scheme, the lack of uniformity can create confusion and mistrust among the public.

It is the responsibility of administrators to ensure that uniforms are uniform and that all personal equipment is standardized as far as possible. All agency personnel should be given an opportunity to express their personal preferences when standards are being adopted, and the standards should be reviewed and possibly amended from time to time. However, once a decision has been made to adopt a standard for a given type of equipment or uniform, all personnel should be expected to comply. This responsibility is not eliminated even when officers are expected to provide some or all of their own equipment; the agency still can specify the types of equipment that are acceptable.

POLICE UNIFORMS

Standard Duty Uniform

In most police agencies, all commissioned officers wear a standard duty uniform except (1) officers assigned to criminal investigations, (2) officers assigned to undercover investigations, and (3) officers whose duties require a specialized uniform. Higher-level administrators usually wear a uniform in smaller agencies, but generally wear civilian clothes in larger agencies; however, these practices are not universal.

Standard police uniforms are available from uniform manufacturers in a vast array of styles and colors. The choice ultimately is left to the administrators of each agency.

Aside from matters of personal taste, the most important consideration in selecting a uniform style is recognizability.[1] The public should be able to identify at a glance, even from a distance, a uniformed police officer, and should be able to distinguish the police officer from a fire fighter, postal carrier, delivery person, bus driver, airline pilot, or any of the other people who wear some kind of uniform. It is also helpful, though not so important, for the public to be able to distinguish among city police officers, sheriff's deputies, and state police officers (see Figure 7.1).

The second important consideration is serviceability. Police work is (at least occasionally) physically strenuous. Sometimes an officer is exposed to dirt, blood, grass stains, and abrasive surfaces such as concrete sidewalks and brick walls. A serviceable uniform should be comfortable to wear and easy to clean. It should not restrict the officer's physical movements, and it should be possible to make at least minor repairs in a neat, nearly undetectable manner. The serviceability of the uniform depends on its style and the fabric of which it is made. A simple, neat style with a minimum of decorative flourishes, tailored out of such fabrics as cotton and polyester, will prove to be most successful.

Figure 7.1. Police officers on patrol carry the "tools of the trade" with them. The uniform itself is an important "tool." (Photo by the authors, courtesy Austin PD.)

The climate in which the uniform will be worn also must be considered. A summer uniform, or a year-round uniform in warm climates, might include short-sleeved shirts. A winter uniform might include a general-purpose jacket, such as the Eisenhower jacket or a modified suit jacket.

The basic components of the standard duty uniform include a shirt, slacks (skirts for female officers are not recommended), a light jacket, a heavy winter jacket, black socks and shoes, and an appropriate cap.

Most urban police agencies continue to use the traditional garrison cap, at least as part of the "formal" or "dress" uniform. Some urban agencies and many suburban agencies have adopted a baseball-type cap that is considerably more practical than the traditional garrison cap. Some sheriff's officers and highway patrol officers wear either a western-style hat or a campaign hat ("Smokey Bear" hat). Particularly in warm climates, hats of any kind have become increasingly optional, and increasingly omitted.

A raincoat or slicker and hat cover complete the basic uniform, to which may be added winter gloves, a winter hat with ear covers, and a reflective jacket or vest for traffic duty at night and in severe weather.

Epaulets (buttoned shoulder flaps), braid, Sam Browne belts (worn diagonally across one shoulder) and other devices that are merely decorative should be omitted from the standard duty uniform, although they might be added to the uniform for special dress occasions (see Figure 7.2). The same thing probably applies to four-in-hand ties that some agencies make a part of the uniform; they serve no useful purpose and may be in the way at times.

A uniform generally requires two belts: one for the slacks, and one to hold a pistol and other working equipment. However, some styles of slacks do not require a belt.

Male and female officers' uniforms should be as nearly alike as possible. Skirts should be avoided not only for the sake of uniformity but also because a skirt simply is not a practical garment for police work. Again, however, a skirt could be part of a female officer's dress uniform.

Figure 7.2. The standard duty uniform can be transformed into a full-dress uniform by adding appropriate accessories. (Photo courtesy Austin PD.)

Each officer should have a minimum of three complete basic uniforms, with one of each type of jacket and bad-weather gear. Five complete uniforms would be even better.

The appearance of a police officer's uniform directly reflects the officer's attitudes and the pride that the agency takes in itself. Each officer must be held responsible for his or her appearance.

A soiled, worn, or damaged uniform should never be worn. If a uniform becomes damaged in the course of an officer's work, he or she should be allowed to return to the police station, or go home if necessary, to change clothes. Supervisors should be responsible for checking the appearance of their subordinates regularly and for correcting any problems that are observed.

Many smaller departments, as we mentioned before, require their personnel to supply all of their own uniforms. This policy is acceptable if the agency's administrators set and enforce standards that are sufficiently detailed to ensure that the uniforms are identical in appearance.[2] Most medium-sized and larger agencies provide all officers with a minimum set of uniforms and pay a monthly uniform allowance that ordinarily covers the cost of cleaning, minor repairs, and the replacement of badly worn or damaged garments.[3]

Officers should be able to purchase additional uniform items as they wish or need to, either directly from the department at cost or from the department's regular suppliers. One advantage to this policy, aside from maintaining standardization, is that the purchase of uniforms in fairly large quantities will help to reduce the cost. A single complete uniform may cost anywhere from $200 to $500; discounts may be available if large quantities are purchased at one time. The argument that officers will take better care of the uniform if they must purchase it themselves can be avoided if supervisors make frequent inspections and enforce a high standard of appearance.[4]

Special-Purpose Uniforms

The standard duty uniform is appropriate for nearly all patrol, traffic enforcement, and office work. However, it is not entirely suitable to some kinds of special duty.

Agencies whose officers are occasionally called on to participate in ceremonial events may require a special dress uniform. The dress uniform, for the sake of economy, may consist of a standard duty uniform to which various decorative touches are added including epaulets, braid at the shoulders, a dress tie or ascot, leather spats, special belts (such as white Sam Browne belts), even a special dress jacket. Some taste and restraint should be exercised so that the dress uniform does not become too gaudy, nor too far removed from the standard uniform so that the officers are not recognizable.

Motorcycle officers and others whose work involves special vehicles or equipment need appropriate uniforms. Motorcycle officers should wear leather boots that cover the shin, for protection of the foot and lower leg. These require slacks with narrow "pipe-stem" legs or jodhpurs (slacks that are loose-fitting above the knee but tight below the knee, originally designed for horse riding). It should go without saying that all motorcycle officers must wear a suitable helmet with goggles or faceplate.

Officers assigned to special weapons and tactics teams (SWAT teams) often wear military-style fatigue coveralls and combat-type boots, garments that are appropriate for any kind of physical activity. Coveralls are also suitable for aircraft officers and might be worn by marine officers.

Plainclothes Officers

Officers who do not wear a standard uniform nevertheless have an obligation to uphold the agency's standards of public appearance. This applies to all criminal investigators, administrators, and others who wear "plain clothes." Undercover agents, naturally, must wear whatever is appropriate to the environment in which they are working, which may mean clothing that is very far removed from ordinary standards of dress.

All other police personnel should be expected to dress in a neat, conventional, and attractive manner. Men ordinarily should wear a business suit or jacket and matching slacks with a dress shirt and tie. Women should wear either an appropriate slacks suit or jacket, blouse, and skirt; in recent years, a type of women's business suit has become fashionable that is appropriate. Colors should be subdued and tasteful.

Arbitrary and unrealistic dress standards should be avoided; there is no need to infringe unnecessarily on individual officers' preferences. However, extreme styles, even though they may be fashionable, should be avoided. Plainclothes officers should in fact wear *plain clothes*. If they want to be noticed, they should wear a uniform. Otherwise, it is to their advantage to be unobtrusive and conservative in appearance.

However, it is also important for plainclothes officers (except undercover agents) to be recognizable as law enforcement agents. One useful device, adopted by the federal Secret Service and other federal law enforcement agencies, is a lapel pin. At a distance it is not noticeable, but to anyone up close it immediately identifies the agent. A pin of about half-inch diameter, bearing a miniature police badge or city seal and the name of the agency, would be sufficient.

In addition, plainclothes officers should carry a full-sized badge in the breast pocket of their jacket, or in a special wallet alongside an identification card bearing their photograph.

Police Insignia

The recognizability of the police uniform depends almost entirely on its general style and color scheme. However, standardized police insignia contribute greatly to recognizability and impart useful information about the individual officer.

The universal device to identify personnel of a law enforcement agency is the badge. The traditional metal badge usually is attached to the front of the hat or cap, and another badge is worn on or above the left breast pocket of the shirt or jacket.

Badges are offered by specialty manufacturers in many styles; the choice of style is purely a matter of taste, except that all officers in an agency should wear exactly the same style.

An officer's rank and duty assignment may be indicated by cloth patches on the sleeve of the shirt or jacket, or by pins on the shirt collar, jacket lapel, or sometimes on the cap. Nearly all police agencies use rank markings similar to those of the U.S. Army, except that not all army ranks are used by most police departments. However, there is nothing to prevent a police department from either eliminating rank markings or adopting its own markings.

Markings are available in various designs to indicate duty assignment or branch of service, such as patrol, traffic, special weapons, tactical squad, and technical services.

Shoulder patches, identifying the agency and perhaps the officer's assignment or branch of service, usually are worn on both shoulders of the jacket or shirt.

Every uniformed officer should wear a name tag, usually in the form of a metal or plastic pin worn above the breast pocket of the shirt or jacket. The name tag is important because it allows a citizen to identify the specific officer with whom he or she has had contact, in case further contact is necessary. The name tag ideally contains the officer's badge number, rank, first initial, and full last name. The minimum would be a name tag bearing only the officer's last name.

In addition to the name tag, or in place of it, some agencies use a photo identification (ID) card, usually worn in a plastic holder attached to the right or left breast pocket. The ID card is valuable, especially for plainclothes officers while they are at the police station. However, the photo ID may not serve as a replacement for the name tag, unless the name on the ID is large enough to be easily read from a distance of more than a couple of feet.

Other insignia, identification marks, and decorative devices tend to clutter up the uniform without adding worthwhile information. Flag patches or lapel pins, for example, may reflect an officer's personal patriotism, but (unless there is evidence to the contrary) most people would assume that all police officers are at least as patriotic as the average citizen. Religious, fraternal, or other insignia should not be worn on a police uniform nor by plainclothes officers because they might give offense unnecessarily to some citizens.

Exceptions certainly should be made for pins or small patches that represent service awards. Many agencies give pins or other insignia for life-saving action, service above and beyond the call of duty, special achievements (such as marksmanship), and so on. These devices contribute to morale and encourage professional attitudes. However, they should be kept to a minimum, and any pins, medals, ribbons, or patches that are authorized for the uniform should be tasteful and discreet. A police uniform, after all, is not a fancy-dress costume. It is a set of working clothes, and anything that detracts from its serviceability could hamper an officer's performance of vital duties.

PROTECTIVE EQUIPMENT

The hazardous nature of police work, especially certain specialized assignments, requires that every reasonable effort be made to protect the officers' lives and safety. In recent years, several types of protective equipment have been developed and gained widespread use for this purpose (see Figures 7.3 and 7.4).

The routine use of protective body armor, or "bulletproof vests," has generated a certain amount of controversy among police officials. The question has been raised as to whether these devices actually provide as much protection as officers may suppose, and whether the wearing of protective equipment may cause officers to become careless or foolhardy.

Figure 7.3. Male body armor. (Photo by the authors, courtesy Austin PD.)

Figure 7.4. Female body armor. (Photo by the authors, courtesy Austin PD.)

These are difficult questions to which no hard and fast answers can be found. However, it is certainly true that no protective device offers an absolute guarantee of safety. Even the best body armor provides only limited protection from certain kinds of hazards.

Before this equipment came into general use, police agencies had developed several procedural rules and policies designed to keep officers safe by avoiding, as much as possible, life-threatening situations. These rules and policies remain by far the best safeguards of officers' lives. Protective equipment cannot substitute for effective, sensible, and cautious police techniques.

Nevertheless, protective equipment can make the difference between life and death in critical situations, provided appropriate equipment is available and is used properly. With these reservations in mind, four kinds of protective equipment can be considered.

For general police use, *body armor* has become increasingly popular. Body armor consists of a sleeveless jacket or vest, ordinarily worn under an officer's shirt or jacket. The armor is made of either nylon or plastic fibers, woven into a tight mesh, covered with fabric. The vest is stiff but somewhat flexible; although advertised as lightweight, most vests are in fact fairly heavy and not very comfortable to wear, especially in warm weather. The most popular type of body armor currently available is made of Kevlar® (a trademark of the E. I. Du Pont de Nemours Company), a type of plastic. Its major advantage is that it can be woven into a mesh that is much more flexible, and feels softer, than other body armor materials.

The purpose of body armor, of course, is to prevent a bullet from penetrating the officer's body. The vest covers only the torso (some vests have a flap that extends over the groin area). Arms, legs, and head are not protected.

Body armor was the first type of police equipment tested by the Technology Assessment Program (TAP), which was started by the International

Association of Chiefs of Police and now administered by the U.S. Department of Justice's National Institute of Justice. In its first report, TAP commented, "no garment manufactured is 'bulletproof'. The term 'bulletproof vest' has been used since crime shows on radio and TV have become popular. Just about anything worn by a police officer can be 'defeated'. When it is defeated, the results are usually disastrous."[5]

According to the TAP report, most body armor is designed to withstand penetration of bullets from .22- and .38-caliber weapons, which account for about 85 percent of the handguns distributed among the public. There is also some evidence that these weapons are most often used in assaults on police officers.[6]

The tests conducted by TAP consist of firing various weapons from a standard distance (depending on the type of weapon) directly at a garment that has been attached to a block of clay, which represents the average density of the human torso. If the bullet deforms the vest or penetrates it beyond a depth of about 1 3/4 inches (44 millimeters, to be precise), the armor is considered to have failed.

Armor that successfully resists penetration is considered acceptable at one of four levels, depending on the type of weapons tested. A Level I vest is tested against .22- and .38-caliber bullets; a Level II vest has successfully withstood bullets of 9-millimeter and .357 Magnum calibers; a Level III vest has been successfully tested against 7.62-millimeter (.308 Winchester) rifles; and a Level IV vest has stood up to a 30.06 armor-piercing bullet. There are also two "sublevels": a Level II-A vest has resisted penetration by .357 Magnum and 9-millimeter bullets but at lower velocities than a Level II vest. A Level III-A vest has been tested against .44 Magnum and 9-millimeter high-velocity weapons.[7]

In the first TAP body armor tests, conducted by independent laboratories in 1978, more than one hundred vests were offered by sixteen manufacturers. Unfortunately, more than half of them failed even at the Level I standard of performance. (Curiously, some vests failed the Level I test but passed at higher levels.) Since then, dozens of new models have been tested in an ongoing evaluation program, and the percentage of passes has increased greatly.

Before any law enforcement agency purchases body armor or declares its use acceptable by its officers, copies of the TAP test reports should be obtained and studied carefully.[8] Officers who wear this equipment routinely should be given thorough instruction in its use and should know precisely what it will and will not do.[9]

Officers should be warned that body armor will not simply deflect a bullet harmlessly. Even the best armor will be deformed more than an inch by many projectiles. This means that the impact of the bullet will be transferred to the wearer's body with about the same force as the round end of a ballpeen hammer swung from a distance of 3 feet. At the very least, the impact will leave a painful bruise. Impact on the rib cage may cause a fracture. Furthermore, the impact is likely to throw the officer off his or her feet. The officer should expect this reaction and be prepared to recover quickly. Only movie superheroes are able to stand up to a hail of bullets without flinching.

A second type of protective equipment is the armored shield. Unlike the vest, the shield is not used routinely, but must be carried whenever the officer feels that danger is imminent. Shields may be made of metal (usually steel) or an impact-resistant plastic such as Lexan. They are used in exactly the same manner that medieval knights used their shields: They are worn or held on the left arm and used to deflect bullets or other projectiles.

In general, a shield provides better protection for the officer's entire body, and it is more likely to resist being penetrated by most projectiles. However, there is no such thing as a perfectly bulletproof shield, and there is always the possibility that a ricocheting bullet will find its way around the shield.

Furthermore, there is the problem of storing the shield where it is readily accessible when it is needed. Usually there is no room in the interior of a patrol car; the shield must go into the trunk, which may be already crowded with other gear.

Finally, because the shield must be carried in or on the left hand, that leaves the officer with only one hand free to use a weapon or other equipment.[10] Because of its limitations, the shield is most useful in special situations such as riot control or terrorist incidents. Even then, it is most effective when it is used in conjunction with body armor for protection against ricochets.

Some police equipment manufacturers offer clipboards and notebooks made of bulletproof or bullet resistant material (usually Lexan). These devices are in no sense a substitute for a full shield or body armor, but there may be instances when an officer could use a high-impact clipboard to deflect a projectile or a hand weapon such as a knife.

As we said earlier, body armor offers no protection at all against one of the most critically vulnerable parts of an officer's body: the head. A shield offers only partial protection to the head, depending on how it is held. The only kind of protection that is effective for the head is a helmet. A suitable protective helmet should cover the entire skull, from the base of the neck to the throat. Most motorcycle-type helmets cover only to the jawline in front and to the nape of the neck in back. The helmets worn by automobile racing drivers (especially drivers in Formula I and other open-cockpit vehicles) provide more complete coverage, but they are relatively expensive.

There are a few helmets designed specifically for police use. They have a transparent faceplate of high-impact plastic that covers the entire face opening of the helmet (coverage of the eyes alone is not adequate) and a broad chin piece covering the entire jaw. A helmet that gives adequate protection is likely to be hot and uncomfortable to wear and may restrict both vision and hearing to some degree. Officers must be trained to use a helmet properly and they must realize its limitations. Even the best helmet will not provide absolute protection against such hazards as a high-powered rifle bullet.

Finally, police officers occasionally may need to use a gas mask, especially if the police themselves are employing some of the nonlethal chemical weapons we will discuss later. Gas masks are used only infrequently; there is little reason for every patrol officer to carry one routinely. Masks vary greatly in their capabilities, but they are all relatively expensive. However, when a gas mask is needed, there is no effective substitute. Masks should be available in sufficient quantities for all officers who may need them, rather than supplying them only to special-duty officers such as members of a SWAT team.[11]

Masks that are connected to an independent air supply (such as an air tank carried by the wearer) are the most effective against any hazard, but also are the most cumbersome and expensive. Filter-type masks are less expensive and far easier to use but offer the least protection and offer no protection at all against some hazards.

Unfortunately, only the largest police agencies can afford to provide a selection of different masks for different applications. Smaller agencies should, at a minimum, make sure that their masks provide at least 30 minutes of protection against the kinds of chemicals they are most likely to encounter including any gas weapons that the agency itself uses.

POLICE WEAPONS

Handguns

Every police officer in the United States carries a handgun of one kind or another. In fact, Great Britain is one of the very few countries in the world where law enforcement officers do not routinely carry handguns as their primary defensive weapon.

The function of the handgun is to protect the officer and innocent citizens from deadly force used by a criminal, by making it possible for the officer to disable the criminal. At times, disabling the criminal results in the criminal's death, an unfortunate but often unavoidable result.

Police officials sometimes propose that officers follow a policy of "shooting to disable" rather than "shooting to kill," but in fact such a policy is impractical in too many instances.

A handgun should not be used by a police officer (or anyone else, for that matter) to stop a fleeing suspect who is not posing an immediate threat, to disable a suspect who is not known to be dangerous, to intimidate any person who is not a criminal subject, or for any other purpose. *Warning shots should not be fired into the air or ground;* when a weapon is fired, it should be for the specific purpose of protecting life from deadly force. These policies should be established, maintained, and firmly enforced by police administrators.

Until recently, the .38-caliber revolver was the most widely used handgun in police service in the United States because of its high reliability and more than adequate effectiveness.[12] It is still widely used but has been replaced in many agencies by 9-millimeter handguns, generally semiautomatic (see Figure 7.5).

Technically, most "automatic" handguns are semiautomatic: they fire one round at a time. A true automatic weapon fires continuously as long as the trigger is pulled. Many officers prefer the semiautomatic because it is able to fire at a rapid rate and an ammunition clip can be reloaded much more quickly than a revolver's cylinder. However, the semiautomatic pistol has a tendency to jam, and misfired rounds must be manually removed before another round can be fired.

Figure 7.5. The 9-mm semiautomatic (left) and the .38 caliber revolver are the two most popular police service handguns in the United States. (Photo by the authors, courtesy Texas Dept. of Public Safety.)

Reloading a revolver can be accomplished rapidly if an *auto-loader* is used. Essentially, the auto-loader is a barrel-shaped device that contains slots for the appropriate number of bullets. When the revolver is empty, its cylinder is opened and the auto-loader is used to inject the bullets into the cylinder all at once.

It is also possible to load a revolver with two or more different types of rounds (such as standard load, Magnum, and soft-nose) and select the round to be used in a given situation.[13]

A short (4-inch) barrel is considered preferable to a longer barrel because the holster is less cumbersome, and because it is less likely that a criminal grappling with an officer at close quarters will be able to grab the barrel and pull the weapon out of the officer's hand. A longer barrel does give greater accuracy, but the difference usually is not great enough to be an important consideration.

We believe that standardization of handguns should be an inflexible policy of the police agency. As we pointed out earlier, there might be situations in which officers would need to share ammunition, which is impossible if they carry different types of handguns. Furthermore, every officer should be trained and competent in the use of every type of weapon used by the agency. Finally, most police agencies supply the ammunition for their officers' weapons, and stocking several different types of ammunition is unnecessarily costly.[14]

Officers should be held responsible for seeing that their weapons, especially their handguns, are cleaned and maintained in perfect working order. Repairs, when needed, might be performed by an agency employee or, if none is available, at the agency's expense.

Every officer should be required to fire practice rounds from his or her handgun, and all other assigned weapons, on a regular basis and should be required to qualify on a practice range (that is, meet specific standards of marksmanship) at least every six months.

Finally, the practice of carrying a concealed second or "backup" weapon should not be encouraged. If proper police techniques are followed, it is unlikely that an officer will be disarmed or unable to use his or her regular weapon. If that unlikely event does occur, it is even less likely that the officer will have the opportunity to reach and use a concealed backup weapon.

Backup weapons occasionally have been used as "throw-down guns"—put in the grasp or near the body of an unarmed suspect who has been shot by an officer, to make it appear that the suspect was armed. This despicable practice cannot be tolerated. If "backup" weapons are permitted, officers should be required to register them with the agency, so that there will be no question about the origin of a weapon found at the scene.[15]

Rifles and Shotguns

Municipal patrol officers rarely have a need for a high-powered, long-range rifle. However, there are situations in which the greater impact and accuracy of a large firearm might be needed, and for those situations a standard-barrel shotgun is appropriate. Most urban police agencies provide patrol officers with a shotgun in their vehicles.[16]

Rural police (sheriff's officers or state troopers) are more likely to need a rifle for its greater accuracy over longer distances. The type of weapon selected is a matter of weighing the various advantages and disadvantages of many different models. However, the choice should be made by the agency's administrators and standardized for the entire agency, whether the agency actually purchases the weapons or they are bought by individual officers.

Few police officers will ever have any reason to use some of the more exotic firearms such as machine guns or armor-piercing rifles. Before any officer is given such weapons, the officer must have extensive training in its proper use.[17]

Nonlethal Weapons

Handguns, shotguns, and rifles all must be classified as lethal weapons because of the high probability that their use will cause the death of the target, even if the officer's intention is merely to disable the subject.

Quite often, lethal force is neither necessary nor desirable, but some force must be used to discourage or disable the subject. For example, a mentally deranged person is as much a hazard to himself or herself as to innocent bystanders, but there is no legal reason to kill the person. A person wanted in connection with a crime may have to be subdued by force, but there is no need or desire on the part of the police to kill the suspect. Citizens involved in a disturbance or demonstration may have to be discouraged or subdued, but lethal force usually is not justified.

In these situations, it would be extremely desirable for the police to have an effective but nonlethal weapon. Unfortunately, none of the nonlethal weapons available to the police has proved to be very effective.

There are four major types of nonlethal weapons, classified by their basic principle of operation: impact, chemical, electrical, and noxious. There are also some miscellaneous devices that do not fit into any of these categories.

The most common impact weapon is the familiar police *baton* (also known as the nightstick or billy club), a wooden or plastic stick usually about 18 to 24 inches in length. There are also more sophisticated baton-type weapons, some based on Asian martial-arts devices, or fitted with a handle perpendicular to the main shaft, and so forth.

Most municipal police departments consider the baton a basic part of the patrol officer's equipment.[18] In fact, the baton has very limited uses. It is properly used either to prod or to push individuals in the direction an officer wishes them to go. A baton also can be used to ward off blows from a fist or from a knife or other hand-held weapon.

The baton should not be used to strike a person. The impact of a baton swung forcefully is almost certain to cause serious, permanent injury. Under no circumstances should a baton ever be swung at a person's head or face, or brought down on a person's back at the base of the neck; such a blow is very likely to be lethal. Many agencies have adopted the policy that a baton may not be used at all unless it is held firmly in both hands and must not be used by poking one end into a person's face or midsection.

Batons filled with metal pellets or reinforcing bars are not permitted by most law enforcement agencies. Batons made of somewhat flexible rubber or plastic are permitted by some agencies, but are subject to the same limitations as wooden batons.[19]

Several types of nonlethal impact projectile weapons have been developed for police use, mostly for the suppression of civil disturbances, but none has proved to be both safe and effective. These weapons include the stun gun, which fires bags of metal pellets; the broomstick gun, which fires short wooden dowels; the ricochet gun, which fires soft plastic pellets that are supposed to be fired toward the ground at the feet of rioters; and the Blake impact gun, which fires soft metal (aluminum) or rubber balls. These devices are very inaccurate, and are as likely to injure innocent bystanders as rioters. They are too inaccurate to be used against individual subjects. Despite being intended as nonlethal weapons, all are likely to cause serious injury or death even when properly used.

The only reasonably effective impact weapon is the water cannon, which sprays a high-powered stream of water. The chances of causing serious injury or death are low unless the subjects struck by the water are forced off their feet or propelled into a barrier such as a plate-glass window. The primary disadvantage of the water cannon is its cost, which is quite high, and the fairly rare occasions when it is likely to be used; it is of little use against individual subjects.

Any ordinary fire department pumper truck can be used as a water cannon, if the fire fighters are willing to work under police supervision or if officers have been trained to operate the equipment. The use of fire trucks for this purpose may be offensive to many citizens, and fire fighters may be reluctant to cooperate because of the resentment toward them that may be generated.

Chemical nonlethal weapons consist of several types of gas that can be disseminated by canisters, grenades, or projectiles from special guns. These weapons are best known by the term *tear gas*.

The two types of chemical most widely used as tear gas are CN (alphachloroacetophenone) and CS (orthochlorobenzal malononitrile). The former is more common; it was developed by the military in World War I and has been used ever since. CS, which was developed in the late 1950s, is now considered much more effective than CN and has the particular advantage of being less likely to cause permanent eye or skin injury. CS gas is also longer lasting because the gas particles tend to cling to clothing and other rough or textured surfaces (including, in some cases, the subjects' hair).

Two other types of chemical weapons, known as DM and HC, are not regarded as suitable for police use, the former because it is extremely dangerous and the latter because it is ineffective.

The method of delivering a chemical or gas may be at least as important as the type of chemical. The most widely used delivery systems are gas grenades and canisters. The former are similar to military incendiary grenades: The gas is contained in a small, barrel-shaped vessel sealed with a lever-type trigger. A pin is pulled to make the trigger operable; the grenade is then hurled, releasing the trigger, and a few seconds later the gas is released. The most serious problem with a gas grenade is that it can be picked up and hurled right back at the police officers.

Canisters or cartridges are fired from a special, large-bore rifle.[20] Some cartridges are designed to explode on impact; others are designed to release a stream of gas. The explosive-type cartridge may cause serious injury and, if shot into a confined area, may start a fire. The other type of canister, like a grenade, can be picked up and thrown back at the user.

The proper use of a gas weapon demands care and skill. Consideration must be given to the wind direction and any other condition that might affect the dispersal of the gas. As much as possible, the user should avoid letting gas drift beyond the target area, where it may injure or annoy innocent citizens. It is also quite likely that the gas will drift or be blown back into the officers' faces. Whenever gas weapons are used, all officers should wear gas masks.

If the target is a building or enclosed area, gas weapons should be used with extra care, precision, and restraint. If one grenade or canister will release enough gas to fill a room, there is no need to use two or three grenades.

Another type of chemical weapon is the aerosol spray, the most familiar of which is Chemical Mace®.[21] The chemical is similar in nature and in effect to CS gas; it is delivered by aiming a spray nozzle on an aerosol can and spraying the target.

Chemical Mace and similar aerosol sprays are intended to disorient and temporarily disable the subject. However, there are several difficulties. First, the spray must be aimed directly into the target's face from a distance of no more than a couple of feet. Second, individuals vary in their sensitivity to the chemical; some people are not bothered by it very much, but others may be permanently blinded by a heavy dose.

Third, because it is unlikely that a subject will stand still and wait to be sprayed, an officer who produces an aerosol can may expect to be met with violent resistance—perhaps more violent than would have been exhibited otherwise. In some cases, subjects have wrestled the can away from the officer or simply deflected the spray back into the officer's face.

Nevertheless, Chemical Mace and similar aerosol sprays can be effective if used with discretion and skill. Once the subject has been sprayed and has reacted to the spray (usually by closing his or her eyes, coughing violently, and trying to cover his or her face), the officer should put the spray can away and take control of the subject by hand, applying handcuffs or other suitable restraint as necessary. Continued spraying of a disabled or unresisting subject is unnecessary and abusive.

As soon as the prisoner is firmly under control, his or her face, neck, arms, hands, and other exposed skin areas should be flushed with water to dilute the spray chemical, which otherwise continues to give an unpleasant stinging sensation. It is especially important to wash the chemical out of the prisoner's eyes and nose to avoid permanent injury.[22]

Two kinds of electrical nonlethal weapons have been used by police agencies. The first of these is the electrified baton, often called a cattle prod. The device consists of a hollow baton or tube that contains electrical batteries and electrical contacts on one end. When the baton is pressed against a subject, an electric shock is transmitted. The shock is not ordinarily dangerous but it is painful. In rare cases, it can be lethal if used on a person with a weak heart or a chronic neurological condition such as epilepsy. Furthermore, it is not a very effective way to subdue a prisoner because the painful shock is likely to make the prisoner more enraged and violent than ever.

A variation on the cattle prod, and sometimes also called a "stun gun" (which confuses it with a projectile weapon), is the hand-held electric device perfected by Nova Technologies and copied by various other manufacturers. The device generates a very high-voltage, low-amperage electric current; when it is pressed against the subject's body and a trigger is pressed, the current temporarily stuns the subject's nervous system, disabling the subject.

This type of device is said to be one of the safest of all nonlethal weapons, although permanent injury or death is possible if the subject happens to have a chronic heart or nervous system problem. The effectiveness of the device is somewhat debatable: Some subjects are less susceptible to it than others, and some recover from the shock more rapidly than others. In any case, the device must be pressed firmly against the subject's body and held there for several seconds, which may be difficult when the subject is resisting violently.

The Taser® is an ingenious device that has been invented as a nonlethal substitute for the pistol. Instead of firing bullets, the specially designed Taser handgun or rifle fires a pair of small metal barbs, not unlike fishhooks. The barbs remain connected to the gun by a pair of lightweight copper wires. When the barbs strike the target, they are supposed to catch on the subject's clothing, hair, or skin. A strong, low-amperage electrical charge is sent through the wires. The charge is strong enough to cause the subject to be paralyzed until the current is turned off.

The disadvantages of the Taser are several. First, the rifle's range is supposed to be up to 500 feet, which means that a rather strong explosive charge is needed to propel the barbs and their trailing wires. If used at closer range, the explosive charge itself could be dangerous or the barbs could be propelled at high enough velocity to penetrate the target's skin and cause serious damage. The barbs are very likely to cause permanent damage if they happen to strike the target's eyes. At longer ranges, more than about 25 feet, the Taser is not very accurate; at long range the barbs may not strike the target with enough impact to catch. Also, the electrical charge may have varying effects on different individuals and could be lethal in some instances. The Taser rifle is no longer being manufactured, but the handgun is still available, and some agencies continue to use the rifle.[23]

Noxious nonlethal weapons include any of several devices that are intended to create such a nuisance that the target will become disoriented or will submit in order to end the annoyance. Some of the noxious-type weapons that have been tested include blindingly bright lights, high-intensity noises, and smoke grenades that release a gas with an obnoxious odor. None of these devices has proved to be very effective against most subjects, and all are likely to be at least as obnoxious to the user (that is, the police officers) as they are to the target subjects. However, on occasion a noxious device (especially one that combines a flash of light with a loud noise) may be useful to distract a barricaded subject, or to create a diversion that will allow officers to move into position to overcome a subject.

Finally, there are several miscellaneous devices that have been proposed as nonlethal weapons.

Animal control authorities have used tranquilizer darts with great success to capture dangerous animals, and it has been suggested that the police use such a device against criminal subjects or rioters. Unfortunately, there is no way to estimate in advance how much tranquilizer will be needed to subdue a violent subject; too little is ineffective, and too much may be lethal. Tranquilizers also are fairly slow acting. These are not serious problems when the target is an animal, but they are great disadvantages when human targets are involved.

Several kinds of foam and liquid agents have been developed for riot control purposes. In most cases, the idea is to make the street, sidewalk, or other surface so slippery that the rioters will be unable to stay on their feet. One product of this sort is called *Instant Banana Peel®*. Some of these agents are quite effective but not terribly useful. If the rioters are unable to stay on their feet, the same will be true for the police officers who are trying to disperse the crowd or take criminal subjects into their control.[24]

In summary, then, all of the so-called nonlethal weapons are either not very effective, or not very useful, or if they are both effective and useful, they are not very nonlethal. By far the most effective "nonlethal weapon" is a well-trained police officer.

POLICE SERVICE EQUIPMENT

In addition to the handgun and baton, one other item of police equipment is nearly universal in American law enforcement: the bracelet-type handcuffs. Handcuffs are among the types of police equipment tested by the Technology Assessment Program.[25]

Officers should be fully trained in the proper use of handcuffs and other restraints. Bracelet-type handcuffs can cause a good deal of pain if they are applied too tightly, and can cause injury to the subject's wrists.[26] Conversely, a subject may be able to slip out of the handcuffs if they are applied too loosely.

Some handcuffs are equipped with a double lock. Officers who have this type of equipment should be trained to use both locks at all times; single-locked cuffs may give way if the prisoner is able to apply sufficient force or leverage.

Whether handcuffs are provided by the agency or officers must supply their own, the agency should establish a standard type that is to be used by all officers, and all should operate with the same key design. This ensures that officers will be able to use one another's cuffs when necessary, and that a lost key will not prevent cuffs from being removed from a subject.

Bracelet-type handcuffs are fairly expensive and can be difficult to use under some circumstances. An alternative that is being used successfully by many police departments is a simple plastic tie, similar to those designed to tie bundles of electrical wires or cables. The plastic tie has a locking device at one end that, once engaged, cannot be disengaged; removing the tie requires the use of a knife or scissors. The tie is cheap, easy to use, not as uncomfortable for the subject as metal handcuffs, and cannot be defeated easily.

Most police officers will need to be equipped with various kinds of specialized tools for their particular assignments. Officers who may be assigned to investigate traffic accidents should have a measuring wheel or tape. The measuring wheel is especially handy because it can be used by one person, it can measure a curved or zig-zag line, and it is not limited in the length of line it can measure. A measuring tape is more accurate and may be less cumbersome to store and carry.

Officers whose duties include investigating crimes should be equipped with an assortment of evidence containers and basic gear such as a fingerprint kit, brushes, tweezers, and a hand-held vacuum cleaner.

Officers assigned to traffic law enforcement should be equipped with a radar speed-measuring unit. Several types of radar units are widely available; the selection of a particular unit depends mainly on the traffic patrol techniques favored by the agency. Probably the most commonly used type is the dashboard-mounted unit, which (depending on the design of the radar transmitter) may be used either in stationary or moving patrol vehicles, and which may be capable of reading the speed of traffic in either direction (that is, oncoming cars or cars approaching from the rear). The dashboard-mounted unit is also much less conspicuous than the other type of bidirectional radar, which is mounted on the side of the vehicle.

Perhaps the most versatile type of radar unit is the handheld or "radar gun" type. The gun can be kept out of sight until a target vehicle approaches, then aimed and triggered to give an immediate reading. This technique makes radar-detection receivers in citizens' cars almost useless, and it avoids the problem of incorrect identification of the target. However, if the radar gun is not used correctly, it will give false speed indications.

Recently there has been some controversy over claims that certain types of radar guns may have caused the officers who used them to develop cancer. As of this writing, the controversy is not resolved, but all of the cases appear to involve officers who may have used the equipment improperly.

One of the most valuable types of equipment in all police work is the camera. Every patrol officer and investigator should be equipped with at least a small snapshot or instant camera for general use at traffic accident and crime scenes. More sophisticated cameras should be used for full investigations of crimes. Some agencies prefer to assign photographic specialists to this duty, although any police officer might be trained to be a competent investigative photographer.

The biggest problem with all photographic equipment is that it is somewhat difficult to use and generally rather expensive. The widely popular videocassette combination camera and recorder, or *camcorder*, is only a little more expensive than a high-quality still camera, but it is infinitely easier to use. The great advantage of the video camcorder is that the tape cassette requires no chemical processing; it can be removed from the camera and viewed on a standard television set. The tape cassettes are inexpensive and can be reused several times.

All patrol vehicles should be equipped with an extensive first-aid kit for immediate care of injured persons when it is impractical to wait for emergency medical personnel and a small fire extinguisher. As a matter of good police practice, officers should not be expected to act as emergency medical technicians or fire fighters. However, they should be prepared, trained, and equipped to render any emergency assistance that may be necessary to protect lives and property until the proper personnel arrive.

The quality of service provided by a law enforcement agency depends, in the final instance, on the quality of the people employed and deployed by the agency. Well-trained and highly motivated officers do not need an elaborate arsenal of exotic, costly equipment. However, they do need a carefully conceived, standardized set of working tools.

It is the responsibility of administrators at every level to see that their officers are properly equipped, whether with uniforms and equipment supplied by the agency, or with gear purchased by the individual officers. Standards must be established and maintained to ensure that all equipment is both appropriate and uniform. This responsibility is just as important a part of the planning and budgeting process as the design of a new building or the purchase of a fleet of vehicles.

REVIEW

1. True or False: Police equipment should be standardized so that all officers can be trained to use it properly.

2. Bulletproof vests (body armor) are valuable because

   (a) They provide a limited degree of protection under certain circumstances.

   (b) They protect an officer from almost any kind of weapon.

   (c) They keep officers from becoming careless.

   (d) They eliminate the need to remember and follow a lot of complicated police procedures.

   (e) They are easier to wear than most other kinds of protective devices.

3. True or False: A police handgun may be used legally only to disable a person who is using or threatening to use deadly force against an officer or citizen.

4. Match the following nonlethal weapons with the basic principle of operation:

(a) Aerosol spray.   (i) Impact.

(b) Water cannon.   (ii) Chemical.

(c) Baton.   (iii) Electrical.

(d) Taser.   (iv) Noxious.

(e) Stun gun.

(f) CS gas.

(g) Cattle prod.

(h) Noise generator.

5. True or False: Photography at an accident or crime scene should be left to a qualified specialist because patrol officers and investigators have too many other duties.

NOTES

[1]D. F. Gunderson, "Credibility and the Police Uniform," in *Journal of Police Science and Administration*, vol. 15, no. 3, September 1987, p. 192.

[2]O. W. Wilson, *Police Planning*, 2nd ed. Springfield, Ill.: Charles C Thomas, 1962, p. 223.

[3]George D. Eastman, ed., *Municipal Police Administration*. Washington, D.C.: International City Management Association, 1969, p. 284.

[4]David A. Varrelman, "Facilities and Materiel," in Bernard L. Garmire, ed., *Local Government Police Management*, 2nd ed. Washington, D.C.: International City Management Association, 1982, pp. 361–62.

[5]*Police Body Armor*, Gaithersburg, Md.: Technology Assessment Program, International Association of Chiefs of Police, December 1978, p. 11.

[6]Robert Little and Max Boylen, "Facing the Gun: The Firearms Threat to Police Officers," in *Journal of Police Science and Administration*, vol. 17, no. 1, March 1990, pp. 49–54.

[7]Daniel E. Frank and Lester D. Shubin, *Selection and Application Guide to Police Body Armor*. Washington, D.C.: National Institute of Justice, 1989, p. 8.

[8]TAP documents, and a catalog of current publications, can be obtained from the TAP at Box 6000, Rockville, MD 20850, or by calling (800) 248-2742.

[9]John Sturner, Sheldon F. Greenburg, and Deborah Y. Faulkner, "Equipment and Facilities," in William A. Geller, ed., *Local Government Police Management*, 3rd ed. Washington, D.C.: International City Management Association, 1991, pp. 406–7.

[10]Wilson, *Police Planning*, p. 222.

[11]Varrelman, "Facilities and Materiel," in Garmire, *Local Government Police Management*, p. 360.

[12]Sturner, Greenburg, and Faulkner, "Equipment and Facilities," in Geller, *Local Government Police Management*, pp. 399–401.

[13]Wilson, *Police Planning*, p. 221.

[14]Sturner, Greenburg, and Faulkner, "Equipment and Facilities," in Geller, *Local Government Police Management*, p. 401.

[15]Tony Lesce, "Back-Up Weapons: Current Thinking," in *Law and Order*, vol. 37, no. 10, October 1989, pp. 32–37.

[16]Sturner, Greenburg, and Faulkner, "Equipment and Facilities," in Geller, *Local Government Police Management*, p. 402.

[17]Ibid., p. 400.

[18]Ibid., p. 405.

[19]T. C. Cox, D. J. Buchholz, and D. J. Wolf, "Blunt Force Head Trauma from Police Impact Weapons," in *Journal of Police Science and Administration*, vol. 15, no. 1, January 1987, pp. 56–61. The article notes that metal flashlights are used by some officers as a baton-type weapon.

[20]Sturner, Greenburg, and Faulkner, "Equipment and Facilities," in Geller, *Local Government Police Management*, p. 404.

[21]Ibid., p. 404.

[22]Donald T. Shanahan, *Patrol Administration*. Boston: Holbrook Press, 1975, pp. 431–32.

[23]Sturner, Greenburg, and Faulkner, "Equipment and Facilities," in Geller, *Local Government Police Management*, p. 405.

[24]Shanahan, *Patrol Administration*, pp. 423–30.

[25]Varrelman, "Facilities and Materiel," in Garmire, *Local Government Police Management*, p. 359.

[26]Joseph E. Scuro, Jr., "Legal Considerations of Handcuffing," in *Texas Police Journal*, vol. 39, no. 6, July 1991, pp. 9–11.

Part Two
Management of Material Resources

POLICE INFORMATION SYSTEMS

Every action taken by a police agency depends on the collection, recording, storage, retrieval, and distribution of information. In an important sense, a law enforcement agency can be regarded primarily as a large information-processing system.[1]

Information about criminal activity comes either from direct *observation* by police officers or, much more often, from citizens who have observed or been victimized by the criminal. This initial *report* of a crime (or, more rarely, of a potential crime that has not yet been committed) becomes the basis for an *investigation* to learn as much as possible about the circumstances of the crime and of the criminal.

The object of the investigation, of course, is to determine the *identity* of the criminal and to assemble enough *facts* about the crime to persuade a judge or jury that the criminal has been accurately identified. At the same time, the identification of the criminal is a necessary step toward his or her apprehension, the legal purpose of which is to ensure that the accused criminal will be presented before a court for trial. *Knowledge* of the criminal's whereabouts is therefore crucial.

Each of the italicized words in the preceding paragraphs represents a type or form of information. In addition, there are many other types of information that concern the police: plans, beliefs, ideas, hunches, policies, and so on. For our purposes, we can define *information* very broadly, as the content and product of human minds.[2]

NATURE OF INFORMATION SYSTEMS

Information held in a single person's mind is useful only to that person and only so long as he or she has access to the memory of it. People are able to expand the usefulness of information by storing it outside of their own memories (for example, in the form of drawings, paintings, or written words) or by transmitting it to other persons (for example, by speech or distribution of the drawings, paintings, or written words). These activities are two basic forms of *communication*: the transfer of information over time or space, or both.[3]

An *information system* is a set of *communications systems* designed to collect, sort, store, retrieve, assemble, and distribute information for some particular purpose. In the case of a law enforcement agency, the purpose is, of course, to detect crimes, identify and apprehend the perpetrators, and bring them to justice.

In this chapter, we discuss the nature and important features of information systems in a police agency. In Chapter 9, we consider communications systems. This division is somewhat arbitrary, since all information systems involve communications to a degree, and all communications systems involve the distribution of information. Modern technologies are making the distinctions between information and communications less clear than they once were.

Unfortunately, law enforcement agencies in the past have not always recognized the importance of establishing a sound, well-planned information system as an integrated whole. Instead, several piecemeal information systems have developed in most agencies, with a good deal of overlap and duplication, significant gaps, and resulting inefficiencies.[4] Inefficiency is evil enough when it merely wastes money or other resources, but when the waste involves an essential resource such as information, the evil is compounded.

The first carefully designed police information systems did not even exist until the end of the nineteenth century, when the concept of scientific investigation began to take hold and, more or less at the same time, legal reforms to improve the criminal justice system placed greater demands on the police to keep accurate records. At the turn of the century, the newly formed International Association of Chiefs of Police attempted without success to establish a uniform system of reporting crimes and the disposition of criminal cases. This system did not actually come about until 1930.[5] Even today, the Uniform Crime Reporting System (UCRS) is not completely uniform and is not used by all law enforcement agencies in the United States.

As late as 1942, O. W. Wilson's suggestion that police departments consolidate their records into a single, centralized file was regarded as a radical notion (although it has come to be the conventional wisdom).[6] Previously, each officer kept his own file of criminal cases, and the only centralized records were those of a primarily administrative nature. Even today, many police agencies do not have a single, central records system to organize and maintain all of the agency's police service and administrative records.[7]

The information and communications systems used in law enforcement developed independently of each other. Wilson and other authorities have insisted since the mid-1930s that central records bureaus and radio rooms should be

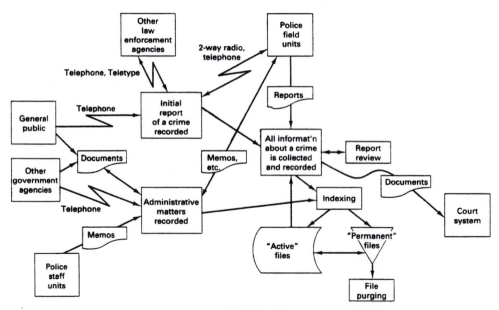

Figure 8.1. An integrated police information system. Square boxes are sources and destinations of information; rectangular boxes are types of information and information processing. "Lightning" lines indicate electronic communications (radio, telephone, and computer); "page" symbol indicates paper documents.

at least adjacent to one another, but that still is not the case in many medium-sized and larger departments. It is usually the case in smaller departments simply because one person is employed as both radio dispatcher and records clerk.

The basic elements and components of an integrated police information system are diagrammed in Figure 8.1. Messages (that is, units of information, regardless of their size or content) enter the police system from several sources: the general public, other governmental agencies, and other law enforcement agencies. Police field units (patrol officers and investigators) and police staff units (administrative and technical support personnel) also generate messages within the system. These messages, from whatever source, concern administrative matters and information regarding crimes (and other police matters). The messages may take any of several forms including written documents (letters, periodicals, memoranda), telephone calls, and teletype messages; communication between the police headquarters and the field units usually is in the form of radio calls or telephone calls.

Incoming messages, of whatever type, are first recorded and then acted on by delivering the message to the proper personnel. In the case of police matters, the field units' major responsibility is to gather more information, which is once again communicated back to a central point for recording (most often in the form of written reports; some agencies now have mobile computer terminals that permit field units to file reports directly into a computerized database).

After a message has been recorded, regardless of its type or form, it must be stored for future use. Some types of messages, such as reports from field units, must be reviewed to make sure that the report is complete, accurate, and in the proper form.

Before any message is stored, it must be *indexed*: A record must be made of the message's location in the storage files. A message then may be stored either in an active file, if it is likely to be retrieved for further use in the near future, or in a permanent file if it is uncertain that the message will be needed soon.

Periodically the permanent files must be "purged," which means that obsolete and useless messages are discarded, to keep the file from becoming overloaded.

Finally, some messages are transmitted from the police system to the court system; these messages are generally in the form of written documents.

In the rest of this chapter, we discuss how these elements can be developed and arranged to meet the administrative and operational needs of a contemporary law enforcement agency.

MANAGEMENT INFORMATION SYSTEMS

The managers of any enterprise must receive a constant stream of information about the condition of the enterprise, the use being made of resources, and the activities of the people involved in it. This is no less true for a police agency than for a steel mill, an army, or a shoe store.

Managers are decision makers. They must have information on which to base their decisions. They must know what resources are available to them, what conditions exist in the organization and its environment, and what the results have been from the last set of decisions they made. Much of this information must come from within the organization itself.

With so much information being bandied about, there is a considerable likelihood that problems will arise. One problem is that the manager may receive inaccurate, obsolete, or otherwise useless information. Another problem is that useful information will be transmitted to the wrong person and not to the person who needs it. Similarly, a manager may receive so much information, all of it useful and necessary, that he or she has no time to digest and act on all of it. These problems exist in almost every organization. Preventing or correcting them can be accomplished only by designing an efficient management information system.

First, consider what kinds of information might be produced by a police agency or introduced to it from the outside. For now, we consider only administrative matters; later we will examine police matters.

The general public might transmit information about any number of administrative matters. A citizen might complain that the police department's budget is too high (or too low!), or that too many (or too few) patrol officers are assigned to a particular neighborhood. Publishers regularly send police departments their magazines, journals, newsletters, books, catalogs, and other literature, either at the request of someone in the agency or unsolicited. Businesses send police departments flyers, brochures, and catalogs of merchandise that the agency might wish to purchase.

Correspondence between the police department and the general public might concern announcements of job openings, promotions, and other personnel actions; negotiations involving the purchase of equipment or property; announcements of newly established police policies that may affect the community; and so on. In fact, almost every aspect of a police agency might result in communications between the agency and some member of the public.

Communications between a police department and other governmental agencies usually involve a smaller range of matters. The agency may receive messages from its parent government concerning newly established policies, procedures, and administrative actions that affect all agencies. There are many communications between the police department and its parent government concerning budget, purchases, personnel, and other transactions. Police departments occasionally receive or transmit correspondence with other governmental agencies (besides the parent government) on miscellaneous matters such as sharing information about the community or about the agency's administrative practices.

Communications with other law enforcement agencies are almost entirely concerned with police matters. Police agencies also sometimes (not as often as they might) share information about administrative practices and policies.

Within the police agency itself, information about administrative matters is generated continuously by every employee. Administrators issue statements of policy, guidelines for particular procedures, and directives that describe actions to be taken by subordinates. Field units produce reports of their activities. The personnel office generates reports of all kinds of personnel actions including job vacancies, appointments, promotions, disciplinary actions, leaves and vacations, tests for promotion, training courses offered and completed, and so on. The accounting office produces reports on budget balances, expenditures, revenues received, and other financial matters. The purchasing office produces estimates of the cost of items to be bought, invitations to bid, specifications, reports of bids, purchase vouchers, and similar documents.

Not every type of administrative message must be transmitted to every administrator. The key factors in designing an effective information system are the answers to three questions: (1) *Who* needs to know? (2) *What* do they need to know? (3) *When* do they need to know it?

Determining the correct answers to these questions is in itself an important management decision. Agency officials should decide what kinds of information are to be transmitted to each person, in what form, and by what means. Or, to look at the problem from another perspective, officials must decide what kinds of information to collect internally and where the information, whether generated internally or externally, is to be sent.

All information received from external sources should go through one central receiving point, where the nature of the message and its destination are recorded. The personnel at the central receiving point (usually the clerks in a mail room) should be given a set of policies to determine where each type of message is to be delivered. Messages that do not fit the standard policies should be reviewed by a supervisor or sent to whichever administrative unit seems most appropriate.

Similarly, information produced within the agency should be recorded and distributed according to an established procedure. Usually the person who generates a message will determine who is to receive it. However, many types of administrative messages will require essentially the same treatment, which should be determined by a standard policy. For example, messages concerning the agency's operating policies should be distributed to all field and operational units in the same form and by the same methods.

Often the form of a message is as meaningful as its contents; for example, a statement of policy distributed to all field units as an "executive order" probably will mean something very different from a statement delivered to only the field supervisors in the form of an "interoffice memorandum."

Again, not all administrative messages should be delivered to all administrative personnel. The reason is not to limit the amount of information available to administrators, but instead to avoid burying them in a mass of unnecessary information. Within reason, all administrative information should be available to anyone who has an interest in seeing it. Some kinds of information, such as records of disciplinary actions against personnel, should be restricted in the interests of individual privacy and security, but those restrictions should be very few. State and federal laws regard most police administrative records as public records that must be made available to any interested citizen. It would be bizarre if an agency prohibited its own employees from examining records that must be shown to anyone who walks in off the street.

The form of nearly all administrative records is the written or printed document. Increasingly, much of this paperwork is being replaced by computerized databases and individual display terminals, but for the present most law enforcement agencies rely on the written word for most administrative messages.

The next most important form of administrative communication is the telephone message. Unfortunately, these communications are almost never recorded, and thus the content is lost to everyone except the parties to the conversation. Very few police agencies adhere to the sensible practice of maintaining detailed telephone logs.

Telephone logs need not be elaborate or complicated. A standard form may be used, permitting several conversations to be recorded on a single sheet. Only the most basic information needs to be recorded: who was called, by whom, the date and time of the call, and the substance of the matters discussed. Each employee should keep his or her own log; completed logs for each day, week, or other appropriate period should be filed in a central place.

Telephone logs should be maintained even in agencies that record incoming calls on magnetic tape. Most agencies keep the tapes only for a limited period; some agencies record only the incoming emergency lines. Few agencies record outgoing calls at all. Logs kept on paper provide the simplest and most convenient way to retrieve information that otherwise exists only in the memories of the people involved in a telephone call. The purpose of maintaining telephone logs is not to check up on employees but simply to ensure that important information is not lost.

Later in this chapter, we examine some of the ways that computers can be used to make a management information system more effective and efficient.

POLICE SERVICE INFORMATION SYSTEMS

The management information systems needed by a police agency are not particularly different from those needed by any other governmental agency. However, a law enforcement agency has an additional set of information needs that concern police matters.

Police service information systems must serve several different, though related, purposes. First, records of police activities are needed by administrators so that they can make appropriate management decisions. Administrative officials must decide how many personnel are needed for each of the various police tasks and functions; this information, of course, provides the basis for planning and budgeting. Administrators also need to know the circumstances and outcomes of police activities so that appropriate policies and procedures can be devised.

Second, police activities are subject to a broad range of legal restrictions and controls. Police officers, in the exercise of their official duties, may impose upon the legal rights and privileges of citizens. In addition, citizens who are involved in disputes with one another may need to know something about actions taken by the police. For example, a driver involved in an automobile accident might decide to sue the other driver, and the official police report of the accident might supply important evidence for the civil suit. In short, records of police activities often have important legal implications, aside from the basic function of helping to control crime.

Finally, records of police activities are indispensable to the basic police role. From the time that the police first become aware that a crime has been committed until the accused perpetrator is turned over to the courts, and sometimes beyond, every action of the police depends largely on the record of what has been done before. A complete record of the police activities in identifying and apprehending an accused criminal is necessary to secure a conviction in a criminal trial. According to some authorities, the difference between successful and unsuccessful prosecution of a suspect often can be traced to the quality of the police reports submitted to the prosecutors.[8]

Each type of police activity or function may require its own information system. For example, if a police agency is required to issue licenses to taxicab drivers, liquor dealers, operators of steam boilers, ambulance services, and so on, separate records must be kept for each type of license issuance. The same is true for noncriminal police services such as inspecting boilers, escorting funeral processions, and operating an emergency medical service. However, we assume that the record-keeping requirements for these non–law enforcement activities are relatively modest, straightforward, and completely separate from crime-related police activities.

Case Report

All crime-related records begin with the *case report* (sometimes called the *incident report* or *complaint report*), the first record that a crime has been, or may have been, committed (see Figure 8.2).

TEXAS DEPARTMENT OF PUBLIC SAFETY
CRIMINAL LAW ENFORCEMENT DIVISION

REPORT OF INVESTIGATION	DATE:
FILE TITLE 1 2 3 4	THIS REPORT IS THE PROPERTY OF TEXAS DPS-CLE DIVISION. NEITHER IT NOR ITS CONTENTS MAY BE DISSEMINATED OUTSIDE THE AGENCY TO WHICH LOANED.

ACTIVE: ___ INACTIVE: ___ CLOSED: ___
RESTRICTED: ___
SUBMITTING INVESTIGATOR
ID-NR: BY:
STATIONED:
APPROVING SUPERVISOR:
ID-NR: BY:
	FILE NR **TYPE** **PROGRAM**
	
	DISSEMINATION CX RELATED FILES
	DIST F1
	AREA F2
	OTHER F3
	        F4
	        F5

RPT-RE:

Figure 8.2. A criminal case report form. (Courtesy Texas Dept. of Public Safety.)

Every official action of the police department (other than the non–law enforcement services already discussed) should be recorded on a standard case report form, even if it does not clearly involve criminal activity; for example, traffic accident investigations, runaway child cases, search and rescue cases, and other noncriminal matters. The case report should be initiated by whichever employee of the agency, whether police officer or civilian telephone operator, first becomes aware of a situation that requires a police response.

Each case report should be given a unique serial number. If serial numbers are issued in chronological sequence from some central control point, it is fairly simple to keep track of the case reports and ensure that none are missing. The serial number should be issued at the moment the case report is initiated; this helps to determine the order in which complaints and requests for service were received. When officers in the field initiate a case based on their own observation, they must immediately report (by radio or telephone) the actions they have taken, so that the dispatcher or telephone operator can initiate a case report and assign it a serial number.[9]

All complaints and cases should be recorded on standard forms. Larger agencies often use a series of different forms for each type of criminal and non-criminal case; there may be separate forms for traffic accidents, traffic violations, homicides, burglaries, bad checks and documents, robberies, missing persons, and so on. Each form can be designed to contain exactly the information that is needed for follow-up police work. If a case involves two or more different types of crimes (for example, a burglary and homicide), both forms should be used but one should be regarded as the master report and only one serial number should be assigned.

Medium-sized and smaller agencies have less need for a multitude of report forms. A single form can be used for all crimes against property, another for crimes against persons, a third for traffic accidents and violations, and a fourth for miscellaneous matters. The use of multipurpose forms not only eliminates the problem of deciding which form to use but also reduces the cost of maintaining an adequate supply of all the forms.[10]

All case reports should be filed in one place, in strict serial-number order. Copies may be retained by the originating officer or distributed to others who need the information (such as investigators). All other reports and documents relating to each case should be filed with the original case report and should bear the same serial number.

The complete case report usually is written by the police officer who is assigned to investigate an incident. This assignment is made by the dispatcher (or telephone operator) or by a supervising official, when the case is not initiated by the police officer in the field. Because most cases are initiated as the result of a complaint received, usually by telephone, in the communications center, a record of the telephone operators' and dispatchers' actions must be kept. The record usually is in the form of a *complaint card* or *dispatch card*, which summarizes the information received from the complainant and indicates when the complaint was received, when an officer was assigned to investigate, and (in brief) what the officer found. The complaint or dispatch card also should bear the case's serial number.[11]

At one time, the common practice in police agencies was to record complaints (whether received by telephone or otherwise) in a log book or "police blotter" kept on the front desk of the police station. This procedure simplifies the issuance of serial numbers, but has little else to recommend it. A few small police departments still follow this traditional practice, but nearly all agencies have adopted a card system instead.

Other Police Service Records

Not all police service records are related to specific cases.

In many agencies, officers are encouraged to question suspicious persons they encounter on the street and to complete a *field interview report* (often a single card). These reports are filed separately from case reports. (However, it should be noted that some states do not permit a police officer to stop people on the street unless the person stopped is specifically suspected of a crime. The rules concerning field interviews vary considerably among the states, and agency policies naturally must conform to state law.)

Evidence and property inventories (see Chapter 10) also should be kept separately, although a copy should be filed with each case to which the evidence or property pertains. The same is true for arrest records.[12]

Personal Identification Files

Most police agencies keep extensive files on individual criminals, suspected criminals, suspicious persons who are not known to be active criminals, and persons whose occupations, place of residence, personal habits, or whatever suggest that they might have some knowledge of criminal activities.

Both the quantity and quality of these records vary from one agency to another; some of the largest files are almost worthless because the records they contain are incomplete, filled with errors, and stuffed with irrelevant trivia. However, a well-managed personal identification file can be immensely valuable to investigators. The file usually contains the following types of records:

Arrest Reports.

Duplicate copies of arrest reports should be maintained, apart from case reports, usually by type of crime. These reports are permanently filed after the arrested person has been either convicted or acquitted in court; until that time, arrest reports should be kept in an active file of persons currently under arrest (whether the person is actually in jail or released on bail).

Modus Operandi Records.

These records consist of descriptions of the methods of operation used by known criminals. They are usually filed by type of crime.

Vehicle File.

A record should be kept of every vehicle that is in any way involved in criminal activity: vehicles owned or used by known criminals, stolen vehicles, and so on.

Accident Reports.

Both those accidents that are actually investigated by the police and those that are reported by the individuals involved in them, but not investigated, should be recorded in a central file.

Fingerprint Files.

Nearly all police agencies maintain a file of the fingerprints of arrested persons and sometimes of other persons who are not known criminals but whose prints are available to the police. In many communities, taxicab drivers, liquor dealers, and citizens engaged in other lawful occupations are required to be fingerprinted by the police before they are issued a license. The value of such noncriminal fingerprint files is very much open to debate, but the fact is that most police agencies maintain these files on the chance that an unknown set of prints might someday happen to match those of someone in the file. Automated fingerprint identification files, which we discuss later in this chapter, may help to make even very large files more useful.

Photograph Files.

Most police agencies also maintain files of photographs of arrested persons, other known or suspected criminals, and various noncriminal citizens such as license applicants. Here again, the use of photographs (usually "mug shots" made when a person is arrested) for "photo lineups" is restricted in some states; nevertheless, virtually all police agencies maintain photo files.

Current Wants and Warrants File.

This file contains the record of every person wanted by the police in connection with a crime (either as a suspect or as a material witness), of persons for whom arrest warrants have been issued by a court, and of vehicles sought by the police (either because they are stolen or because they are thought to have been used in a crime). Only current, outstanding warrants should be kept in an active file. Once the person or vehicle has been found, or there is reason to think that the person or vehicle is no longer wanted, the record should be removed and placed in a permanent file (such as the case report, if that is appropriate).

Indexing

Earlier we said that records in a police information system must be *indexed*, to create a record of the original record's location in the files. Otherwise, finding the original record when it is needed may involve a time-consuming and possibly futile search.

The index files may contain nothing more than 3- x 5-inch index cards. One set of cards should contain the *name* of every person who comes to the attention of the police: suspected and known criminals, victims, witnesses, complainants, persons interviewed in the field, and so on. A second set of cards should be made for every *location* involved in a police matter: crime scenes, residences and known hangouts of suspected and known criminals, traffic accident sites, and so forth. A third set should contain descriptions of *vehicles* involved in police matters.

Each index card should contain one name, location, or vehicle description and a reference to the case report or other complete record to which the card is related. A new card should be made for each entry.

Name cards should be filed alphabetically by last name. Vehicle cards should be filed either by license number or, in states that issue a new license each year, by the manufacturer's vehicle identification number or the state-issued title number. The card also should contain the make, model, original color, and physical description of the vehicle.

Location cards should be filed in the same order that addresses are listed in the postal ZIP Code directory: named buildings in alphabetical order, followed by named streets in alphabetical order (with addresses on the same street in numerical order), followed by numbered streets. The names of businesses may be filed in alphabetical order and cross-indexed to the street address.

A good, well-maintained indexing system is one of the most crucial elements of an effective police information system. Unfortunately, it is also one of the elements most often overlooked or slighted by smaller agencies whose personnel are too busy to take the time to make index cards.

Summaries and Bulletins

Most police agencies that have twenty-five or more employees compile and distribute a *daily bulletin* that contains a summary of the previous day's police activities. Smaller agencies often find that a weekly bulletin is sufficient. The bulletin keeps everyone informed of important events and of crime trends. It also provides a record for administrative and other purposes of the level of police activities.

The bulletin should be compiled and duplicated in time to distribute it to the first shift as it comes on duty each morning, usually between 6:00 and 8:00 A.M. Copies should be distributed to all officers at each shift's roll call and should be placed on the desk of all personnel who work standard office hours.

The contents of the bulletin may vary from one agency to another, but certain items should always be included; they are the following:

- A list and brief description of all wanted persons, with an indication of whether a warrant has been issued.
- A list and description of all wanted vehicles.
- A summary of the previous day's (or week's) police activities: complaints received, requests for service, calls handled, reports filed by type of crime, arrests or other clearances (disposition of cases) by type of case, and so on.

The bulletin also might contain some or all of the following items:

- A brief analysis of current crime trends; what types of crimes are increasing in frequency, where particular types of crimes are occurring, and so on.
- A description of significant unsolved crimes, wanted persons, or other matters that deserve special attention.
- Announcements of new administrative policies, guidelines, directives, and so on.
- Announcements of personnel actions of general interest such as promotions, transfers, new appointments, and retirements.

The bulletin is not the place for personal messages or news of social events such as the Police League softball schedule. These matters are more appropriately handled in a separate employee newsletter or by posting on bulletin boards. The bulletin should be treated as an official record.

The daily or weekly bulletin also serves as the source for a monthly compilation of statistical information about police activities and crime trends. For that reason, the crime information in the bulletin should be arranged according to the UCRS. The monthly summaries, prepared from the daily or weekly bulletins, serve in turn as the source for quarterly, semiannual, and annual summary reports that are used for administrative purposes such as planning and budgeting.

Uniform Crime Reporting System

The UCRS was established in 1909 by the IACP, but was not widely used until the early 1930s when the FBI agreed to administer it and to organize its own files according to the UCRS categories. Since then, the FBI has received regular reports from some 15,000 U.S. agencies, all on a voluntary basis. These reports are compiled, added to the FBI's own records, and published by the Department of Justice each year.[13]

The Uniform Crime Reports (UCRs) are often criticized, and indeed there may be much about them to question. However, they are also the only statistics maintained on a nationwide, continuous, and reasonably uniform basis. They may not be the best statistics possible, but they are the best ones available.

All agencies should submit reports on a regular basis, following the guidelines developed by the FBI and the IACP committee that monitors the UCRS. The more agencies that submit reports in a standardized manner, the more likely that the overall statistics will reflect at least the general pattern of criminal and police activity across the nation.

INFORMATION TECHNOLOGIES

Enormous strides have been made in information and communications technologies during the present century, yet law enforcement agencies have been even slower than most governmental bureaucracies in adopting the newest advances. New technologies are often costly, of course, but the main reason for the reluctance of police administrators to improve their information systems seems to be not a lack of funds but a lack of knowledge about how new technologies can be applied to police work.

The information technologies most widely used by the police are writing (including printing, xerographic copying, and facsimile), photography, the telephone, and the radio (more properly called the radiotelephone). The teletype is also used to some extent by many police agencies, although not always as well as it could be.

Unquestionably the most important advance in information technology in the present century has been the extraordinary development and spread of computers. In the mid-1980s, most state police agencies and a few large metropolitan police departments used computers to store administrative records and, in some cases, to compile and store police service records (especially large files of wants and warrants, criminal histories, and personal identifications).[14] Few medium-sized and smaller police departments could afford, or felt the need for, such costly equipment. Less than a decade later, it is a rare police agency that does not use computers extensively for all of those purposes and much more.[15]

But before we focus our attention on computers, let us consider how law enforcement agencies use more conventional information technologies.

Writing

Writing on paper is by far the most prevalent of all information technologies. Paper documents are easy and inexpensive to create, duplicate, distribute, store, and retrieve.

The greatest disadvantage to paper documents is their relatively low *packing density*: the volume of information that can be recorded in a given amount of space. Standardized forms and codes help, but paper documents still tend to be bulky and awkward to store. Paper documents are also vulnerable to loss from fire, moisture, vermin, and the mere passage of time.

Furthermore, writing involves the use of "natural" language. Language is not always precise. Words may have various meanings and can be used inaccurately. Both writing and reading are fairly time-consuming activities. Most adult Americans are literate, but the ability to use language well varies from one person to the next.

Photography

Photography is used by most police agencies to record the conditions that are found at crime and accident scenes and sometimes to record police activities (for example, to record the procedures used by the police to control a civil disturbance).

The great advantage of photography is that quite a bit of information can be stored in a relatively small space; compared with writing, photography has a much higher packing density. Furthermore, photography records visual information in much the same way that it would appear to a person actually present at the time a photograph was made, so it is generally assumed that photographs represent an accurate account of a scene.

The principal disadvantages of photography are that the equipment is relatively expensive and requires a degree of skill to operate it properly. Some police agencies insist that only specially trained personnel should be given the responsibility for making crime-scene photographs, where skill in using the equipment is especially important. In other agencies, all officers who have crime-investigating duties are given some training in photography, because even poorly made photographs taken as early in the investigative process as possible are much better than either no photographs at all or photographs made at a later time when a specialist happens to be available.

Finally, photographs, like written documents, can be inaccurate or misleading. Still photographs represent an event at an instant of time, and the activities shown in the photograph may be misinterpreted if the viewer does not know what went before or after. Because a photograph shows only a two-dimensional view of a three-dimensional scene or event, information about the relative positions of objects in the scene also may be misrepresented. These problems can be partially overcome by the skill of the photographer or by taking a series of photographs at different instants in a sequence of events, or at different angles in a scene.

Motion picture photography, by its nature, represents a sequence of events more accurately, because most or all of the sequence can be shown. However, it is still possible to represent an event in a misleading way, either by the choice of angles at which the scene is recorded or by stopping and starting the camera selectively. It is worth noting that most of the special effects in television and theatrical motion pictures are obtained in just this way, to mislead the viewer (for dramatic or creative reasons).

Furthermore, even when there is no intention to mislead viewers, motion picture photography requires a relatively high degree of skill, and motion picture film is very expensive.

Video represents another form of photography that is considerably less expensive and more suited to the demands of police work.[16] Modern video equipment includes video camcorders that are relatively easy to operate. The video tape, housed in a plastic cassette, is inexpensive and can be reused dozens or even hundreds of times. Most important, the video image does not need to be developed, as motion picture film does; the video tape can be viewed immediately after it has been made. We discuss some of the uses of video recording in Chapter 9.

Telephones

Telephones are generally considered as part of a communications system, so we defer to Chapter 9 a detailed discussion of their use in police agencies.

New telephone technologies can be integrated with information systems in various ways. An ordinary telephone answering machine, for example, greatly increases the likelihood that important information, received by telephone, will be recorded for future reference. It has been the practice in most police departments for many years to record all incoming calls for police service (that is, calls received over emergency lines), but it has not been feasible to do so for all telephone calls. Today the technology to record all incoming and outgoing calls does exist; whether it is necessary and desirable to use the technology in this way is a matter for administrative consideration.

Another new technology combines the telephone with written documents: *facsimile machines* are widely available to transmit written material from one place to another (actually making a copy of the original document). The facsimile machine, or "fax," has quickly become a standard fixture in police records rooms, especially in agencies that have personnel located in precinct or neighborhood stations scattered around the city.

Radiotelephones

The convergence of information and communications technologies is nowhere more obvious than in the case of radio systems. For that reason, again, we put off a discussion of police radios until Chapter 9.

Teletype

Teletype is an automated telegraph system. The teletype service of most interest to law enforcement agencies is the National Law Enforcement Telecommunications System (NLETS), a teletype network operated by the U.S. Department of Justice. It connects all state and federal law enforcement agencies, including some military agencies.

Each state police agency serves as the "gateway" to NLETS for local law enforcement agencies. This means that a teletype terminal in a municipal police department can be connected to the NLETS center at the state police headquarters to transmit messages to or from other agencies in the same state, agencies in other states, or federal agencies.

NLETS can be used simply to exchange messages between police agencies. For example, a local police department may use the system to distribute information about a fugitive suspect or to inquire about whether the police in another city have information concerning certain stolen property. More important, the NLETS network provides direct electronic access to state and federal computer databases of law enforcement information. Personal identification information, criminal histories, reports of wanted persons and vehicles, and vast amounts of similar law enforcement information can be obtained through the NLETS network.[17]

COMPUTERS

Before about 1970, computers were so expensive and difficult to operate that only the very largest police agencies could even consider having one.[18] Since then, the size and cost of the equipment have declined steadily at the same time that its capabilities and ease of use have advanced. Now there are computer systems readily available in almost any size and at almost any price.

As recently as 1985, a clear distinction could be made among three general types of computers: *mainframe* computer systems, *minicomputers*, and *microcomputers* (now more often called *personal computers*). The distinction was based on the overall size, speed, and capacity of each type.[19]

A mainframe system typically consisted of several separate units, each contained in its own cabinet, connected together by cables. One cabinet (often the size of a large refrigerator-freezer) contained the *central processing unit* (CPU). This unit consisted of one or more processors, the "engine" that drives a computer system, which we discuss in more detail shortly (see Figure 8.3). Other cabinets contained various kinds of memory devices and switching systems. Information could be entered into and retrieved from the computer by using magnetic tape, punched paper tape, punched cards, or keyboards and display monitors.

A minicomputer system essentially was a miniaturized version of the mainframe system. The CPU, and usually some memory and a switching system, were contained in a single cabinet, often about the size of a clothes washer. Additional memory devices, such as magnetic tape drives and mag-

Figure 8.3. A mainframe computer system's central processor. It may look unimpressive, but this CPU can process four tasks simultaneously at a rate of 76 million instructions per second, and has access to nearly 190 gigabytes of stored information (roughly equal to 15,500 four-drawer file cabinets of paper records). (Photo by the authors, courtesy Texas Dept. of Public Safety.)

netic disk drives, could be connected to the CPU. Information usually was entered and retrieved by the use of keyboard-and-monitor terminals; teletypewriter terminals also could be used, and sometimes punched cards or punched paper tape.

Microcomputers were first developed and marketed in the 1960s, but the early models were so primitive and difficult to operate that they had little commercial use. Practical microcomputers began to appear in the 1970s; by 1980, they had grown in their capabilities and ease of use to the point that they began to appear in many business offices. When the International Business Machines (IBM) Corporation, which had been the world's biggest manufacturer of mainframe computers and had successfully developed a line of minicomputers, introduced its first microcomputers, what had been an important trend suddenly became a full-fledged revolution.

A microcomputer consists of a processor built on a single solid-state chip, about the size of a large postage stamp, connected to a limited amount of active memory and a switching system, all in a single cabinet smaller than a briefcase. The cabinet usually contains two or three magnetic memory devices. A keyboard is used to put information into the computer, and a display monitor is used to retrieve information. Most microcomputers also are connected to a printer, for a paper copy of information.

Early microcomputers took up several square feet of desktop space. Later models, some with the keyboard and monitor built into the computer body, were reduced to the size of a large portable typewriter, then to the size of a looseleaf notebook. Today there are hand-held models.

What is most significant about microcomputers is not merely their size, but their ever-increasing capabilities. Today's desktop microcomputers are as fast, and can manipulate as much information, as the best mainframe computers of the 1960s. Furthermore, the computing capability that might have cost several million dollars in 1970 can be obtained now for a couple of thousand dollars (see Figure 8.4).

Figure 8.4. A personal computer or microcomputer. A PC can be used by itself or can be connected to other PCs, a minicomputer network, or a mainframe system. (Photo by the authors, courtesy Texas Dept. of Public Safety.)

Some Computer Basics

It is not necessary to know exactly how a computer works to use one, any more than it is necessary to be an automotive engineer to drive a car. However, some understanding of computers is essential to every manager or administrator.

The capabilities of any computer system are judged primarily by the system's speed and capacity. The standard measure of a computer's speed is millions of instructions per second: the rate at which the computer's processor operates.

Microcomputers (and sometimes minicomputers) are often rated according to their processors' *clock rate*, in millions of cycles per second, or *megahertz* (MHz). The two measures are related but not directly; in other words, a processor with a clock rate of 33 MHz does not perform 33 million instructions per second, because the latter is determined by several factors. In fact, clock rate alone is only a rough approximation of the speed at which a given computer works.

The capacity of a computer concerns both the volume of information it can manipulate at one time and the volume of information it can store. The former depends mostly on the size of the computer's active memory, or *random access memory* (RAM). While a computer is being used, both the program it is following and the data it is using are stored in RAM. In most computers, however, information stored in RAM disappears when the power is turned off. Information therefore must be stored in some other device; we describe such devices in a moment.

The capacity of both RAM and storage memory is measured in bytes. A *byte* is a unit of information, actually a numerical code consisting of eight binary digits, or *bits*.* One byte represents a single character, such as a letter of the alphabet or a numeral, or a command to the computer. The RAM capacity of

*Strictly speaking, a byte may contain any number of bits. However, the standard code used by virtually all microcomputers and minicomputers, and by many mainframe systems, uses an 8-bit byte.

most microcomputers is expressed in thousands of bytes, or *kilobytes* (KB), or millions of bytes (*megabytes*, or MB). Some minicomputers and most mainframe computers have memory capacity expressed as billions of bytes, or *gigabytes*. To give you some idea of how much these figures represent, each chapter in this text contains between 50 and 100 KB of information.

The essential difference between a mainframe computer, a minicomputer, and a microcomputer is the speed and capacity of its CPU and RAM. The distinction is no longer as clearcut as it once was: some microcomputers have about the same capabilities as the less expensive minicomputers, and some minicomputers have faster CPUs and more RAM than a small mainframe system.

Processors

The processor is the key component of any computer. In mainframe systems and minicomputers, the processor consists of several large integrated circuits connected together. In microcomputers, the processor is a single integrated circuit.

It is the processor that does the computing in a computer. Even the most sophisticated processors have limited abilities to perform a small number of operations, such as adding two numbers together, subtracting one number from another, or routing a byte of information to a particular destination. What gives processors their power is the enormous speed at which they can combine these few operations and repeat them. For example, to multiply, say, 15 times 12, the processor simply adds 15 twelve times.

Modern mainframe computer processors often are designed to permit *parallel processing*, which means that two or more operations can be carried out simultaneously. The processor automatically divides a problem into several parts, solves each part separately, then combines the solutions into one.

One reason for the explosive growth in the development of microcomputers has been the rapid evolution of their processors. When IBM began to build microcomputers, the company and its principal supplier of processors agreed that the processors would be available to anyone else who wanted to build similar computers. Dozens of other companies quickly began to produce "IBM-compatible" microcomputers based on the same processors. This "open systems" approach encouraged the rapid acceptance of the IBM design. As a result, IBM and the processor manufacturer were able to support the rapid development of more sophisticated designs. As of the early 1990s, approximately six to eight IBM-compatible computers were being sold for every non–IBM-type computer.[20]

Memory

Aside from processor design, the most crucial component of a computer is its active memory.

Mainframe systems and minicomputers generally have several gigabytes of RAM. This enormous active memory permits the use of extremely complex *programs*, the detailed step-by-step instructions that control a computer's operations. The early IBM-type personal computers often were equipped with as little as 4 KB of RAM, which severely limited their usefulness. Today, top-of-the-line personal computers have 10 MB of RAM, and sometimes more. This is considerably less than most minicomputers, but designers have found ways to get around this limitation by clever design of programs and the use of "external" memory.

"External" memory means any memory device connected to a computer other than the active RAM. In personal computers, "external" memory devices include magnetic disk drives that are often housed in the same cabinet as the processor and RAM; they are "external" only in a technical sense. Most personal computers also can be connected to memory devices, such as magnetic tape drives and laser disk drives, contained in separate boxes. External memory for minicomputers and mainframe systems usually includes magnetic disk drives and magnetic tape drives.

The magnetic disk drives used in personal computers are of two types: "hard" disks (sometimes called "fixed disks") and "floppy" disks. In a hard disk drive, a metal disk coated with magnetic material is spun by a motor. Magnetic heads are positioned above and below the disk. The entire drive may be concealed inside the microcomputer's cabinet. Hard disk drives typically are capable of storing anywhere from 10 to 200 MB of data; their capacity seems to increase by a factor of two every year.

"Floppy" disks are plastic disks coated with magnetic material (essentially the same material used in magnetic audio and video recording tape); the disk is held in a paper or plastic envelope. The disk drives are accessible from the front of the computer cabinet so that the disks can be inserted and removed easily. Each floppy disk may hold up to about 1 1/2 MB of data, but virtually unlimited amounts of data can be stored on a number of disks and "swapped" in and out of the computer as necessary. For this reason, floppy disks are especially useful to store information that is used infrequently, and to keep a permanent "back-up" copy of the information on a hard disk.

New storage devices, including memory chips that can be inserted into a slot in some hand-held computers, may replace floppy disks in the future.

Monitors

A computer is not very useful unless you can enter information into it and retrieve information from it easily. Early computers required an operator to "translate" data and instructions into switch positions, and the only device to "read out" the results of the computer's processing was a row of blinking lights.

In the 1950s, teletype machines came to be a standard "input-output" device for computers. The operator entered information using the teletype keyboard; the results were printed out by the computer on the teletype machine.

Finally, in the 1960s, most mainframe systems and minicomputers used electronic keyboards for input and television-like monitors for output (with various kinds of printers used for permanent copies of the computer's output).

Television-like monitors are also used by personal computers. Special circuits in the computer translate information from the processor into a video signal similar to, but differing in some details from, the signal used by an ordinary television set. Whether the display is black-and-white or in color depends on the type of video circuit installed in the computer and on the type of monitor connected to it.

The design of color monitors has evolved over the years; in general, the newest video circuits and monitors offer better *resolution* (fine detail) and a greater range of colors than did earlier models. When a computer is going to be used more or less continuously, resolution is especially important, since a high-resolution image is much less fatiguing to the operator's eyes than a blurry, low-resolution image.

Printers

Computer printers have evolved along with all the other components of computer systems. The teletype terminals used as output devices in the 1950s and 1960s were slow, noisy, and crude. During the 1970s, two new types of printers were developed: dot-matrix printers and daisywheel printers. Both were faster and quieter than teletype, and both produced much more attractive print.

Daisywheel printers are now virtually obsolete; only a few models are still being made. Briefly, a daisywheel printer uses a print wheel on which the characters are embossed at the ends of metal "petals" (from which the name is derived). The wheel spins continuously; a hammer taps the proper character, causing the image to be transferred to paper. Electronic daisywheel typewriters are still very popular.

In a dot-matrix printer, the printing element is a *print head* that contains several small rods or pins. Early models had 9-pin heads; the better models today have 24-pin heads. An electronic signal from the computer causes some of the pins to stick out a fraction of an inch as the head taps against an ink ribbon; thus the pattern formed by the pins is transferred to the paper. The best 24-pin dot matrix printers can form images that are hard to distinguish from high-quality type. Dot-matrix printers are relatively inexpensive, fast, and reasonably quiet.

Two other types of printers developed in the 1980s have become increasingly popular. *Ink-jet* printers work much like dot-matrix printers, but instead of pins striking an ink ribbon, the ink-jet machines have a printing head that contains a number of tiny nozzles that spray ink directly onto the paper. *Laser* printers use technology similar to that of xerographic copiers: a low-powered laser forms the desired image on an electrostatic drum, causing tiny graphite particles to cling to the drum; the particles are then transferred to paper. Both ink-jet and laser printers are capable of much higher speeds and more attractive images than dot-matrix printers and are virtually silent in operation.

Finally, color printers are now available in dot-matrix, ink-jet, and laser models, as well as earlier (and cruder) colored pen printers (more accurately called *plotters*).

Connecting Computers

Mainframe computers and minicomputer systems generally are capable of serving hundreds of individual *terminals* or *workstations*: input-output devices such as keyboards and monitors. The terminals can be connected to the CPU by electrical cables, *fiber-optic* cables, telephone lines, or even radio signals. Such systems are designed from the outset to permit many different users to have access to the computer at the same time, and to use different programs simultaneously.

Personal computers are designed primarily as separate, "stand-alone" devices: that is, the computer and its associated keyboard-monitor, printer, and other devices are intended to be used by one person at a time. However, it is not only possible but increasingly common for personal computers to be connected to one another, and to be connected to minicomputers and mainframe systems.[21]

The simplest way to connect two computers is through telephone lines. A device called a *modem* is used to convert the computer's digital electronic signal into a series of tones (like those produced by a touch-tone telephone pad) for transmission over the telephone system, and then to convert the tone signal back into a digital signal. Virtually any personal computer can be attached by modem to the telephone system; it is then simply a matter of placing an ordinary telephone call to any other modem-equipped computer. Special software is used to control the communications process.

Modems and the telephone system are adequate to connect computers on an irregular basis, and are essential to connect an individual personal computer to any of several commercial computer data services. However, this type of connection has some disadvantages.[22] The telephone system is designed mainly for voice signals; its limitations impose a fairly low "speed limit" on computer signals. Because the signals must travel through a multitude of electronic devices—amplifiers, relays, switches, and so forth—from one point to another, errors and losses of information can occur. Finally, and of special concern to law enforcement administrators, extreme measures must be taken to avoid unauthorized access to or disruption of computer signals.

Where personal computers need to be connected together most of the time, especially if they are being used to access a common database or to exchange programs, any of several types of specialized computer networks can be used. Such networks are known generally as *local area networks* (LANs). There are several common types of LANs and many variations. All involve the installation of special circuits in each microcomputer.[*] Any of several types of cable are then used to connect the machines together. Special software must be installed in each computer. In some LANs, one computer (often a minicomputer, for its greater speed and capacity) is used to control the network. With a well-designed LAN, up to several hundred personal computers can be connected, enabling any one computer to access any other's memory for data and to share programs.[23]

---

[*]It should be noted that Apple's Macintosh computers come equipped with the circuitry necessary to use one type of LAN; in fact, several Macintosh computers can be connected together with the cable supplied with them. See Jon Zilber, "The Macintosh: The Choice for Graphics," in *PC Magazine*, vol. 11, no. 9, May 12, 1992, p. 141.

Software

We have already mentioned software several times, which suggests the difficulty of even discussing computers without referring to it. In principle, a working computer system consists of two main parts: the hardware, or machinery and electronic devices; and the software, the information that a computer computes. Software, in turn, includes two elements: *programs*, the instructions that a computer follows to perform useful work; and *data*, the information that is to be—or has been—computed.

But it is not quite as simple as that.

First of all, there are several layers of software in even the simplest computer system. A computer's processor—whether in a microcomputer or a gigantic mainframe—is essentially just a glorified electrical switch, capable of doing nothing more than turning various circuits on or off. Which circuits it turns on and which it turns off is determined by its instruction set, part of which is built into the processor's circuits themselves, and part of which is contained in a separate memory device (called a *read-only memory*). These instructions are expressed as stored binary digits, or bits, in what computer designers call "machine code."

It would be possible to operate a computer by directly entering instructions in machine code. That is how the early, crude microcomputers worked, and at one time that is how the earliest large computers worked. As you can imagine, however, it would be an unbearably tedious process.

Instead, the processor and the other main elements of a computer are controlled by a second layer of instructions: the *operating system*. The operating system translates a long list of coded instructions into machine code, and in some cases directly controls parts of the computer, such as the magnetic disk drives. In fact, the most widely used operating system for IBM-type personal computers is called the *Disk Operating System* (DOS), although it does much more than operate the disk drives. The video circuits that translate information into visual images, the circuits that operate printers, and the circuits that connect a personal computer to a network are all adjuncts to the operating system.

The operating system can be used to operate a computer directly. In a microcomputer system, for example, it is possible to enter instructions and data from the keyboard and have the results displayed on a monitor or printed on a printer. However, the number of instruction codes in an operating system is fairly limited. For example, the version of DOS released in 1987 contained fewer than one hundred commands, about forty of which were likely to be used frequently.[24] Several of those commands actually consisted of *subprograms* or *routines*: combinations of simpler instructions that are to be performed in a particular sequence. It is also possible to develop one's own lists of instructions—that is, programs—using only the commands in the operating system.

Most people, however, are interested in using computers to do useful work. They do not have the time, skill, or inclination to write their own programs. A program that is designed to perform a particular type of useful work is known as an *application program*. A simple application program may contain a few dozen commands to be performed in a specific sequence. More sophisticated programs, such as the word-processing program we used to write this text, contain hundreds of thousands of commands, and are divided into dozens or hundreds of routines and subprograms.

Before leaving this general introduction to the subject of software, one more "layer" should be mentioned: computer languages. A *language* is a type of program that translates a particular set of instructions into machine code. Operating systems and application programs are written in a particular language. That is, they use the set of instructions that a given language contains. DOS, for example, is written in a computer language known as *BASIC*.[25]

Computers in Law Enforcement

This introduction cannot begin to suggest all of the ways that computers might be used by a police agency. As we said at the beginning of this chapter, a law enforcement agency is essentially an information-processing enterprise. Computers are general-purpose information-processing machines whose speed and enormous capacity are ideal for the critical functions of police service.[26]

Mainframe computer systems usually require custom-designed application programs, although a few commercially designed programs are available for the more popular mainframe systems. A somewhat wider variety of "off-the-shelf" programs are available for many minicomputers, and the computers' manufacturers usually offer a selection of common application programs for their machines.

Commercial software developers and publishers offer thousands of application programs for personal computers.[27] Indeed, choosing the most appropriate programs for a personal computer can be difficult because of the huge variety available. Fortunately, several magazines regularly publish reviews comparing the features and advantages of new programs, and there are consulting services that, for a fee, provide extensive evaluations of specific programs or program types.[28]

The four types of general application programs that are likely to be of interest to a police agency are word processing programs, desktop publishing programs, database programs, and accounting programs. In addition, there are several specialized types of application programs intended for law enforcement use.

Word Processing

A *word-processing program* enables the user to produce written documents including letters, memoranda, reports, and anything else that consists primarily of text. Most of a word-processing program's features are designed to make it easy to edit and rearrange a document in the computer before printing it.

For example, most word processors can insert a character or a word into the middle of a sentence, moving the rest of the sentence over automatically. At the end of each line, the program determines whether the next word will fit within the margins; if not, the word is moved to the beginning of the next line (a feature called "word wrap"). Margins, tab stops, and other characteristics can be decided ahead of time and can be changed at any time; usually the program will reformat an entire document if the margins are changed.

Many word processors can move, copy, or delete an entire block of text designated by the operator. Blocks of text that are used routinely (sometimes called "boilerplate") can be stored in the computer's memory and inserted into a document by merely pressing a few keys. Some word processors can store a mailing list in memory and automatically generate mailing labels or address envelopes from the list. Some programs can even insert selected names or other information from a stored list into new documents.

In a law enforcement agency, a word processing program that offers a fairly extensive array of text-editing features will soon become indispensable for preparing correspondence, memoranda, policy announcements, and case reports. If the word processor includes automatic subprograms to check spelling and punctuation, the quality of case reports should rise dramatically.

Desktop Publishing Programs

Many police departments might find a basic *desktop publishing program* useful. This type of program is intended for the preparation of newsletters, brochures, pamphlets, posters, and other documents that combine text with graphic elements.

Most desktop publishing programs contain a limited number of text-editing features like those of word-processing programs. Many desktop publishing programs also are designed to use word-processing programs as the source of textual matter; the text can be "imported" or copied into the desktop publishing program and combined with graphic elements (such as fancy type, borders, shaded boxes, and illustrations).

Database Programs

A *database program* (sometimes called a *file management program*) is designed to store, organize, and manipulate large quantities of information, especially information of a common type. For example, a database program can store thousands of names and addresses, along with related information such as telephone numbers, job titles, and brief notes. Selected lists then can be compiled from the master list and arranged in, say, alphabetical order, or postal code order, or by some other criterion.

In a database program, the user first establishes the database by deciding what kind of information is to be stored and in what format. The information is divided into individual records (such as the name, address, etc., for one person) containing a certain number of *fields*; the person's name might be one field, street address a second field, city a third field, and so on. The program then can select records according to the contents of one field (all persons in a given city, for example) and compile a list of the records, sorted by alphabetical or numerical order.

All of the types of police service records we discussed earlier in this chapter—case reports, identification files, and the rest—can be stored, organized, and manipulated by appropriate database programs much more easily and efficiently than any paper-based system can accomplish. The use of a database program usually eliminates the need for a card-based indexing system because indexing is an inherent feature of most database programs.[29] Records can be entered into the system as rapidly as someone can type the information into a keyboard. The program automatically keeps the file properly organized, and records can be retrieved simply by entering a request at the keyboard. Even files containing hundreds of thousands of records can be sorted and a needed record can be retrieved in a few seconds.

There are a few cautions to be considered before rushing out to buy a database program. First, a great deal of thought must be given to the kind of information to be stored and how it is to be organized: that is, what is to be included in each record, and how the fields are to be arranged within each record. A poorly designed database is difficult and slow to use. There are professional consultants who can help in designing a database. There are also some commercially published designs, specifically intended for use by police agencies, that can be used with one or more of the popular database programs.

Second, law enforcement databases grow rapidly. A police agency in a small town that handles an average of 10 calls per day will have accumulated at least 3650 case reports in a year, more than 18,000 in 5 years, nearly 40,000 in 10 years—plus all of the other types of files we discussed earlier. A medium-sized police department might easily amass a hundred thousand files *each year*. If a single record contains, say, 500 bytes of information (equivalent to about one-half page of typewritten material), one year's worth of records will require 50 MB of computer memory, or at least half the capacity of a personal computer's hard disk (or approximately 35 to 45 floppy disks).

Except in very small agencies, we would suggest that a minicomputer be used for police service database files; in larger agencies, a mainframe system may be required. Bear in mind, however, that relatively inexpensive personal computers still can be used for other applications, and can be connected by LAN or telephone modem to the database when needed.

A third consideration is that the information in case reports and other written documents will have to be entered into the database on a continuing basis, usually by typing in the information at a keyboard terminal. Furthermore, the existing paper files will have to be entered into the system—a task that may be overwhelming. Otherwise, the agency's records personnel will be operating what amounts to two systems: the "old" paper files and the "new" computer files. In many cases, both systems will have to be searched to find a needed file.

There are electronic *scanners* that can "read" a typed (or sometimes even a handwritten) document and convert the information into computer code. Unfortunately, the scanners presently available have neither the speed nor the accuracy that are desired for use in transferring paper files into a police records system. Still, even a relatively slow scanner whose output must be checked and corrected manually might be preferable to the excruciating task of transferring thousands of files manually.

Finally, all of the personnel who will have access to the system must be trained in its proper use. This may include not only records clerks but virtually all of the agency's officers. Of course, the easier the system is to use, the less time that will be required for training.[30]

Because database systems can become so large and complex, computer program designers have begun to use a special set of techniques, known as *artificial intelligence* (AI) to organize and manage the files. An AI program typically relies on an *expert system*, a series of subprograms that guide the user through the files to locate or to analyze the information that is sought.[31]

For example, the computer program may present a series of questions. As the operator answers each question, the computer narrows the area within the database that is to be searched.[32] Some AI programs are capable of performing extremely complicated searches. Others can project trends, produce training simulations, and develop profiles of criminals.[33]

Because of the enormous memory requirements of large database systems, new technologies may be needed. Optical disks, which use a laser beam to record information by literally burning a tiny pit in the metallic surface of the disk, and a lower-powered laser beam to read the information, can store much larger quantities of data than magnetic disks of comparable size. The optical disk, once the recording has been made, is virtually indestructible.[34]

In the next chapter, we consider some of the ways that database files can be used in a police communications system.

Accounting Programs

Word-processing, desktop publishing, and database programs all have important applications in both police service records systems and in management information systems. Just to round out this discussion, we point out that accounting software is available that can improve an agency's bookkeeping procedures.

Many small police departments and even some medium-sized agencies rely on their parent government to maintain their financial records, simply because the police department does not have personnel with the appropriate skills. In such cases, a good, fairly simple accounting program that can be operated by the police department's records clerk, for example, would enable the police administrator to regain control over the agency's finances.

Special Application Programs for Law Enforcement

Several specialized application programs have been designed for use by police agencies. Some are intended for mainframe or minicomputer systems, to take advantage of those computers' speed and capacity. However, some specialized law enforcement programs can be operated on personal computer systems, and in any case microcomputers can be connected to a minicomputer and used as workstations.

Some of these specialized programs involve communications and are discussed in more detail in Chapter 9. Here we simply list and describe them briefly.

Computer-Assisted Dispatching (CAD).

It has long been recognized that one of the most difficult and crucial jobs in a police agency is that of the dispatcher, who must keep track of all of the available field units and decide quickly how to assign someone to respond to each call for service. In effect, the dispatcher is the commander of the agency's field forces (see Figure 8.5).

CAD programs generally maintain a display of all field units, indicating their current status (available, en route to a call, busy, and so forth) and, in some systems, their location.[35] The program also receives information, usually from the telephone operator's terminal, regarding each call for service and displays the information on the dispatcher's terminal. Some programs automatically indicate the nearest available field unit. A few programs automatically forward that information to the field unit, in effect dispatching the unit (unless the dispatcher manually overrides the computer).

Map and Address Files.

Special computer databases can be designed to contain map displays of the agency's service area. The operator can enter an address and the appropriate map will be displayed, perhaps with a highlight or other indicator of the exact location of the desired address. Map and address files are part of the software available for advanced 911 telephone systems, but this type of program can be established separately.

Figure 8.5. A computer-aided dispatching station. The dispatcher receives information from the 911 operator on the computer screen, which also displays the units currently available and those that are currently "busy." The computer automatically "proposes" the assignment of an available unit to each call; the operator either accepts or replaces the proposed assignment, which is then transmitted to the unit's mobile data terminal. (Photo by the authors, courtesy Austin PD.)

Court Records Management.

Agencies such as sheriff's departments that are involved in generating and processing court records (such as summonses and warrants) are likely to find this type of program extremely useful. One program even includes a module to keep records of sheriffs' sales (of forfeited property) and includes a word-processing module.[36]

Disaster Response Management.

Some disaster management programs are designed primarily to assist in developing disaster plans. Others are useful when a disaster occurs, for such purposes as calling up key personnel, recording damage assessments, and handling logistical records. Where a small police department is responsible for part or all of a community's disaster response, this type of software may prove invaluable.

Inventory Management.

Several companies have developed inventory management programs with the specific needs of law enforcement agencies in mind. Commodities inventories, capital equipment inventories, found property, and case-related property control features are available in various programs.

Jail Management.

Again, where the law enforcement agency has responsibility for operating a jail, an accurate and efficient record-keeping system can be enormously beneficial. Some of the programs of this type are designed to keep track of each inmate's location, status, visitors lists, medical records, court appearance dates, and such related information as activity schedules, and even food service records.

Case Tracking and Management.

Several computer programs are now available to keep crime investigation, accident investigation, and complaint records from initiation to closing. Some programs automatically produce daily, weekly, and other periodic management reports (cases initiated, complaints received and disposition, status of open cases, and so forth), usually in a form that is compatible with the FBI's UCRS. There are also specialized computer programs that contribute directly to casework, such as programs that generate crime-scene sketches,[37] some that produce a graphic reconstruction of motor vehicle accidents[38], a criminalistics lab management system,[39] and other case-management functions.

Personnel Records Management.

Several programs have been designed to meet the specific needs of law enforcement personnel managers. These programs produce and maintain records of personnel applications, testing, selection, training, certification, appointments, promotions, leave, vacations, salaries, commendations and reprimands, and so forth. There are also programs that produce and maintain records of personnel benefit systems, such as health insurance, pension contributions, and the like.

Mobile Data Terminals (MDTs).

Small, compact computer terminals can be installed in vehicles, such as patrol cars, to enable field officers to communicate directly with the agency's computer system. The MDTs vary in design; some include only a compact keyboard and a printer (similar to the printer in a desktop calculator), while others include a display monitor. The MDTs also vary considerably in the number and type of functions they can perform. However, all require special software in the agency's main computer to control the system's operations.[40] One significant advantage of MDTs is that transmissions are in digital form, rather than voice, and therefore difficult for unauthorized persons to intercept and decode.

Interagency Networks.

Earlier in the chapter we mentioned the NLETS computer network that allows personnel of any agency to communicate with any other local, state, or federal law enforcement agency. The NLETS system is controlled by computers in each state's police headquarters and by a master controller in Arizona.

Smaller, localized computer networks can be established as well. For example, several suburban police departments could establish an interagency network to exchange information about current wants and warrants, or to coordinate field operations.[41] Such a network would require connecting the main computer of each agency, with one computer (equipped with special software) serving as the network controller.

Automatic Print Identification.

Most police agencies maintain large files of fingerprint samples, usually on paper cards. Some large agencies' files contain millions of cards. When an unknown set of prints must be identified, searching through a card file can be an enormous task. Several print identification programs have been developed including one commissioned by the FBI.

These programs are capable of comparing thousands of prints per second in search of possible matches. None of the programs has proved to be perfectly accurate and dependable, but progress in this area is being made. Similar search-and-match programs may be developed in the future for photograph files ("mug shots") and perhaps DNA comparison files.[42]

This brief list certainly does not exhaust the number of ways that computers are being used in law enforcement today. We introduce some other specialized application programs in Chapter 9.

Perhaps the biggest drawback to the introduction of computers in law enforcement is that the number of available programs is so large and varied, it is almost irresistibly tempting to buy a program for every conceivable purpose. Unfortunately, few agencies can afford to do so. Instead, the administrator—with a good deal of discussion with operating personnel and middle managers—must select carefully the most needed and useful software, compatible with the agency's existing hardware (or with hardware to be purchased specifically for the new software). The administrator also must follow through to see that the software is properly installed, that all personnel are trained to use it correctly, and that in fact the software is used for its intended purpose.

Police administrators must remember that the ultimate "system" begins and ends with people—particularly the men and women of the police agency. The information system must meet their needs; it must give them fast, accurate, and convenient access to the information they need to do their jobs.

Furthermore, no law enforcement agency can afford to operate in a vacuum. Even rural police departments increasingly must coordinate their activities with those of their neighbors. Computer technology, properly used, can enable police agencies not only to collect, analyze, and retrieve information, but—perhaps for the first time—to share it as well.[43]

REVIEW

1. True or False: A law enforcement agency is primarily an information system.

2. The UCRS is

(a) A voluntary system for collecting crime statistics from participating agencies.

(b) A system for recording crimes that have been reported by uniformed police officers.

(c) A collection of statistics on cases investigated by the FBI.

(d) A method of comparing state laws so that similar crimes will have the same definitions.

(e) A computerized database of criminal case histories.

3. Records must be kept on all police activities because

(a) They are needed to make administrative decisions.

(b) The services rendered by the police usually have legal significance.

(c) They often influence the agency's future activities.

(d) All of the above.

(e) None of the above.

4. _____ must be maintained to assist in locating records once they have been filed.

(a) Personal identification records.

(b) Documents.

(c) Case reports.

(d) Indexes.

(e) Summaries.

5. The general application programs that are most likely to be used in law enforcement agencies are

(a) Accounting.

(b) DOS.

(c) Word processing.

(d) Subroutine.

(e) Database.

(f) Desktop publishing.

(g) CAD.

NOTES

[1]Charles A. Gruber, Jerry Eugene Mechling, and Glenn L. Pierce, "Information Management," in William A. Geller, ed., *Local Government Police Management*, 3rd ed. Washington, D.C.: International City Management Association, 1991, pp. 308–9.

[2]D. F. Gunderson and Robert Hopper, *Communication and Law Enforcement*. New York: Harper and Row, 1984, pp. 2–5.

[3]Sam Souryal, *Police Administration and Management*. St. Paul: West Publishing, 1977, p. 322.

[4]Kent W. Colton, *Police and Computer Technology*. Washington, D.C.: U.S. Dept. of Justice, National Institute of Law Enforcement and Criminal Justice, 1979, p. 2.

[5]V. A. Leonard, *The Police Records System*. Springfield, Ill.: Charles C Thomas, 1970, pp. 37–42.

[6]O. W. Wilson, *Police Records and Their Installation*. Chicago: Public Administration Service, 1942.

[7]Richard E. McDonnell, "Information Management," in Bernard L. Garmire, ed., *Local Government Police Management*, 2nd ed. Washington, D.C.: International City Management Association, 1982, pp. 311–13.

[8]Peter W. Greenwood, Jan M. Chaiken, and Joan Petersilia, *The Criminal Investigation Process*. Lexington, Mass.: D.C. Heath, 1977, pp. 182–90.

[9]McDonnell, "Information Management," in Garmire, *Local Government Police Management*, p. 318.

[10]Ibid., pp. 318–19.

[11]Donald G. Hanna and John R. Kleberg, *A Police Records System for the Small Department*, 2nd ed. Springfield, Ill.: Charles C Thomas, 1974, pp. 12–16.

[12]Richard N. Holden, *Modern Police Management*. Englewood Cliffs, N.J.: Prentice Hall, 1986, pp. 177–78.

[13]United States Department of Justice, Federal Bureau of Investigation, *Uniform Crime Reports for the United States*. Washington, D.C.: U.S. Government Printing Office, published annually.

[14]McDonnell, "Information Management," in Garmire, *Local Government Police Management*, pp. 313–14.

[15]Holden, *Modern Police Management*, pp. 176–77.

[16]Joseph Missonellie and James S. D'Angelo, *Television and Law Enforcement*. Springfield, Ill.: Charles C Thomas, 1984.

[17]"Focus on NCIC: Identifying the Unidentified," in *FBI Law Enforcement Bulletin*, vol. 59, no. 8, August 1990, pp. 22–24. NCIC is the National Crime Information Center, the computerized database maintained by the FBI and accessed through NLETS.

[18]Kent W. Colton, *Police Computer Technology*. Springfield, Ill.: Charles C Thomas, 1978, p. 19.

[19]Joseph A. Waldron, Carol S. Sutton, and Terry F. Buss, *Computers in Criminal Justice*. Cincinnati: Pilgrimage Books, 1983, pp. 3–6.

[20]Jim Seymour, "Platforms: How the PC Stacks Up," in *PC Magazine*, vol. 11, no. 9, May 12, 1992, p. 115.

[21]Waldron, Sutton, and Buss, *Computers in Criminal Justice*, p. 7.

[22]Frank J. Derfler, Jr., "Maximum Modems," in *PC Magazine*, vol. 11, no. 5, March 17, 1992, pp. 285–339.

[23]Frank J. Derfler, Jr., "Connectivity Simplified: An Introduction to the Ways of Networking," in *PC Magazine*, vol. 11, nos. 6 and 7, March 31 and April 14, 1992.

[24]Al Stevens, *Teach Yourself DOS*. Portland, Ore.: Management Information Source, 1989, pp. 241–44.

[25]For more information about this language, see James S. Coan, *Basic BASIC*. Rochelle Park, N.J.: Hayden Book Co., 1970.

[26]Gary W. Sykes, "Automation, Management, and the Police Role: The New Reformers?" in *Journal of Police Science and Administration*, vol. 15, no. 1, March 1986, pp. 24–29.

[27]Waldron, Sutton, and Buss, *Computers in Criminal Justice*, pp. 28–39.

[28]One of the better sources of information about software for law enforcement purposes is the Fulcrum Group, a consulting company that publishes *The Guide to Public Safety Software*, which lists hundreds of programs for every type of computer. The Guide is updated quarterly. For more information, contact The Fulcrum Group, Inc., Post Office Box 335, Schwenksville, PA 19473.

[29]McDonnell, "Information Management," in Garmire, *Local Government Police Management*, pp. 325–26.

[30]Ibid., p. 317.

[31]Edward C. Ratledge and Joan E. Jacoby, *Handbook on Artificial Intelligence and Expert Systems in Law Enforcement*. New York: Greenwood Press, 1989.

[32]A. B. Badiru, J. M. Karasz, and Bob T. Holloway, "AREST: Armed Robbery Eidetic Suspect Typing Expert System," in *Journal of Police Science and Administration*, vol. 16, no. 3, September 1988, pp. 210–15.

[33]Ratledge and Jacoby, *Artificial Intelligence*, pp. 2–9.

[34]Daniel L. Arkenaw, "Records Management in the 1990s," in *FBI Law Enforcement Bulletin*, vol. 59, no. 6, June 1990, pp. 16–18.

[35]Gruber, Mechling, and Pierce, "Information Management," in Geller, *Local Government Police Management*, pp. 311–12.

[36]*Business Office Sheriff's System*, United Systems Technologies, Inc., 3021 Gateway Ave., Suite 290, Irving, TX 75063.

[37]*Compuscene*, published by Visatex Corp., 1745 Dell Avenue, Campbell, CA 95008, is one such program; there are several others.

[38]*Computerized Accident Reconstruction*, also published by Visatex Corp. (see n. 37); again, there are several other programs that perform the same function.

[39]*Labman*, published by Integrated Computer Concepts, 3250 N. Arlington Heights Road, Suite 210, Arlington Heights, IL 60004.

[40]Gruber, Mechling, and Pierce, "Information Management," in Geller, *Local Government Police Management*, pp. 312–13.

[41]*Multi-Community Command and Control Systems in Law Enforcement*, Washington, D.C.: U.S. Dept. of Justice, Law Enforcement Assistance Administration, National Criminal Justice Information and Statistics Service, 1978.

[42]Bruce J. Brotman and Rhonda K. Pavel, "Identification: A Move Toward the Future," in *FBI Law Enforcement Bulletin*, vol. 60, no. 7, July 1991, pp. 1–6.

[43]Stephen Bardige, "Technology Boosts Law Enforcement," in *American City and County*, February 1990, p. 14.

POLICE COMMUNICATIONS SYSTEMS

In Chapter 8, we discussed the enormous importance of information in law enforcement. We pointed out that a police agency is, in a sense, an information-processing enterprise whose primary task is to collect, record, store, retrieve, and act on information. The essential role of the police communications system is to make that information available and useful to all of the agency's personnel.[1]

Communication is the process of transferring information from one place to another, or from one time to another. The means by which this process is carried out is the *communications medium*, which can be a simple device such as writing on paper, a complex technological system of devices such as a telephone network, or a natural mechanism such as speech. Ultimately, all technological media must simulate one of the natural mechanisms available to human beings.

A *communications system* (see Figure 9.1) consists of four major elements.

Encoding.

Information must be converted into whatever form the system requires. For example, the brain converts information into nerve signals that control muscles in the throat and mouth to produce the sounds of intelligible speech. In a telephone system, these sounds must be encoded into electrical signals.

Transmission.

The information must be moved from one place to another, or from one time to another. In speech, transmission is through the air. In a telephone system, transmission is through a pair of wires and a multitude of signal-routing and maintaining devices: relays, switches, amplifiers, and so forth.

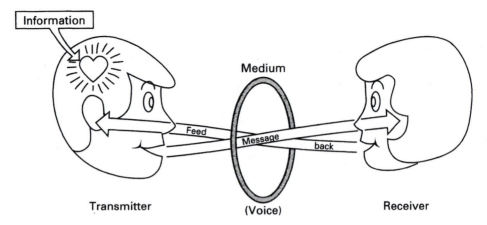

Figure 9.1. A communications system: the information, the transmitter, the medium of transmission, the message, the receiver, and feedback.

Decoding.

When information is received at the intended place or time, it must be converted back into a form that the human brain can process. The sounds of speech, for example, are received by the ear and converted into an electrochemical signal, carried over the auditory nerve to the brain. In a telephone system, the receiver converts an electrical signal into a sound.

Feedback.

For a communications system to work properly, there must be some means for the sender of information to know that it has been accurately received. In speech, this is accomplished by the response of the person who receives and acknowledges the sounds. In a telephone system, feedback is accomplished by the fact that the system permits two-way conversations.

Spoken language is certainly the oldest and most important form of communication, not only in law enforcement but in human society generally. If police officers could simply exchange information by talking to one another, nothing else would be needed. But the nature of police work requires individuals to work at different times and places, and even in small agencies requires the exchange of so many kinds of information among so many people that more elaborate communications media are required.

The nature of law enforcement's mission also imposes some other requirements on a police communications system.[2]

Dependability.

Above all, the system must work properly at all times, or as close to that ideal as any mechanical system can come. A system that often fails to work, that introduces distortions or errors into the information being transmitted, or that does not have sufficient capacity to handle the volume of information being transmitted, is an intolerable hindrance.

Security.

A related requirement is that the system must be protected from being disrupted, whether deliberately by someone who intends to interfere with police operations, or accidentally such as by natural phenomena.

Confidentiality.

Security is important not only to ensure that the system works but to prevent sensitive information from falling into the wrong hands. The information that police officers routinely exchange involves such matters as accusations of criminal behavior that should not become public knowledge, whether they are true or false, until competent evidence is fully and properly presented in court.

Accessibility.

A police communications system must be available to all of the police personnel who have need of it, at the time and place where it is needed. A sophisticated radio system is not very useful to an officer who, for whatever reason, does not have a radio equipped with the proper channels. Similarly, an officer who has the right radio equipment is not well served if the channel is so busy that an emergency message cannot be transmitted.

Speed.

Even in routine matters, police officers need to exchange information as quickly as possible. In an emergency, communication must be immediate.

Easy to Use.

Communications systems may be technologically complex and sophisticated, but they should be easy for the officer to operate. Most communications technologies require some training in their operation, but, to borrow a phrase that computer designers use to describe their products, the system should be as "user friendly" as possible. An ideal communications system should be virtually "transparent": The mechanics of the system should not be apparent to the user.

Police communications systems have been fragmented in the past partly for lack of a suitable technology. The first telegraphic call boxes were put into use by American police departments in the 1880s. They were further developed and refined over the next forty years until they finally were replaced with telephone call boxes (see Figure 9.2).

The latter were first introduced about 1890 but were not widely used because they were expensive and unreliable. However, police stations did install telephones to receive calls from the public in the late 1880s.

The most significant advance in police communications was the development of the mobile radio in the 1920s. Other communication technologies have been added from time to time. The old telegraphic call boxes disappeared in the 1920s, but telegraphic communication, in the form of teletype, was reborn about 1930. In the late 1960s NLETS, a teletype network dedicated specifically to law enforcement use, finally was established. At about the same time, state police

Figure 9.2. Telephone call boxes were the only practical means of communicating between precinct or departmental headquarters and the beat patrol officer from the late nineteenth century until the 1950s. (Courtesy of The Bettman Archive)

agencies and the larger metropolitan police departments began to establish computerized information-processing systems, and these were tied together over the national teletype system.[3]

POLICE COMMUNICATIONS TECHNOLOGIES TODAY

In Chapter 8, we briefly mentioned the convergence of information and communications technologies. This is an important trend with implications for law enforcement management.

As recently as two decades ago, information and communications technologies were separate and distinct. For example, telephones, television, and computers had very little in common. Today, nearly all computers use television-like display screens and can be connected over the telephone network to exchange data.

Many new television sets contain computer-like circuits to process and enhance the video signal, and one TV manufacturer built a telephone into some sets. Video cameras and small television display screens can be connected directly to the telephone network, enabling people to see as well as talk to each other. Some computers can be used to record video information, manipulate video images in various ways, and produce animated video images with sound.

These are but a few examples of technological convergence: different technologies that are becoming more and more alike, and more and more a part of one another.

In the past, developing an integrated information and communications system would have been extremely difficult simply because the various technologies were incompatible. Today it is possible, and in the future it should be even easier, to create a single system that uses a variety of related technologies, working together to serve the needs of police agencies.

In the following pages, we treat several communications technologies as if they were still separate and distinct, but you should also consider ways that they can be combined into an integrated system.

Written Documents

It is not likely that paper documents will disappear from police agencies anytime soon. The convenience, economy, and universal acceptability of written matter ensures that this very old technology will continue to serve the bulk of information and communications needs for the foreseeable future.

Communicating written documents involves their physical movement from one place to another, or their storage and later retrieval (movement from one time to another). The former process usually means the transportation of documents by a messenger or through an agency such as the U.S. Postal Service.

However, there are two new technologies that can greatly increase the speed with which written documents are communicated. We have already mentioned *facsimile communication,* or *fax:* the transmission of documents over the telephone system, which actually results in the production of a copy of the document at the receiving end. The other new technology is *electronic mail.*

Fax machines are relatively inexpensive and easy to operate, and can be connected to any telephone line. Transmitting a document is merely a matter of establishing the telephone connection and feeding the original document into the machine. A copy is produced in the receiving machine within a few seconds.

Most fax machines are designed to both transmit and receive documents. A fax machine that is permanently connected to a telephone line can be dialed directly, and some machines can transmit and receive documents automatically without an operator's attention. Businesses use these machines to transmit routine documents at night, when telephone rates are lowest and the machine is not busy with more urgent matters.

The most serious drawback to fax communication is the relatively low quality of the image most machines produce. Some are better than others, and it is desirable to use equipment that produces the best possible image. For some applications, however, high image quality is essential. Many police agencies use facsimile machines to transmit fingerprint images, photographs, and other non-written material between precinct or neighborhood stations and the agency headquarters, or between agencies. For this purpose, a low-quality image may be a serious handicap.

Electronic mail, sometimes called *E-mail,* is the transmission, storage, and display of documents in a computer system. In this case, there may never be a document on paper. Documents such as memoranda, notes, reports, and letters can be created in one computer, probably using a word-processing program, then transmitted electronically to another computer, or more often a central computer network server. The message is stored in the receiving computer's memory. Whenever the intended recipient wishes to receive his or her E-mail, the accumulated messages can be called up on a computer display monitor, and only those that need to be kept for future reference might be printed on paper.

E-mail programs vary considerably in their capabilities. The personnel who are authorized to transmit and receive E-mail must be trained to use the system, and may have to be reminded occasionally to check their "electronic mailbox" for messages, which otherwise can accumulate at an alarming rate, using up the computer system's memory capacity and making it unavailable for other purposes.

Telephones

The minimum equipment for police communications consists of a one-line telephone and a single-channel radiotelephone. When the telephone operator receives a request for police service, the information is then retransmitted orally over the radiotelephone to the nearest available field officer. The telephone operator or the dispatcher and the field officer also may exchange additional information over the radio.

The principal advantage of the telephone is that it is readily accessible to almost everyone. In the United States, approximately 99 percent of all homes and businesses contain at least one telephone, and the nationwide telephone system permits almost anyone to place a call within a few seconds from any one telephone to any other. Furthermore, a great deal of information can be conveyed rapidly by means of a telephone conversation.

However, the telephone system is not without some disadvantages.

- Telephones are used primarily for oral communication, and the spoken word can be misunderstood or misinterpreted, although persons engaged in a telephone conversation can speak back and forth repeatedly until both parties are satisfied that the correct information has been conveyed. This problem is made somewhat worse by the relatively poor quality of the sound signals carried by the telephone system and by the interference that sometimes affects telephone circuits.

- The person receiving a telephone call usually has no means of determining the true source of the call. People sometimes intentionally use the telephone to misrepresent their identity or location or to convey false information, such as false alarms.

- Telephone technology generally permits only a limited number of telephone calls to be received at one time over a particular set of equipment. Thus, the equipment may be busy with one set of calls while another, possibly more urgent, call is blocked. It is highly desirable, in all but the very smallest police agencies (those with fewer than five employees), to have separate external telephone lines for emergency and administrative calls. Telephone systems are available that permit the administrative lines to be connected to specific internal lines (stations, as they are called by telephone engineers, or what most people call extensions). Emergency calls should always be terminated at the switchboard.[4]

Some of the most elaborate police telephone systems are the 911 systems. American and some foreign telephone companies have agreed to use the numbers, 9-1-1, exclusively for emergency calls, and special equipment has been designed for 911 systems. In a community that has a 911 system, that number is published as the sole number for all kinds of emergency services: police, fire department, ambulance service, and so on.

At a 911 Center, the operators first determine what kind of emergency service is needed by each caller. Depending on how the 911 Center is set up, the operators may handle all incoming calls by initiating a dispatch card and forwarding the information to the appropriate agency, or the incoming call may be switched to separate operators or dispatchers for each agency (see Figure 9.3).

Many 911 systems serve a countywide or larger area that contains several law enforcement and other emergency service agencies. The 911 operator must determine which police department, fire department, or ambulance service is to be notified for each call.

A 911 system demands a high degree of cooperation among all the participating agencies, whether the 911 Center operators actually dispatch all calls or simply transfer calls to the various agencies (see Figure 9.4). In some cases, considerable cost savings can be achieved by using the 911 Center as a centralized point for both receiving calls and dispatching. However, a 911 system can be the source of endless confusion if it is not managed properly.

Some 911 systems include a *calling number display* feature. This means that the telephone number from which each call is being placed is displayed on the operator's console. False reports may be detected if the operator asks for the caller's number and the number given is different from the number displayed. Also, if the call is interrupted, the operator may be able to trace the location from which the call came, by referring to a cross-index of telephone numbers.

An even more sophisticated 911 system includes *automatic location display*, which does just what the name suggests: the street address from which an incoming call is being placed is displayed on the operator's console.

Figure 9.3. A 911 operator station. The operator receives calls from the public and either routes each call to the appropriate agency or, if the call involves the city police, takes the information and transmits it by computer to the appropriate dispatcher. (Photo by the authors, courtesy Austin PD.)

Figure 9.4. The emergency medical service operates its own dispatching system within the 911 Center because of its specialized need for certain information. All of the EMS dispatchers are certified paramedics. (Photo by the authors, courtesy Austin Dept. of Emergency Medical Services and Austin PD.)

This system requires a computerized database containing all of the street addresses in the community and their corresponding telephone numbers, and the database must be continuously updated. Establishing the database can be a monumental task, not just in metropolitan areas, but even more so in rural counties where many roads have never been given names and many buildings have no specific address.

One of the most important concerns in developing a police telephone system, whether or not it is part of a 911 system, is the system's capacity. It is not possible, in theory or in practice, to build a telephone system that is capable of handling any number of calls that might be received.

Telephone systems are designed to have enough incoming lines so that ordinarily no caller will receive a busy signal during *normal peak periods*. A normal peak period is the time of day that is busiest under ordinary circumstances. However, there are bound to be times when all incoming lines are jammed, such as during a period of severe weather or a community-wide disaster.

Some telephone switchboards can be designed to include an overload feature that automatically switches emergency calls to other lines, such as the agency's administrative lines, or lines to another agency, when all emergency lines are busy. The newer switchboards also are capable of *queuing* (which is pronounced like "cueing") incoming calls, so that when all lines are being used, other callers are not given a busy signal but are made to wait until the next non-busy line is available.[5]

In addition to these protective devices, a police telephone system should have at least one completely separate line, with an unpublished number, that can be used for urgent outgoing calls when all the other lines are busy. The number for the separate line also might be given to selected police officials and a very few authorized persons such as the city manager or mayor to use for urgent incoming calls.

Dedicated lines, popularly known as hotlines, also should be established among the various public safety and emergency service agencies in the community. Dedicated telephone lines are fairly expensive; a much less expensive alternative might be the use of voice channels over a cable television system. We describe this possibility more fully later in this chapter.

Telephone engineers and independent communications system consultants are the best sources of information about the design of a suitable police telephone system and its integration with the rest of the agency's information and communications systems. The engineers can conduct studies of telephone usage to determine the most efficient number of lines for normal peak periods and whether special overload protection devices are appropriate. The local telephone service company is not the only source of telephone equipment. Switchboards and other equipment, some of it extremely sophisticated, can be purchased from many different manufacturers and dealers.

Radiotelephones

Radio systems designed specifically for police service are offered by several manufacturers. These radio systems vary in two important ways: the *frequency bandwidth*, the portion of the radio frequency spectrum at which they operate, and the number of channels available. There are other differences among the various police radio systems, including a variety of special features, but bandwidth and channel capacity are the most basic.

The Federal Communications Commission (FCC) decides what uses can be made of various parts of the radio frequency spectrum in the United States and allocates frequency bandwidth to different uses. At present, four bandwidths are available for general police use:

Low VHF (very high frequency) band	about 40 MHz
High VHF band	about 155 MHz
Low UHF (ultra-high frequency) band	about 455 MHz
High UHF band	about 800 MHz

The different bandwidths have very different operating characteristics. A low-band VHF system requires the least electrical power to transmit a signal over a given area, but it is also the most subject to interference from other radio systems and natural phenomena. A UHF system requires much more power to cover the same area, but the signals tend to penetrate most obstacles. However, UHF signals are sometimes absorbed or reflected by heavy rainfall, fog, snow, or even dense foliage.

One advantage of a UHF system is that the mobile antennas are very short, generally only a foot long, and resemble the antennas used for popular cellular telephones. The police radio antenna, therefore, is not obvious on unmarked police vehicles.[6]

A police radio system consists of these major components.

Base Station.

The main radio transmitter and receiver, ordinarily located at the agency's communications center, often in its headquarters facility. In some large metropolitan areas, each district or precinct has its own base station and, in effect, its own independent radio system. A better system is to establish a single dispatching center through which all communications must flow. In any case, there should be a *backup base station* at a different location, in case the principal base station is unavailable because of some catastrophe. Both the main base station and the backup base should be equipped with emergency power batteries and generators, what communications engineers call an *uninterruptible power supply*.

Mobile Radios.

The mobile radios used in police vehicles are properly called transceivers, because they are combined transmitters and receivers. In some radio systems, the mobile units receive on one frequency and transmit on another. This kind of system, technically known as a *half-duplex system*, is necessary where a repeater is used, which we will discuss in a moment.

In a *full-duplex* system, the base station and the mobile units are equipped to transmit and receive simultaneously. A system that does not use a repeater, in which both the base and mobile radios transmit and receive on a single frequency, is a *simplex* system.

It is very desirable for every mobile unit to be able to receive and transmit over all radio channels used by the agency. When an agency has several channels for different purposes, mobile units that are limited to only certain channels are relatively inexpensive. However, the problems of coordinating personnel from several different districts or elements of the agency are compounded when they are not able to communicate with one another over a common channel.[7]

Hand-held units.

All field personnel, and most headquarters personnel, should be equipped with hand-held, portable transceivers, popularly called *walkie-talkies* (see Figure 9.5). Again, it is very desirable for all portable transceivers to be able to operate on every channel used by the agency. Newer models can be equipped with a small combination speaker-microphone that can be clipped to the officer's shirt collar, jacket lapel, or pocket. Some models also can be equipped with a small earphone. These devices allow officers to hear radio transmissions even in a noisy environment, without turning up the radio's volume to the point of annoying other people or compromising the confidentiality of communications.

Repeaters.

Repeaters, or relay transmitters, are used when it is impractical or impossible for a single central transmitter to cover the agency's entire geographic territory. Usually a central transmitter is located at or near the highest point in town, which may be some distance from the base station, with either a separate radio link or a wire connection between them.

It is usually desirable for the central transmitter to be close to the geographic center of the area to be served. However, the transmitter may be unable

Figure 9.5. One of the most significant reasons for the revitalization of foot patrol is the development of powerful, efficient, lightweight portable two-way radios. (Photo by the authors, courtesy Austin PD.)

to reach mobile units in remote parts of the community. Repeaters can be used to ensure that communications are uninterrupted no matter where the mobile units happen to be. Each repeater receives the radio signals from mobile and handheld units, amplifies the signals, and retransmits them on a second frequency. Two frequencies are necessary or the retransmissions would interfere with the signals being received.

A police radio system may require any number of radio channels. A very small agency might be able to share one channel with other nearby agencies, if the total radio traffic for all of them is quite low. Using a shared frequency encourages the various agencies to work cooperatively and can mean considerable cost savings if a single dispatching center and central transmitter are used by all.

As the number of officers in the field increases, so does the volume of radio traffic. It does not take long for a single-channel system to become overloaded. There is no general rule on the number of field units that can be accommodated by a single-channel system, because that depends also on the number of routine messages generated by each unit and on how quickly the dispatcher is able to dispose of each exchange of messages. Adding more channels to a radio system is costly, and often requires replacing the entire system. The first step when a channel is approaching its traffic-handling capacity is to improve the efficiency of communications by using codes for routine messages, such as the well-known ten codes, and by standardizing the form of many messages.[8]

Ideally, a dispatcher should be able to handle all communications with at least twenty-five patrol units or as many as fifty nonpatrol units (investigators, special units, and administrators). Thus, an agency with one hundred officers on duty at one time, half of whom are on patrol, would need a minimum of three channels; four channels would be preferable.

However, the volume of radio traffic is not the only consideration. If the patrol units are dispersed over a relatively large geographical area and if it is fairly common for police activity to be taking place simultaneously at several different locations, more channels may be needed to avoid confusion and to reduce the complexity of the dispatcher's job.[9] A few agencies have attempted to use two or more dispatchers on each channel, but this practice generally adds to the confusion.

The standard practice in most police agencies is to assign channels to different geographical areas, usually the patrol districts. At least one channel should be set aside for use by investigators, administrators, supervisors, and other nonpatrol units. This channel also might be used by patrol officers who are engaged in lengthy conversations with the dispatcher or with one another (for example, during a search operation). Also, it is a common, and extremely desirable, practice in many metropolitan areas to have one channel that can be used by all law enforcement and other public safety agencies in the area.

Thus a police department might require eight or more radio channels; for example,

Channel 1	Patrol district A
Channel 2	Patrol district B
Channel 3	Patrol districts C and D
Channel 4	Patrol district E
Channel 5	Patrol supervisors
Channel 6	Investigators and administrative
Channel 7	Tactical and special units
Channel 8	Metro (all public safety agencies)

Ordinarily, the FCC will grant a license to a police agency for whatever number of radio channels the agency requires. However, the number of channels in each bandwidth is limited. If nearby agencies are assigned the same channels they will interfere with one another.

Interference is especially a problem on the VHF channels, and is a growing problem on low-band UHF. The more channels an agency needs, the more difficult it may be to find sufficient bandwidth to accommodate them. Also, the greater the channel capacity of a radio system, the more expensive the equipment will be. Thus it is desirable not to overdesign the radio system by building in more channel capacity than is really needed.

It is not necessary for all channels to be in the same band. Patrol channels might be in, say, the VHF low band, while administrative and special-unit channels are on high-band UHF. Each type of communication is served by a radio system that is particularly suited to its needs. For example, communications with patrol units may require more geographic range than communications with administrative and special units, so VHF low band may be more economical.

One disadvantage in this arrangement is that operating two or more separate radio systems may be more expensive than operating one system, even when the total number of channels remains the same. Most mobile and hand-held radios are limited to a single bandwidth, although there are some relatively expensive radios that are capable of operating over both the VHF and UHF bandwidths. Otherwise, the patrol units would not have access to the administrative and special-unit channels, and vice versa.

One way to increase channel capacity without wasting bandwidth is to use *digital voice signaling*, a technique in which sounds are converted into digital signals, like the electronic signals used by computers. The digital signals are then transmitted over a radio channel, and reconverted by the receiver into sounds. Digital signals by their nature require less frequency bandwidth than conventional (*analog*) signals;[10] they are almost invulnerable to interference. It is possible to combine several digital signals together without distorting the information. Digital signaling is inherent in a system that uses mobile data terminals, which we discuss shortly; digital voice signaling is a logical counterpart.[11]

Digital voice systems are also inherently "scrambled"; the conversations cannot be understood by anyone who intercepts them, accidentally or deliberately, unless the person has the same type of equipment as the police. There are also analog "scrambling" systems that, when they work properly, are almost completely invulnerable, but they are also rather expensive, and some are not very reliable.

Many law enforcement agencies have found that digital *pagers* can be extremely useful as a supplement to the radiotelephone system. Depending on their design, pagers can be used simply to alert someone that a message is waiting for them, or to convey a brief message, such as a telephone number that the person should call. For many investigators, criminalistics technicians, and administrators, a pager is sufficient for nearly all routine communications. By using pagers, an agency can free its more costly and scarce radio channels.

All police telephone and radio systems should be connected to a recording system. Tape recording systems designed specifically for this purpose are available from several manufacturers. Some have a number of separate tracks for each telephone line and radio channel. Even a small agency can afford to install ordinary voice-actuated tape recorders on its telephone line and radio base station. The tapes should be kept for at least one week, and longer if any questions arise about any of the conversations that have been recorded.[12]

The selection of telephone and radio equipment is crucial to the development of a sound, efficient police communications system. Thanks to the revolution in communications technology that has occurred in the past twenty years, an incredible variety of equipment is available. This does not make the task of choosing the best, most appropriate system any easier.

Some help is available from the Technology Assessment Program of the U.S. Department of Justice's National Institute of Justice, which regularly evaluates radio equipment and publishes its findings.[13] In addition, there are independent communications consultants who, for a fee, will research the available technology and design a system for a specific situation. Consultants' services are not cheap, but they might save more than they cost in the long run, if they succeed in designing a system that meets a police agency's immediate needs, allows for future expansion, and promotes efficient police operations.

Computers

We have already discussed some of the general applications of computers to a police agency's information needs. Here we consider some specific applications of computers to a police communications system.

Computer-Assisted Dispatching.

One of the most important uses of the computer in modern police service is CAD. Several types of CAD systems have been developed with many variations. In a typical CAD system, the telephone operators and dispatchers use computer terminals connected to either a mainframe computer or minicomputer. The terminal itself can be a microcomputer.[14]

When a complaint or request for service is received by telephone, the telephone operator enters the information into the computer. The computer verifies the information (if no such address exists, the computer will let the operator know this), then automatically displays the essential information on the appropriate dispatcher's terminal. In some CAD systems, the computer determines which patrol or other field unit should be assigned to the call. If the dispatcher agrees with the assignment, merely pressing a button causes the information to be transmitted to the appropriate mobile unit. If the dispatcher disagrees, which should occur rarely, the computer's assignment can be overridden manually.[15]

Everything we have described so far can take place in much less time than it takes to describe it. The computer keeps a permanent record of every call handled, as well as detailed information about the activities of each field unit. This information can be retrieved in the form of lists, summaries, or, depending on the computer system, graphs and charts.[16]

The more elaborate CAD systems require mobile data terminals in each police vehicle. An MDT is a miniature computer terminal. The terminal usually includes a compact keyboard and either a small display screen or a small printer, or sometimes both (see Figure 9.6). The MDTs are connected to the central computer by digital radio signals.[17]

When a call is transmitted to a mobile unit, the information is stored in the MDT. If the officer is not in the vehicle or is temporarily unavailable, the information waits until the officer acknowledges it. If no acknowledgment is received within a certain time, the computer notifies the dispatcher and proposes another assignment. Also, field personnel can communicate directly with the computer to obtain additional information, such as current wants and warrants on a person or license number, or to file a summary preliminary report of an investigation.[18]

A well-designed CAD system can do much more than just speed up the dispatching process. It can connect every field unit to the agency's central data base and, through it, to regional, state, and national files of police information. At the same time, the CAD system connects every element of the police department into a single nerve system through which every kind of administrative and police service information can flow.

An elaborate CAD system requires at least a large minicomputer; some are based on small mainframe computers. Automatic dispatching and mobile data

Figure 9.6. A mobile data terminal in a patrol car gives the officer instantaneous access to information. (Photo by the authors, courtesy Austin PD.)

terminals are not essential. Many smaller police agencies can get by with a mini-computer or even a microcomputer to assist the dispatcher, without the more elaborate features. The computer can be used to store information about each call, entered manually by the telephone operator and dispatcher, and to assist the dispatcher by looking up addresses, criminal histories, wants and warrants, and other data base information. The dispatcher makes case assignments and communicates with field units by radiotelephone.[19]

Automatic Vehicle Location (AVL).

One of the difficulties faced by police dispatchers is that they ordinarily have no way of knowing exactly where field units are located at any given time. When a call for service is received, or there is some emergency such as a pursuit in progress, the dispatcher either must make a "blind assignment," calling for any unit in the vicinity to respond, or must guess as to which unit is likely to be closest and available. During busy periods, precious time can be wasted in determining where everyone is. AVL systems have been developed to alleviate this problem somewhat.[20]

Several types of AVL systems have been developed. In some systems, each vehicle is equipped with a special transceiver, and one or more AVL transmitters are placed around the service area. The AVL transmitters automatically send out a "polling" signal periodically, usually more than once every second. The signal contains a code that causes the mobile unit's transceivers to send a responding signal. These response signals are received at two or more dedicated receivers. A computer compares the relative strength of the response signal as it is received at each AVL receiver.

In principle, the relative strength of the signal indicates the distance between the mobile unit and the receiver. The computer then is able to calculate the mobile unit's location by triangulation: If you know the distance from an unknown point (the mobile unit) to two or more known points (the AVL receivers), the unknown point is at the intersection of the two circles whose radii are the known distances.[21]

The principal disadvantage of this system is that it requires a good deal of expensive equipment and produces results of uncertain reliability. Several factors other than distance can affect the strength of a radio signal. Most AVL systems are designed to "poll" each mobile unit repeatedly and use the average of several calculations to compute the unit's location. Even so, these calculations are subject to considerable error, especially when a unit is in motion.

A more reliable system uses many low-powered receivers scattered throughout the service area. Each mobile unit contains a low-powered transmitter that broadcasts a continuous coded signal. As a mobile unit passes a receiver, the receiver picks up the signal and "reports" it to a central computer. In principle, then, every mobile unit can be located according to its proximity to one of the receivers. A similar system uses low-powered transmitter-receivers that transmit a continuous "polling" signal, rather than continuously broadcasting transmitters in the mobile units.

This system, too, has its drawbacks. One problem is that a mobile unit might be stationary at a location that is not close to a receiver, in which case the mobile unit "disappears" from the system. Two or more mobile units passing near a receiver can confuse it. At best, this system can only show the mobile units' locations within a general area around each receiver. To improve the system's "resolution"—that is, to display more specific locations—the number of receivers must be increased. This type of AVL system is extremely expensive because of the huge number of receivers that must be installed and maintained throughout the service area to produce useful information. However, it is generally more reliable than the triangulation system.

A third type of AVL system uses a small computer in each mobile unit to calculate the unit's location by keeping track of its movement. The computer periodically transmits a "report" to the central base station. This kind of "dead-reckoning" system is extremely expensive and unreliable.

The newest type of AVL system overcomes most of the limitations of the three conventional systems we have described, by relying on communications satellites. There are several commercial versions that use commercial satellites, and a system operated by the U.S. Air Force using its own satellites. All function similarly: a continuous signal is transmitted from the satellite, whose position in the sky is accurately known. Special receivers in each vehicle receive the satellite signal, and from its strength, direction, or slight variation in frequency, compute the vehicle's position in relation to the satellite. Depending on the type of equipment used, the vehicle's position can be simply displayed on a monitor in the vehicle, or transmitted over a separate channel to the agency's base station.

A dependable automatic vehicle location system is a tremendous aid to dispatchers. In studies of AVL systems, it has been found that dispatching efficiency, measured in terms of response time, is not significantly improved, but that the AVL system is extremely valuable in tactical situations, such as a pursuit or a civil disturbance, and in situations in which an officer's safety is in jeopardy.

Television

Law enforcement agencies have slowly begun to develop innovative and effective uses of television and its ancillary technologies, video tape recording and cable television.

There are several rather obvious uses of broadcast television for purposes such as public information and community relations.[22] Video tape recording also has been used extensively in police training programs, which we discuss in a little more detail in Chapter 12.[23] There are some other uses of television that are less obvious.

Surveillance Systems.

In the late 1960s the Detroit Police Department installed remote-controlled television cameras on several bridges over a major expressway. The cameras fed pictures to monitors at the police headquarters, enabling an officer to observe traffic flow and operate a system of lane control signals to direct traffic.[24]

Some experiments were conducted at about the same time in the use of television cameras for surveillance of public areas such as parks, beaches, and shopping malls. There were protests from some citizens concerned with possible invasions of privacy, but the courts ruled that the presence of cameras in a public place was no more invasive than the presence of an actual police officer. Nevertheless, these experiments usually did not last long, partly because of the public outcry over the "Big Brother" implications and partly because of the high cost of leasing special transmission lines to carry signals from the cameras to a central monitoring point.[25]

Television is used increasingly for surveillance in and around police buildings, jails, and other locations where security is a major concern.

Visual Arrest Records.

Video tape recordings are now made routinely of persons at the time of arrest, especially for driving while intoxicated or driving under the influence of drugs.[26] These recordings can be made at the police station, to record the arrested person's appearance and behavior.[27] Recently, compact video camcorders have been installed in patrol vehicles to record events in the field, especially traffic stops. In a widely publicized case in Texas, a dashboard-mounted camcorder recorded the beating death of a county constable at the hands of two men who had been stopped for a traffic violation. The perpetrators were caught and convicted largely because of the videotaped evidence.

Court Presentations.

On occasion, witnesses and even defendants have been represented in court by video tape recordings of their testimony, if it was impractical for them to appear in person. Video tape recordings of crime and accident scenes have been allowed as evidence in some court cases and are often valuable to investigators even if they are not presented as evidence.[28]

Video Teleconferencing.

State police agencies and some large metropolitan police departments have used video teleconferencing as a substitute or supplement for meetings of far-flung employees. *Video teleconferencing* means that individuals at two or more locations communicate by the transmission of both visual and audio information.

Video teleconferencing is fairly expensive, mostly because of the cost of transmitting two-way video signals over the telephone or other communications networks. Nevertheless, the cost savings over "live" meetings can be substantial. More important, it may be possible for individuals at different locations to "meet" by teleconference more frequently and more conveniently than meetings in person would permit.

Cable Television.

Cable television systems offer a means of transmitting television and other signals that can be extraordinarily cost effective.

Most people think of a cable system purely as a means for distributing entertainment programs to homes. However, nearly all cable systems built during the past few years, and many older systems, have one or more channels reserved for use by local government. These *access channels* can be used by law enforcement agencies for a variety of purposes. Crime prevention and other public information programs can be transmitted to the general public at little or no cost. Training programs can be transmitted to personnel at several locations at once, or even to officers' homes, so that they can receive training during their off-duty hours.

Similarly there may be cable channels available for law enforcement use that cannot be received by the public. Some large cable systems have *institutional channels*, or completely separate institutional networks, designed for this purpose. The institutional channels often can be used for two-way transmissions, or at least one-way video and two-way audio, which means that they can be used not only for interactive training programs but also for teleconferencing.

Most cable television systems also can be used to carry almost any kind of electronic signal including voice (in effect, a citywide intercom system) and data. The nonvideo channels can be used to connect computer terminals between the police headquarters and precinct stations, or between the police department and other governmental agencies; to connect remote surveillance cameras and other surveillance devices to a central monitoring point; to synchronize and control traffic signals throughout the city from a central computer; and to transmit teletype messages among several law enforcement agencies in the area. These applications represent only a handful of suggestions. The potential for police use of cable television is almost limitless.

POLICE-PUBLIC COMMUNICATIONS SYSTEMS

In this chapter we have been concerned primarily with communications systems within a police agency. We also have referred to communications between the police and other governmental agencies.

Police agencies have another important communications need: communications between the agency and the public it serves. Of course, there is a good deal of this type of communication. Citizens call in complaints or requests for police service over the telephone, or come in person to the police station to file a complaint or to give information. Field officers routinely communicate with individual citizens, and sometimes with citizens' groups, on the street. Some police agencies make effective use of the mass media (newspapers, radio, and television) to provide the public with information and to solicit information from the public.

With one exception, the telephone, all these forms of police-public communication are usually initiated by the police. When the police department wishes to inform the public or wishes to receive information from the public, the department does the communicating, by whatever means it finds convenient.

Additional channels of communication can be opened between the police and the public, to the benefit of both. For example, it is not difficult and certainly not expensive to encourage the operators of radio-equipped mobile fleets to assist the police by reporting traffic accidents, hazardous conditions, or suspicious incidents. In many cities there are dozens of radio-equipped fleets on the streets at almost all times: buses, taxicabs, delivery trucks, construction crews, utility crews, and individuals with mobile radiotelephones. The operators of these fleets can be contacted and asked for assistance. The police department should conduct a training session for the fleet drivers and dispatchers, explaining what kinds of things they should look for and how to report them.

In hundreds of cities, volunteer citizens' groups have been organized specifically to provide emergency communications by Citizens Band (CB) and other widely used radio systems. The FCC has reserved CB Channel 9 specifically for emergency messages and assistance to travelers. All of the frequencies in the amateur radio service can be used for emergency messages.

National organizations such as the American Radio Relay League and REACT International have been established to promote the proper use of amateur and CB radio for emergency communications, and to work with local police and other public safety authorities.

The local chapters or affiliates of these national organizations tend to be quite autonomous. Some are much more effective than others. Police authorities should welcome the interest of civic-minded volunteers and should strongly encourage their activities. Contact should be made as early as possible with the groups' organizers or leaders to establish a cooperative working relationship.

If possible, the police should participate in the training of volunteer emergency monitors and should work with the volunteers to establish reporting procedures, liaison for the communication of problems or grievances, and so forth.

All police agency personnel, especially in the communications center, should be fully informed about the groups' purposes and agreed-on operating procedures.

If no volunteer emergency radio group already exists, the police might sponsor one of their own. Where there is already a police reserve or police buddy program, it might be expanded to incorporate emergency radio monitoring. Not all CB and amateur radio operators will respond favorably, but in almost every community there are plenty of civic-minded citizens, willing and able to serve their community, if they are only asked.

EFFECTIVE COMMUNICATIONS

The emphasis in this chapter and in Chapter 8 has been on the technologies used by law enforcement agencies to collect, store, retrieve, and communicate information. As we have said, the extraordinary development of these technologies has created an opportunity for police agencies, regardless of their size, to establish highly effective, integrated information-communications systems. However, there is one part of the system we have said little about, and yet it is surely the most crucial part of all: the agency's personnel.

We once asked the chief of a medium-sized police department what he looks for in prospective recruits. His answer was surprising and thought provoking. He said, "I want people who like to talk." He went on to explain that, in his view, nearly all of a police officer's job involves talking to people: interviewing complainants, victims, witnesses, suspects, and just ordinary citizens going about their affairs. An officer who is articulate and who is able to "speak the language" of the people he or she encounters will be more successful and effective than an officer who relies on the authority of the badge and the threat of the gun. Persuasion, he said, is the most important tool a police officer can use.[29]

There are dozens of books presenting various theories of effective communication. We are aware of no evidence that any one theory is superior to the others. There does seem to be a consensus that effective communicators are also good listeners and are generally sympathetic to their intended audience. It is difficult to communicate effectively when there is hostility, suspicion, or indifference between the sender and the receiver.

REVIEW

1. The four major elements of a communications system are

    (a) The sender.

    (b) Encoding.

    (c) Transmission.

    (d) Written documents.

    (e) Decoding.

(f) Feedback.

(g) The receiver.

2. "Technological convergence" refers to

(a) The tendency of information and communications systems to become increasingly alike and part of one another.

(b) The tendency of police agencies to rely excessively on a particular technology.

(c) The increasing expense of maintaining a variety of information and communications systems.

(d) The importance of channeling all information through a single reception point.

(e) The tendency of new technologies to replace old ones.

3. "Queueing" is used

(a) To prevent telephone lines from being jammed with too many calls.

(b) To determine the normal peak period for telephone traffic.

(c) To hold incoming calls in sequence, when all circuits are busy, so that callers do not hear a "busy" signal.

(d) To ensure that hotlines are available for urgent outgoing calls.

(e) To determine which incoming calls are most urgent.

4. True or False: In a police radio system, it is important for all channels to be in the same frequency bandwidth.

5. AVL systems have not become common in American police agencies because

(a) There is little need for dispatchers and supervisors to know the precise location of every vehicle.

(b) Until recently, AVL systems were extremely expensive and unreliable.

(c) Taxpayers resent AVL as a potential intrusion on their privacy.

(d) It is the officer's responsibility to know where he or she is at all times.

(e) AVL equipment is too bulky and fragile for use in American police cruisers.

NOTES

[1] Richard N. Holden, *Modern Police Management*. Englewood Cliffs, N.J.: Prentice Hall, 1986, p. 178.

[2] Thomas F. Adams, *Police Field Operations*. Englewood Cliffs, N.J.: Prentice Hall, 1985, pp. 106–8.

[3] Alan Burton, *Police Communications*. Springfield, Ill.: Charles C Thomas, 1973, pp. 28–38.

[4] Roger W. Reinke, *Design and Operation of Police Communications Systems*. Washington, D.C.: International Association of Chiefs of Police, 1964, p. 3.

[5] Reinke, *Police Communications Systems*, pp. 10–11.

[6] Reinke, *Police Communications Systems*, p. 19.

[7] See, for example, "Police Practices: Pennsylvania's Interdepartmental Communications System," in *FBI Law Enforcement Bulletin*, vol. 59, no. 9, September 1990, pp. 18-19.

[8] Adams, *Police Field Operations*, pp. 124–28.

[9]Reinke, *Police Communications Systems*, pp. 12–13.

[10]R. L. Sohn, J. E. Abraham, W. G. Leflang, R. D. Kennedy, J. H. Wilson, and R. M. Gurfield, *Application of Mobile Digital Communications in Law Enforcement*. Washington, D.C.: U.S. Dept. of Justice, National Criminal Justice Information and Statistics Service, 1978.

[11]However, see Karen Layne, "Unanticipated Consequences of the Provision of Information: The Experience of the LVMPD," in *Journal of Police Science and Administration*, vol. 17, no. 1, March 1990, pp. 20–31.

[12]Reinke, *Police Communications Systems*, p. 16.

[13]The Technology Assessment Program can be contacted at the National Institute of Justice, TAP Information Center, Box 6000, Rockville, MD 20850, or by calling (301) 251-5060.

[14]Holden, *Modern Police Management*, p. 178.

[15]Charles A. Gruber, Jerry Eugene Mechling, and Glenn L. Pierce, "Information Management," in William A. Geller, ed., *Local Government Police Management*, 3rd ed. Washington, D.C.: International City Management Association, 1991, pp. 311–12.

[16]R. L. Sohn, R. M. Gurfield, E. A. Garcia, and J. E. Fielding, *Application of Computer-Aided Dispatch in Law Enforcement*. Washington, D.C.: U.S. Department of Justice, National Criminal Justice Information and Statistics Service, 1978.

[17]Gruber, Mechling, and Pierce, "Information Management," in Geller, *Local Government Police Management*, pp. 312–13.

[18]Adams, *Police Field Operations*, pp. 128–29.

[19]Gruber, Mechling, and Pierce, "Information Management," in Geller, *Local Government Police Management*, pp. 311–13.

[20]Ibid., p. 313.

[21]G. R. Hansen and W. G. Leflang, *Application of Automatic Vehicle Location in Law Enforcement*. Washington, D.C.: U.S. Department of Justice, National Criminal Justice Information and Statistics Service, 1978, pp. 9–12.

[22]Joe Missonellie and James S. D'Angelo, *Television and Law Enforcement*. Springfield, Ill.: Charles C Thomas, 1984, pp. 142–44.

[23]Ibid., pp. 123–27.

[24]V. A. Leonard, *The Police Communications System*. Springfield, Ill.: Charles C Thomas, 1970, p. 41.

[25]Missonellie and D'Angelo, *Television and Law Enforcement*, pp. 116–22, 128–31.

[26]Michael Giacoppo, "The Expanding Role of Videotape in Court," in *FBI Law Enforcement Bulletin*, vol. 60, no. 11, November 1991, pp. 1–5.

[27]Missonellie and D'Angelo, *Television and Law Enforcement*, pp. 97–115.

[28]Giacoppo, "Videotape in Court," pp. 1–3.

[29]Personal interview with Leonard Hancock, Chief, Temple (Texas) Police Department, October 10, 1977.

POLICE PROPERTY AND EVIDENCE MANAGEMENT

All the things we have discussed in the past four chapters have one characteristic in common: They must be purchased. Buildings, furniture, equipment, vehicles and their accessories, uniforms, weapons, personal equipment, records forms, computers, gasoline and oil for police vehicles, and food for jail prisoners all must be bought from someone.

Once items have been purchased, they must be cared for and used properly. When they are no longer useful or needed, they must be disposed of.

The police also act as custodians of several kinds of property that do not actually belong to them: property that has been found, or whose ownership is unknown or disputed; stolen property that has been recovered; evidence connected with a crime; and the personal belongings of prisoners. As custodians, the police have a legal responsibility to care for such property. Especially in the case of evidence the police have a professional interest in keeping it secure.

PROPERTY ACQUISITION

The purchasing of materials and services is a continuing, indeed never ending, function of police administration. Usually about one-fourth of a police agency's annual budget is allocated to items that are to be purchased (the rest being devoted to personnel costs, rent, utilities, and other fixed costs).

In private business and in government, purchasing is a highly specialized field.[1] Purchasing agents or managers often have advanced degrees in business administration, special training in purchasing, and years of experience as purchasing clerks or buyers of specific kinds of merchandise. Unfortunately, this level of expertise is not always available to law enforcement agencies.

Some very large police departments have established a centralized purchasing unit (sometimes called the procurement section, materiel acquisition section, or supply section). Medium-sized and moderately large agencies—those with between one hundred and five hundred employees—often have no specialized purchasing unit, but instead rely entirely on the parent government to provide this function. In smaller agencies, purchases are made by individual administrators or entirely by the chief executive.

The first step in establishing an efficient purchasing system is to designate some one individual as the agency's purchasing agent.[2] In small agencies, this person may have other duties. However, it is preferable for the person designated as *purchasing agent* to be someone other than the person responsible for supervising the budget and keeping the agency's accounts.

There are four major steps in the purchasing process, all of which require the participation of the purchasing agent and of other agency administrators.

Requisition.

Every purchase begins when someone in the agency has a need for some item or service that must be acquired from outside. This need is expressed in a standard form, a *purchase requisition*, that describes what is to be obtained.[3]

Selection.

Because most items and services could be obtained from several sources, and often at different prices, an effort must be made to determine the best source for whatever is needed and then to obtain the best possible price for the item or service.

Transaction.

The purchase itself is a legal transaction, a commitment of the agency's funds in exchange for goods or services to be delivered.

Receipt.

Finally, the goods must be received by the agency and either accepted or, if they are not correct, rejected; or the service must be performed to the agency's satisfaction.

The purchase requisition might be prepared by anyone in an agency, subject to whatever limits are imposed by the agency's policies. When the requisition is submitted to the purchasing agent, the item or service to be bought must be described so thoroughly that there is little chance that the wrong thing will be bought. It is the purchasing agent's responsibility to get only what the requisition specifies. If the requisition says merely, "One dozen 1/2-inch-long screws," it is not the purchasing agent's fault if the screws that are bought have the wrong thickness or thread pattern.

Because the requisition is so important, most agencies require that it be approved by at least one supervisor or administrator before it is sent to the purchasing agent (see Figure 10.1).

LAS VEGAS METROPOLITAN POLICE DEPARTMENT
REQUEST FOR PURCHASE REQUISITION

CHECK TYPE:
☐ If over $500, must have attached memo listing **verbal** quotes from three vendors.
☐ If over $2,500, but less than $10,000, must have attached **written** quotes from three vendors. If preferred vendor is not lowest quote, justification must be attached.

TO:

    DIRECTOR OF FINANCIAL SERVICES

        VENDOR   **DATE** _____

NAME   _____

ADDRESS   _____

TELEPHONE _____ PRICES QUOTED BY _____ DATE OF QUOTE _____

REQUESTING BUREAU _____ BUREAU# _____

REQUESTED DELIVERY DATE _____ ☐ DELIVER ☐ PICK UP
        DO NOT USE "ASAP"

FILL IN ALL AREAS; IF MORE ITEMS USE SECOND FORM

QTY.	UNIT	CATALOG/ MODEL#	ITEM	UNIT COST	TOTAL COST

SHIPPING INSTRUCTIONS: SHIPPING (if not included in price)

SPECIAL INSTRUCTIONS: **GRAND TOTAL**

BUDGETED ITEM ☐ YES ☐ NO   IF NO, ATTACH JUSTIFICATION

REQUESTED BY (EXT.)   DIVISION/OFFICE COMMANDER

BUREAU COMMANDER   ASSISTANT SHERIFF

BUDGET APPROVAL	**ACCOUNTING PROCESSING**
ACCOUNT _____	POR# _____
BY _____ DATE _____	ENTERED _____ COM _____
F.Y. _____	BUYER _____

Figure 10.1. A purchase requisition form. (Courtesy Las Vegas Metropolitan PD.)

The purchasing agent's most obvious value comes in the selection part of the process. An experienced and capable purchasing agent should be familiar with most of the usual sources for the goods or services needed, and should be able to select the source that is capable of delivering what is needed in the appropriate quantity and quality at the best possible price.

The purchasing agent also must know how to complete the transaction in a manner that is legally correct and that best achieves the agency's goal of obtaining goods and services in a timely fashion, at the least cost for the quality of goods desired.

Finally, the purchasing agent's responsibilities end when the goods are delivered and checked to be sure that they are in proper order, or when the service has been performed to the agency's satisfaction. However, a competent purchasing agent also wants to know whether the agency's personnel are satisfied with the items that have been bought, whether expensive equipment performs as it is supposed to, whether the vendor provides prompt and satisfactory maintenance services, whether the warranty is honored, and so on.

Requisition

Each purchase requisition should carry out some part of the plans contained and reflected in the agency's budget. If a need arises that was not foreseen, and for which the budget has no provision, a decision must be made to amend the budget before the purchase is made. As a general rule, no purchase should be made unless the funds for it have been provided in the budget.[4]

The most important part of any purchase requisition is the *specification* of what is needed. As we have said, the specification should be sufficiently complete that there is no question about what is needed. On the other hand, the specification should not be so detailed and restrictive that there is only one possible source for an item, because that limits the purchasing agent's chances of seeking the best possible price through competitive bidding.

Specifications should concern only those details of an item's functional characteristics or its performance that have a direct bearing on the use that is to be made of it. Details about appearance, size, construction, or other matters should not be included unless they are relevant to the item's intended use. Brand names and manufacturers' trade names should be used only to illustrate the type or quality of item desired, unless there is some good reason that only an item from a specific manufacturer is suitable (for example, if the item to be bought is a part that must be assembled with some equipment that the agency already owns).[5]

Preparing specifications can be a complicated matter, especially when complex and expensive equipment is to be purchased, such as police vehicles. The expertise of an experienced professional purchasing agent can save an agency a good deal of money, not to mention the unpleasant consequences of acquiring unsatisfactory equipment.

There also may be help from an unexpected source. A commercial vendor has compiled a computer database, using information from the Technology

Assessment Program of the U.S. National Institute of Justice (see Chapter 6), based on the police vehicle tests conducted by the Michigan State Police. The database, called *AutoBid*, is designed specifically to help police agencies prepare bids for vehicle purchases. The police administrator or purchasing agent can choose from "menus" of features and specifications, selecting those that are relevant to the agency's needs. The program then indicates which vehicles meet the specifications and prepares the bid form automatically.[6]

Selection

Police officers must be able to rely on the equipment and materials purchased for their use. At the same time, police departments do not have unlimited funds at their disposal; they must get the most out of every dollar they spend. The best way to do this is to make every purchase from the *lowest qualified bidder*.[7] This does not always mean that the lowest price must be used as the only criterion for selection.

Somewhat different procedures are used for different kinds of purchases. The kinds of purchases that occur most often in police agencies are commodities purchases, item orders, and service contracts.

Commodities are goods that are used more or less continuously. They tend to be highly standardized items, usually bought in large quantities. Some examples would be vehicle fuel and motor oil, office supplies and stationery, ammunition, cleaning and maintenance supplies, food for jail prisoners, and photographic supplies. It would be ridiculous to require a separate requisition for a day's supply of gasoline, wait for bids to be taken, and then take a purchase order to the chosen gas station. Instead, fuel and other commodities are purchased in sufficiently large quantities to meet the agency's needs during some period, often as much as a year.

Commodities usually are delivered periodically, as they are consumed, so that the agency does not have to store the entire year's supply at once. For example, if the department has 10 vehicles, each of which consumes on the average 25 gallons of gasoline per day, the department would need a total of 3500 gallons per week, or 182,000 gallons per year. The department might have adequate storage for a week's supply, plus a reserve for higher-than-normal usage or late delivery, and require the vendor to deliver 3500 gallons each week.

Item orders are purchases of individual items or small groups of items on an irregular basis as the goods are actually needed. Many purchases are for single items that may vary in cost from a few cents to several hundred thousand dollars. Each item order is handled as a unique case, as if nothing of the sort had ever been purchased before, except that some goods (such as uniforms) may be standardized and obtained routinely from the same vendor.

Services are work performed by individuals who are not employees of the agency. Police departments may have many occasions to purchase services from individuals. For example, an attorney might be paid a monthly fee, or retainer, for advice about the legal aspects of police activities. Similarly, a radio technician might be paid on a contract basis to maintain the agency's radios. Usually services are purchased from nonemployees when there is not enough work to justify hiring someone on a regular basis or when the volume of work varies so greatly and irregularly that it would be inefficient to keep enough employees on the payroll.

In some cases, purchased services do not involve the work of people, but of machines. For example, some businesses offer the use of small computers, xerographic copiers, and other specialized office equipment on a service basis. Instead of buying or leasing the machines, the purchaser merely buys the use of them. The cost might be calculated on the basis of the number of hours (or minutes) the computer is used during a month, or the number of copies made, and so on.

A distinction should be made between a service contract and a *maintenance contract*. The former is an agreement to pay a certain amount for the services of an individual (or, as we have just seen, for a machine) to do certain work. A maintenance contract is an agreement to pay in advance a predetermined fee for any repairs or maintenance services that might be needed on a particular piece or type of equipment. Unfortunately, some businesses refer to their maintenance contracts as service contracts, which adds to the confusion.

For any of the three types of purchases, the selection process begins with the compiling of a *source list*, or bidders' list. Experienced purchasing agents maintain a source list for each type of merchandise or service that the agency routinely buys. Lists are constantly updated, with new businesses added and out-of-date information removed. Vendors who have previously delivered unsatisfactory merchandise or services may be removed from the list, either temporarily or permanently. The list is compiled from as many sources of information as the purchasing agent has available: from trade advertising, direct mail advertising, salespersons' calls, telephone directories, trade association directories, and so on.[8]

Once a source list has been compiled, the purchasing agent is prepared to invite *bids* or price *quotations* on the item or items to be purchased (see Figure 10.2).[9] The invitation may be on a standardized form, containing the description given in the purchase requisition, plus any other conditions of delivery, agency policies regarding discounts, and so forth. In some cases, the invitation may be given in the form of a letter. Anyone interested in bidding on the purchase is required to reply by a certain time and date.[10]

For items of relatively small value, especially when they are reasonably standardized or when precise specifications are not necessary, quotations may be solicited by telephone and written bids may not be required.[11]

Commodity purchases are almost always made on the basis of written bids because the size of the purchase usually is quite large. Some governments require commodity purchases to be made only after invitations to bidders have been published in newspapers (often in the classified advertising columns) or in official publications. For example, the federal government uses several of its own publications to solicit bids, including the *Federal Register* and the U.S. Department of Commerce's *Business and Commerce Daily*.[12]

If the item is unusually expensive, complicated, or involves services as well as goods, a two-step purchasing process may be used. For example, if a police department decides to buy a new computer system, there are a vast number of

INVITATION TO BID
CITY OF FORT SMITH
Mail Sealed Bids To→ PURCHASING AGENT
P.O. BOX 1908
FORT SMITH, ARKANSAS 72901

Place _____PURCHASING_____

Date _____

Bid Opening Date 10 A.M. _____ 19 ____

ON THE OUTSIDE OF SEALED BID ENVELOPE SHOW:

(1) RETURN ADDRESS OF FIRM

(2) BID OPENING DATE

(3) BID ON 4-DOOR SEDAN

Bidder agrees to comply with all conditions below and on reverse side of this bid. Purchases made for the City of Fort Smith are exempt from Federal Excise Tax. Do NOT include on your bid. Unit prices and extensions are required. If a trade discount is shown on bid, it should be deducted and net line extension shown. Bidder guarantees product offered will meet or exceed specifications identified in this bid invitation.

BIDDER MUST FILL IN & SIGN	
NAME OF FIRM	
STREET ADDRESS	
CITY	STATE ZIP
AUTHORIZED SIGNATURE	DATE
TITLE	
QUOTE F.O.B. DESTINATION	

ITEM NO.	DESCRIPTION AND/OR SPECIFICATIONS	QUANTITY	MFG & MODEL	UNIT PRICE	EXTENSION
1	Automobiles as per specifications				
	GENERAL: Standard four (4) door sedans with center post. All vehicles bid shall be of a general quality level equivalent to the Chevrolet Impala, Ford L.T.D. II, Pontiac Catalina, or Dodge Monaco. Sedans are to be furnished with all interior and exterior trim and appointments listed by the manufacturer in printed specifications and literature as "Standard Equipment". Vehicles must meet all Federal Safety Emission Regulations.				
	ENGINE: 350 cubic inch displacement, 8 cylinder.				
	TRANSMISSION: Automatic shift specifically recommended by manufacturer for the engine being offered. Shall include a first-gear lock out feature.				
	WHEELBASE: 116 inch				
	SUSPENSION: Heavy-duty designed and engineered for pursuit-type work. Should include heavy-duty front and rear sway bar, torsion bar, heavy duty rear springs and heavy duty shock absorbers 1 3/16" front and 1 3/8" rear.				
	STEERING: Power				
	BRAKES: Power disc front metallic-impregnated, drum rear.				
	ELECTRICAL: 12 volt with 61-ampere alternator and 80-ampere/hour battery, dual horns and radio noise suppression equipment.				
	AIR CONDITIONING: Factory installed integrated with heater-defroster.				
	GLASS: Tinted.				
	TIRES: Five (5) radial ply BSW (Includes mounted spare).				
	SPEEDOMETER: Calibrated, (0–100 MPH) in 2 mph increments.				
	ACCESSORIES: Glove and trunk compartment lights. Interior dome lights NOT to light when doors are opened. Vinyl upholstery – Colors to be compatible with vehicle color.				

By the signature hereon affixed, the bidder hereby certifies that neither the bidder nor the firm, corporation, partnership or institution represented by the bidder, or anyone acting for such firm, corporation or institution has violated the Federal Antitrust Laws, nor communicated directly or indirectly the bid made to any competitor or any other person engaged in such line of business.

DELIVERY IN _____ DAYS

CASH DISCOUNT

_____ % _____

Figure 10.2. A bid invitation form. (Courtesy Fort Smith, Arkansas, PD.)

computer systems that might fit the agency's needs, and the department might not have anyone among its employees who would be competent to design such a system and specify the components to be purchased. Instead, a *request for proposals* (*RFP*) should be issued. The RFP consists of a statement of the general sort of thing that is desired and, in as much detail as possible, what it is expected to accomplish. Interested bidders respond with a detailed proposal, including the design of the system, the required components, and an estimated price.[13]

The responses to an RFP may represent a number of different approaches to the department's needs. The selection process focuses not on the price, but on the quality of the proposed system design: Will it do what the department wants to do, and is it the most efficient and economical of the various designs submitted? Sometimes specific features of one design will be very attractive, while another vendor's overall design comes closer to meeting the agency's needs. Ultimately, the agency (usually with the participation of every administrator who will be affected by the purchase) must choose the vendor whose design comes closest to meeting those needs.

The second step, then, is the negotiation of a contract with the chosen vendor. Changes in the design might be made to incorporate new features (including some that may have been proposed by another vendor) or to delete features that the agency decides are unnecessary. The contract often will specify conditions such as the time of delivery, how the equipment is to be installed, training that is to be provided by the vendor for the agency's employees, and how repair service is to be provided under a warranty. Finally, a fixed price is negotiated and written into the contract.

Transaction

Once bids have been received and the lowest qualified bidder has been selected, or a contract has been negotiated, the actual purchase is made, usually by issuing a *purchase order* to the successful bidder.[14] The merchandise may not be actually delivered until sometime later. However, a purchase order is a legal, enforceable contract; if either the buyer or the seller fails to complete the transaction, the contract is broken and the other party has recourse to the courts.[15]

Receipt

When the purchased goods are delivered, a responsible person in authority must examine them to be sure that they are the items that were ordered and that they are in proper condition. For items of a routine nature, the delivery inspection may be performed by a clerk. For expensive or unusual items, the inspection should be performed by the purchasing agent or by the person who will actually use the item.

If everything is in proper order, a delivery receipt is given and the items are accepted. If anything is not in order, the delivery should be refused. Once the agency has accepted delivery, the department may be obligated to pay for the item even if it later turns out to be defective or unsuitable. Furthermore, once the item has been accepted, it is legally the property of the department and the vendor is not responsible for any losses that might occur as a result of theft, fire, weather, and so on. Thus, as soon as an item has been accepted, it should be moved to the place where it is to be used or, if this is not possible, it should be put into safe storage.

Merchandise, especially commodities that are delivered periodically, may be delivered with an *invoice*, the supplier's claim for payment. Otherwise, the invoice is sent directly to the purchasing agent. All invoices must be checked against the agency's delivery records to make sure that the items in the invoice were in fact received and accepted. Any deviation from the terms of the purchase order must be settled before payment is made (and preferably before the item is put to use).

When the invoice is found to be correct, it is forwarded to the department's accounting unit, where it should be checked again by comparison with the purchase order and, if possible, the original requisition.[16]

These routine checks are more important than may be apparent. Careless handling of purchases is one of the largest sources of waste and corruption in both government and private industry. Swindlers frequently mail out large batches of falsified invoices, knowing that careless businesses and government agencies will pay the bill without checking.

Legal Considerations in Police Purchasing

Aside from the legal implications of purchasing we have already discussed, there are some particular legal considerations that apply specifically to police agencies.

The purchasing process usually involves a great deal of money, which means that there is always the possibility of abuse, corruption, or waste. To avoid these problems, legislators and agency administrators devote considerable attention to purchasing procedures. Federal, state, and local laws dictate the kinds of things that may be bought by public agencies; laws and administrative rules specify the exact procedures that must be followed in purchasing them.

Most cities and counties have established their own standard purchasing procedures. Standardized forms must be used for purchase requisitions, bid invitations, requests for proposals, purchase orders, and so on. Naturally, every police agency must follow the laws of its parent government. For that reason, we have not gone into great detail about these procedures, which may vary from place to place.

There also may be *flow-through requirements*. When an agency receives funds from a source other than its parent government, or when the parent government receives funds from another source and passes them on to the police department, any requirements imposed by the original source of the funds may "flow through" and affect the way those funds are used.

For example, the federal government may provide money to a state criminal justice planning agency, which in turn may award a grant to a local police department. The federal grant might have some restrictions as to how the funds may be used or may carry some conditions on the purchasing procedures that may be used. These restrictions or conditions apply to the local police department, even though it was not the direct recipient of the federal money.

Flow-through requirements can be extremely complicated, especially when funds pass through two or more levels of government. Often each level adds its own specific restrictions or conditions.[17]

There are also federal and state laws designed to use purchasing procedures for various social purposes. Governments spend a great deal of money, so it is only reasonable to use some of it to achieve the general goals of society. For example, laws or rules may require agencies to purchase goods from minority-owned businesses whenever possible or from businesses in economically depressed areas. Many local governments have formal or informal policies of buying from businesses in the community. Other laws and administrative rules may require purchased goods to meet special qualifications to protect the natural environment, to reduce energy consumption, and so on.[18]

The real problem for a police administrator is to know what rules apply to a particular purchase, especially since the social welfare requirements also may flow through from higher to lower levels of government. Even if the parent government has an exceptionally well-organized central purchasing office, the police department should have at least one person on its own staff who is familiar with the special requirements of police purchasing.

Donated Equipment and Funds

One other special problem affects police agencies more than most other governmental agencies. Public-spirited citizens occasionally offer to donate equipment, services, or money to their local police department. In some communities, there are established charitable organizations whose primary purpose is to support the local police department by supplementing the funds that are available from local government.

Police officers and administrators often feel uneasy about accepting such contributions. There is a genuine concern that accepting anything from outsiders might compromise the agency's integrity. In some states, there are laws that specify whether and how a police agency may accept donations.

No police department has more money than it needs. Police departments, especially in smaller communities, are chronically strapped for funds. They do not have enough money for all of the necessities, much less the extras that might make police work more enjoyable, more efficient, or even safer.

Unless there are very strict laws to the contrary, police agencies should accept donations, but only on certain conditions.

The first condition is that the donation should not be from a single individual. Donations should be accepted only from organizations that have qualified under state law or the rules of the U.S. Internal Revenue Service for designation as nonprofit organizations. Private foundations or trusts are also acceptable donors, if they are administered by a board of trustees composed of reputable persons.

The second condition is that the donor must attach no restrictions to the contribution. Equipment or merchandise must become the permanent property of the agency, to be used in any way the agency sees fit.

The third condition is that the donation should be publicly acknowledged. Some donors prefer to remain anonymous, and may have practical reasons for doing so, but a police agency should not accept any anonymous donation nor any donation from a known source who insists that the contribution not be made public. The police do not need to seek publicity for every donation, but the fact of the donation, the amount, and the source should be considered public information.

The fourth condition is that the acceptance of a donation must not obligate the agency to do (or fail to do) anything. Some people, out of sincere and honest motives, contribute to the police in the hope of promoting the enforcement of particular laws. For example, funds might be contributed to establish an antiobscenity project. The funds should not be accepted unless the police have already decided to establish such a project. Even after the funds have been accepted, the police agency must be under no obligation to continue the project once proper law enforcement objectives have been achieved.[19]

INVENTORY AND PROPERTY CONTROL

Once goods have been purchased for use by a law enforcement agency, the agency has a responsibility to its parent government, the taxpayers who support it, and to its own personnel to make sure that the goods are properly maintained and used. Waste and unnecessary loss are probably no more common in law enforcement agencies than in any other kind of organization, but there is always room for improvement.

Many police agencies have adopted the policy that each employee is responsible for each piece of equipment assigned to him or her. If a piece of equipment, whether it is a handgun, an automobile, or a computer terminal, is lost or damaged through the employee's negligence, the cost of repair or replacement is supposed to be charged against the employee's salary.

In practice, however, this policy is unenforceable. Negligence is often difficult to prove when equipment is lost or damaged in the heat of police action. Even when negligence is clearly established, assessing the cost against an officer's pay (which is usually not very generous to begin with) often results in an excessively severe penalty.

A fairer and more realistic policy is to treat negligence or abuse of equipment as a disciplinary matter. All personnel should receive thorough training in the use and care of their equipment; supervisors should observe any instances of misuse and take appropriate corrective action.[20]

Inventory Controls

From an administrative perspective, the control of police property is largely a matter of establishing and maintaining proper records of the assignment and use of materials.[21]

Each nonconsumable item (such as furniture) should be marked with the agency's name and a unique serial number. The marking should be done at the time the item is received and accepted. Once an item has been marked, a permanent record should be established, showing when and from whom the item was purchased, along with whatever other information (cost, physical description, and so on) may be desired. The record should indicate to whom the item has been assigned for use or where it is located.[22]

At least once a year, a complete physical inventory of all furniture and equipment should be taken. If the inventory records have been computerized, the computer should be able to provide a printed list of the numbered items, their description, and their assigned location. A clerk then takes the list to each unit of the agency and checks to make sure that every numbered item is actually present. Equipment in use on the streets or elsewhere also must be checked, either by having a clerk go to each location where the equipment should be found or by having the equipment brought to the headquarters for checking. The physical inventory is time-consuming and inconvenient, but it is a necessary procedure to make sure that equipment is not lost or abused.

Consumable items must be handled differently. All commodities should be placed in storage when they are delivered, and there should be an individual responsible for each storage location or facility. When consumable supplies are withdrawn to be used, the person taking them out of storage must leave with the storage supervisor a receipt, showing what has been taken out. From time to time, the storage supervisor should report to the accounting office (or to the purchasing agent) the total amount of supplies that have been withdrawn and how much remains in inventory.

Losses of consumable supplies are most likely to occur after they have been withdrawn from storage. For this reason, the accounting office or the purchasing agent should periodically review all supply receipts to determine whether any units of the agency, or even individual employees, are taking out unreasonable quantities of supplies.

Property Maintenance

Equipment cannot be used properly unless it is in proper working condition. Unfortunately, property maintenance is an area of chronic neglect in many agencies, especially the smaller ones whose budgets are most limited. Once a piece of equipment has been acquired, it is used until it stops working for whatever reason, and then it is simply discarded or put in storage until funds can be found to repair it.

Any machine, any device that has working parts, requires some sort of maintenance, if only to keep it clean. Weapons, communications equipment, laboratory and photographic equipment, and vehicles all require fairly elaborate maintenance. Information on correct maintenance procedures usually is supplied by manufacturers or can be found in commercially published technical manuals.

The responsibility for maintenance should be clearly assigned. In general, the first responsibility rests with the individual officers and employees who are assigned the use of each type of equipment. It is their responsibility to keep their equipment clean and to report promptly any signs of malfunctions or excessive wear. Employees should not be expected to make repairs themselves, nor to pay for repairs caused by normal use or accidents that are not attributed to their own negligence.

Periodic preventive maintenance and repairs are the responsibility of the agency. If the agency is too small to afford full-time repair technicians, maintenance contracts may be negotiated with qualified technicians who may perform repairs either at the department's facilities or in their own shops.

Disposition of Obsolete and Unneeded Equipment

No matter how thorough the maintenance effort, eventually every item of equipment will reach the end of its useful life. Police equipment often becomes outdated and useless to the agency even before it stops working.

When each piece of equipment is obtained, an estimate of its useful life should be made. Nothing lasts forever, and most manufacturers can provide a reasonable estimate of the useful life of their products.

It may seem that the most economical way to use equipment is to "run it into the ground," that is, to use it until its useful life has been exhausted, or until it has broken down to the point that it can no longer be repaired. Actually, that is the least economical practice. It is much better to keep an item in use only until

1. Its maintenance costs are equal to or greater than the amortized cost of a new item.
2. It has reached the end of its designed life expectancy but is still in good working condition.
3. A new model is available that, because of its greater efficiency, will cost less to operate or maintain than the item it will replace.

For example, consider the following:

1. The agency has a copying machine that cost $5000 when purchased eight years ago. At the beginning of the ninth year, the machine has required only routine servicing. However, during the ninth year it begins to break down frequently, and each breakdown means an average repair bill of $100. A replacement machine will cost $6000 and may be expected to last five years. Thus, the new machine's amortized cost would be $1200 per year. Because the old machine is costing at least $1200 per year to keep in operation, the new machine should be bought.
2. The agency has a fleet of 10 vehicles, all purchased three years ago at a cost of $18,000 each. All have about 100,000 miles on their odometers and have been maintained in good working condition. If they are sold now, as used cars, they are likely to bring $6000 each. If they are kept another year, and used for an average of 30,000 miles each, they

are likely to bring only $2000 apiece (assuming that none of them begins to break down, which probably is not a valid assumption).If the cars are sold now, they will have cost an average of $4000 per year (the original cost of $18,000, less the resale value of $6000, divided by three years). If they are kept another year, they still will have cost $4000 per year ($18,000, less $2000 resale value, divided by 4 years). So there is no economic advantage in keeping them another year. They should be sold.

3. The agency has a three-channel radio system, bought and installed 20 years ago at a cost of $50,000 for the base station equipment and $1200 per mobile unit. Today the agency has 60 mobile units. The three-channel system is still in good working order, but the three channels are badly overloaded with traffic; there is frequent interference from other transmitters on adjacent channels; and the system's design does not permit mobile units to hear one another, so mobile-to-mobile messages must be relayed through the dispatcher.

A new radio system with four channels, at a higher frequency that is less susceptible to interference, and with mobile-to-mobile capability, will cost $100,000 for the base station and $1000 per mobile unit, or a total of $160,000. The old radio system cannot be sold because there is no market for obsolete equipment.

In this case, the decision is not entirely economic. There is no doubt that the new system will reduce the volume of radio traffic on each channel, reduce time lost because of interference, and increase the efficiency of both dispatching and patrol operations. The end result will be better service to the public. These considerations alone indicate that the old radio system should be replaced.

Once a decision has been made to replace outdated or unneeded equipment, a long time may pass before the replacement actually occurs. Most police agencies operate on an annual budget; funds for replacement equipment must be written into the next year's budget. Even after the budget has been approved, the purchasing procedure may take several months to complete. From the time a decision is made to purchase new equipment until the equipment is actually delivered may be anywhere from a year to two years.

For that reason, equipment replacement should be done on a scheduled basis. Equipment does not have to be kept until it is falling apart; it is much better to replace it while it is still in good working order and can be sold for a fair price. Therefore, all equipment should be scheduled for replacement when it is first acquired. Many kinds of equipment that are used continuously, for example, vehicles, should be replaced on a regular basis. For example, a police department might replace one-third of its fleet each year, so that no vehicle is kept longer than three years.

Equipment that is being replaced should not be simply discarded. If possible, it should be sold for its full, fair market value, either through a negotiated sale (perhaps with another law enforcement agency) or through a public auction. Even equipment that is not in good condition may be sold for its parts or its scrap value. Often the money made from the sale of obsolete equipment is enough to greatly reduce the cost of the replacement equipment.[23]

CONTROL OF NONPOLICE PROPERTY

Law enforcement agencies, more than any other governmental agency, often must act as custodians of property that belongs to someone else. This *custodianship* is a legal status: The responsibility for the care of someone else's property is not merely a matter of good administrative practice, it is a legal obligation.

There are two broad classes of nonpolice property that a police agency must be concerned with: evidence and nonevidence.

Evidence

The legal meaning of *evidence* is, quite simply, anything that may have some bearing on the facts of a crime or other legal issue. More precise definitions are available in state statutes and in texts on criminal investigation. For our purposes, we are concerned with any tangible item that has been collected in connection with the investigation of a crime, and that must be kept either for laboratory examination or for presentation in court.

We are not presently concerned with the handling of evidence by investigators, laboratory technicians, or other police personnel, but instead with the administrative aspects of custodianship of evidence. In this regard, the proper handling of evidence ought to be simple.

The first requirement is that adequate, appropriate storage facilities must be provided. Second, one or more officers must be assigned the responsibility of caring for all evidence.[24]

Evidence may be in any form, size, or shape, which makes proper storage difficult. Some evidence might be in the form of a few human hairs and a bit of dust; another item might be a bicycle, a television set, or an automobile. Obviously, the same storage facilities will not be appropriate for all.

Most evidence should be stored in a single room equipped with shelves and cabinets, and floor space for bulky items. Each item received as evidence must be clearly marked as to its source, the case number to which it applies, and the identity of the officer investigating the case. Small items should be packed in boxes or envelopes, clearly marked on the outside.

The shelves and cabinets in the evidence storage room should be divided by partitions into separate bins, with wire or solid doors on the front of each bin and, best of all, a lock on each bin. All evidence that pertains to a single case should be kept in one bin if possible (see Figure 10.3).

Specialized facilities must be available for some types of evidence. Some materials may need to be refrigerated until (or after) they have been examined in the lab. Drugs and weapons must be kept in a safe or vault. Very

Figure 10.3. A police property room. Each item of evidence is placed in a labeled box, or the item itself is marked with the case number to which it pertains. Proper storage and control of evidence property is essential to the legal processes of criminal law enforcement. (Photo by the authors, courtesy Austin PD.)

bulky items such as a refrigerator or a piece of furniture must be stored on the floor, but they too must be protected against contamination and tampering. A few items (for example, an automobile) cannot be brought into the police building; they should be kept in a garage or at least in a secured area under constant surveillance.

A few police departments have adopted the policy of providing each officer with a small locker in which evidence for his or her cases may be kept. This procedure has few advantages and many disadvantages. Evidence from several cases may be mixed together, and the lockers usually are not large enough for even moderately bulky items (such as a suitcase or a stereo). Centralized storage under the supervision of a designated officer is a more practical approach.

When an item of evidence is delivered to the storage supervisor, a permanent record should be established. The record should include a description of the item, the case number, where the evidence was found, and the signature of the investigating officer. The investigator should keep a copy of the record, and another copy should be kept at the storage facility.

Each time an item is removed from storage, that fact should be noted on the permanent record. The evidence custodian also should obtain a signed receipt from the person taking the item out, showing where it is being taken and for what purpose. These records may be needed in court, to demonstrate that the chain of custody for each piece of evidence has been properly maintained, and that there is no likelihood that the evidence has been contaminated.[25]

Medium-sized and larger agencies may accumulate a substantial quantity of evidence and crime-related property, awaiting lab examination, court proceedings, or final disposition. Some of this material is dangerous (such as weapons), some is valuable (such as money or jewelry), and some is both (such as drugs). It is a sensible precaution for someone other than the property room staff to conduct a periodic inventory of evidence, checking to see that records are complete and that every item listed in the current records is, in fact, accounted for.

Nonevidence Property

Items that have been found and turned in to the police, the personal property of prisoners, stolen property that has been recovered, and items whose ownership is unknown or disputed all must be stored until they can be returned to their proper owners. Again, the police have legal custodianship of this property.[26]

The storage of nonevidence property is not as demanding as it is for evidence. Locked storage bins are not necessary, provided the property is kept in a single room under the control of a responsible officer. The agency does have an obligation to protect the property from loss or damage. This could mean that some items such as food or prescription drugs must be refrigerated, and items of considerable value such as jewelry should be kept in a safe or vault.

Again, a permanent record should be kept of each item of nonevidence property held by the police, indicating its source and any information available about its ownership. No property should be removed for any purpose without a notation being made on the record and a receipt being given by the person taking the property.

Police agencies traditionally have been eager to return nonevidence property to its rightful owners. Unfortunately, sometimes they are too eager. Anything delivered to the police as found property, recovered stolen property, or property of questionable ownership must be turned over to the rightful owner only after that person has established his or her claim to it. If the police turn property over to the wrong person, the department could be sued by the rightful owner. Great caution must be exercised.

The best procedure in many cases is to turn the matter over to the courts. Most states have reasonably simple procedures for the disposition of property whose ownership is uncertain or disputed. Ordinarily, the decision to release the property to the person claiming ownership may be made by a magistrate. Following these procedures will relieve the police of responsibility if an honest mistake is made.

Some law enforcement agencies, especially sheriff's departments, are legally authorized to sell or otherwise dispose of certain property, such as unclaimed found property. In some states, the sheriff is required by law to conduct an auction to sell real estate and other property that has been forfeited for nonpayment of taxes or for some other reason. The laws governing these sales vary considerably from state to state. Naturally, any law enforcement administrator who has such a responsibility must become thoroughly familiar with the applicable laws, and must take care to see that the auctions or other sales are conducted properly.

REVIEW

1. The four major parts of the purchasing process are:

   (a) Selection.   (e) Requisition.

   (b) Receipt.   (f) Specifications.

   (c) Transaction.   (g) Advertising.

   (d) Evaluation.   (h) Acceptance.

2. _____ are goods that are used more or less continuously.

   (a) Services.

   (b) Nonconsumables.

   (c) Commodities.

   (d) Items.

   (e) Bidders.

3. A request for proposals might be used in the purchase of

   (a) Commodities.

   (b) Items.

   (c) Services.

   (d) All of the above.

   (e) None of the above.

4. A purchase order is

   (a) A request for the vendor to give the price for goods.

   (b) A contract.

   (c) A receipt for the delivery of merchandise.

   (d) A payment for goods.

   (e) A statement of administrative policies.

5. A nonconsumable item such as equipment should be replaced as soon as

   (a) It no longer works and cannot be repaired.

   (b) It is more than three years old.

   (c) It can be sold for profit.

   (d) It has reached the end of its design life.

   (e) None of the above.

REFERENCES

[1]Robert L. Janson, *Purchasing Agent's Desk Book*. Englewood Cliffs, N.J.: Prentice Hall, 1980, pp. 26–27.

[2]O. W. Wilson, *Police Planning*, 2nd ed. Springfield, Ill.: Charles C Thomas, 1962, p. 65.

[3]Paul V. Farrell, ed., *Aljian's Purchasing Handbook*, 4th ed. New York: McGraw-Hill, 1982, pp. 5–3 to 5–7.

[4]Jackson E. Ramsey, *Budgeting Basics*. New York: Franklin Watts, 1985.

[5]Farrell, *Aljian's Purchasing Handbook*, pp. 3–29 to 3–30.

[6]"Police Vehicle Selection," in *FBI Law Enforcement Bulletin*, vol. 59, no. 6, June 1990, p. 24.

[7]Farrell, *Aljian's Purchasing Handbook*, pp. 1–5 to 1–6.

[8]Ibid., pp. 6–10 to 6–23.

[9]Janson, *Purchasing Agent's Desk Book*, pp. 91–94.

[10]Farrell, *Aljian's Purchasing Handbook*, pp. 5–7 to 5–10.

[11]Ibid., p. 5–51.

[12]Ibid., pp. 20–23 to 20–24.

[13]Ibid., pp. 5–7 to 5–10.

[14]David Farmer, ed., *Purchasing Management Handbook*. Brookfield, Vt.: Gower Publishing Company, 1985, pp. 247–50.

[15]Farrell, *Aljian's Purchasing Handbook*, pp. 5–10 to 5–41.

[16]Ibid., pp. 5–48 to 5–49.

[17]Ibid., pp. 20–39 to 20–41.

[18]Ibid., p. 6–31.

[19]David A. Varrelman, "Facilities and Materiel," in Bernard L. Garmire, ed., *Local Government Police Management*, 2nd ed. Washington, D.C.: International City Management Association, 1982, p. 350.

[20]Ibid., pp. 347–48.

[21]Wilson, *Police Planning*, p. 58.

[22]John Sturner, Sheldon F. Greenburg, and Deborah Y. Faulkner, "Equipment and Facilities," in William A. Geller, ed., *Local Government Police Management*, 3rd ed. Washington, D.C.: International City Management Association, 1991, p. 410.

[23]Varrelman, "Facilities and Materiel," in Garmire, *Local Government Police Management*, pp. 353–54.

[24]Sturner, Greenburg, and Faulkner, "Equipment and Facilities," in Geller, *Local Government Police Management*, p. 407.

[25]George D. Eastman, ed., *Municipal Police Administration*. Washington, D.C.: International City Management Association, 1969, pp. 270–74.

[26]Sturner, Greenburg, and Faulkner, "Equipment and Facilities," in Geller, *Local Government Police Management*, pp. 408–9.

Part Three
Management of Human Resources

PERSONNEL RECRUITMENT
AND SELECTION

In Part II, we discussed the physical resources that police administrators must manage: the buildings, vehicles, and various kinds of equipment. Here we discuss the management of the most fundamental resource of a police agency: its people.

Since law enforcement has existed as a separate discipline and governmental service for more than a hundred years, and policing began well before the modern era of sophisticated equipment and motor vehicles, you might suppose that all the important issues in managing a police agency's human resources were settled long ago. However, that is far from the case. The proper management of police personnel is easily the biggest source of controversy within the law enforcement profession, and between the police establishment and the community it serves.

How do you attract the best-qualified individuals to become police officers? For that matter, what does "best-qualified" mean? How do you select, from among those who apply for a police job, those who are most likely to be successful? Once you have made a choice, how do you determine whether those selected are, in fact, successful? How much training should new recruits be given, and what kinds of training? Do experienced officers also need training, and, if so, what kinds and how often? How closely should police officers be supervised to make sure that they are doing their jobs properly? What should be done if there is evidence that they are not doing their jobs properly? What can be done to prevent and correct instances of abuse of police authority or of outright police corruption?

These are only a few of the thorny issues that confront police administrators today. On each issue, nearly everyone has an opinion, but those opinions are not always backed up by facts.

The management of human resources is clearly the greatest challenge in police administration. Unfortunately, as John Furcon points out,

> The percentage of the annual police budget allocated to personnel salaries and benefits usually ranges from 75 to 90 percent of the total....The percentage of the budget related to vehicle fleet purchases is ordinarily far less. Ironically, however, we find that the amount spent on vehicle operation and maintenance is many times the amount expended for personnel selection and development and training activities.[1]

In a world in which, generally speaking, you get what you pay for, Furcon's comment may reflect a disastrous misplacement of priorities. A police department *is* the people who wear its uniform and carry its badge. How they are recruited and selected determines the basic quality of the agency and of the services it provides to its community.

RECRUITMENT

Most police agencies in the United States have no formal recruiting program at all. Many police administrators reason, "Why should we go to the trouble and expense of recruiting when we have very few positions open, and usually there are more applicants than jobs available?"

The problem with this approach is (1) the number of applicants at a given time might not meet the agency's immediate personnel needs, and (2) the quality of applicants may not correspond at all with the agency's requirements.

A few very large agencies, especially those that enjoy an excellent reputation in their communities, have many more applicants than they need. Nevertheless, they conduct a continuous recruiting program to attract new applicants. The reason is simply to ensure that not only is the number of applicants adequate, but that applicants of the highest possible quality are available at all times. Furthermore, experience has shown that it is more efficient and less costly in the long run to maintain a constant recruiting effort, even when there are no vacancies at the moment, than it is to start and stop recruiting drives over short periods.

Attracting qualified applicants for police vacancies has become increasingly difficult and will only get more difficult in the future. Part of the reason is simple demographics: the general population of the United States is, on the average, older than it has been in the recent past, and it is becoming increasingly older. Thus there are relatively fewer young people to recruit.

At the same time, the proportion of the population that is legally defined as "minority" (a term we will discuss shortly) is increasing. According to population experts, by sometime early in the next century the "majority" (Caucasians of European ancestry) will comprise something less than one-half of the population. The increase in minority population comes not only from persons born in the United States but also partly from new immigration: people who often do not speak English and who may have little knowledge or understanding of American traditions and customs.[2]

A sophisticated recruiting program is designed to locate, identify, and attract individuals who have the personal characteristics, basic skills, and motivation to become outstanding police officers. These traits are not widely distributed among the general population; finding these individuals demands forethought and diligence. As sociologist Jennie Farley points out, "*Where* employers look for applicants may well determine who they find."[3] She could have added that *how* employers look is just as important.

Before any recruiting program is established, the administrator must be able to answer the question, "What kind of people do we want?" The answer may seem obvious, but it is not. Defining the ideal police officer has proved to be very difficult despite the best efforts of hundreds of police officials and social scientists.

If one were recruiting, say, basketball players, the task would be comparatively simple: Give each prospective applicant a ball, put him or her on the court with several players of known ability, and see how the recruit performs. The Boston Symphony Orchestra has used the technique of having applicants play their instruments while a panel of judges, hidden behind an opaque screen, listen to and evaluate the performance.[4] But in both cases, the standards for selecting basketball players or symphony musicians are relatively clear-cut. The job of police officer is so complex and subtle that it defies any simple analysis.

Entry-Level Personnel

Nearly all American law enforcement agencies have adopted the same basic personnel system as the military services. With rare exceptions, all new personnel enter at the lowest level and are assigned general duties that, in theory, demand only modest skills and experience. Those who perform the entry-level duties successfully may be promoted, sooner or later, to higher-level positions that involve either supervision of the entry-level personnel or more specialized skills.

In other words, all new police officers are assigned to patrol. Those who prove to be good patrol officers have the opportunity to become, in turn, sergeants, lieutenants, captains, or detectives, communications specialists, crime-prevention specialists, and so on. It is very rare for an American police agency to hire someone for any position other than patrol officer.

Ironically, the job of patrol officer is actually the most demanding and difficult of all law enforcement jobs. No other police officers are likely to be exposed to the endless variety of challenges that face patrol officers on the street. Yet the system demands that the best, most effective patrol officers move up to positions that are, in some ways, less demanding (although they involve broader responsibilities). By implication, officers who remain in the patrol force throughout their careers are often —perhaps unfairly—regarded as failures.

This system places a tremendous burden on the recruitment and selection process. If new police officers were put in jobs that involved relatively modest demands, they could be given the time and opportunity to prove themselves and to learn what will be expected of them before they were put on the streets.

Instead, new officers must be nearly at the peak of their capabilities from their first day on the job. It is simply too dangerous—to the recruit, other police officers, and the community—to hire someone as a police officer whose capabilities are uncertain.

Under these circumstances, recruiters must cast their net as widely as possible, hoping to bring the greatest possible number of applicants into the selection process. Because the criteria for selecting successful patrol officers are so difficult to determine, the recruiter can never be certain which individuals are likely to succeed.

The techniques used in police recruiting are not especially different from those used in private industry, the military, and other governmental services. Posters, pamphlets, booklets, and other printed materials are distributed wherever young people are likely to see them. Police personnel speak at high school assemblies, church youth group meetings, and anywhere else prospective recruits might congregate. Radio and television *public service announcements* ("commercials" that are broadcast at no cost to the agency) and appearances on radio and television talk shows also may be effective. All these activities are designed simply to call attention to the fact that the police department is seeking applicants.

The recruiting materials and literature must be developed with care (see Figure 11.1). They must be attractive to their intended audience, of course. Dull and unimaginative posters or public service announcements are not likely to attract anyone.

At the same time, recruiting materials must not mislead prospective recruits into false expectations about police work. Jack L. Kuykendall and Peter

Figure 11.1. Recruiting literature must be not only attractive but also informative and accurate. (Courtesy Los Angeles PD, Las Vegas Metropolitan PD, Macon, Georgia, PD, and Mesa, Arizona, PD.)

C. Unsinger have observed, "All too often, applicants get images from the [entertainment] media of what law enforcement is all about, and they are quite shocked when they are informed, or discover, what being a policeman [sic] actually involves."[5]

Consequently, recruiting materials and presentations by recruiters must not only attract qualified applicants, but must serve to correct some of the misconceptions that the public has. People whose knowledge of law enforcement is gained mostly by watching television crime shows will be astonished to learn that police officers are not encouraged to beat up informants, shoot suspects on sight, race at breakneck speed in their patrol vehicles, trick suspects into confessing their crimes, or flirt with citizens who have been stopped for traffic violations.

Prospective applicants should be told forthrightly that police work is occasionally dangerous, usually tedious, often frustrating, and sometimes emotionally devastating. It demands great physical stamina, mental alertness, an exceptionally reliable memory, constant attention to minute details, and nearly unshakable emotional stability. It is never glamorous. Once in a while it is extraordinarily, deeply satisfying.

Recruiters must be candid not to discourage qualified applicants but to discourage those who might apply for the wrong reasons. By the same token, recruiters must be careful never to mislead prospective applicants about matters such as pay, job security, fringe benefits, or the chances for promotion. Human resources expert Robert L. Desatrik warns, "During the recruiting process an organization will foster unrealistic expectations....These bright young people will look around at their competition, perceive that they can reach the top more quickly elsewhere, and resign....Conversely, the ineffective performers will stay as long as possible, knowing that they cannot compete in the open market."[6]

Police recruiters must realize that they are in fact competing with all other employers, both those in governmental services and private industry. Unfortunately, the competition frequently will offer higher starting wages, more attractive fringe benefits and working conditions, and at least equally good chances for advancement. Recruiters may be tempted to emphasize, even exaggerate, whatever slight advantages they have, to the point of misleading recruits who will later be disillusioned and resentful when they learn that they have been misled.

What can recruiters offer that will attract qualified candidates? They should be able to offer wages and other benefits that are not unreasonable for the level of education and experience that are required. More important, recruiters can appeal to the idealism and desire for service to the community that are widespread among young people.

Perhaps the most important factor is the general reputation of the police agency. An agency that enjoys a well-deserved reputation for effectiveness, integrity, and dedication to serving the public will be far more successful in its recruiting than an agency with an unfavorable image. However, it is not sufficient for an agency to cultivate a good image by papering over the truth. In the long run, an agency that wants to enjoy a good reputation must be in fact effective, incorruptible, and dedicated to public service.

Recruiters should attempt to maintain a pool of applicants that will meet the agency's foreseeable personnel needs over at least the next year, even when there are no current vacancies. It is not unreasonable to estimate that of every ten applicants, only one will be chosen for entry-level training, and that of every ten trainees only five or six will complete the training and be accepted as permanent employees. Thus for every vacancy to be filled in the future, the recruiter needs to attract twenty or more applicants.[7]

Recruiters must be aware that their ultimate success lies in the future. They must not become so concerned with filling the agency's immediate vacancies that they neglect longer-term needs. For example, many police agencies now require recruits to have a minimum of two years of college; a few agencies require a college degree. But this does not mean that recruiters should concentrate entirely on college students as prospective applicants. They should devote a part of their time to recruiting among high school students, if only to let potential future recruits know what requirements they will have to meet. Even grade school and junior high school students should be encouraged to consider careers in law enforcement and to prepare themselves in high school and college.

Recruiters should not assume that they are the only members of the agency who are responsible for finding qualified applicants. Several studies have shown that most applicants for police vacancies learned of the opportunity from friends or relatives who were already police officers.[8] Thus every member of the agency has a role to play in recruiting.

College Students

Should a college degree be required of all recruits? The question is controversial among police authorities. The prevailing opinion seems to be that a degree is not essential, but that, all else being equal, a police officer's job performance increases in direct proportion to the amount of his or her formal education.[9] One study has even suggested that college-educated police officers are more likely to make "ethically correct decisions"; that is, they are less likely to act in ways that are illegal or improper.[10] Consequently, many agencies now require at least a year or two of college.[11]

College education is relatively specialized. Are some college degrees more valuable than others? This, too, is a controversial question. Some police officials place a high premium on academic training in law enforcement, police science, or criminal justice. Other officials take just the opposite approach. They believe that students with degrees in law enforcement will have to be detrained of overly idealistic and unrealistic attitudes. Obviously, the problem is that some officials lack confidence in the academic program![12] Nevertheless, it is a fact that some police officials prefer graduates with degrees in the liberal arts, sciences, or other fields.

Whatever an agency's preferences may be, recruiting on college campuses will be a major part of the overall recruitment effort. The effort should begin with standard techniques such as the distribution of posters and literature, but that is only the beginning.

Most colleges have a placement office whose specific responsibility is to help students in finding suitable employment after graduation. The police recruiter should work closely with the placement office staff in identifying promising students. Whenever possible, the recruiter should meet with individual instructors whose students are likely prospects.

On-campus recruiting should be a two-step process. First, students who might be interested in law enforcement as a career must be identified. Second, interested students must be screened to be sure that they have the appropriate qualifications for the recruiter's agency, and then they should be encouraged to submit an application.

The first step can be accomplished with the help of the placement office and by such techniques as operating a booth in a high-traffic area, perhaps in the student union. The second step requires personal contact between the recruiter and promising students. This contact might be initiated by a letter from the recruiter to all students who have been identified as prospects. The letter might invite the student to submit a preapplication, or to meet the recruiter at a designated time and place for a brief interview. Students could be invited to a social gathering where they would have the opportunity to meet recruiters for informal discussions, after which individual interviews would be held.

Most of the recruiter's time will be spent answering students' questions. At the same time, the recruiter should evaluate each student in terms of the agency's basic requirements. Students who are clearly unsuitable for law enforcement careers should be candidly discouraged from wasting time that could be better spent preparing for another career. Students who are marginally qualified or whose qualifications are uncertain should be encouraged to take the steps necessary to correct their deficiencies. Students whose qualifications are superior should be identified for follow-up contacts.

It is especially important for promising prospects to receive follow-up attention. One brief on-campus visit or interview is not enough. At the very least, prospective recruits should receive a letter, a week or two after the initial visit, thanking them for their time and encouraging them to keep in touch with the agency. The most promising prospects should be contacted by letter or, if possible, by telephone every few months, to remind them of the agency's interest. They might be encouraged to enroll in the agency's volunteer program or intern program, if this exists. Follow-up contacts should continue until the student formally applies for a police job or indicates that he or she is no longer interested.

In addition, students who have been identified as promising recruits should be asked to refer the recruiters to any of their friends who might be interested in a police career. These referrals are not only an excellent way to meet other good prospects, but they give the recruiter additional background information about the initial prospect.

Many police departments have organized volunteer programs that are especially appealing to college-age young people who are interested in a career in law enforcement. Some agencies have *police cadets*, unpaid volunteers who assist commissioned officers in a variety of routine tasks; *police aides*, either part-time or full-time paid employees who perform clerical and public service duties;

Explorer Scouts, affiliated with the Boy Scouts of America, who carry out special public service projects; or *community service officers*, noncommissioned personnel (usually paid, full-time employees) who assist the police in crime prevention, community relations, and social welfare projects. College-age prospective recruits should be invited to participate in any of these programs as a valuable introduction to law enforcement.

Recruiters must be aware of a special problem if their agency prefers applicants with some college. According to most studies, minority individuals have less access to a college education than do members of the majority population.[13] Consequently, a college education requirement tends to have a discriminatory effect. Recruiters should make a special effort to identify minority college students who may be interested in law enforcement, and particular attention should be paid to schools that have a concentration of minority students.

Minority Applicants

Strictly speaking, a person is a member of an ethnic minority if he or she belongs to a racial, nationality, or language group that makes up something less than half the population of the community. In that sense, all Americans are members of ethnic minorities, because no one group accounts for more than half of the total population.

However, *minority* also has a specific legal meaning. For reasons of history and social policy, the courts and legislatures have defined minority to include certain specific groups who have been the victims of various kinds of discrimination in the past. Those groups are African-Americans, Hispanics, Native Americans (including the native tribes of Alaska), and Asians. We use the term *minority* to mean these specific ethnic groups.

It is illegal to discriminate against any applicant or employee because of his or her gender, age, race, creed, color, national origin, native language, marital status, or other characteristics that are irrelevant to the job. In fact, by the end of the 1970s there were more than 130 federal laws and regulations concerning personnel practices in industry and public employment, and most of those laws and regulations were designed to prevent discrimination.[14]

Furthermore, it is poor administrative practice to allow discrimination to persist. Members of minority communities often are extremely sensitive to these conditions and reject the authority of any agency that does not recognize and protect their rights as citizens.

It is not enough for the police administrator to declare, "We will no longer discriminate against minority individuals. From now on, anyone who meets the standards we have set for new employees will be accepted." The problem is that the standards themselves may be discriminatory, intentionally or not. For example, many police agencies at one time required all recruits to be not less than 5 feet, 10 inches tall; some required recruits to be at least 6 feet tall. Of course, not many women could meet that requirement, and very few Hispanic or Asian men could do so. Thus, the requirement discriminated against those groups.

The fact that a height requirement discriminated against women and some men would not, in itself, justify changing the standards, if there were some evidence that height had a direct bearing on a person's ability to be a police officer.

During the past two decades, every requirement used by police agencies and other employers to screen applicants has been tested in court or by the administrative procedures of the federal *Equal Employment Opportunity Commission (EEOC)*. Those requirements or standards that are found to be clearly and fairly relevant to the work of the police officer have been allowed to stand. However, requirements that are purely arbitrary or that have no bearing on a person's ability to be a competent police officer (such as height requirements) have been eliminated or modified to remove their discriminatory effect.

Simply removing unfair standards and treating all applicants equally may not be enough. Because of discriminatory practices in the past, ethnic minorities are strongly *underrepresented* in many police agencies. That is, there are fewer minority police officers than there would be if there had been no discrimination, and there are fewer than there should be in proportion to the number of minority individuals in the general population. Ethnic minorities are especially underrepresented at the higher administrative levels of most police departments.

By the late 1960s, the courts and legislatures began to realize that passively ending discrimination would do little to correct the problem of underrepresentation. A positive effort would be required.

This positive effort can be expressed in the form of an *affirmative action plan*, a statement of the methods that an agency will use to correct minority underrepresentation.[15] Some affirmative action plans have been developed as a result of court orders and have had to be submitted to the EEOC for approval. However, many agencies have developed voluntary plans, which also may be submitted to EEOC for acceptance. Later, if an individual complains that an agency's personnel policies are discriminatory, the existence of an EEOC-accepted plan may be sufficient to demonstrate the agency's good intentions.

Once a plan has been developed, submitted to the EEOC, and accepted, it must be carried out. The test of any plan is the results that it produces. An affirmative action plan should produce a clear increase in the proportion of minority individuals at all ranks.

These results usually are measured against a set of specific goals that are expressed in terms of the percentages of employees of different ethnic backgrounds. For example, an agency might begin with a personnel roster that consists of 92 percent white males, 5 percent African-American males, 1 percent Hispanic males, and 2 percent females. Its affirmative action plan might call for the following percentages over a five-year period:

| | Male (%) | | | Female (%) |
Year	White	African-American	Hispanic	
1	88	6	2	4
2	80	10	4	6
3	75	12	8	15
4	65	15	10	20
5	60	15	12	23

These goals would bring the agency's personnel complement more into line with the proportional minority membership in the community (assuming that the single category of female includes minority women in appropriate proportions).

Actually, women are not a minority but a slight majority of the general population. They still would be underrepresented at the end of the five-year period in this plan, because only 23 percent of the employees, not 51 percent, would be female if the goals were achieved. However, out of the total adult female population, generally a little more than half are employed or seeking employment. Consequently, these goals may be realistic. (Note that the figures shown here are intended merely for purposes of illustration and do not represent the actual plan of any particular agency.)

The use of proportional goals often has been misinterpreted to mean that a police agency is required to meet specific quotas. In fact, a few court decisions have imposed proportional goals as quotas, as a means of evaluating an employer's good faith and efforts to correct underrepresentation. Ordinarily, the percentages are simply goals toward which an agency has committed its efforts.

No police agency has been required to hire only minority individuals or women, or to hire them in preference to all qualified white males. However, some agencies have been required by courts to hire minority or female applicants in preference to *equally qualified* white males until the representation goals have been achieved. If the current goal is to raise the percentage of African-American males from, say, 5 to 10 percent, then African-American male applicants would be given preference over white males who have essentially the same qualifications, until the goal is met; after that, all applicants should be considered equally.[16]

Affirmative action plans do not consist only of goals for proportional representation. Some EEOC-approved plans do not even contain specific goals. However, a plan must state that all arbitrary standards of selection have been eliminated and that every effort is made to seek, identify, and attract qualified minority applicants.

That brings us back to the subject of recruiting. In principle, the recruitment of minority applicants is no different from the recruitment of anyone else. Recruiters use essentially the same techniques: literature distributed to places where young people are likely to find it, speeches to community organizations, and, when possible, the use of the mass media (see Figure 11.2).

An extra effort must be made to bring the recruiting message to the minority population. Recruiters who are actively seeking minority applicants must be willing and able literally to "speak the language" of their prospects. A recruiter who speaks Spanish reasonably well is more likely to be successful recruiting Hispanic young people than a recruiter who does not know the language at all.

Even more important than proficiency in another language is the recruiter's understanding of the minority population's culture, social environment, attitudes, and values. Individuals from different ethnic groups have different atitudes toward police officers and careers in policing. Recruiters who do not personally share their targets' social background must at least be trained to understand it.[17]

Figure 11.2. Despite progress in recent decades, minorities continue to be underrepresented in many police agencies and therefore are a special concern of recruiters. (Photo courtesy Texas Dept. of Public Safety.)

Personal contact with the minority community is vital. It is not enough for the recruiter to make a speech at a high school with a high percentage of minority students. The recruiter also must meet with the school's administrators and teachers, asking them to identify students who might be good prospects. Similarly, the recruiter should contact coaches, ministers, and leaders of civic organizations, especially those that are youth-oriented, such as the Scouts, YMCA and YWCA, Boys and Girls Clubs, and so on.

The police recruiter also must face a difficult fact: The most outstanding minority young people may not be interested in a police career. Virtually all employers are actively seeking minority applicants, which gives the exceptional young person many opportunities. A career in law enforcement, with its chronically low salaries, mediocre fringe benefits, and generally unfavorable working conditions, may not be very appealing. Also, most law enforcement agencies have a far worse reputation in the minority community than they do in the community at large, whether or not the poor reputation is deserved.

There is nothing the recruiter can do about low salaries and poor working conditions. There is very little the recruiter can do about an agency's poor reputation, except to try to correct it by giving factual information in response to misconceptions and distortions. The recruiter must not try to ignore the problem or counteract it with misleading statements, exaggerations, or false propaganda. In particular, the recruiter must be careful not to overstate the agency's progress in carrying out its affirmative action plan. The recruiter should candidly admit the extent of progress in raising minority representation and in opening opportunities for advancement.[18]

Women

Women have been employed by American police departments since about 1845, when New York City first hired women as police matrons. However, until the late 1960s women police officers were employed only for a limited range of duties: They were used as matrons in women's prisons or jails, as plainclothes officers in juvenile, crime prevention, and other specialized units, and occasionally as undercover agents in vice and organized crime units. Many were assigned to essentially social-service, non–law enforcement duties. Because of their limited roles and lack of exposure to general law enforcement duties, women had virtually no opportunity for advancement in police careers.[19]

As an outgrowth of the civil rights and women's movements, women began to speak out against these limitations during the mid-1960s. In 1968, the Indianapolis Police Department became the first in the United States to employ women as general-duty police officers.[20] In 1972, the U.S. Supreme Court declared that job discrimination based on gender was unconstitutional, in a case that did not involve a police department. Congress passed the Equal Rights Amendment, which was never ratified by the states, but nevertheless influenced federal and state legislation and both private and public employment practices.

The FBI and the U.S. Secret Service immediately began to hire women as regular agents. The metropolitan police in Washington, D.C., St. Louis, and New York City, and the Pennsylvania State Police began to employ women as general-duty police officers.[21] By the mid-1970s, most urban police departments had at least a few women police officers. In short, nondiscriminatory employment policies apply to women as well as to minority men.

Assigning women to general patrol duties and making available to them advancement into the higher ranks of law enforcement is not without problems. According to sociologist Susan Ehrlich Martin, about one-half of the male officers in the Washington, D.C., Metro Police Department were dogmatically opposed to working with female officers, regardless of what the law might say. Female officers themselves had great difficulty in adapting to the demands of police patrol work and in relating to their male and female colleagues.[22]

More recent studies have shown that these problems have not gone away, despite the fact that research indicates either no significant differences between male and female officers' performance, or, in some cases, that women have actually performed better than men.[23]

Other than that, recruiting women is not substantially different from recruiting men. As with the recruitment of minority males, the recruiter must make a special effort to go where the candidates are likely to be found. This may include youth organizations that cater to young women, such as the Girl Scouts, Campfire Girls, and YWCA, or women's colleges. However, women are most likely to be found in the same places as men: high schools, youth groups, churches, colleges, and universities. When addressing such groups, the recruiter should make a special effort to make it clear that women are sought as prospective recruits.

Some recruiters are baffled by the "problem" of minority women: Should they be treated as women or as members of a minority, or both? Or are they a special category distinct from either one? Probably the best approach is to treat minority women first as members of a minority, because they are likely to identify with the minority group and its culture and values. At the same time, the recruiter must address the candidates' special concerns about the treatment of women.

Ultimately, every prospect must be treated as an individual. The recruiter must learn what his or her interests, concerns, and capabilities are, regardless of gender or ethnic background. After all, the recruiter's job is to attract the most promising individuals, not the most promising "categories."

Experienced Personnel

One of the peculiarities of American law enforcement is the absolute refusal of most agencies to consider applicants from other police departments for higher-level positions.[24]

A study of police personnel practices, conducted under the auspices of the Police Foundation in 1972, found that about one-fourth of all American police departments have a firm policy of not accepting transfers from other agencies. Among the three-fourths of departments that have no such policy, transfers were still exceedingly rare; less than one-third of the agencies that responded to the survey had ever hired a police officer with previous police experience.

Even when agencies do employ officers with previous experience, almost always the new employee must start over as a basic trainee or patrol officer. Out of some 670 American police agencies, only 9 had hired an officer with previous experience for any position above the rank of police officer; 4 of those 9 were small police departments whose chief previously was employed in a nearby urban agency.[25]

Why is it so difficult for police officers to transfer their hard-won skills and experience from one agency to another?

Most agencies that have policies against such *lateral entry* claim that there are laws or civil service regulations against it. However, this merely shifts the blame without explaining why these laws or regulations exist.

Another common explanation is that experience in one agency has no particular value in another agency, where administrative policies, law enforcement techniques, and the social environment are likely to be completely different. Some police officials insist that an officer with experience in another agency must be completely "retrained" before he or she is competent in their agency.

It is virtually impossible to prove or disprove the validity of these claims. The differences in administrative policies and law enforcement techniques from one agency to another, especially if both are in the same state, do not appear to be overwhelming. True, the social environment in large urban agencies is different from that in small suburban or rural agencies. Nevertheless, it is hard to imagine that a competent, experienced officer could not adapt to the changed environment. People in other professions adapt to different working conditions when they change jobs.

To put it another way, it seems at least equally likely that the training and experience gained by a police officer in one setting would enable him or her to adapt to a new environment more rapidly than a brand new, totally inexperienced recruit will.

The problem may not lie in differences in working conditions, but instead in differences in the standards of recruit selection and training among different agencies. A personnel officer in Agency A has no way of knowing whether an applicant from Agency B has gone through the same kind of selection process and training that Agency A would have given its own recruits. Until recently, a majority of small police agencies have given their recruits no training at all.

The solution to this problem may rest partly in the certification programs that are now in effect in all but a few states. *Certification* simply means that an individual must meet certain basic minimum standards and must complete a prescribed training program in order to be employed as a police officer anywhere in the state. There is no obvious reason for a police agency to prohibit certified police officers from other agencies from competing for positions.

However, there are still some states with no certification program at all, and no two state programs are exactly alike. Thus, an officer qualified in one state may not meet even the minimum qualifications in another state. Until there is a greater degree of standardization from state to state, and police officials generally accept certification as a valid means of establishing an officer's credentials, it is likely that lateral transfers will continue to be rare.

SELECTION

Recruiters and police administrators have no control over who applies for a job in a police department. Individuals who are obviously unsuitable may be discouraged from submitting an application, but it is not always obvious whether an applicant is qualified or not. Thus an applicant's suitability for employment must be determined through some process of evaluation.

Even if there were only one position to be filled and only one applicant for that position, some kind of evaluation process would be necessary to ensure that the applicant is qualified for the position. In fact, there are usually several applicants for every vacancy, and therefore it is necessary to determine which applicant is best suited to the position. This is done through a process of selection.

At one time, police departments were free to use whatever selection criteria they liked. Some of the criteria were entirely arbitrary; they were based solely on the personal prejudices and whims of the agencies' administrators. Other criteria were based on vague notions and unproven theories about what makes a good patrol officer.

A survey conducted in 1989 of thirty-six state and twenty-six municipal police departments in the United States revealed a wide array of selection criteria. Virtually every police agency surveyed used some sort of written test to evaluate the applicants' cognitive abilities; most agencies used tests prepared by the Civil Service agency for their state or city. The validity of such "locally designed" tests is very much open to question.[26]

About three-fourths of the agencies used a written "personality" test, usually the well-known *Minnesota Multiphasic Personality Inventory*. One-third of the agencies required applicants to complete lengthy "biographical data forms" even though some of the questions asked on the forms could infringe on the applicants' civil rights, and even though the information demanded on the forms probably has no predictive validity. Similarly, forty-nine of the sixty-two agencies required applicants to complete physical strength or agility tests, although these are commonly understood to discriminate against some applicants, and the job-relatedness of such tests has not been established.[27]

More than half of the agencies surveyed in 1989 used some sort of situational or simulation tests. About half used polygraph examinations, and about the same number used some type of psychological assessment or screening.

All of the agencies required a background investigation before an applicant would be accepted for employment, and all of the agencies required applicants to pass a medical exam.[28]

At first glance, each of the criteria listed above seems fairly reasonable. However, *none* of these standards was used by all police agencies. Some agencies had much less specific standards; others had much stricter criteria.

Most agencies used the same selection criteria for all applicants. This automatically meant that women and many minority males were disqualified because they could not meet the height, weight, and physical agility standards. The courts have ruled that an employer can use any standard in screening or evaluating prospective employees, even if the standard discriminates against some individuals, *provided* the employer can prove that the standard is a reasonably accurate method of determining whether a person can perform the work that the employees would be expected to perform. Criteria that are not clearly job-related cannot be used, even if there is no proof that they are discriminatory.[29]

When police departments began trying to implement the court rulings, they discovered a serious problem: No one knew exactly what qualities in an applicant are most likely to predict the applicant's future success as a police officer. In fact, as police officials and social scientists attempted to develop new, job-related selection criteria, they found that they could not agree on even a simple definition of what makes a successful police officer, or of what "successful police officer" means. Does it mean someone who makes a lot of arrests? Or does it mean someone whose preventive patrol techniques are so good that a high volume of arrests is not needed?[30]

A police department must screen out those applicants who are unfit in the sense that they are incapable of becoming competent police officers. At the same time, the department is interested in screening in or selecting those applicants who are most likely to become not just competent, but superior police officers. Any standards or criteria used to screen out or screen in applicants must be clearly related to the job of the police officer, and they should have *predictive validity*. That is, they should accurately determine whether an individual is likely to become a competent police officer.

All this must be accomplished despite the fact that there is no universally agreed-on definition of what a "competent police officer" is or does, or of what qualities in an applicant are likely to produce competence.[31]

Despite this dilemma, police departments must continue to hire new personnel, and they must continue to select from the applicants available to them. If there are no perfect methods of selection, the only recourse is to use the best available methods (even though imperfect) and continually seek to improve them (see Figure 11.3).

APPLICATION FOR EMPLOYMENT
CITY AND COUNTY OF MONTGOMERY
"An EQUAL Opportunity Employer"
PERSONNEL DEPARTMENT
City Hall
Montgomery, Alabama 36102

	Accepted	Rejected
Citz.		
Ed.		
Exp.		
Res.		
Other		

Title of Position:

INSTRUCTIONS:
ALL BLANKS MUST BE
FILLED IN COMPLETELY

Name: (Type or print name)

Mr.
Mrs. _____ Race _____ Age ___
Miss   First   Middle   Last

Address _____ Tel. No. _____
     House No.   Street   City   State   Zip

How long have you lived in Alabama immediately prior to date of application _____
                             Yr.   Mo.

Are you a U.S. citizen? _____

Date of Birth _____ Place of Birth _____
                                   City   County   State

PERSONAL DATA: Height ___ ft. ___ in. Weight _____ lbs. What is the condition of your health? _____

Marital Status: Single ☐; Married ☐; Separated ☐; Widowed ☐; Number of dependents under 18. _____

Do you have any physical handicaps? _____ . If so, attach a description to this application. Do you object to having your present employer questioned about your work? _____ . Have you ever been discharged or forced to resign from a position? _____ . If so, attach a complete explanation to this application. Have you ever been convicted of any law violation other than a minor traffic violation? _____ If so, give name and location of court, date, nature of charge and disposition. _____

SOCIAL SECURITY NO. _____

EDUCATION	Circle Highest Grade You Completed		Date Completed
Grammar and High School	1 2 3 4 5 6 7 8 9 10 11 12		
College or University (name of schools)	1 2 3 4 5 6 7	Degrees	
	Major		
Business, Trade or Correspondence School			
	Courses Studied		
List your professional certificate or license			

List three reliable persons, not relatives or employers, who know you well enough to give information about you:

	Address	Occupation

Figure 11.3. An employment application form. (Courtesy Montgomery, Alabama, PD.)

WORK HISTORY

Beginning with your PRESENT or most recent employment, list in REVERSE ORDER periods of employment. Each time you changed jobs or your title changed should be listed as a separate period. Give complete information, especially about the kind of work you did. (Use extra sheet if necessary.) Applicant must be specific and accurate in stating their experience and training for this position.

EMPLOYMENT RECORD: List all employment				
Employment Dates	Occupation and Description of Duties	Employer's Name and Address	Salary Received	Reason For Leaving
FROM				
TO				
TOTAL MOS.				
FROM				
TO				
TOTAL MOS.				
FROM				
TO				
TOTAL MOS.				
FROM				
TO				
TOTAL MOS.				
FROM				
TO				
TOTAL MOS.				
FROM				
TO				
TOTAL MOS.				
FROM				
TO				
TOTAL MOS.				
FROM				
TO				
TOTAL MOS.				

Show other experience by using additional sheets.

I hereby certify that all statements made hereon and attached hereto are true and correct to the best of my knowledge. Any false statement may be cause for denying me the right to examination or employment.

Date _____   Signature _____

Figure 11.3. Continued.

Multiple-Hurdle Procedure

The traditional method of selecting recruits from a group of applicants involves the use of a series of separate tests. An applicant's failure to pass any one test immediately disqualifies the person from further consideration.

First, applicants are informed of the agency's basic minimum qualifications: factors such as minimum age, minimum educational attainment, height and weight standards, and residency requirements. These are all factors that an applicant can judge for himself or herself. If the agency requires a minimum height of 5 feet, 8 inches, a person who is 5 feet, 4 inches tall will not bother to apply.

All applicants who meet the basic minimum standards are then put through a series of tests. The tests actually used vary from agency to agency, and the order in which the tests are applied also varies. However, a typical sequence might begin with a written test that the applicant must pass with a specific minimum score. Those who fail the written test are immediately dismissed. Those who pass the written test are then given a thorough physical examination; those who pass it are required to complete a physical agility test. After that, some agencies use either a written psychological test or applicants are interviewed by a clinical psychologist or psychiatrist. Each test or "hurdle" produces a clear, pass-or-fail decision: those who fail any one test are not allowed to continue.

The last two steps in the multiple-hurdle process usually are an oral interview with anywhere from one to seven police officials and a background investigation conducted by the agency's investigative personnel. We will have more to say about these two steps later.

The biggest advantage of the multiple-hurdle procedure is its efficiency. The various hurdles can be arranged in a sequence according to their relative cost and difficulty. Administering the written intelligence test is relatively inexpensive and simple, while conducting a background investigation is expensive and time-consuming. The more costly steps are not wasted on applicants who are certain to be disqualified anyway. Second, each unsuccessful applicant knows (or should know) where he or she failed and, in some cases, may be able to correct the deficiency and successfully take the tests another time.

However, there are several disadvantages to the multiple-hurdle procedure. Applicants vary in their ability to take certain kinds of tests, especially written tests, regardless of their potential capabilities. An applicant might fail a test on a bad day that he or she would have passed on another day. Usually the multiple-hurdle procedure makes no provision for second chances; an applicant who fails any one test must start over.

Some of the tests that are commonly used in multiple-hurdle procedures are of questionable value. As late as 1975, many police departments continued to use written intelligence tests that were originally developed for the testing of U.S. Army recruits in World War I. There is very little evidence that even the newest intelligence and psychological tests have any direct relationship to the job of patrol officer, although there is no question that patrol officers must have above-average intelligence and must be emotionally stable.

Psychological interviews, and even the medical examination performed by a physician, may involve subjective judgments that have no proven relationship to an individual's ability to perform the job of police officer. For example, some physicians would automatically fail an applicant who has a heart murmur. Others would require that the applicant undergo lengthy and expensive cardiovascular testing to determine whether the heart murmur poses a health threat. Still other physicians would consider a slight heart murmur in an otherwise healthy young person to be irrelevant.

Finally, the multiple-hurdle process ignores the widely held belief that no single set of personal characteristics can be used to define the ideal patrol officer. Some successful patrol officers are more intelligent than others, some are more physically vigorous than others, and some are more emotionally stable than others. Certainly, a person who is significantly deficient in any of those three areas would not be fit for consideration. However, a person might be only marginally qualified in one or more areas, yet so highly qualified in another area that the deficiencies are not that important.

An arbitrary pass-or-fail cutoff point on consecutive tests does not allow an applicant to compensate for a minor deficiency in one area. Furthermore, most people, given proper training and guidance, can improve their intellectual, physical, and emotional performance. Young people, especially those from a relatively deprived social and economic background, may show deficiencies that can be corrected in a well-conceived and properly executed training program (see Figure 11.4).

Multiple Assessment Procedure

Some of the disadvantages of the multiple-hurdle process can be avoided by eliminating the use of a single set of pass-or-fail standards and, instead, considering the overall performance of each applicant on a series of tests.

Often the tests are the same ones used in the multiple-hurdle procedure. However, each applicant receives a numerical score on each test, and the scores may be weighted according to their relative importance. For example, the scores on a physical agility test might range from 0 to 10 and the scores on a written intelligence test might range from 0 to 25. After an applicant has completed all the tests, his or her scores are added together. Sometimes each applicant's average score is computed. In any case, all of the applicants who complete the test series are ranked according to their total or average scores.

This procedure allows an applicant to make up for a low score on one test by getting a high score on another test. If weighting formulas are used, greater emphasis can be placed on the characteristics that are presumed to be most significant.

After all applicants have been ranked according to their scores, oral interviews are conducted beginning with the highest-ranked applicant and continuing down the rankings as far as necessary to fill the available vacancies. Usually more applicants are interviewed than will be hired, since some recruits are certain to drop out during training.

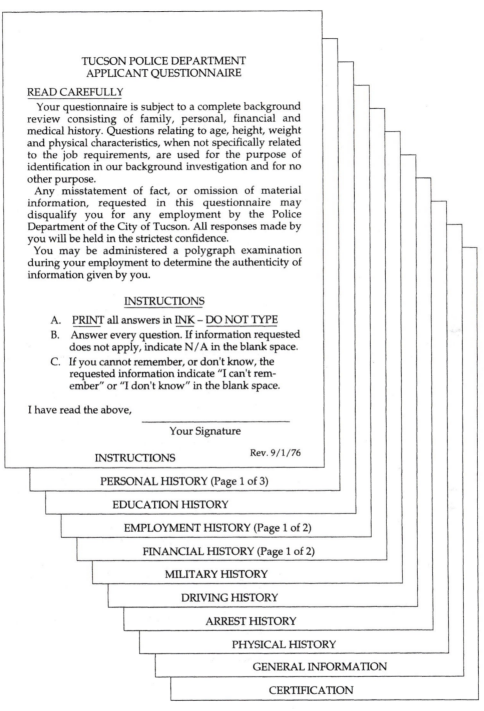

TUCSON POLICE DEPARTMENT
APPLICANT QUESTIONNAIRE

READ CAREFULLY

Your questionnaire is subject to a complete background review consisting of family, personal, financial and medical history. Questions relating to age, height, weight and physical characteristics, when not specifically related to the job requirements, are used for the purpose of identification in our background investigation and for no other purpose.

Any misstatement of fact, or omission of material information, requested in this questionnaire may disqualify you for any employment by the Police Department of the City of Tucson. All responses made by you will be held in the strictest confidence.

You may be administered a polygraph examination during your employment to determine the authenticity of information given by you.

INSTRUCTIONS

A. PRINT all answers in INK – DO NOT TYPE
B. Answer every question. If information requested does not apply, indicate N/A in the blank space.
C. If you cannot remember, or don't know, the requested information indicate "I can't remember" or "I don't know" in the blank space.

I have read the above, _____

Your Signature

INSTRUCTIONS   Rev. 9/1/76

PERSONAL HISTORY (Page 1 of 3)

EDUCATION HISTORY

EMPLOYMENT HISTORY (Page 1 of 2)

FINANCIAL HISTORY (Page 1 of 2)

MILITARY HISTORY

DRIVING HISTORY

ARREST HISTORY

PHYSICAL HISTORY

GENERAL INFORMATION

CERTIFICATION

Figure 11.4. Applicants to the Tucson PD must complete this extensive application form. (Courtesy Tucson PD.)

Some agencies also have a cutoff point below which they will not go. Only applicants ranked in the top half, or top third, are interviewed even if this leaves some unfilled vacancies.

Assessment Center Method

The newest selection procedure is an expanded form of the multiple-assessment procedure; it is called the *assessment center method*. Usually all applicants are brought to a single location and put through a series of assessment tests. However, according to psychologist Robert J. Filer, "An assessment center is a method...; it is not a place."[32]

Many of the same testing procedures and techniques that are used in the multiple-hurdle and -assessment procedures are also used in the assessment center method. In addition, applicants are required to complete several *performance tests* in which they are expected to demonstrate their ability to perform tasks similar to those actually performed by police officers. The performance tests often take the form of simulated police activities. Other simulations involve situational testing. Filer describes a situational test used by some Colorado police departments:

> The applicant [was] brought to the testing room and given a gun belt to wear. He or she was then briefly instructed in handcuffing and frisking procedure and was handed a card on which minimal instructions were typed. An example would be, "You are driving on patrol in the downtown area when you notice a young man (approximately 25 years of age) prying at a parking meter with a screwdriver. It is 4:45 P.M. Do your duty."
>
> When the applicant had read and understood the card, he or she entered the room in which [a police officer playing the part of] an irate citizen was kicking and prying at the parking meter. The "theme" that the [actor] followed in this situation was that he had been looking for a parking place for 15 minutes and that, when he finally found one, his nickel jammed in the meter....Once the applicant had completed one situation, he was handed instructions for the next until all three situations were completed.[33]

An important aspect of the assessment center method is that each applicant is rated not by one evaluator, but by several, and the applicant's final score on performance tests is a composite of the several evaluators' ratings. Written tests and other testing procedures with specific, fixed scores also may be used, and applicants' scores may be averaged or weighted in various ways. However, the idea behind the assessment center method is to treat the applicant as a whole person and to evaluate a number of possible areas of strengths and weaknesses, rather than merely to set arbitrary pass-or-fail criteria.[34]

The advantage of the assessment center method is that it allows applicants to be rated on their overall performance on both objective tests and situational tests. The use of several evaluators whose ratings are combined in some manner helps to overcome the possibility that a single evaluator might be biased for or against an applicant.

However, the assessment center method is, as you might expect, a very costly procedure. The testing process is more time-consuming than either of the other procedures we have discussed, and it requires teams of well-trained evaluators. There are no well-established standards against which to measure the applicants' performance on simulations and situational tests.

Nevertheless, some of the assessment center techniques appear to be promising. As more experience with this procedure is gained, it should be possible to determine whether applicants who successfully enter the police ranks through an assessment center turn out to be competent police officers.

Psychological Tests and Screening

The 1989 study by Ash, Slora, and Britton indicated that many police agencies use some sort of psychological examination or assessment of applicants, in an effort to determine the applicants' overall emotional stability and suitability for police work.

This has become an area of considerable controversy, especially when the testing takes the form of clinical examination by a psychologist or psychiatrist. One study reports that such screening, especially when it includes both written tests and clinical interviews, has a high degree of predictive validity; that is, the screening indicates which applicants are likely to be successful police officers.[35] But another study concludes that psychological screening has "unknown validity" and that research does not support its use.[36]

It is undeniably important to learn whether an applicant is emotionally mature and capable of withstanding the psychological stresses of police work. Unfortunately, the tests that are currently being used seem to be more successful in screening out the unsuitable applicants than in screening in the superior ones.[37] That being the case, it is hardly reasonable to use any psychological test on a "pass-fail" basis. The information provided by psychological testing (whether written tests or clinical interviews are used) should be taken into consideration but only as one factor among many in judging an applicant's suitability.[38]

Oral Interviews

Nearly all police agencies use oral interviews as the last or next-to-last step in the selection process.

The rationale for the oral interview is that every new police officer should be acceptable in overall personality and behavior. This unavoidably involves a subjective judgment on the part of the officers who conduct the interview, and often it is a judgment made in a surprisingly short period. When many applicants are being processed, interviews may last only three to five minutes, which means that the judgments are based on no more than superficial impressions.

There is some disturbing evidence that the members of oral interview boards have a strong tendency to give favorable ratings to applicants who appear to be most like themselves.[39] Thus the composition of the board may predetermine the outcome: A board made up exclusively of white males is less likely to give favorable ratings to females or minority males, and vice versa.

The oral interview is only as valuable as the skill of the interviewers can make it. For that reason, the members of an oral interview board should be chosen carefully, rather than being assigned at random or on the basis of such factors as official title or seniority. The board should consist of five to seven members including the personnel officer, one or two senior administrators, one or two midlevel supervisors (sergeants or lieutenants), and one or two experienced patrol officers chosen from different units. A small board cannot represent all elements of the agency, while a board that is too large will merely confuse and intimidate the applicants.

Interview board members should meet well in advance of the interview period to discuss and agree on what they intend to accomplish. Board members should compile a list of questions, perhaps fifteen or twenty, that will be put to each candidate.

The questions should not be random or arbitrary. Questions that call for a simple "yes-no" answer should be used only to confirm basic minimum qualifications or background data. Most of the questions should call for a more open-ended, exploratory response from the candidate: questions that ask, "Why do you...?" or "How would you...?" or "What would you do if...?" Follow-up questions should be used if the applicant's responses are incomplete or ambiguous or if the applicant simply fails to address the original question.[40]

Questions must not concern matters such as the applicant's ethnic background, nationality, marital status, sexual preferences, or other matters that are not clearly job related and that could imply discrimination on the part of the agency.[41]

Finally, applicants should be given an opportunity to ask questions, too, and the interviewers should answer them as fully as possible. Often the questions asked by an applicant are as revealing as the answers to the board's questions. For example, an applicant who is overly concerned with salary, prompt promotions, fringe benefits, and pension plans may be more interested in a secure job than in serving the public.

It is essential for the interview board to be provided in advance with copies of each candidate's background records, including the original application form and the results of all of the tests that each candidate has completed. Board members should have ample time to study the background packages, so that each applicant is not met as a stranger.

Immediately after each interview, the board should be given time to discuss the members' impressions. Each board member should keep a score sheet on which to record ratings of the applicant during the interview. During the discussion period, board members should decide on their final ratings and make a group judgment on the applicant.

Background Investigations

Those applicants who have passed every test and been approved by the oral interview board should be given a thorough background investigation. In larger agencies, this task is often assigned to a special unit composed of experienced investigators assigned to the personnel office. In smaller agencies, the investigations may have to be performed by regular detectives whenever they are able to find the time.

The background investigation should be a two-step process. The first step consists of a routine screening. Local, state, and federal criminal records files such as the *National Crime Information Center* and drivers' license records are checked to determine whether the applicant has an unfavorable history. Nearly all agencies disqualify an applicant with a past felony conviction, certain misdemeanor convictions, or an excessive number of traffic violations.[42]

Many agencies also conduct a credit check, on the theory that individuals who are unable to manage their personal finances are likely to be irresponsible. However, this information must be judged carefully. Credit reporting agencies are not always scrupulous about maintaining accurate records, and an unfavorable record could be due to special circumstances (a serious illness in the family or some other misfortune) that were beyond the individual's control.

If nothing unsatisfactory turns up in the routine screening, a candidate is subject to a more complete investigation. Neighbors, relatives, former employers, and other potential information sources are interviewed to learn as much as possible about the candidate's personal background and reputation. A standard set of interview questions should be developed to focus the investigator's attention on matters that have some bearing on the applicant's ability to be a competent police officer. Questions that could imply improper discrimination, or plain snooping into matters that are none of the agency's concern, must be avoided.

A formal, written report of the background investigation should be prepared and placed in the applicant's permanent file, on which the final selection decision will be made. It is also useful to have the applicant review the investigator's report and to allow the applicant to submit corrections or objections if necessary.

Whatever selection procedures are used, eventually someone must decide whether an applicant is to be admitted to employment or not. In a multiple-hurdle procedure, this decision may be more or less automatic; those who survive all the hurdles are accepted for training. In a multiple-assessment or assessment-center procedure, after oral interviews and background investigations have been completed, the final decision may be a matter of accepting all applicants whose ratings or rankings are above a certain cutoff point. In any case, if the selection process is relatively automatic, the personnel office should inform all candidates, including the unsuccessful ones, as quickly as possible.

In many larger agencies, where the number of qualified applicants exceeds the current vacancies, all applicants who have met the minimum requirements are ranked on an *eligibility list* according to their total scores. The final decision then is made by a selection board, which may or may not consist of the same officers as the oral interview board. The selection board sometimes is given the authority to request additional tests or investigations in questionable or marginal cases.

The stakes in recruiting and personnel selection are high; they concern nothing less than the ultimate success of the police enterprise.[43] Criminal justice professor Peter Horne warns, "If a poor or meaningless selection process exists, then low-caliber officers will be recruited and they will perform poorly in the field for, perhaps, a twenty or thirty year career."[44] For the police administrator, recruitment and selection are the means by which the most important resource of all—the human resource—is acquired.

REVIEW

1. Many police departments recruit continuously because

(a) Most applicants are not qualified.

(b) They are not able to fill their vacancies.

(c) An unexpected vacancy could occur at any time.

(d) It is more efficient than an occasional, short-term campaign.

(e) That is the main duty of recruiting officers.

2. True or False: Recruiting officers often must make a special effort to correct the misconceptions about law enforcement that people get from the entertainment media.

3. An affirmative action plan is

(a) A court order requiring an employer to stop discriminating against minorities.

(b) A plan to increase the proportion of minority individuals and women in the agency's work force.

(c) A set of quotas specifying the number of minority individuals and women who must be hired.

(d) A plan to eliminate all standards that favor minority applicants.

(e) A plan to reduce the number of white male employees in an agency.

4. True or False: Equal Employment Opportunity laws require all female and minority applicants to be regarded as qualified until proven otherwise.

5. In principle, all three of the selection procedures described in this chapter are supposed to

(a) Be equally cost efficient.

(b) Prevent unqualified persons from applying for law enforcement jobs.

(c) Ensure that all recruits receive adequate training.

(d) Provide adequate numbers of recruits to fill all of the agency's vacancies.

(e) Be job related and have predictive validity.

NOTES

[1]John Furcon, "An Overview of Police Selection," in Charles D. Spielberger, ed., *Police Selection and Evaluation*. Washington, D.C.: Hemisphere, 1979, p. 7.

[2]Ralph S. Osborn, "Police Recruitment," in *FBI Law Enforcement Bulletin*, vol. 16, no. 6, June 1992, p. 21.

[3]Jennie Farley, *Affirmative Action and the Woman Worker*. New York: American Management Association, 1979, p. 44.

[4]Ibid., pp. 57–58.

[5]Jack L. Kuykendall and Peter C. Unsinger, *Community Police Administration*. Chicago: Nelson-Hall, 1975, p. 248.

[6]Robert L. Desatrik, *The Expanding Role of the Human Resources Manager*. New York: American Management Association, 1979, p. 92.

[7]Ibid., p. 23.

[8]Osborn, "Police Recruitment," p. 23.

[9]Lawrence W. Sherman and the National Advisory Commission on Higher Education for Police Officers, *The Quality of Police Education*. San Francisco: Jossey-Bass, 1978, p. 2; Mitchell Tyre and Susan Braunstein, "Higher Education and Ethical Policing," in *FBI Law Enforcement Bulletin*, vol. 61, no. 6, June 1992, pp. 7–8.

[10]Tyre and Braunstein, "Ethical Policing," pp. 8–9.

[11]David L. Carter and Allen D. Sapp, "College Education and Policing," in *FBI Law Enforcement Bulletin*, vol. 61, no. 1, January 1992, p. 10.

[12]Ibid., p. 14.

[13]Ibid., p. 11.

[14]Farley, *Affirmative Action*, p. 1.

[15]J. Edward Kellough, "Integration in the Public Workplace," in *Public Administration Review*, vol. 50, no. 5, September–October 1990, pp. 557–65.

[16]Farley, *Affirmative Action*, p. 59.

[17]Ibid., p. 38.

[18]Ibid., pp. 109–10.

[19]Joseph Balkin, "Why Policemen Don't Like Policewomen," in *Journal of Police Science and Administration*, vol. 16, no. 1, March 1988, pp. 29–30.

[20]Peter Horne, *Women in Law Enforcement*, 2nd ed. Springfield, Ill.: Charles C Thomas, 1980, p. vii.

[21]Ibid., p. 32.

[22]Susan Ehrlich Martin, *Breaking and Entering: Policewomen on Patrol*. Berkeley: University of California Press, 1980, p. 49.

[23]Sean A. Grennan, "Findings on the Role of Officer Gender in Violent Citizen Encounters," in *Journal of Police Science and Administration*, vol. 15, no. 1, March 1987, pp. 78–85; Balkin, "Why Policemen Don't Like Policewomen," pp. 30–32.

[24]Geoffrey P. Alpert and Roger C. Dunham, *Policing Urban America*, 2nd ed. Prospect Heights, Ill.: Waveland Press, 1992, p. 43.

[25]Terry Eisenberg, Deborah Kent, and Charles J. Wall, *Police Personnel Practices in State and Local Governments*. Washington, D.C.: The Police Foundation, 1973, p. 27; Edward A. Thibault, Lawrence M. Lynch, and R. Bruce McBride, *Proactive Police Management*. Englewood Cliffs, N.J.: Prentice Hall, 1985, pp. 233–35.

[26]Vance McLaughlin and Robert Bing, "Selection, Training, and Discipline of Police Officers," in Dennis Jay Kenney, ed., *Police and Policing*. New York: Praeger, 1989, pp. 28–29.

[27]Robin Inwald and Dennis Jay Kenney, "Psychological Testing of Police Candidates," in Kenney, ibid., pp. 34–41.

[28]Philip Ash, Karen B. Slora, and Cynthia F. Britton, "Police Agency Officer Selection Practices," in *Journal of Police Science and Administration*, vol. 17, no. 4, December 1990, pp. 258–69. This study is in part a replication and updating of a similar survey conducted by the Police Foundation in 1972 and reported in Eisenberg, Kent, and Wall, *Police Personnel Practices*.

[29]Charles D. Spielberger, John C. Ward, and Harry C. Spaulding, "A Model for the Selection of Law Enforcement Officers," in Spielberger, *Police Selection and Evaluation*. Washington, D.C.: Hemisphere, 1979, pp. 11–12.

[30]Melany E. Baehr and Arnold B. Oppenheim, "Job Analysis in Police Selection Research," in Spielberger, ibid., pp. 33–59.

[31]Joseph M. Fabricatore, "Preentry Assessment and Training," in Spielberger, ibid., p. 79.

[32]Robert J. Filer, "The Assessment Center Method in the Selection of Law Enforcement Officers," in Spielberger, ibid., p. 212.

[33]Ibid., p. 224.

[34]Ronald G. Lynch, *The Police Manager*, 3rd ed. New York: Random House, 1986, pp. 180–95.

[35]Dierdre Hiatt and George E. Hargrave, "Predicting Job Performance Problems with Psychological Screening," in *Journal of Police Science and Administration*, vol. 16, no. 2, June 1988, pp. 122–25.

[36]William O. Dwyer, Erich P. Prien, and J. L. Bernard, "Psychological Screening of Law Enforcement Officers: A Case for Job Relatedness," in *Journal of Police Science and Administration*, vol. 17, no. 3, September 1990, pp. 176–82.

[37]Robert D. Meier, Richard E. Farmer, and Davis Maxwell, "Psychological Screening of Police Candidates," in *Journal of Police Science and Administration*, vol. 15, no. 3, September 1987, pp. 210–15.

[38]Alan W. Benner, "Psychological Screening of Police Applicants," in Roger C. Dunham and Geoffrey P. Alpert, eds., *Critical Issues in Policing*. Prospect Heights, Ill.: Waveland Press, 1989, pp. 72–86.

[39]Steve Falkenberg, Larry K. Gaines, and Terry C. Cox, "The Oral Interview Board: What Does It Measure," in *Journal of Police Science and Administration*, vol. 17, no. 1, March 1990, pp. 32–39.

[40]Thibault, Lynch, and McBride, *Proactive Police Management*, pp. 236–37.

[41]Wayne F. Cascio and Leslie J. Real, "The Civil Service Exam Has Been Passed: Now What?" in Spielberger, *Police Selection and Evaluation*, pp. 124–30.

[42]Kuykendall and Unsinger, *Community Police Administration*, pp. 254–55.

[43]Vance McLaughlin and Robert L. Bing III, "Law Enforcement Personnel Selection: A Commentary," in *Journal of Police Science and Administration*, vol. 15, no. 4, December 1987, pp. 271–76.

[44]Horne, *Women in Law Enforcement*, pp. 52–53.

TRAINING AND CAREER DEVELOPMENT

A little more than twenty years ago, the IACP discovered that only about 15 percent of all U.S. police agencies provided their recruits with any kind of training, or even required them to receive any training before they were put to work.[1] Police officials recruited and selected young men (rarely young women at that time), gave them a badge and a gun, and hoped that they would survive long enough to learn what they were supposed to do.

Unfortunately, that is still the case for some small police departments, especially in rural areas, in states that have no mandatory certification program. Even where certification programs do exist, some states permit recruits to be hired without training and to be assigned to patrol duties for up to a year while they complete the certification requirements.

The lack of training for new police officers usually is explained as a matter of economics. It is expensive and time-consuming to provide recruits with the training they need. There also is a persistent notion in the minds of some police officials, politicians, and ordinary citizens that policing is a rather menial task, easily learned by anyone with normal intelligence and a few weeks of practical experience. The opposite idea—that policing is an extremely complex discipline requiring practitioners with a substantial body of knowledge, technical skills, and professional attitudes—is dismissed by some people as a ploy to justify more money for law enforcement.

Our own view, which should come as no surprise, is that adequate training is absolutely vital to the success of any law enforcement agency. No one is a natural-born police officer, equipped from birth with the knowledge, skills, and attitudes needed to perform this most demanding and difficult of jobs. We consider it astonishing that some states demand more education and training of their barbers and undertakers than of their police officers.

Not only do newly recruited police officers need to be adequately trained before they begin work, but all police officers, regardless of their rank or amount of experience, must continuously improve their abilities through training and education. Education and training are the principal means by which an individual police officer can advance his or her career, learning new specialized skills or broader responsibilities.

Modern police administrators simply cannot avoid the need to provide adequate training for all of their employees: new recruits, experienced officers, and even noncommissioned personnel. Rising public expectations of law enforcement, in a society that seems to be changing more rapidly all the time, create a demand for more effective and sophisticated policing, which in turn requires more and better training of personnel.[2]

But what kinds of knowledge, skills, and attitudes do police officers need to learn? Who should provide the training (given the fact that it is costly and time-consuming) and under what conditions? What are the most effective and efficient ways to provide training? Finally, what should be the relationship between advanced training and career advancement: Should promotions be offered only to those who have completed prescribed training courses, or should other factors also be considered? How can administrators determine which of their officers are most likely to be successful in higher positions that require more specialized skills or leadership abilities?

TRAINING

Perhaps the first question that needs to be answered is, What do police officers need to know when they begin working as patrol officers and, later, as they seek advancement to more specialized positions or positions of broader responsibility?

This is hardly a new question. Law enforcement authorities have been asking it, and proposing answers, ever since the need for training began to be recognized about a century ago. In 1965, the IACP advocated a 200-hour basic training course that included academic study of the criminal justice system, judicial procedure, the laws of evidence and arrest, and the organization of law enforcement agencies; and skills training in patrol procedures, the use of firearms, and emergency vehicle operation, among other topics.[3]

The IACP's curriculum has not been universally adopted. Aside from those agencies that offer no recruit training at all, basic training programs vary from one or two weeks in length (40 to 80 hours) to as much as eleven weeks (440 hours), and in a few cases even more.[4] Not only do training programs vary in length, they vary even more widely in content, despite the standardization that is supposed to be imposed through state certification programs.[5]

In principle, any basic training program should be designed to fulfill one objective, succinctly stated by William J. Bopp: "to develop in officers those skills necessary to successfully perform a new job."[6] To accomplish this objective, according to Bopp, "Trainers must be thoroughly familiar with the tasks policemen are expected to perform."[7] In short, the content of the training program should be derived from a close examination of the work that is actually done by police officers.

But that is more easily done in theory than in practice. As we discussed in Chapter 11, the work of the patrol officer is so complex and varied that it all but defies analysis.

Basic training for new patrol officers generally occurs in two phases. The first phase, preservice training, may take place either before or after a person has been hired by a police agency. The content of the training includes the knowledge and skills that are virtually universal in policing: a basic understanding of criminal law and the criminal justice system; standard procedures in crime prevention, patrol, criminal investigation, traffic law enforcement, and other specialized areas of police science; and similar matters.

Preservice training may be obtained either in a police academy operated by a law enforcement agency or in a community college or university. Many police administrators believe that preservice training should be part of a broader educational experience in the social role of policing and in the liberal arts as a whole. Other police officials tend to be skeptical of the value of college programs in law enforcement. In a 1987 study, it was discovered that police chiefs, sheriffs, and college teachers of criminal justice courses were almost entirely in agreement on what needed to be taught. Nevertheless, the police chiefs and sheriffs insisted that the college programs were "too theoretical."[8]

The second phase of preservice training consists of the specific policies and procedures of the agency in which the person has been employed. The content of this orientation training might include local criminal ordinances, administrative and operational structure of the agency, and local policies concerning personnel matters, duty assignments, radio procedures, and dozens of other items. By its nature, orientation training must be provided by each agency, in an academy if one exists, or in some other manner in smaller agencies.

Who Should Provide Training

The responsibility for securing adequate training and education must be shared by the individual officers or prospective officers and the police agency. In the past, this responsibility has been borne almost entirely by the individual officers. Most police agencies today recognize and accept their share of the burden.

There are three possible sources of training and education: individual study (such as the reading of books about law enforcement practices), academic programs conducted by colleges and other educational institutions, and training programs conducted by individual police agencies. In some states, regional councils of governments operate regional academies to train personnel from several small agencies that cannot afford to operate their own academies.

The value of individual study should not be underestimated. Young people who are interested in pursuing a career in law enforcement should be encouraged to learn as much about the field as they can by reading books on the subject (preferably sound academic works rather than just popular fiction). In addition, as we mentioned in Chapter 11, some police agencies have organized volunteer programs for young people that give them limited but valuable experience in police-related activities.

Individual study is just as important for active law enforcement personnel. Every police officer should seek to learn more, to acquire new skills and polish old ones, and to increase his or her ability to perform. However, individual study in law enforcement faces the same limitations as it would in, say, the field of medicine: Reading about brain surgery is not the same as performing an operation. Furthermore, a person pursuing a course of individual study usually lacks the clear objectives and guidance that an organized training program would provide.

Academic programs conducted by two-year community colleges, or by baccalaureate colleges or universities, represent a relatively recent development in the field of law enforcement. Academic programs in criminology (the study of the causes and consequences of crime) have existed since the end of the last century, usually within departments of sociology or psychology. Similarly, some universities have offered programs in criminal justice for many years, usually within the departments of political science or law. However, few colleges have offered programs intended specifically for prospective or practicing law enforcement personnel, until the mid-1970s, when federal funds became available for this purpose; several hundred colleges created such programs. Unfortunately, when the federal funding stopped, many of the programs stopped. Nevertheless, some college-level programs have continued because there is little question that they meet a vital need.

The third source of police training is the police academy. All state police agencies and most large municipal police departments operate their own academies. As we mentioned, in some cases several police agencies are served by a single regional academy, either operated jointly by the participating agencies or funded through a council of governments or a separate state agency (see Figure 12.1).

Generally, a police agency's primary responsibility is to provide basic training for all of the agency's new personnel. In addition, most academies conduct periodic seminars, workshops, and short courses (training programs that last a week or two) for current personnel.[9]

Methods of Training

This text is not the place for an extended discussion of training methods and techniques; information on this subject is readily available elsewhere. However, a police administrator does need to know enough about training methods to permit appropriate planning.

Any training program should begin with an evaluation of the agency's existing operations to determine what kinds of training are needed and for

Figure 12.1. Nearly all state police agencies and many metropolitan police departments operate their own academies to train recruits and to conduct in-service training programs, seminars, and workshops. (Photo by the authors, courtesy Texas Dept. of Public Safety.)

whom. Training is not (or should not be) an end in itself, but instead a means of improving the performance of police personnel. Thus, the first step in establishing a sound training program is to examine the performance of all personnel, determine their weaknesses, and decide which of those weaknesses may be corrected through training.

For example, if it is found that too many traffic citations are being dismissed by the courts, the reason may be that patrol officers are making technical mistakes in their traffic arrest and citations procedures. A training program then could be organized to correct the error. Or, to take another example, if an evaluation of the agency's current operations reveals the need for a specialized unit to investigate organized crime, a training program ought to be developed to prepare the personnel assigned to the new unit. Merely establishing the new unit and assigning personnel to it will not satisfy the need.[10]

A training program does not necessarily involve bringing together a number of people in a classroom under the direction of an instructor. In fact, training programs can be organized and conducted in several different ways (see Figure 12.2).

Preservice or recruit training usually includes classroom instruction plus laboratory sessions for practice in skills. Inservice training of current personnel may take the form of classroom instruction, workshops, or seminars lasting from an hour or two to several days. Inservice training also may be accomplished partly through the distribution of printed training bulletins, memoranda, policy statements, and so forth. Short training sessions may be conducted at roll call at the beginning of each patrol shift.

It also may not be necessary for all personnel to receive training at the same time and place. Videotape and inexpensive means of distributing video programs

Figure 12.2. Classroom training is only one part of a thorough recruit training program. (Photo by the authors, courtesy Austin PD.)

by cable and microwave permit a great deal of instructional material to be prepared in advance and delivered to whoever needs it, wherever they may be. Videotape also offers the advantage that a student who misses a lecture or demonstration, or who simply wants to review the material, can view it at any time.

The FBI's National Law Enforcement Academy, based in Quantico, Virginia, now offers videotaped training programs and two-way video conferences, using the Law Enforcement Satellite Network, to any police agency. In most cases, the programs are free of charge, but the participating agency must have the appropriate equipment, which may cost several thousand dollars.[11]

Each type or form of training has its advantages and disadvantages. Some of the most common training methods are discussed in Table 12.1.

Table 12.1 is not intended to be exhaustive, but instead to illustrate the number of different training techniques that can be used, to suggest their limitations, and to indicate the most beneficial uses of each technique.[12]

Another training technique that is not listed in the table, but that has become increasingly popular, is the *simulation*: any kind of activity that represents a portion of real life.

Simulations designed for educational or training purposes, like most popular games, have the advantage of involving the students or players in an activity that, if the simulation is properly designed, accurately represents some kind of real-life activity.

A *role-playing simulation* involves assigning two or more students to act out roles in a certain situation. One example of a role-playing exercise was described in Chapter 11, where a student is instructed to conduct a field interrogation and, if necessary, an arrest of a suspicious person who appears to be breaking into a parking meter.

Any number of situations can be used as the basis for role-playing exercises.[13] For example, two students might play the roles of a husband and wife engaged in a heated argument, and two other students could be assigned to act as patrol officers responding to a disturbance call. The "husband" and "wife" are

TABLE 12.1 COMMON TRAINING METHODS

Method	Description	Advantages, Disadvantages
Lecture	Instructor presents information to a group (class)	Least expensive method of presenting information to a large group; requires relatively little preparation by instructor. Should be used only if all students need the same information and have essentially the same background and preparation, and if the lecture is followed up with practice or application of knowledge. Success depends heavily on ability of the lecturer to get and keep students' interest.
Television, film, etc.	Audiovisual material shown to group or individual	Can be a highly individualized form of instruction; any size group, from one to several hundred, may view a television program or videotape recording, film, slide program, etc.; the best method for showing a procedure, technique, or anything that cannot be brought into a classroom. Relatively expensive; requires careful preparation by instructor, prompt follow-up; content of material, if purchased from external source, may not fit students' needs precisely.
Conference or seminar method	Students (with or without instructor) discuss subject	Excellent for group problem solving, exploration of issues, modification of attitudes; involvement of students in discussion reinforces learning. Without a skilled instructor, students may learn the wrong information or attitudes. Discussions may be time-consuming and unproductive unless carefully led by the instructor. Limited to small groups (ten to fifteen students) or some will have no chance to participate.
Demonstration method	Instructor shows students	The best way to teach a technique or process. Each student must have a chance to practice immediately after demonstration. Good follow-up to a lecture or use of audiovisual media. Often time-consuming; may require costly specialized training facility or materials for demonstration; requires careful preparation by the instructor. All students must be able to view and participate in the demonstration.
Practice or laboratory method	Students practice techniques or procedures	Most effective form of instruction for learning procedures, techniques. Should be used as follow-up to lecture, demonstration, or use of audiovisual materials. May be costly if many students are given a chance to practice; may require special equipment, facility, or materials. Limited number of students can be supervised by one instructor.

likely to gain some insight into the strong feelings that law-abiding citizens experience when outsiders intrude into their private business. The students assigned to act as patrol officers should learn what is likely to happen in a real-life situation of this type.

The exercise can be made even more valuable if (1) the four students exchange roles after a while; (2) the exercise is recorded on videotape and played back so that the role-playing students can see themselves objectively; and (3) the exercise is carefully discussed afterward by the instructor and the rest of the class.

Real-life situations and activities can be simulated by computer programs. Most video games are actually computer-based simulation exercises designed for entertainment rather than education. However, computer-based simulations have been developed for training purposes in other fields, notably the training of aircraft pilots. Their potential in law enforcement training is almost limitless.

Whatever methods are used, the instructors and the administrators responsible for training need to be aware of current research and innovations in education and training. Much has been learned in recent years about the most effective ways to motivate and teach adults. Methods that have been used in the past, such as the "stress training" that is a traditional feature of military recruit training programs, have been shown to be not just ineffective but actually counterproductive; they inhibit learning rather than promote it.[14]

Police training administrators also must recognize that the changes in the nature of the police work force may dictate changes in training methods. Methods that might have been appropriate when virtually all recruits were middle-class white males with at most a high school education may be inappropriate when many recruits are female, members of ethnic minorities, and college educated.[15]

Field Training

Long before there were any formal training programs in law enforcement, it was the common practice to assign newly hired personnel to patrol a beat for a few weeks in the company of an experienced officer. Even after formal training programs came into being, this practice of on-the-job training continued. Police administrators readily conceded that a formal training program, no matter how well it might be organized and conducted, did not adequately prepare a rookie officer for life on the street.

On-the-job training, more frequently known as *field training*, continues to be the standard practice in most police agencies today. A recruit who has completed the formal training program, either in an academy or in a college, is first hired on a probationary basis for a period of anywhere from one month to two years; more typical probationary periods are six to twelve months.[16]

During the probationary period, an officer is subject to being dismissed at any time, usually without notice and without much, if any, right to appeal. Even if the probationary employee is not dismissed, a permanent position is not guaranteed, but depends on an evaluation of the rookie officer's overall performance during the probationary period. The recommendations of the rookie's immediate supervisor usually carry a great deal of weight. Thus a probationary officer is under considerable pressure to satisfy the expectations of his or her supervisors.

There is a long-standing tradition in policing, according to which each rookie officer is told by the experienced officer (*field training officer*) to whom he or she is assigned, "You can forget all that stuff they told you in the academy, because I am going to show you how things are really done." The implication that formal training is irrelevant, unrealistic, and worthless serves to undermine the rookie's confidence and leaves the new officer vulnerable to confusion. Some rookies, eager to please their supervisors, take the advice literally and promptly forget much of what they have learned about sound police practices.[17]

It may well be that some formal training programs are not as relevant and realistic as they ought to be. However, that does not justify dismissing them so lightly. A far better approach, of course, would be to involve experienced officers, at all ranks, in critically evaluating and improving the formal training.

In fact, the experienced officers who are responsible for the field training of rookies often have no personal knowledge of what is currently being taught in the academy. The experienced officer may have graduated from recruit training five years or more in the past and be completely unaware of major changes in the formal training program since then. Quite often the procedures used by the field training officer are nearly identical to those taught in recruit training.

In many police agencies the assignment of probationary employees to field training officers is left to the discretion of unit supervisors (sergeants or lieutenants) who may not realize what an important matter this is. The influence of a field training officer—good or bad—can color a rookie officer's entire career. Clearly, field training officers should be chosen with great care and should be given adequate preparation. They should be exemplary officers whose job performance is as close as possible to the ideal, and they should be informed in detail about the content of the recruit training program.

The field training phase itself should not be left to chance. Clear, definite training objectives should be established, and the field training officers should know exactly what is expected of them.

Because a major part of field training is the evaluation of the probationary officer's performance, the field training officers should be trained to conduct proper evaluations. Field training should not contradict or undermine the formal recruit training program; instead, field training should be an extension and continuation of the formal program.[18]

Another aspect of field training deserves comment here. The personnel in any police agency naturally form a social group with its own culture: a set of knowledge, ideas, values, and ways of doing things that are, to a greater or lesser degree, shared among all the members. Any newcomer to the group must learn this culture in order to be accepted as a member. Acceptance into the group is vital to the newcomer, since the failure to win acceptance means being shut out of the group's internal, informal systems of communication and social interaction.

The process of learning a group's culture and gaining acceptance is called *socialization*.[19] This process is important in all stable social groups, not just in law enforcement. It actually begins during the recruitment process: Applicants who want to be accepted as recruits attempt to demonstrate the behaviors that, as far as they can tell, will lead to acceptance into the police society. Individuals who do not appear to be capable of learning acceptable behaviors are simply not admitted as recruits.

The socialization process continues during recruit training, as the recruits learn more about the specific behaviors that are expected of them and the ideas and attitudes that experienced police officers display. The traditional advice to the rookie to "forget everything you learned in the academy" is so devastating because it amounts to saying, "The knowledge, ideas, values, and attitudes you have tried to master will not help you gain acceptance in the society of real police officers." The field training period is the "final exam" in the socialization process, and this is one exam that every rookie officer must pass.

Socialization is not necessarily a conscious, deliberate process. Often neither the new recruit nor the experienced officer is fully aware that this process is occurring. Recruits are often confused about what is expected of them, what they must do to gain acceptance. Members of the group may reject the newcomer not because of any fault of the recruit, but because of prejudices, fear that the newcomer is a threat to their own jobs, or some other hidden cause.[20]

Administrators and supervisors must be aware of the socialization process and of its significance. They cannot guarantee that every recruit will be accepted into the agency's society, nor should they try to do so. However, they can try to ensure that the process is fair. Discrimination against whole classes of newcomers—such as women or minority members—should not be tolerated; each recruit is entitled to be judged as an individual and given a fair chance to win acceptance.

If it appears that experienced personnel are unreasonably hostile toward or fearful of newcomers, the causes should be discovered and corrected. Often it is helpful to make the field training officers themselves aware of, and sensitive to, the socialization process. They should understand that a rookie not only needs to demonstrate technical competence as a professional law enforcement officer, but also must fit into the social world of policing. The field training officer, knowingly or otherwise, can either encourage or hinder this vital process.[21]

CAREER DEVELOPMENT

In any bureaucratic organization, both responsibility and rewards are said to flow upward. The person at the top of the organizational pyramid has more of both than anyone else in the organization, and the persons at the bottom have less of both than anyone else.

The members of a bureaucratic organization are supposed to be assigned to their positions according to their individual competence, or merit. In principle, every member of an organization should be the one person who is most capable of performing the specific tasks assigned to him or her.

The fact that this is not always the case does not invalidate the principle. People sometimes attain high positions of responsibility and rewards for which they are not competent. When this happens, whatever the reason may be, the organization as a whole suffers a loss of efficiency and effectiveness. Thus it is very much in the interests of the organization to see that every position is filled by the most competent person.

At the same time, most people want to receive all the rewards they can, and thus it is in the interests of every individual to achieve the highest position that he or she is competent to fill. Competence is not an inborn, permanent quality of an individual. It is at least partly acquired over a period, through the individual's experiences, including formal and informal training. As a person gains experience, he or she should become competent to hold higher positions in the organization, up to each individual's natural limits of talent, intelligence, and so on.

Thus the organization gains by making the best use of each member's abilities and the individual gains by holding the highest position for which he or she is competent. Conversely, both the organization and its members suffer when people are prevented from reaching the highest position for which they are competent, especially if some of the higher positions are held by individuals who are less competent.

These principles apply as surely to a police department as to any other government agency, business corporation, or other bureaucracy. To put them into practical terms, police administrators have an obligation to see that both the needs of the agency and the needs of the individuals within it are served by making the best possible use of the agency's human resources. Every member of the agency should be assigned to the one job for which he or she is best suited. Every member of the agency should be continuously preparing to advance to the next higher job for which he or she is capable of becoming competent.[22]

If promotions to higher positions of authority are given solely on the basis of seniority (the length of time a person has worked for the agency) or, worse yet, on the basis of the personal prejudices of some higher official, not only will the organization's efficiency suffer but lower-ranking personnel will become dissatisfied when they realize that their own abilities are not being properly recognized and rewarded. Dissatisfaction leads to poor performance throughout the agency because dissatisfied personnel feel no desire to do their jobs well.

In extreme cases, dissatisfied personnel may engage in destructive acts in a deliberate or unconscious effort to sabotage the agency that they perceive to be unfair.[23] The importance of this principle can hardly be overstated. As Susan Ehrlich Martin points out,

> Those individuals who perceive opportunities (although they may be quite limited in reality) will work for them; those who feel blocked (although opportunities may actually exist) will seek alternative sources of satisfaction.[24]

The alternatives may include devoting little or no effort to one's work, or treating the agency's clients (in the case of a police department, the general public) with contempt and hostility, or looking for more satisfying work elsewhere.

Opportunities for promotion are not the only concern of employees. All people need to feel satisfaction in their work: that they are being appropriately rewarded, and that their contributions are recognized and appreciated. One of the most important discoveries of the human relations school of management (see Chapter 2) was that people who feel satisfaction in their work generally are more productive. Traditional, mechanical approaches to increasing productivity—which usually have amounted to finding ways to make people work harder—ignored this vital principle. Managers of successful enterprises, including law enforcement agencies, realize today that part of their responsibility includes helping all employees to find satisfaction in their work.[25]

Later, we will see how this responsibility can be carried out as part of an overall career development program. First, however, we look at the more traditional approach to career development, with its emphasis on evaluating employees' performance and establishing systematic procedures for promotions.

Performance Evaluation

If the most competent employee is to be assigned to each position, then it is necessary to evaluate the competence of every employee.[26]

Those who are performing their present jobs successfully are presumed to be most capable of handling a higher job requiring more specialized skills or broader responsibilities equally well, and are also presumably entitled to be rewarded for their efforts. The two methods most widely used in law enforcement to evaluate employees' performance are known as work product measurement and trait rating.

Work product measurement simply means counting the amount of work an employee produces. This method works very well in organizations where employees produce tangible products or perform the same kind of operation repetitively.

For example, in a factory the assembly-line workers' output might be the number of products assembled in an hour or a day. Even if an employee performs a variety of different tasks, work product measurement is practical if each type of task can be readily measured. A clerical worker's output might consist of so many letters typed, files filed, telephone calls received and transferred, and so forth per day.

But how can the work of a police officer be measured? Should patrol officers be evaluated on the basis of the number of arrests they make, traffic citations issued, field interviews conducted, and so on? In fact, these items have been used by many police departments.[27]

The problem is that there is no real basis for comparing one officer's work with another's, using these "products." One officer might be assigned to a district with little crime while another is assigned to a high-crime area. Even within a single area, the level of police activity may vary from one time of day to another. Thus the "product" of officers on different shifts would not be comparable.

It is almost impossible to measure the intangible, but no less important, work of a police officer in patrolling a beat, counselling citizens, observing traffic and suspicious activities, and many similar activities. Furthermore, merely counting a police officer's products says nothing about the quality of the officer's work. How many of the officer's arrests resulted in convictions? How much of the officer's patrolling prevented crimes in the first place? These factors are ignored in most work product measurement systems.[28]

A *trait-rating system* attempts to look beyond the measurable work and examine the quality of an officer's job performance in terms of the traits, or personal characteristics, a successful officer is supposed to display.[29] In a trait-rating system, supervisors are given a list of the traits that are supposed to be exhibited by police officers. The supervisor is then required to give a numerical rating to each officer for each of the listed traits. The ratings "points" are then added up to arrive at an overall rating for the officer.

The first problem in any trait-rating system is that there is no universal agreement on what traits a police officer ought to display. The trait lists developed by various police agencies show a remarkable diversity. Some trait lists include specific tasks or behaviors, such as, "The officer performs at least twenty-five field interviews per week," or "The officer responds to each radio call within four minutes." Other traits concern general characteristics of behavior: "The officer treats all members of the public in a courteous and dignified manner," or "The officer presents a neat, well-groomed appearance." Still other traits may concern matters that are not readily observable: "The officer is conscientious and diligent in his or her work," or "The officer has good judgment." Under the best of circumstances, it would be hard for any supervisor to assign valid ratings to every officer on each of these traits.[30] (Figures 12.3 and 12.4 show two variations of trait-rating evaluation forms.)

In fact, the situation is even more problematic. Most police officers, especially on patrol, work alone or in the company of their regular partner most of the time. A supervisor has few opportunities to observe officers while they are engaged in actual police duties. Thus, the supervisors' ratings are based not on actual, direct observation but on feelings, impressions, and comments of other officers, and, consciously or unconsciously, personal biases and favoritism.[31]

A few police agencies, recognizing the serious limitations of both work product measurement and trait rating, have tried to develop sound evaluation systems by combining the two methods. Supervisors are required to complete a periodic evaluation of each officer, usually every six months or once each year, that considers both the quantity of product (arrests, traffic citations, and so on) an officer has produced and an assessment of the officer's personal traits.

This combination of methods may be better than either method alone, but the original problem still is not solved. The work performed by police officers, especially on patrol, is so complex and varies so greatly from time to time and from place to place that it is nearly impossible to measure in an accurate, objective way.[32]

Another variation of trait rating also has been used with some success: *peer rating*. This merely means that each officer fills out a trait-rating sheet for every other officer in his or her unit. Peer ratings are intended to overcome some of the problems inherent in supervisory ratings. Presumably, peers (that is, other officers of the same rank or level) have more opportunity to observe their fellow officers at work, and therefore their ratings of one another are more likely to be based on actual performance.[33]

CHEYENNE POLICE DEPARTMENT
EMPLOYEE EVALUATION REPORT

NAME: Position: Rating Period:

	Exceeds Standards ☐	Meets Standards ☐	Requires Improvement ☐	Unsatisfactory ☐
KNOWLEDGE OF WORK: Comments:				
QUALITY OF WORK: Comments:	Exceeds Standards ☐	Meets Standards ☐	Requires Improvement ☐	Unsatisfactory ☐
QUANTITY OF WORK: Comments:	Exceeds Standards ☐	Meets Standards ☐	Requires Improvement ☐	Unsatisfactory ☐
ATTITUDE & WORK INTEREST: Comments:	Exceeds Standards ☐	Meets Standards ☐	Requires Improvement ☐	Unsatisfactory ☐
RELATIONS WITH OTHER EMPLOYEES: Comments:	Exceeds Standards ☐	Meets Standards ☐	Requires Improvement ☐	Unsatisfactory ☐
INITIATIVE: Comments:	Exceeds Standards ☐	Meets Standards ☐	Requires Improvement ☐	Unsatisfactory ☐
PROMOTION POTENTIAL: Comments:	Exceeds Standards ☐	Meets Standards ☐	Requires Improvement ☐	Unsatisfactory ☐
SUPERVISORY ABILITY: (when applicable) Comments:	Exceeds Standards ☐	Meets Standards ☐	Requires Improvement ☐	Unsatisfactory ☐
OVERALL PERFORMANCE: Comments:	Exceeds Standards ☐	Meets Standards ☐	Requires Improvement ☐	Unsatisfactory ☐

	Signature	Position	Date
RATER:			
REVIEWER: Comments:	Signature	Position	Date
REVIEWER: Comments:	Signature	Position	Date
EMPLOYEE: Comments:	Signature	Position	Date

White – Personnel   (See Instructions on Reverse Side)
Blue – Division Copy
Pink – Individual

Figure 12.3. A trait-rating personnel evaluation form. (Courtesy Cheyenne, Wyoming, PD.)

Please return to coordinator by _____

CITY OF PORTSMOUTH
Performance Evaluation and Counseling Form
PATROL OFFICER

EMPLOYEE NAME	INGRADE PROGRESSION DATE	DEPARTMENT AND DEPARTMENT NUMBER
CLASS SPECIFICATION	EMPLOYEE NUMBER	SOCIAL SECURITY NUMBER
EVALUATION DATE	EMPLOYMENT DATE	SUPERVISOR/REVIEWER

☐ Probationary Employee 1st 6 Month Review　　☐ Probationary Employee 2nd Review for Permanent Status & Ingrade Progression　　☐ Permanent Employee Annual Review 120 Days Prior to Continuous Service Date　　Other (Please Specify) ☐ _____

INSTRUCTIONS

Using the numerical scale below, compare the performance of the employee being rated against the performance criteria listed for each factor. Select the number which best indicates your perception of that individual's performance on each of the criterion and enter it in the box provided. Then enter a number indicating a composite, or overall evaluation for the factor. Your complete evaluation should not necessarily reflect an average of the criteria rating since some criterion are more important than others. Examples of past performance **must** be cited.

EVALUATION SCALE

O　E　|　M　|　B　|　U
9　8　7　6　5　4　3　2　1

(O)	Outstanding	– Exemplary performance far exceeding performance criteria.
(E)	Exceeds Expectation	– Performance which exceeds the level supervisor normally expects.
(M)	Meets Expectation	– Generally meets supervisor's expectation on performance criteria.
(B)	Below Expectation	– Erratic performance on criteria, falling short of the normally expected . . . requires remedial attention.
(U)	Unsatisfactory	– Unacceptable performance which must receive immediate attention.
(NA)	Not Applicable	– Evaluation of the factor or criterion is inappropriate for the employee being rated.

FACTOR A: SUPPORT OF UNIT'S OBJECTIVES: PLANNING/TEAMWORK.
Performance Criteria:

	Works with supervisor and other officers in planning and building an effective team.
	Officer's objectives, talents and efforts are directed toward the needs of the department and accomplishment of unit's goals.
	Improved methods are suggested and readily tried to improve effectiveness and solve traffic/crime problems.
	New and additional assignments are accepted and performed.
	Composite Evaluation for Factor.

Cite examples of past performance to support your evaluation. ————

FACTOR B: RESPONSE TO RADIO CALLS AND ASSIGNMENTS.
Performance Criteria:

	Radio calls are not missed without valid cause.
	Response is made promptly, safely, and appropriately.
	Further action is rarely required in minor cases.
	Assistance provided is appropriate to the need or problem.
	Proper radio procedures are followed.
	Composite Evaluation for Factor.

Cite examples of past performance to support your evaluation. ————

FACTOR C: INVESTIGATION OF CRIMES OR TRAFFIC ACCIDENTS.
Performance Criteria:

	Appropriate investigative steps are taken.
	Investigations are thorough, and each step properly documented.
	Evidence is properly preserved, collected, and thoroughly described.
	Other divisions or agencies are properly notified as required.
	Composite Evaluation for Factor.

Cite examples of past performance to support your evaluation. ————

FACTOR D: APPREHENSION, ARREST OF CRIMINAL SUSPECTS.
Performance Criteria:

	All necessary information is included in arrest reports and case files.
	No unnecessary force is used.
	Uses correct procedures in effecting an arrest.
	Composite Evaluation for Factor.

Cite examples of past performance to support your evaluation. ————

Figure 12.4. A more elaborate personnel evaluation form with detailed, job-related criteria. (Courtesy Portsmouth, Virginia, PD.)

FACTOR E: ENFORCEMENT OF TRAFFIC REGULATIONS.
Performance Criteria:

☐	Traffic tickets are issued.
☐	Actions taken are appropriate to the offense.
☐	Citations are rarely returned for correction.
☐	Traffic is controlled as required.
☐	Composite Evaluation for Factor.

Cite examples of past performance to support your evaluation. ————————

FACTOR F: PERFORMANCE UNDER STRESSFUL, EMERGENCY OR UNUSUAL CONDITIONS.
Performance Criteria:

☐	No serious deviations from expected performance are demonstrated under unusual circumstances.
☐	Demonstrates ability to take command of emergency situations.
☐	Composure is maintained under stress.
☐	No major errors identified by supervisor in post-operation review.
☐	Composite Evaluation for Factor.

Cite examples of past performance to support your evaluation. ————————

FACTOR G: RELATIONS WITH CITIZENS AND THE COMMUNITY.
Performance Criteria:

☐	Does not antagonize or insult citizens.
☐	Knows and is responsible to community problems.
☐	Courtesy is demonstrated in citizen contacts.
☐	Anger and verbal abuse from citizens does not adversely affect performance.
☐	Composite Evaluation for Factor.

Cite examples of past performance to support your evaluation. ————————

FACTOR H: WORKING RELATIONSHIPS AND COOPERATION WITH OTHER PERSONNEL.
Performance Criteria:

☐	Valid complaints are not received from fellow officers or supervisory personnel.
☐	Problems in personal relationships with other personnel do not impair work relationship.
☐	Readily assists/backs other officers.
☐	Trains and guides less experienced officers.
☐	Composite Evaluation for Factor.

Cite examples of past performance to support your evaluation. ————————

FACTOR I: PREPARATION AND PRESENTATION OF REQUIRED REPORTS AND INFORMATION.
Performance Criteria:

☐	Field interview reports are used.
☐	Reports are legible, concise, grammatically correct, submitted at agreed upon time, with required information.
☐	Reports and information are rarely returned for correction.
☐	Evidence is preserved and thoroughly described.
☐	Composite Evaluation for Factor.

Cite examples of past performance to support your evaluation. ————————

FACTOR J: OPERATION, MAINTENANCE AND CARE OF DEPARTMENTAL EQUIPMENT.
Performance Criteria:

☐	Vehicles are not abused through poor driving habits.
☐	Specified operating and safety procedures are followed in the use and maintenance of equipment.
☐	Weapons are checked for cleanliness and serviceability.
☐	Automobiles are returned clean (interior) and serviced for next shift.
☐	Equipment wear, malfunctions, damages are identified and reported.
☐	Composite Evaluation for Factor.

Cite examples of past performance to support your evaluation. ————————

FACTOR K: CONFORMANCE TO WORK SCHEDULES, ASSIGNMENTS AND INSTRUCTIONS.
Performance Criteria:

☐	Instructions are followed and assignments completed on schedule.
☐	Work does not have to be closely supervised.
☐	Deviations from instructions and schedules are explained satisfactory to supervisor.
☐	Unassigned time is effectively utilized, i.e., maintain citizen contact, proactive patrol.
☐	Special attention is given to high accident or crime areas.
☐	Composite Evaluation for Factor.

Cite examples of past performance to support your evaluation. ————————

FACTOR L: CONFORMANCE TO DEPARTMENT POLICIES, REGULARITY OF ATTENDANCE AND PUNCTUALITY.
Performance Criteria:

☐	Policies, rules and regulations are followed as prescribed.
☐	Appearance meets departmental specifications.
☐	No unnecessary delays in starting work at specified time.
☐	No abuse of meal periods, coffee breaks, quitting time, or other special absences.
☐	Supervisor is given proper notice in advance of absences.
☐	Composite Evaluation for Factor.

Cite examples of past performance to support your evaluation. ————————

Figure 12.4. Continued.

FACTOR M: OTHER FACTORS IMPORTANT TO SUPERVISOR. (List any unique duties performed by employee not apparent in Class Specification.)

Performance Criteria.

Cite examples of past performance to support your evaluation.

OVERALL PERFORMANCE RATING

Based upon preceding evaluations, but not necessarily an average of the factors since some are more important than others, carefully read the criteria for each of the performance levels and check the term which best describes the employee's overall performance for the evaluation period.

	Outstanding:	Exemplary overall performance deserving special recognition normally occuring in less than 5% of the workforce.
	Exceeds Expectation:	Performance exceeding the supervisor's expectation on nearly all performance factors.
	Meets Expectation:	Performance generally meeting supervisor's expectation on most performance criteria.
	Below Expectation:	Erratic performance falling short of that expected on most factors. Special review recommended in 60 days. USE OF THIS OVERALL RATING REQUIRES COMPLETION OF REMEDIAL ACTIVITIES SECTION BELOW. This rating is not sufficient to deny Ingrade Progression.
	Unsatisfactory:	Unacceptable performance: Ingrade Progression for Permanent Employees may be withheld only upon submission of two unsatisfactory ratings, in accordance with the Pay and Classification Plan. Probationary Employees may be terminated immediately. USE OF THIS OVERALL RATING REQUIRES COMPLETION OF THE REMEDIAL ACTIVITIES SECTION BELOW.

Remedial Activities: Actions which supervisor and employee have agreed upon to correct performance evaluation Below Expectation or Unsatisfactory.

Development Activities: Action which supervisor and employee have agreed upon to further develop employee capabilities and to prepare for greater responsibility.

RATER'S SIGNATURE   DATE:

COMMENTS OF EMPLOYEE:

SIGNATURE OF EMPLOYEE: DATE:

(Signature indicates only that appraisal has been reviewed with employee.)

SIGNATURE OF RATER'S SUPERVISOR: DATE:

Figure 12.4. Continued.

Despite all the difficulties we have mentioned, performance evaluation is an indispensable part of police administration. Both work product measurements and trait ratings can be valuable tools in planning, especially in identifying problem areas and making decisions about the deployment of human resources. A trait-rating system of some kind is nearly unavoidable if employees are to be evaluated for their current performance and their potential for promotion.[34]

However, neither of these tools is perfect. The value of a trait-rating system depends first on the quality of the trait lists; the traits to be rated must be selected carefully, to reflect actual, observable performance and not vaguely defined personal qualities. Peer ratings, in addition to supervisory ratings, make a trait-rating system even more useful.[35]

Promotional Systems

If it is so difficult to evaluate the performance of individual officers, how can police administrators select those officers who are the best qualified to be promoted?

In the early days of American law enforcement, this posed no problem at all, since assignments and promotions were based almost entirely on political considerations. One of the most important, and most bitterly contested, of the many reforms along the way toward police professionalism was the establishment of *merit systems* for the selection and promotion of officers. A merit system, as the name suggests, simply means that some orderly procedure is used to select recruits and award promotions.

A sound merit system should be objective and free of personal biases. It should allow all qualified candidates to compete fairly for every available position.

Most merit systems are administered not by the police department itself, but by an independent *Civil Service Commission* or similar agency, which may serve the same function for other governmental departments and agencies. The federal Office of Personnel Management (formerly the U.S. Civil Service Commission), for example, supervises the hiring, placement, and promotion of all federal employees in the executive branch (except the military) and independent federal agencies, except those in the highest administrative positions who are appointed directly by the president. Most state and local Civil Service Commissions are organized along similar lines, although there is a great deal of variation from one agency to the next.[36]

When Civil Service merit systems were first introduced, it was widely believed that the most important criterion in choosing candidates for promotion was knowledge about the job to be filled. It seemed that the simplest and fairest way to award promotions would be to require all candidates to pass a test covering the areas of knowledge that a job involved; whoever obtained the highest score on the examination would be the best qualified for the position. Thus the competitive Civil Service examination became the basis for most merit systems, and this procedure is almost universally used today.[37]

Competitive examinations have the advantage of being, in principle, fair and objective. As long as everyone takes the same test and the tests are scored accurately, it is a simple matter to determine which candidate has the best score and is therefore presumed to be best qualified. But there are several problems with this procedure that are not immediately apparent.

The first problem concerns the composition of the examinations. What questions should be asked? What knowledge is required for a person to be successful in a given job? Who should formulate the questions? In many cases, the task of developing the questions has been left to police administrators, who may have no experience or talent in analyzing the requirements of each position and framing questions that are clear, fair, and accurate in assessing the candidates' abilities.

Over the years, the science of developing standard tests, including merit system tests, has progressed to a considerable degree of sophistication. The commercially developed tests used in college entrance examinations, for example, are the product of thousands of hours of work by specialists in psychological testing. Many merit system examinations are made up by small committees or by individual administrators who have no special qualifications for this task.

Several professional organizations, including the IACP, have attempted to improve the quality of merit system tests by developing standard tests that may be used by local police agencies. Some state certification agencies also provide promotional tests for use by local police agencies and Civil Service Commissions. There are also various commercially developed tests intended for this purpose.

Local police and Civil Service officials frequently object that these standard tests are not suitable because they do not take into account local conditions, policies, or requirements. Many police administrators prefer to use a test that has been developed locally, even though its validity (whether it actually measures what it is supposed to be testing) and reliability (whether it consistently produces the same results) may be questionable.[38]

A second problem is even more fundamental: Can any written examination fairly determine whether an individual is capable of performing a given job? It may be true that every job involves applying a given body of knowledge. But the fact that a person has the required knowledge does not necessarily mean that the person can apply it successfully.

It is not clear that any written examination can measure all of the skills that may be required to perform a job successfully. Furthermore, people vary in their ability to do well on written tests. Some people might possess the knowledge and skills to do a job very well, but still receive a low score on a written exam. Others may score high on the exam because of their test-taking skill even though they are not really better qualified than some other candidate.

For all of these reasons, the use of competitive written examinations as the basis for merit system promotions has come under a growing cloud of controversy and skepticism.[39] But if competitive written examinations are not the answer, how can administrators choose the best-qualified candidates for promotion?

Most law enforcement agencies use a variety of methods, in combination, to award promotions. Typically, officers who wish to be considered for promotion must first establish their eligibility. Usually they must have been employed in their present position for a minimum period of time, anywhere from one to eight years.[40] Then they must pass a written exam, usually with a minimum score of 70 percent. Some agencies also permit only those officers with the greatest seniority in their current rank to compete for promotions. Oral examinations and ratings by supervisors are also used by some agencies to establish eligibility.[41]

Once candidates have established their eligibility for promotion, a different set of criteria may be used to rank the candidates from "most qualified" to "least qualified." Many agencies continue to use candidates' scores on written exams as one of the bases for ranking. Oral exams and supervisors' ratings are also widely used for ranking. Seniority in the candidates' present rank and length of service are also considered. Finally, two other factors—the candidates' educational achievements and participation in the department's inservice training program or equivalent advanced training received elsewhere—are considered by a few agencies.

For most American law enforcement agencies, the promotional system and the recruit-selection system are very similar; the methods used to screen applicants for entry into police employment are used again whenever an employee is being considered for promotion.

Career Development Systems

There is an alternative to the conventional system of promotion we have just discussed. However, it is considerably more difficult to establish and maintain.

A few police departments, mostly large ones, have developed a comprehensive *career development system* that attempts to overcome the drawbacks of the usual promotional system while better serving the needs of both the agency and its employees. One such career development system was developed in the early 1970s by the Los Angeles County Sheriff's Department and reported in a monograph published by the Law Enforcement Assistance Administration.[42]

Briefly, the essential steps in establishing a career development system are as follows:

1. Each position in the agency must be analyzed and a thorough description must be developed, stating the skills and knowledge required for the position, and defining the minimum criteria for assignment to it. The analysis may be performed by administrators or, better yet, may be based on information provided by officers who are already serving in each position.[43]

2. An assessment must be made of each employee's attitudes, capabilities, desires, skills, and current performance. These assessments should be based on information from a number of sources: personnel records, supervisors' ratings, scores on standard tests, and responses to questionnaires completed by the employees themselves.

3. In principle, whenever a vacancy occurs in any position in the agency, the employee records are checked to determine which employees have the experience, skills, and capabilities required. The idea is to match the agency's needs with the human resources available. In all but the smallest agencies, this may require some kind of computerized matching system.

The object of the career development system is not merely to match personnel with vacancies, but to encourage the best use of human resources. Development is the key word here. Each employee must be allowed, and in fact assisted, to reach his or her maximum potential. This is done by treating personnel not merely as jobholders, but as individuals who, over the length of their careers, should be able to progress from one position to another as they gain experience, skill, and knowledge.

The agency has certain needs, defined by the positions that must be filled now and in the future. It is therefore to the agency's advantage to ensure that its personnel acquire the abilities they will need to fill those positions.

This is where the employees' attitudes and desires are important. As we noted earlier, most people want to achieve the highest position they are capable of handling successfully, which will enable them to receive the greatest possible rewards. However, all employees do not necessarily aspire to the same positions. Some patrol officers may be interested in becoming detectives, while others may be more interested in achieving higher ranks in patrol or specializing in traffic enforcement, crime prevention, or some other particular field. The career development system should allow each employee to fulfill his or her individual goals.[44]

The basic mechanism for this purpose is *career counseling*. Experienced officials (usually supervisors, although sometimes specialists in personnel are assigned this responsibility) should discuss with each employee his or her aspirations and should develop a *career development plan* that will enable the employee to become qualified for whatever positions he or she would like to fill in the future.

The career development plan for an individual might involve specialized training, or it might involve transferring into a specialized unit (while remaining at the same rank) to gain additional experience. For example, a patrol officer who is interested in traffic law enforcement might be assigned to an inservice training course in that specialty and then reassigned to the traffic patrol unit as soon as a vacancy occurs. Another patrol officer who is more interested in becoming a detective might be encouraged to enroll in a college-level course in criminal investigation, followed by an academy training course in criminalistics.

A career development plan represents an agreement between the employee and the agency. Like any agreement, it is subject to modification if future conditions require it. For example, a patrol officer in 1994 might indicate an interest in becoming a narcotics investigator in the future and might develop a career plan accordingly; however, in 1998 it may be apparent to both the officer and his supervisor that the need for narcotics investigators is declining. By the same token, in 1994 an officer might express an interest in traffic law enforcement, but find in 1996 that she is more interested in community relations and crime prevention. In all such cases, it makes sense for both the employee and the agency to modify the career development plan.

Career counseling also is an opportunity for a supervisor to discuss an employee's work performance in a nonthreatening, nonpunitive setting. Whatever method is used to evaluate employees' performances, the results should be discussed, confidentially, with each employee. Those who are performing well should be encouraged to continue, of course. Those whose performance is not satisfactory should be counselled on ways to improve. If additional training or a different work assignment might bring about better job performance, those measures can be made part of the employee's career development plan. At the same time, the employee should be told exactly what degree or amount of improvement is expected in a certain period. The responsibility for improvement is placed on the employee in return for the agency's agreement to assist the employee in carrying out the career development plan.

Career planning is only half of a complete career development program. The other half is *job enrichment*, which means that an effort is made to make every job in an organization more intrinsically satisfying and rewarding. People do not enjoy doing routine, meaningless tasks in an unpleasant work setting. According to Robert McGregor's Theory Y (discussed in Chapter 2), most people want to perform jobs that involve meaningful, responsible work; they want to have a substantial degree of control over their own working practices; and they want to know that the results of their work are important.

It might seem, at first glance, that job enrichment is totally unnecessary in a police agency. Surely the vital duties of police officers are meaningful, responsible, and require a high degree of individual control. Unfortunately, that is not always true, especially at the patrol level.

In the more traditional police agencies, patrol officers are given only the most routine and meaningless duties to perform, such as patrolling endlessly in an area where the most significant criminal act is likely to be jaywalking. Even in high-crime areas, if patrol officers are limited to merely taking reports from complainants and turning cases over to detectives for investigation, much of the patrol officers' time is wasted on trivial tasks.

Throughout a police agency, for both commissioned officers and civilian employees, much can be done to make everyday work more interesting and satisfying. Supervisors and their subordinates must approach this aspect of career development together, in partnership, to identify areas where employees find their work unsatisfying and to make whatever changes are necessary and feasible.

The Culver City (California) Police Department has adopted an unusual method of encouraging job enrichment: the periodic rotation of all personnel assignments. The department's four captains, who head the agency's four major bureaus, exchange jobs every eighteen months. Lieutenants and sergeants also rotate from one assignment to another every eighteen months. Patrol officers rotate at varying intervals, ranging from every twelve months to every thirty-six months, depending on their current assignment. A patrol officer also may request a transfer at any time to an open position.

The agency reports that the rotation system increases both morale and efficiency by exposing every officer to a variety of challenging assignments, and prevents personnel from stagnating in one position too long.[45]

A sound, functional career development system does not come into existence merely because someone wants it to. A great deal of effort must be spent in planning and installing the system. Once the system is in operation, energy must be devoted to making it work, day in and day out, from one year to the next.

Training opportunities must exist, either within the agency or through cooperative arrangements with other organizations, not only for new recruits but also for experienced personnel. Performance evaluations must be as thorough, accurate, and fair as possible. Career counseling must be oriented toward satisfying the needs, both immediate and future, of both the agency and its employees. If these criteria are met, the career development system can be one of the most valuable of all the administrative tools for making effective use of human resources.

REVIEW

1. Which of the following items should be included in basic recruit training?
   (a) Content of the state criminal code.
   (b) Criminal and civil court procedures.
   (c) Duties and authority of the various units of the police agency.
   (d) Personnel matters such as salaries and benefits.
   (e) Driving skills.
   (f) All of the above.
   (g) None of the above.

2. True or False: It is not practical for small agencies to provide training opportunities for their personnel.

3. The final exam in the recruit's socialization process usually occurs during
   (a) Academy training.
   (b) The recruit selection process.
   (c) Inservice training.
   (d) The probationary period.
   (e) Role-playing exercises.

4. The members of a bureaucratic organization are supposed to be assigned to their positions according to
   (a) Merit.
   (b) Seniority.
   (c) Probation.
   (d) Performance evaluation.
   (e) Natural ability.

5. The purpose of a career development system is to
   (a) Ensure that the agency's personnel needs are met, both immediately and in the future.
   (b) Assist individual employees in taking advantage of the career opportunities that are open to them.
   (c) Avoid the need for costly, time-consuming training programs.
   (d) All of the above.
   (e) None of the above.

NOTES

[1]Jack L. Kuykendall and Peter C. Unsinger, *Community Police Administration*. Chicago: Nelson-Hall, 1975, p. 259.

[2]Michael Brave, "Liability in Law Enforcement," in *American City and County*, vol. 107, no. 6, May 1992, p. 10; Gary M. Post, "Police Recruits: Training Tomorrow's Workforce," in *FBI Law Enforcement Bulletin*, vol. 61, no. 3, March 1992, p. 19.

[3]Kuykendall and Unsinger, *Community Police Administration*, p. 261.

[4]Thomas Shaw, "The Evolution of Police Recruit Training," in *FBI Law Enforcement Bulletin*, vol. 61, no. 1, January 1992, p. 3.

[5]Geoffrey P. Alpert and Roger C. Dunham, *Policing Urban America*, 2nd ed. Prospect Heights, Ill.: Waveland Press, 1992, p. 53.

[6]William J. Bopp, *Police Personnel Administration*. Boston: Holbrook Press, 1974, p. 183.

[7]Ibid., p. 181.

[8]Robert J. Meadows, "Beliefs of Law Enforcement Administrators and Criminal Justice Educators Toward the Needed Skill Competencies in Entry-level Police Training Curricula," in *Journal of Police Science and Administration*, vol. 15, no. 1, March 1987, pp. 4–6.

[9]Larry D. Armstrong and Clinton O. Longenecker, "Police Management Training," in *FBI Law Enforcement Bulletin*, vol. 61, no. 1, January 1992, pp. 22–26.

[10]Bopp, *Police Personnel Administration*, pp. 181–83.

[11]Ginny Field, "The FBI Academy," in *FBI Law Enforcement Bulletin*, vol. 61, no. 1, January 1992, p. 20.

[12]John C. Klotter and Joseph Rosenfeld, *Criminal Justice Instructional Techniques*. Springfield, Ill.: Charles C Thomas, 1979.

[13]Joseph M. Fabricatore, "Preentry Assessment and Training," in Charles D. Spielberger, ed., *Police Selection and Evaluation*. Washington, D.C.: Hemisphere, 1979, p. 82.

[14]Post, "Police Recruits," pp. 19–24.

[15]Shaw, "Police Recruit Training," pp. 2–6.

[16]Michael S. McCampbell, "Field Training for Police Officers: State of the Art," in Roger C. Dunham and Geoffrey P. Alpert, eds., *Critical Issues in Policing*. Prospect Heights, Ill.: Waveland Press, 1989, pp. 111–20.

[17]Philip Bonifacio, *The Psychological Effects of Police Work*. New York: Plenum Press, 1991, pp. 30–32.

[18]Bopp, *Police Personnel Administration*, pp. 192–93.

[19]John J. Broderick, *Police in a Time of Change*. Prospect Heights, Ill.: Waveland Press, 1987, pp. 34–35, 139–40.

[20]Bonifacio, *Police Work*, pp. 37–42.

[21]Lawrence Sherman, "Learning Police Ethics," in M. C. Braswell, B. R. McCarthy, and B. J. McCarthy, *Justice, Crime and Ethics*. Cincinnati: Anderson Press, 1991, pp. 97–113.

[22]Law Enforcement Assistance Administration, *Career Development for Law Enforcement*. Washington, D.C.: U.S. Department of Justice, 1973, pp. 3–4.

[23]Ron A. DiBattista, "Creating New Approaches to Recognize and Deter Sabotage," in *Public Personnel Management*, vol. 20, no. 3, February 1991, pp. 347–52.

[24]Susan Ehrlich Martin, *Breaking and Entering: Policewomen on Patrol*. Berkeley: University of California Press, 1980, p. 13.

[25]Edwin B. Flippo and Gary M. Munsinger, *Management*, 5th ed. Boston: Allyn and Bacon, 1982, pp. 268–72.

[26]N. F. Iannone, *Supervision of Police Personnel*, 4th ed. Englewood Cliffs, N.J.: Prentice Hall, 1987, pp. 235–39.

[27]James L. Farr and Frank J. Landy, "The Development and Use of Supervisory and Peer Scales for Police Performance Appraisal," in Spielberger, *Police Selection and Evaluation*, pp. 61–62.

[28]Stephen D. Mastrofski and Robert C. Wadman, "Personnel and Agency Performance

Measurement," in William A. Geller, ed., *Local Government Police Management*, 3rd ed. Washington, D.C.: International City Management Association, 1991, pp. 368–70.

[29]For example, see Charles B. Saunders, Jr., *Upgrading the American Police*. Washington, D.C.: The Brookings Institution, 1970, pp. 19–21, for a list of traits of the "ideal police officer."

[30]Michael K. Brown, *Working the Street*. New York: Russell Sage Foundation, 1988, pp. 115–17.

[31]Ibid., pp. 119–21.

[32]Iannone, *Police Personnel*, pp. 239–42.

[33]Mastrofski and Wadman, "Personnel and Agency Performance Measurement," in Geller, *Local Government Police Management*, pp. 372–73.

[34]George T. Felkenes, "Police Performance Appraisal," in Calvin J. Swank and James A. Conser, eds., *The Police Personnel System*. New York: Wiley, 1983, pp. 252–74.

[35]Ernest C. Froemel, "Objective and Subjective Measures of Police Officer Performance," in Spielberger, *Police Selection and Evaluation*, p. 87.

[36]George W. Greisinger, Jeffrey S. Slovak, and Joseph J. Molkup, *Civil Service Systems: Their Impact on Police Administration*. Washington, D.C.: U.S. Department of Justice (Law Enforcement Assistance Administration), 1979.

[37]Dennis E. Nowicki, Gary W. Sykes, and Terry Eisenberg, "Human Resource Management," in Geller, *Local Government Police Management*, p. 285.

[38]James M. Poland, "The Written Examination," in Swank and Conser, *The Police Personnel System*, pp. 77–90; Paul M. Whisenand and George E. Rush, *Supervising Police Personnel: Back to the Basics*. Englewood Cliffs, N.J.: Prentice Hall, 1988, pp. 192–94.

[39]William A. Caitlin, "Promotion and Career Development," in Swank and Conser, *The Police Personnel System*, p. 279.

[40]Nowicki, Sykes, and Eisenberg, "Human Resource Management," in Geller, *Local Government Police Management*, p. 284.

[41]Caitlin, "Promotion and Career Development," in Swank and Conser, *The Police Personnel System*, pp. 279–80.

[42]LEAA, *Career Development*.

[43]Whisenand and Rush, *Supervising Police Personnel*, pp. 194–96; Mastrofski and Wadman, "Personnel and Agency Performance Measurement," in Geller, *Local Government Police Management*, pp. 365–68.

[44]Caitlin, "Promotion and Career Development," in Swank and Conser, *The Police Personnel System*, pp. 285–87.

[45]Tom Gabor, "Rotation: Is It Organizationally Sound?" in *FBI Law Enforcement Bulletin*, vol. 61, no. 4, April 1992, pp. 16–19.

LEADERSHIP AND DISCIPLINE

LEADERSHIP

One of the most crucial characteristics of any police agency is the quality of its leadership. Throughout the course of the police professionalization movement, great emphasis has been placed on improving the quality of leadership exercised by administrators and supervisors, from the highest executive level all the way down to the first-level supervisors (usually sergeants).

This concern for the quality of police leadership is a direct result of the ambiguous nature of law enforcement tasks. If the duties of police officers were simple, clear-cut, and routine, it would be relatively easy to instruct officers in their duties and check to see that those duties were carried out properly. But, as we have said, the duties of police officers are complex and diverse and often require a good deal of individual judgment and discretion.[1]

Furthermore, most police officers perform their duties independently, without immediate supervision. These conditions place a considerable burden on the supervisors to exercise a reasonable degree of control, to promote conformity to the agency's rules and policies, and to obtain reliable information about their employees' performance, by relying on techniques of leadership rather than the more straightforward techniques of management.[2]

What exactly is *leadership* and how does it differ from management? Sociologist Keith Davis defines leadership as "the ability to persuade others to seek defined objectives enthusiastically."[3] The key word here is "persuade."

A manager does not need to persuade employees to do their work properly. The manager's authority includes the right to require employees to perform or, if they fail to do so, to face various consequences. As long as the manager is able to determine whether in fact the employees are performing as expected, persuasion is not needed; authority alone is sufficient.

However, if it is not possible for the manager to exercise such direct supervision, the only alternative is to persuade the employees that it is in their best interests to perform well.[4] And that is exactly the position in which police managers find themselves.[5]

Davis's definition includes two other key concepts: "defined objectives" and "enthusiastically." As we will see, one of the most important duties of a leader is to ensure that employees' objectives are well defined, so that they will know what they are supposed to do.

Whether employees perform their jobs enthusiastically or not may seem unimportant at first. As long as the job gets done, what does it matter whether the employees are enthusiastic? But on further reflection, enthusiasm indicates the extent to which employees are motivated to perform their jobs to the best of their ability. Where direct supervision is possible, usually it is enough for employees to perform up to a given, often arbitrary standard. But where a job requires initiative and individual judgment, as police work does, the employees' enthusiasm and desire to perform *better* than the minimum standard assumes greater importance.

The importance of leadership is made dramatically clear when the consequences of poor leadership are considered. As Richard J. Lundman points out, "misconduct by police patrol officers is a complex phenomenon involving . . . at least two elements: illegitimate opportunities and informal departmental operating rules that encourage or tolerate officers who take advantage of these opportunities."[6] Indeed, wherever corruption, abuse of authority, and other police misconduct exist, the problem can be traced to insensitive, inconsistent, and incompetent leadership.[7]

Unfortunately, merely recognizing the importance of leadership does not guarantee that superior leaders will occupy positions of authority in a police agency, or, for that matter, in any other organization. Leadership is an intangible quality attached to individuals. For more than a century, social psychologists and management experts have been trying to define exactly what makes for superior leadership and how superior leaders might be identified.

Functions of Leadership

What exactly do leaders do to exercise leadership? As we have already said, leadership involves several key elements: persuasion, defining of objectives, and promoting enthusiasm.

Each of these elements may require a wide range of specific activities. In Chapter 2, we discussed some of the attempts by social scientists and management theorists to determine how and why some managers are successful leaders while others are not. If any generalization can be made, it is that particular methods or techniques are likely to be successful only for certain leaders, with certain groups of subordinates, under certain conditions.[8] No one technique or group of techniques seems to work consistently for all leaders all of the time.

According to Roy R. Roberg, the subordinates' expectations largely determine what leadership behavior will be successful.[9] If employees expect their leader to make forceful decisions, issue orders, and demand strict compliance with arbitrary rules or standards, then a leader who behaves in that fashion will be successful. On the other hand, if employees expect a leader to allow them to participate in making decisions and to give them considerable latitude in carrying out their tasks, then a more authoritarian leader is likely to encounter resentment and resistance.[10]

One of the most important functions of a leader, according to the theory of Terence R. Mitchell and James R. Larson, Jr., is to clarify the organization's objectives and the path by which those objectives may be reached.[11] Mitchell and Larson, like most contemporary social psychologists, assume that employees generally want to perform their jobs well because that is most likely to earn them significant rewards (both tangible, such as higher pay, and intangible, such as personal satisfaction). However, employees are often frustrated when the organization's objectives are unclear, or when employees are uncertain about what they are expected to do to reach those goals. Thus it is up to the leaders to make sure that all employees understand the objectives and how to achieve them.

We can see this principle in police agencies. Almost any law enforcement agency will subscribe to some global goals, such as enforcing the law or preventing crime or treating all citizens equally and fairly. But these goals remain little more than noble sentiments until they are translated into specific policies, operating procedures, and programs.

The clear fact is that no police agency enforces all laws equally. Some laws are essentially unenforceable, either because the agency lacks sufficient resources, or because there is no public support for enforcement. Others are enforced only sporadically (for example, traffic laws), and still others are enforced only when there is a public demand for enforcement (for example, some vice laws).

If the chief of a police department maintains publicly that the department always enforces the law, although every patrol officer knows that some laws are generally unenforced but is never sure which laws those are, the agency's objectives are ambiguous and confused.

Similarly, no one believes that the police can ever prevent all crimes from occurring. For a police administrator to claim piously that preventing crime is a major objective of the department, without specifying what kinds of crime are to be prevented and by what methods, merely leads to cynicism and disillusionment on the part of lower-ranking officers.

The complexity of law enforcement creates a great need for leaders to clarify goals and objectives and to be sure that their employees know what is expected of them.

At the same time, employees need to know what rewards they will receive when they attain the agency's goals. Again, these rewards are not necessarily tangible. The approval of fellow officers, praise from superiors, and even the self-satisfying sense of a "job well done" are powerful rewards. By way of contrast, nothing is more discouraging than for an officer to feel that he or she has done a good job or reached some important goal and then be criticized for some trivial, incidental fault.

Leadership Styles

In 1939, three social psychologists conducted a study in which they attempted to learn what kinds of leadership behavior were most effective. The study was designed to test three contrasting leadership styles, which the researchers called autocratic, democratic, and laissez-faire (a French term meaning, approximately, "whatever they like").

Individuals trained to exercise leadership in one of these three styles were assigned at random to carefully matched groups of students, who were then given a relatively straightforward set of tasks to complete. Afterward, the groups' efforts and behavior were compared. The researchers found that, in trial after trial, the group with a democratic leader had the best performance, and that group members expressed the greatest degree of satisfaction with their efforts. The groups with an autocratic leader (one who made all decisions for the group and assigned specific tasks arbitrarily) sometimes performed well and sometimes poorly; group members, no matter how well they had performed, generally expressed resentment and hostility toward their leader and toward one another. The groups with a laissez-faire leader generally performed poorly and the members expressed a great deal of frustration and confusion.[12]

This kind of study has been criticized on the grounds that it depends on having groups that are carefully selected and matched to be representative of the general population. Real-life groups are not always average, and the characteristics of a particular group have a bearing on a leader's success.[13]

For example, as we pointed out earlier, a group that expects its leader to be more or less dictatorial might perform best under an autocratic leader, and might be very frustrated under a democratic leader who expected group members to contribute to decision making and to display initiative in carrying out their tasks.[14]

Other theorists have questioned whether there are only three important styles of leadership.[15] Mitchell and Larson, for example, have identified four leadership styles and have defined them somewhat differently from the three-leaders categories.

1. *Directive.* This style, similar to the conventional autocratic category, is characterized by a leader who makes most of a group's decisions, assigns specific tasks to group members, and enforces conformity to rules and performance standards.

2. *Supportive.* This style, approximately comparable to the laissez-faire style, is characterized by a leader who is concerned solely with the welfare and feelings of group members, who allows group members to make all decisions about policies and performance standards, and who generally exercises no active leadership.

3. *Participative.* This style, similar to the traditional democratic style, is characterized by a leader whose chief concern is to involve group members in decision making.

4. *Achievement oriented.* This style is characterized by a leader who is concerned solely with achieving the organization's goals; who therefore sets and enforces performance standards and generally enforces the organization's rules and policies, but otherwise leaves group members free to perform tasks in any manner they like.[16]

No one of the four leadership styles is necessarily superior to the others. An effective leader may be successful with any of these styles, depending on the characteristics of the group and of the situation in which they must perform. Thus a leader, to be consistently successful with all groups and in all situations, must be capable of adopting whatever style is appropriate to the circumstances.[17]

One of the most important findings of research on leadership is that the degree of satisfaction felt by group members is an important factor in determining the group's performance.[18] If group members have confidence in their leader, accept the leader's style as appropriate to their needs, believe that the organization's objectives are appropriate and attainable, and believe that they will be properly rewarded for their efforts, they are likely to be successful even in the face of many obstacles.

Conversely, if group members lack confidence in their leader, feel that the leader's style is inappropriate, feel that the goals set for them are inappropriate or unattainable, and feel that there is no consistent relationship between their efforts and the rewards available to them, performance is almost certain to be poor.

Thus one of the most important functions of the leader, no matter what his or her style might be, is to monitor the group's attitudes and, insofar as possible, correct any of the conditions that are likely to undermine the group's performance.[19] A leader must be especially sensitive and responsive to any grievances expressed by subordinates.

Leadership and Employee Grievances

A *grievance* is simply an expression of dissatisfaction. Since it has long been recognized that dissatisfied employees do not perform well, most modern organizations have attempted to develop procedures to discover and correct the sources of employee grievances wherever possible.[20]

There are two main types of employee grievances: those that are general among employees and those that pertain to a particular individual employee. General grievances are most often the result of working conditions that apply more or less equally to a group of employees, although it is not necessarily the case that all employees are affected or feel dissatisfaction. Individual grievances are more often the result of the personal interactions between the dissatisfied employee and his or her immediate supervisor.

Individual grievances should be, and often are, the easiest to correct. An employee may feel that he or she is being treated unfairly for any number of reasons. Quite often, the supervisor is not even aware that these feelings exist. Unfortunately, employees, especially those in the lowest ranks, often feel powerless to do anything about their dissatisfaction. They are afraid to complain to the supervisor, who might retaliate in some way.

These fears are not always unjustified; if the supervisor is discriminating against the employee, consciously or otherwise, the employee's complaints could make matters worse rather than better. Unless the employee has a reasonable opportunity to take the grievance to a higher authority, the situation is likely to become progressively worse and may affect not only the dissatisfied employee, but other employees who are aware of the unresolved grievance.[21]

Individual grievances can be a particularly serious problem in a police agency, where employees perform much of their work independently and the organization's success depends so heavily on their individual initiative. A dissatisfied patrol officer or detective is likely to do as little work as possible and to feel little loyalty toward the organization that (in the officer's mind) is treating him or her shabbily.[22] Police supervisors, particularly at the first-line level, must be acutely sensitive to this kind of problem and must be ready at all times to deal with it in a constructive and positive manner.

The first step is to adopt clear policies regarding both the formal and informal resolution of individual grievances.[23] Any employee who feels that he or she is being treated unfairly by a supervisor should have an unlimited right to complain to the supervisor's superior, without any fear of retaliation.

When presented with a complaint, the superior must treat the subordinate with courteous attention. If possible, the problem should be resolved informally. The superior officer should investigate the circumstances of the complaint and determine whether, in fact, the supervisor has acted improperly toward the subordinate. If so, usually this is a violation of the agency's policies, and it would be up to the superior to advise the supervisor that such behavior is unacceptable. The superior also must see that the improper behavior stops and that there is no retaliation against the complaining subordinate.

Sometimes the superior's investigation will show that there has been no intentional discrimination, no violation of agency policy, or that the problem falls into the gray area of personal misunderstandings or personality conflicts. In that case, the usual remedy is to bring the supervisor and the subordinate together for a frank discussion of the problem.

More often than not, the offending supervisor is unaware that the problem exists or attributes it entirely to the fault of the subordinate. By bringing the conflict out into the open, both parties may be able to see the other's point of view and adjust their behavior accordingly.

Again, the superior should not let the matter drop or insist that the two parties "settle this among yourselves." The superior should follow up after a reasonable period of time to see that the problem has been resolved. If it has not, further action might be required, such as specific instructions to the supervisor to correct his or her offensive behavior or a transfer of the subordinate to some other unit of the agency.

These informal procedures should be sufficient to resolve most individual grievances.[24] In those few cases that cannot be resolved by the supervisor's immediate superior, it may be necessary for the dissatisfied employee to invoke a formal grievance procedure. Usually, formal procedures are needed only when a grievance is generally shared among a large group of employees throughout the agency. However, some contracts between the police officers' union or bargaining unit and the parent government specifically require formal grievance procedures in individual cases.[25]

General grievances are most often directed at a particular policy or rule of the agency, or at some specific condition or practice. For example, women patrol officers may object to a policy that requires them to be assigned only to two-person patrol units, since this policy implies that they are less competent than male officers (and, since in many agencies there are comparatively few two-person patrol units, also limits their job opportunities).[26] Or a group of minority officers might claim that they are being discriminated against by supervisors who assign them only to the most disagreeable and undesirable duties.

Whatever the cause or nature of a general grievance might be, usually it does not involve the relationship between group members and a particular supervisor, and usually no one manager can resolve the problem. Instead, it must be addressed as a matter of policy throughout the agency or the subunit that is directly involved. For that reason, a formal grievance procedure is necessary.[27]

Most large police departments have established formal procedures for both individual and general grievances. In agencies that have a negotiated contract or collective-bargaining agreement with an employee union or unionlike organization, grievance procedures usually are spelled out in considerable detail.[28]

However, many small and medium-sized agencies have no formal procedures, and often the executives of smaller agencies fail to see a need for them. The need exists even in the smallest agency because employees who find no appropriate outlet for their grievances invariably become dissatisfied, resentful, and frustrated.

A formal procedure does not have to be so formalized that it becomes a bureaucratic ritual. We consider the important elements of a formal grievance policy to be (1) the employees must present the grievance as a group; (2) the grievance must be stated in writing; (3) all interested parties (including those who currently benefit from the rule or policy in question) must have an opportunity to respond to the complaint; and (4) there must be a clear, definite resolution.

Even individual grievances should require a group presentation if they cannot be resolved informally. For example, suppose a patrol officer feels that she is being discriminated against by her sergeant. First she should be required to attempt to resolve the problem by appealing to her lieutenant. If the lieutenant's informal efforts are unsuccessful or unsatisfactory, the patrol officer should then be allowed to draft a complaint to higher authorities, but should *not* be permitted to submit the complaint alone.

If the agency has a union or unionlike organization, it should submit the complaint on the patrol officer's behalf. If there is no appropriate employee organization, the officer should be required to have her complaint cosigned by several of her colleagues who are aware of the problem and who support the officer's position. The minimum number of cosigners should be established by agency policy; it should be low enough to ensure that employees have a fair opportunity to bring a complaint, but high enough to discourage trivial and frivolous complaints. Naturally, there should be a clear, enforceable policy to prevent any retaliation against the cosigners.

The written grievance statement or complaint should include the specific nature of the problem (whether it is an agency policy or an individual supervisor-subordinate conflict) and what action the complainants would like taken. Any evidence in support of the complainants' position should be included.

Once the written statement has been submitted, the grievance procedure should concentrate entirely on the problem stated in the complaint, and neither the complainants nor other interested parties should be permitted to bring up unrelated matters.

Formal grievance procedures often involve a hearing before a board composed of agency officials. In larger agencies, the hearing board might be the Civil Service Commission or some other permanently established entity. In smaller agencies where formal grievances are infrequent, it is usually necessary to establish a special hearing board. The board should be composed primarily of high-ranking officials, preferably those who have no immediate interest in the policy at question. The agency's chief executive usually should not be a member of the hearing board, since he or she may be called on to review the board's decision.

Within reason, the board should include officers from all of the agency's organizational elements. For example, it might include the chiefs of each of the three major divisions or bureaus. It is also desirable for the board to include at least one member whose rank is the same as that of the complainants.

The hearing procedure does not have to be conducted with all the formality of a court trial, but it should be conducted in a dignified manner. The complainants should have ample opportunity to make their case including the presentation of any documents, witnesses, or other evidence. Any other interested parties such as officers (of whatever rank) who are opposed to the complainants should then be given an equal opportunity to present their point of view. The entire proceeding should be recorded, at least on audiotape if a stenographic record is impractical.

Usually it is not a good idea for a hearing board to be given the authority to make a final decision. After the hearing, the board should consider the evidence that has been presented and, if necessary, seek additional evidence. Then the board should prepare a recommendation to the agency's chief executive. The action, or *relief*, requested by the complainants should not be the only option; the board may find an alternative that is more appropriate. The board's recommendation should be made in writing, with a statement summarizing the evidence that has been presented to them, and state what actions the board feels are necessary and appropriate to resolve the grievance.

The board's recommendation must be acted on within a reasonable period of time by the agency's chief executive, and the action should be publicly announced. If the chief executive decides not to follow the hearing board's recommendations, the reasons for that decision should be clearly stated. The complainants should be given a reasonable opportunity to submit any additional evidence or arguments that may have been overlooked earlier but not simply to repeat the same arguments indefinitely.

The procedures we have suggested do not guarantee that every grievance will be resolved to everyone's complete satisfaction. Some problems or conflicts simply cannot be resolved, given the limited resources available to police agencies and the complex, often conflicting demands made on them. At least, employees should feel that the procedure has given them a fair chance to air their problems and that, even though they have not gotten exactly what they wanted, the final decision is just and reasonable.

Employee grievances are not the only source of problems that test a leader's abilities. In every formal organization there are various rules, policies, and procedures that all members or employees are expected to observe. Rules in an organization, like laws in a society, are worthless unless they are enforced, unless all members are required to obey them and violators are faced with appropriate consequences.

In a police organization, many of the rules are especially important since they are necessary to maintain the agency's integrity and the confidence of the public and, in some cases, to protect the lives of citizens and of police officers themselves. As in society as a whole, enforcement of the rules must depend to a great degree on the voluntary compliance of all members. The extent of voluntary compliance, in turn, is a product of the degree, kind, and quality of discipline that exists throughout the agency's membership.

DISCIPLINE

Too often, discipline is considered purely in terms of conformity with a strict set of rules and standards and the punishments that may result if the rules and standards are violated. This negative concept of discipline, taken to an extreme, reduces the members of an organization to the level of unthinking, fear-ridden sheep.

A more positive concept of discipline is not only possible, but highly preferable.[29] The very word *discipline* is derived from the same Latin word as *disciple* (follower, or, more specifically, pupil). In its original sense, discipline does not mean blind obedience; it means following the leadership of someone in authority, someone who occupies the role of a teacher.[30]

Discipline is necessary in any organization because it is impractical and unrealistic to expect every member to have enough knowledge and understanding to recognize, and act on, the organization's overall interests. Some members occupy positions of authority and leadership precisely because (in principle, if not in fact) they are most capable of making decisions, setting policies, and assigning tasks that will guide all members to achieve the organization's ultimate goals. Subordinate members are therefore obliged to follow a leader's directions not merely because the leader occupies a position of authority, but rather because the subordinate should be loyal to the organization and its goals.

Most of the rules and policies of an organization concern routine, day-to-day events. The rules themselves are often rather arbitrary. For example, there is no inherent reason that it is better for automobiles to be driven on the right half of the road rather than on the left; the choice is arbitrary, and in fact there are many places (notably Great Britain and Japan) where automobiles are driven on the left. However, once the choice has been made and a rule established, everyone must obey it, or there would be chaos on the streets.

Similarly, there are a great many ways that patrol officers could be assigned to their beats. However, in every police department one particular method of assignment is established and individual officers are not free to ignore the official routine and choose whatever beat they like (unless that is the official routine).

Other rules and policies are designed to deal with unusual events in which there may not be time or opportunity for each member to gather information, consider its effects on the organization as a whole, and decide independently on the proper course of action. Instead, the organization's leaders have tried to anticipate these events and have decided on the best way to deal with them. When such an event occurs, members of the organization are expected to respond according to the rules. If the leaders have correctly anticipated the proper course of action, the organization's interests will have been protected.

For example, making a traffic stop exposes a patrol officer to a variety of dangers, including the possibility that the motorist will produce a gun and shoot the officer. This is clearly not in the interests of the police department (much less those of the officer). Therefore, most police departments have adopted specific rules on how traffic stops are to be made.

These rules are designed to reduce the officer's vulnerability and give the officer time to recover if the situation suddenly becomes dangerous. An officer who carelessly violates the traffic-stop rules is not only increasing his or her risk unnecessarily, but, in a sense, behaving disloyally toward the department.

Finally, some rules and policies are established to protect an organization against the chance that some of its members might set out to harm the organization itself.

For example, a police department collects a great deal of information, in the course of its usual work, that could be of great value to criminals. A police officer might inform the operators of a drug-smuggling ring that they were about to be raided by investigators. No doubt the smugglers would be willing to pay well for this information, which would enable them to frustrate the police agency's legitimate work. To avoid this possibility, the department's officials establish policies concerning who may have access to sensitive information and strict rules that forbid stealing and selling the information to anyone outside of the agency.

These few examples merely illustrate the different kinds of rules and policies that exist in police agencies and, for that matter, in all sorts of organizations. The point is that *obedience to the rules is not a goal in itself*. The rules exist because they are necessary to the proper functioning of the organization. Obedience to the rules should be a product of the members' loyalty to the organization as a whole and their desire to further the organization's ultimate goals.[31]

Blind, unquestioning obedience is not always required. Sometimes rules produce the wrong results, or they are inappropriate to the particular circumstances of a given event. In such cases, the members' proper concern for the organization's highest goals may lead them to question and even disobey a particular rule.

When this happens, it is important for leaders to recognize the subordinates' true motives. In a police agency, where individual initiative is demanded, an employee who acts out of a genuine desire to serve the agency's interests ought to be encouraged, even if the employee's actions do not conform strictly to the agency's rules and even if, in a particular instance, the employee's actions turn out to be mistaken or fruitless.[32]

Violations of the agency's rules that are a result of thoughtlessness, carelessness, ignorance, or malice cannot be tolerated. Some rules involve interests of the agency that are so overriding in importance that violations cannot be condoned even if the employee's intentions were good.[33]

Thus, for a police administrator, maintaining proper discipline is much more than merely a matter of setting up a system of rules and policies and enforcing them rigidly. Police officers are expected and encouraged to show initiative and to exercise discretion in interpreting and applying the rules.

When it appears that a rule or policy has been violated, the administrator must consider the intentions and motives of the offending officer and must weigh the violation against the larger interests of the agency. A few rules may be so vital to the agency's interests that no violation is permissible, no matter what the officer's intentions. However, all members of the agency should know which rules are inviolable and under what circumstances they apply.

Finally, as much as possible, the enforcement of rules must be consistent. If some officers are permitted to violate the rules routinely, but others are punished for violating similar rules, respect and loyalty toward the agency and its leaders will be undermined.

Sources of Discipline Problems

Assuming that the personnel of a police agency have been selected for their intelligence, competence, and strong desire to serve the goals of law enforcement, most discipline problems are likely to result from poor management rather than the officers' malice.[34]

The three most common causes of poor discipline are these:

Excessive or Poorly Formulated Rules.

If there are so many rules, policies, procedures, guidelines, orders, and so on that no one can possibly be expected to know, understand, and remember all of them, the inevitable result will be confusion and frequent violations. Rules and policies should be kept to the absolute minimum necessary to maintain smooth operations and to protect the agency's interests.

Similarly, if rules and policies are so poorly stated that no one understands them, or there is uncertainty about when a particular rule applies, or conflicting rules exist for the same circumstances, proper adherence to the rules may be impossible. Administrators should examine and reconsider *all* agency rules, policies, and procedures periodically, should eliminate those that are not absolutely needed, and should revise those that may be confusing or contradictory.

Inadequate Training.

The English courts, centuries ago, established the principle that "ignorance of the law is no excuse." This principle is observed throughout the American system of justice, but it is balanced by another principle: that no one can be expected to comply with a law until *due notice* has been given. For example, a legislature must publish the laws it has adopted; no one can be accused of violating an unpublished, secret law.

The same principles must apply to agency rules and procedures. It is the responsibility of administrators to see that all personnel are aware of the rules and policies that have been adopted, and how the rules are to be applied. The more important the rule, the more thoroughly personnel must be trained in its content and application. It is not enough simply to provide new recruits with a "policy handbook" or a "manual of orders" and expect them to memorize it. Recruits should be trained to know all of the agency's rules, and all personnel should be retrained periodically, or whenever significant new rules are established.

Poor Supervision and Leadership.

Proper supervision involves assisting personnel in carrying out their tasks and checking to make sure that work is performed properly. If supervisors do not carry out their duties diligently and consistently, they have no opportunity to correct their subordinates' mistakes, bad habits, or misinterpretations of the agency's rules and policies.

As we said earlier, the fact that most police officers operate independently and more or less in isolation, rather than under the direct control of supervisors, places a considerable burden on the supervisors. They cannot observe everything their subordinates do. However, they can and must observe subordinates in the field as much as possible, review the subordinates' official reports of actions taken, and occasionally ask subordinates to report informally on how they are doing their work.

Furthermore, supervisors can exercise leadership by setting a good example of conformity with the agency's rules and standards and by expecting their subordinates to perform well. A careless, lazy, or incompetent supervisor inevitably will have careless, lazy, and unproductive subordinates.

The remedy for these common problems is fairly obvious. Excessive or poorly formulated rules can be revised or, if possible, eliminated. Personnel can be trained to know the rules, know why they exist, and know when to apply them. Poor leaders can be retrained to make them better leaders, or, if necessary, replaced.[35]

Serious Discipline Problems

There is a certain type of human personality that is sometimes attracted to law enforcement as a career. This kind of person tends to interpret every event and every relationship with another person, even the most fleeting ones, in terms of a possible threat to the person's self-image, values, and desires. Such a person tends to be highly aggressive, even combative in the face of the slightest provocation. The Spanish word *macho* is often used to describe this kind of person.[36]

Individuals with this personality type may be excellent police officers. They approach their work with a great deal of energy, dedication, and determination to succeed. Unfortunately, they sometimes carry their zeal to such an extreme degree that they become overly aggressive, careless about following sound police practices, and dangerous to themselves as well as to the public they are pledged to protect. Every encounter with a citizen becomes a confrontation, a test of wills in which the officer's authority and legal power may be used to force the citizen to submit. An arrest or search for evidence may be pursued with no regard for commonsense rules of safety and proper legal procedure.

The macho personality is only one example of what we might call the *insubordinate personality* type. More generally, insubordinate personalities are those that cause a person to be incapable of accepting and observing proper discipline. A person with this type of personality resists all rules and orders, often regarding them as attacks on his or her rights as an individual.

The causes of the insubordinate personality do not concern us here; that is more a subject for a psychology text. For the police administrator, the insubordinate personality can be a serious problem. A single officer with this type of personality can be a disruptive influence on other personnel and, through improper actions, may undermine the integrity of the whole agency. Several officers who are insubordinate, especially if they are concentrated in key areas such as a high-crime-area patrol district or an investigative unit, can wreak havoc on the agency's ability to perform its vital tasks and on the public confidence that every police agency must have.

Correcting this problem is not easy. Ideally, individuals with insubordinate personality types should be identified and screened out before they enter a police agency. Unfortunately, even with extensive and sophisticated psychological testing procedures, the screening process is never foolproof—especially because police agencies want to attract individuals who are dedicated, aggressive, and assertive, qualities that are displayed by people who might be a problem at some later time.[37]

A better opportunity to identify and screen out the insubordinate personality exists during the training process. Recruits who are unable to accept direction, who are careless about learning and following correct police procedures, or who are excessively aggressive and heedless of sensible safety precautions should not be allowed to enter the police service.

Even so, individuals with insubordinate personalities may get through the most rigorous screening process, or their personality characteristics may become more extreme after they earn a badge. At that point, there are three possible remedies, depending on the seriousness of the individual problem.

- The insubordinate officer can be retrained to accept proper discipline.
- The officer can be required to undergo psychological treatment (which may be very costly and time-consuming) to correct the personality defect.
- If all else fails, the officer can be released from employment. Usually this last step requires a history of violations of agency rules and policies, or a violation so serious and flagrant that dismissal is justified.

Individual officers who are persistently unable to conform to the agency's rules may be not only a problem in themselves, but they may encourage other personnel to disregard proper discipline. In a sense, the insubordinate personality is a kind of "infection" that can weaken or destroy an entire agency. The ultimate result can be systematic corruption of the agency.

Systematic Corruption

Nearly every metropolitan police force in the United States has experienced periodic police scandals in which police officers have been accused of corruption.

Invariably these scandals have followed a pattern of events. First there is an arrest of a single officer or a small group, often as a result of an internal investigation started by the agency's administrators or as a result of a citizen's complaint investigated by public officials outside of the agency (such as the district attorney or the city council). The investigation quickly widens. Other citizens complain that they have been the victims of corrupt police practices. The news media and politicians carry out their own investigations.

Sometimes a special investigative body is formed, made up of elected officials, citizens who have impeccable reputations, and perhaps one or two experts from outside of the community. More police officers, usually those at lower ranks and middle-management levels, are arrested, indicted, and drummed out of the agency. Top administrators are swept out (usually they are permitted to resign or take early retirement) and replaced with reform-minded newcomers.

The agency is quickly reorganized, during which more managers and administrators are either terminated or transferred to less sensitive duties. After a few months of frantic activity, the community's political leaders announce that all the bad apples have been eliminated and the agency has been purified.[38]

Unfortunately, history indicates that merely changing the administrative structure and replacing a few top officials will not prevent corruption from recurring. Systematic corruption—as opposed to the presence of a few insubordinate individuals—is a product of the attitudes and values held by the members of the entire police agency.

In 1972, one of the periodic police scandals we have described erupted in New York City. A special investigative body was formed, known as the *Knapp Commission* after its chairman, William Knapp. But unlike most such bodies in the past, the Knapp Commission was given enough authority and investigative resources to go beyond a superficial study of what had gone wrong in the New

York Police Department. Some of the commission's findings, which were published and widely circulated, make clear the sources of systematic corruption.

The commission found that "group loyalty, stubbornness, hostility and pride" among members of the police force create the circumstances in which corruption can flourish. Even though members of the force are aware that corruption exists, they refuse to do anything about it because of misplaced loyalty to one another, hostility toward the outside world (that is, the general public), and determination not to permit the agency's reputation to be tarnished.

Thus, even high-level administrators and officers who are not themselves corrupt ignore the existence of corruption around them. These circumstances are clearly ideal for those who wish to take advantage of them. New recruits are "invited" to take part by corrupt senior officers; those who decline the invitation find themselves distrusted and shut out of the police society. Those who complain to a superior are ignored or treated as "trouble makers."[39]

The findings of the Knapp Commission have been confirmed by several other studies of police corruption in both large and small departments in every part of the nation. Clearly, the *"bad-apple" theory*, that corruption is caused by a few individual officers, has been thoroughly discredited. In every police agency, Richard Lundman observes, "Police work is exceptionally rich in opportunities for corruption."[40]

If the police culture—the set of values and attitudes held by members of the police society—permits corrupt behavior to be ignored and to go unpunished, it will become increasingly widespread and nearly impossible to uproot. It is not enough for administrators to treat particular instances of corruption, such as an officer who is caught accepting a bribe, as mere examples of "bad apples." Administrators must promote a positive police culture, through their personal example and through vigorous enforcement of agency rules.

Ways to Deal with Discipline Problems

Effective administration of a police agency requires a strong and positive attitude toward discipline. Administrators, managers, and supervisors—everyone in the agency above the lowest ranks—must demonstrate by personal example their wholehearted commitment to the agency's goals and their willingness to conform to the agency's rules, policies, and procedures.

At the same time, the police administrator must recognize that perfect discipline is impossible. Some personnel will commit violations of the agency's rules for one or more of the reasons we have cited: ignorance, carelessness, misunderstanding or misinterpretation of the rules, or, in the more serious cases, malice or defiance.

Violations cannot be ignored or condoned without damaging the agency's morale and public reputation. However, excessively harsh disciplinary procedures can rob officers of the initiative and sense of personal responsibility they must have to be effective.

What is needed is a range of disciplinary procedures that can be used flexibly, to fit the circumstances of each type of violation, but with enough consistency to maintain a sense of fundamental fairness for all personnel.[41]

A distinction must be made among the various types of disciplinary violations, and among the different degrees of seriousness of violations. The types of violations might be classified in any number of ways, but generally a police agency has four broad classes of rules, policies, and procedures: those that concern purely internal matters of routine administration; those that concern protecting the integrity and security of the agency and its personnel; those that concern proper police conduct in the delivery of law enforcement services; and those that concern clearly illegal acts that might be committed by police officers. Within each of these broad classes, there are degrees of seriousness.

The following list presents examples of actions by police officers that might be violations of specific agency rules, policies, or procedures. The list is by no means exhaustive; in most agencies, it could be expanded considerably. However, it illustrates both the different types of violations that may occur and the different degrees of seriousness. Within each group of violations, examples are listed in their order of seriousness, from minor to major.

1. Violations of internal procedure
   (a) Reporting late for assigned duty
   (b) Poor grooming; improper wearing of uniform
   (c) Abuse of privileges such as misuse of sick leave or vacation time, trading assignments without permission, and so on
   (d) Delayed, incomplete, or inaccurate filing of an administrative report
   (e) Failure to report for duty without excuse
2. Violations of agency integrity or security
   (a) Improper disclosure of confidential police information
   (b) Failure to keep a prisoner in secure custody
   (c) Careless handling or abuse of police equipment, such as weapons
   (d) Misuse or theft of evidence, prisoners' property, or police equipment
3. Improper conduct in performance of police duties
   (a) Discourteous treatment of a citizen (whether victim, witness, suspect, person requesting assistance, or whatever)
   (b) Failure to obtain a complete report of an incident (crime, traffic accident, etc.)
   (c) Delayed, incomplete, or inaccurate report of an incident involving police service
   (d) Disregard for safe procedure in carrying out a police service
   (e) Sleeping, drinking, and so forth, while on duty
   (f) Failure to perform an arrest, interrogation, or search for evidence in a legally proper manner
   (g) Ignoring or condoning an apparent crime known to the officer
4. Illegal acts of police officers

(a) Acceptance of anything of value from a citizen without proper authorization
(b) Physical abuse of a citizen (whether or not suspected of a crime)
(c) Solicitation of a bribe
(d) Falsification of a report; solicitation of false testimony from a witness, victim, suspect, etc.
(e) Commission of any act that is generally a criminal violation (theft, burglary, etc.)

Again, this list is not intended to be complete, but merely to serve as a list of examples. Depending on local laws and regulations, several of these violations might be regarded as criminal. For instance, physically abusing a suspect may be a crime as well as a serious violation of agency policy.

The disciplinary process requires two steps: (1) violations must be detected, and (2) corrective action must be taken. The process also should operate in two different modes, informal and formal, with regard to both detection and correction.

For example, the informal detection of violations is one of the principal functions of supervisors. By observing their personnel in all phases of their work, supervisors should become aware of instances of improper conduct.

As we have already noted, the fact that much police work occurs out of the supervisors' immediate sight does not (or should not) prevent supervisors from observing their subordinates. Supervisors may use indirect means such as reviewing their officers' reports or questioning officers about their own behavior and that of their colleagues. In addition, supervisors should devote as much time as possible to accompanying individual officers in the field, primarily to observe them at work.

Informal correction should occur as soon as possible after a violation has been detected.[42] Supervisors have a number of methods of informal correction at their disposal, ranging from the simple reprimand or "chewing out" to such techniques as denying privileges (such as refusing to approve an officer's request for vacation time), assigning the violator to a less desirable duty (perhaps transferring a motor officer to a foot patrol beat), and so on.

Supervisors should not forget that retraining may be the most appropriate response to minor violations, if they are the result of ignorance or carelessness. The retraining may consist of nothing more than the supervisor's explaining the proper procedures or assigning the violator to an appropriate in-service training class.

More formal methods of both detection and correction are required for serious violations.

Every police agency, even the smallest, should have at least one person specifically designated to conduct internal investigations. In very small agencies (those with fewer than 10 commissioned officers), this person is almost always the chief executive. In small- to medium-sized agencies (up to about 50 commissioned officers), someone other than the chief should be designated as the agency's *inspector*. This person may have other duties as well; often the inspector is the head of a major division.

In the role of inspector, the administrator is responsible for investigating any complaints of improper police conduct that come to the agency's attention and reporting the findings of those investigations to the chief executive. In addition, the inspector should take positive steps such as reviewing official reports, questioning supervisors and individual officers, and observing officers in the field whenever possible to discover any improper conduct that might not be the subject of a complaint.

Larger agencies require a full-time, formal investigative unit under the command of an inspector who has no other duties, and with as many additional personnel as the size of the agency requires. The internal investigations unit should be attached directly to the chief executive's office. The head of the unit should hold rank equivalent to captain or higher, and all members of the unit should hold rank equivalent to lieutenant or above. However, it may be preferable to use such titles as inspector and chief inspector (unless these terms are used elsewhere in the agency) to emphasize the unit's independence from the agency's administrative hierarchy.[43]

Again, the internal investigations or inspections unit should not only investigate any complaints that are brought to the agency but also should take steps to discover any improper conduct that otherwise would not become known.

In all cases, the results of each investigation carried out internally should be reported first to the chief executive and then, with the chief's approval, to the officers involved and their immediate superiors. In large agencies where many investigations might be conducted, reports of routine matters need not be made to the chief, but only to the officers involved and their superiors.

For minor violations, corrective action may be left to the discretion of the violator's supervisor. Supervisors should be able to apply relatively minor disciplinary measures on their own initiative. These measures might include placing a written reprimand in the officer's permanent personnel file, assigning the violator to extra duty without extra pay, suspending the officer without pay for not more than three days, or transferring the officer temporarily to a less desirable assignment (see Figure 13.1).

Supervisors also might be given the authority to apply somewhat more severe measures with the approval of their superiors. Examples might be suspension without pay for up to one week, permanent transfer to another unit, temporary reduction in rank for up to six months, or revocation of a probationary promotion. In most agencies, officers who have been promoted are considered to hold their new rank on a probationary basis for a period ranging from three months to one year.

The disciplinary measures we have just discussed are formal in the sense that they are applied only following an independent investigation and written report. Still more severe disciplinary measures require an even more formalized process to protect the rights of the individual officers as well as the interests of the agency.[44]

Whenever an investigation produces evidence that an officer has committed a serious violation of the law or of agency rules, the complete report should be submitted to the chief executive only. The accused violator and his or her

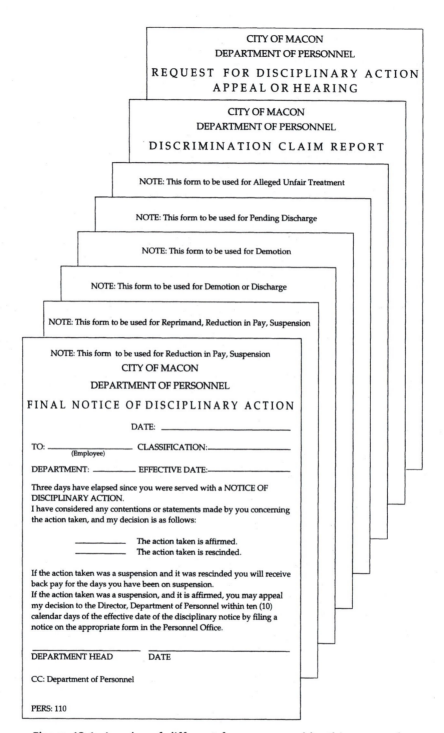

CITY OF MACON
DEPARTMENT OF PERSONNEL

REQUEST FOR DISCIPLINARY ACTION
APPEAL OR HEARING

CITY OF MACON
DEPARTMENT OF PERSONNEL

DISCRIMINATION CLAIM REPORT

NOTE: This form to be used for Alleged Unfair Treatment

NOTE: This form to be used for Pending Discharge

NOTE: This form to be used for Demotion

NOTE: This form to be used for Demotion or Discharge

NOTE: This form to be used for Reprimand, Reduction in Pay, Suspension

NOTE: This form to be used for Reduction in Pay, Suspension

CITY OF MACON

DEPARTMENT OF PERSONNEL

FINAL NOTICE OF DISCIPLINARY ACTION

DATE: _____

TO: _____ CLASSIFICATION: _____
       (Employee)

DEPARTMENT: _____ EFFECTIVE DATE: _____

Three days have elapsed since you were served with a NOTICE OF
DISCIPLINARY ACTION.
I have considered any contentions or statements made by you concerning
the action taken, and my decision is as follows:

_____ The action taken is affirmed.
_____ The action taken is rescinded.

If the action taken was a suspension and it was rescinded you will receive
back pay for the days you have been on suspension.
If the action taken was a suspension, and it is affirmed, you may appeal
my decision to the Director, Department of Personnel within ten (10)
calendar days of the effective date of the disciplinary notice by filing a
notice on the appropriate form in the Personnel Office.

_____   _____
DEPARTMENT HEAD   DATE

CC: Department of Personnel

PERS: 110

Figure 13.1. A series of different forms are used by this agency for
each step in the disciplinary process. (Courtesy Macon, Georgia, PD.)

superiors should be given notice of the nature of the accusation and of the possible consequences. The officer should be suspended with pay until a final determination is made (see Figure 13.2).

Some agencies use disciplinary hearings only to determine the punishment for a major violation, once guilt has been determined by the internal investigations unit. However, police officers are just as entitled to due process as any other citizen, and a fairer procedure is to submit the entire matter to a hearing.

A disciplinary hearing board should be composed of at least three high-ranking officers who have no direct interest in or knowledge of the subject of the hearing. In smaller agencies, this will be very difficult to achieve, and it may be desirable to include at least one member of the board who is not an official of the agency. For example, an assistant district attorney, or a ranking officer from another agency, or even a citizen who has a high reputation in the community might be appointed to the hearing board along with two ranking officials of the agency.

A disciplinary hearing is not a criminal trial, and there is no legal requirement that the procedures and rules of evidence for a criminal trial be observed. Nevertheless, the hearing must be conducted in a fair and impartial manner. The accusation and evidence should be presented by the head of the internal investigations unit, and the accused should be allowed to challenge the evidence, to question the inspectors who have brought the accusation, and to present witnesses or evidence on his or her own behalf. Some agencies permit an accused officer to be represented by an attorney; many agencies do not permit legal representation or place some restrictions on the attorney's role.

The purpose of a disciplinary hearing is (1) to determine whether the accused is guilty of a violation (except in those agencies that allow a hearing only to determine punishment); and (2) if the violation is proven, to assess punishment. In some agencies, especially smaller ones, the findings of a disciplinary hearing board are subject to the approval of the chief executive; where this is the case, however, the chief executive thereby assumes full legal responsibility for the consequences. The hearing board (with or without the final approval of the chief) should be able to assess punishments ranging from suspension without pay for up to six months (in some states, the law permits suspension for up to one year), to permanent reduction in rank or, in the most severe cases, dismissal from the force.[45]

If the disciplinary board finds the officer guilty of a criminal act, the board's finding should be forwarded to the district attorney for possible prosecution.

In some agencies, the process we have just described is subject to state law and Civil Service rules, which may call for more elaborately formal procedures and may place various restrictions on the composition of the hearing board and the possible outcomes. In other agencies, the contract between the police department (or its parent government) and the employees' union or unionlike association contains detailed provisions for the handling of disciplinary matters. However, in general the purpose of a formal disciplinary procedure is to ensure that all serious accusations of police misconduct are thoroughly investigated and that the final disposition is fair to the accused officer, to the agency, and to the public.[46]

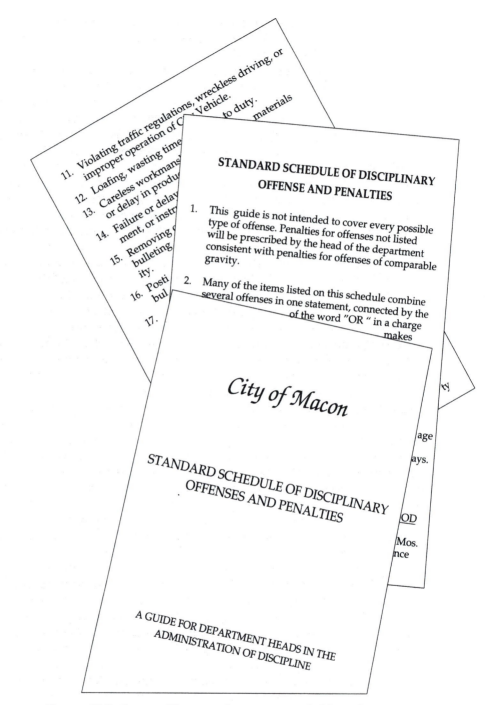

11. Violating traffic regulations, wreckless driving, or
    improper operation of City Vehicle.
    _____ to duty.
12. Loafing, wasting time _____ materials
13. Careless workmans _____
    or delay in produ _____
14. Failure or dela _____
    ment, or instr _____
15. Removing _____
    bulleting _____
    ity.
16. Posti _____
    bul _____
17. _____

**STANDARD SCHEDULE OF DISCIPLINARY
OFFENSE AND PENALTIES**

1. This guide is not intended to cover every possible
   type of offense. Penalties for offenses not listed
   will be prescribed by the head of the department
   consistent with penalties for offenses of comparable
   gravity.

2. Many of the items listed on this schedule combine
   several offenses in one statement, connected by the
   _____ of the word "OR" in a charge
   _____ makes

City of Macon

STANDARD SCHEDULE OF DISCIPLINARY
OFFENSES AND PENALTIES

A GUIDE FOR DEPARTMENT HEADS IN THE
ADMINISTRATION OF DISCIPLINE

Figure 13.2. Every officer receives a copy of this policy manual describing in detail the disciplinary procedures. (Courtesy Macon, Georgia, PD.)

Disciplinary procedures are a matter of great concern to all police agencies. Many police officials feel uncomfortable even considering or discussing these matters. There is a widespread belief that serious misconduct is rare (the bad-apple theory) and that any public awareness of these rare incidents will undermine the public's confidence in the whole agency.

The truth, however, may be just the opposite. The public's confidence is reinforced, not undermined, when the community knows that police misconduct will not be ignored or condoned, that police officers are expected to maintain the highest standards of integrity and honesty, and that the agency has a vigorous, fair, and consistent system of maintaining discipline.

REVIEW

1. One of the most important functions of any leader is to
    (a) Clarify the organization's objectives.
    (b) Make forceful decisions and issue clear orders.
    (c) Allow all subordinates a high degree of autonomy.
    (d) Reward employees for their performance.
    (e) Prevent subordinates from making mistakes.
2. The traditional three leadership styles are
    (a) Participatory.
    (b) Democratic.
    (c) Supportive.
    (d) Laissez-faire.
    (e) Autocratic.
    (f) Managerial.
3. True or False: All employee grievances should be handled through a formal hearing process to ensure that they are properly resolved.
4. The most common causes of discipline problems are
    (a) Poor rules, ignorance, and lack of leadership.
    (b) Poor rules, overaggressiveness, and unrealistic attitudes.
    (c) Carelessness, ignorance, and egotistical attitudes.
    (d) Lack of leadership, lack of training, and lack of adequate resources.
    (e) None of the above.
5. The inspections unit of a police agency should be responsible for
    (a) Ensuring that all personnel are properly equipped and wear proper uniforms.
    (b) Detecting and investigating all instances of improper police behavior.
    (c) Conducting hearings into accusations of improper police behavior.
    (d) Investigating only those complaints of improper police behavior that are submitted in writing by citizens.
    (e) Ensuring that all disciplinary problems are handled within the agency, without attracting unwarranted public notice.

NOTES

[1]Herman Goldstein, *Problem-Oriented Policing*. New York: McGraw-Hill, 1990, p. 16.

[2]Michael K. Brown, *Working the Street*. New York: Russell Sage Foundation, 1988, pp. 119–21.

[3]Keith Davis, *Human Behavior at Work*, 6th ed. New York: McGraw-Hill, 1981, p. 124; Geoffrey P. Alpert and Roger C. Dunham, *Policing Urban America*, 2nd ed. Prospect Heights, Ill.: Waveland Press, 1992, p. 81.

[4]Harold J. Leavitt and Homa Bahrami, *Managerial Psychology*, 5th ed. Chicago: University of Chicago Press, 1988, pp. 122–30.

[5]Roy R. Roberg, *Police Management and Organizational Behavior: A Contingency Approach*. St. Paul, Minn.: West Publishing, 1979, pp. 158–59.

[6]Richard J. Lundman, *Police Behavior: A Sociological Perspective*. New York: Oxford, 1980, p. 10.

[7]Alpert and Dunham, *Policing Urban America*, pp. 82–83.

[8]James G. Hunt and Lars L. Larson, eds., *Contingency Approaches to Leadership*. Carbondale, Ill.: Southern Illinois University Press, 1974. See especially pp. 189–98.

[9]Roberg, *Police Management*, p. 165.

[10]Steven A. Goren, *Productive Management*. Acton, Mass.: Brick House Publishing, 1990, p. 24.

[11]Terence R. Mitchell and James R. Larson Jr., *People in Organizations*, 3rd ed. New York: McGraw-Hill, 1987, pp. 433–60.

[12]Kurt Lewin, Ronald Lippitt, and Ralph White, "Patterns of Aggressive Behavior in Experimentally Created 'Social Climates,'" *Journal of Psychology*, 1939, pp. 271–99; White and Lippitt, *Autocracy and Democracy*. New York: Harper and Row, 1960; Davis, *Human Behavior*, pp. 135–37.

[13]Robert L. Katz, "Retrospective Commentary," in *Harvard Business Review*, September-October 1974, pp. 101–102, summarizes the reactions to the "three-leaders" studies.

[14]Fred E. Fiedler, *Leadership*. New York: General Learning Press, 1971.

[15]Frank J. Landy and Don A. Trumbo, *Psychology of Work Behavior*. Homewood, Ill.: Dorsey Press, 1976, pp. 365–406.

[16]Mitchell and Larson, *People in Organizations*, pp. 351–54.

[17]Arthur W. Sherman, Jr., George W. Bohlander, and Herbert J. Chruden, *Managing Human Resources*, 8th ed. Cincinnati: South-Western Publishing, 1988, pp. 355–58.

[18]Harry Levinson, *Ready, Fire, Aim*. Cambridge, Mass.: The Levinson Institute, 1986, pp. 92–94; also see pp. 3, 19–21.

[19]Ronald G. Lynch, *The Police Manager*, 3rd ed. New York: Random House, 1986, pp. 17–19.

[20]N. F. Iannone, *Supervision of Police Personnel*, 4th ed. Englewood Cliffs, N.J.: Prentice Hall, 1987, p. 75.

[21]Ron A. DiBattista, "Creating New Approaches to Recognize and Deter Sabotage," in *Public Personnel Management*, vol. 20, no. 3, Fall 1991, pp. 347–52.

[22]Sherman, Bohlander, and Chruden, *Managing Human Resources*, p. 306.

[23]Iannone, *Supervision of Police Personnel*, pp. 76–77.

[24]Michael J. Duane, "To Grieve or Not to Grieve: Why Reduce It to Writing?" in *Public Personnel Management*, vol. 20, no. 1, Spring 1991, pp. 83–90.

[25]Sherman, Bohlander, and Chruden, *Managing Human Resources*, p. 464.

[26]Nancy K. Grant, Carole G. Garrison, and Kenneth McCormick, "Perceived Utilization, Job Satisfaction, and Advancement of Police Women," in *Public Personnel Management*, vol. 19, no. 2, Summer 1990, pp. 147–61.

[27]Iannone, *Supervision of Police Personnel*, pp. 68–75.

[28]Dennis E. Nowicki, Gary W. Sykes, and Terry Eisenberg, "Human Resource Management," in William A. Geller, ed., *Local Government Police Management*, 3rd ed. Washington, D.C.: International City Management Association, 1991, pp. 303–04.

[29]Michael R. Carrell, Frank E. Kuzmits, and Norbert F. Elbert, *Personnel/Human Resource Management*, 4th ed. New York: MacMillan, 1992, p. 634.

[30]Vance McLaughlin and Robert Bing, "Selection, Training, and Discipline of Police Officers," in Dennis Jay Kenney, ed., *Police and Policing*. New York: Praeger, 1989, pp. 30–31.

[31]Carrell, Kuzmits, and Elbert, *Personnel/Human Resource Management*, pp. 632–44.

[32]Levinson, *Ready, Fire, Aim*, pp. 50–56; Goren, *Productive Management*, pp. 33–35.

[33]Sherman, Bohlander, and Chruden, *Managing Human Resources*, pp. 368–70.

[34]Warren G. Bennis and Burt Nanus, *Leaders: The Strategies for Taking Charge*. New York: Harper and Row, 1985, p. 21.

[35]Carrell, Kuzmits, and Elbert, *Personnel/Human Resource Management*, pp. 629–32.

[36]For a thorough discussion of the psychology of police officers, see Philip Bonifacio, *The Psychological Effects of Police Work*. New York: Plenum Press, 1991, especially pp. 140–60; Thomas Barker and David L. Carter, *Police Deviance*. Cincinnati: Pilgrimage Press, 1986, pp. 154–55; Mary Jeanette Hageman, *Police-Community Relations*. Beverly Hills, Calif.: Sage Publications, 1985, pp. 54–55.

[37]Robin Inwald and Dennis Jay Kenney, "Psychological Testing of Police Candidates," in Kenney, *Police and Policing*, pp. 34–41.

[38]Frank Schmalleger, *Criminal Justice Today*. Englewood Cliffs, N.J.: Prentice Hall, 1991, pp. 191–95. See also Barker and Carter's discussion of the Knapp Commission Report on corruption in the New York Police Department, *Police Deviance*, pp. 22–28.

[39]Commission to Investigate Allegations of Police Corruption in New York City (William Knapp, Chairman), *Report*. New York: George Braziller, 1972.

[40]Lundman, *Police Behavior*, p. 225.

[41]Sherman, Bohlander, and Chruden, *Managing Human Resources*, pp. 442–58.

[42]Carrell, Kuzmits, and Elbert, *Personnel/Human Resource Management*, p. 630.

[43]Alpert and Dunham, *Policing Urban America*, pp. 114–16.

[44]Edward A. Thibault, Lawrence M. Lynch, and R. Bruce McBride, *Proactive Police Management*. Englewood Cliffs, N.J.: Prentice Hall, 1985, pp. 194–95.

[45]Carrell, Kuzmits, and Elbert, *Personnel/Human Resource Management*, pp. 645–50, discuss the legal ramifications of dismissal.

[46]Barker and Carter, *Police Deviance*, pp. 16–19.

ADMINISTRATION OF PATROL OPERATIONS

We have devoted a large part of this text to a discussion of the organization, selection, training, and management of personnel because they are the crucial, indispensable element of a law enforcement agency. Efficient and comfortable facilities, suitable vehicles, and appropriate equipment are all important, but in the final analysis policing is almost entirely dependent on people: the officers and civilian employees of a police agency.

Personnel are also a costly resource. Salaries and personnel benefits generally make up 75 percent or more of a police department's budget. Administrators are responsible for seeing that these human resources are used efficiently, to ensure that as much useful work as possible is obtained for every dollar spent. This does not mean that the agency's personnel should be treated like plantation slaves, under constant pressure to work harder. It does mean that all employees should have a continual stream of work to do, at a pace that allows them to perform at their highest capacity, without either undue strain and pressure or excessive periods of wasted time.

Police administrators have a peculiar problem in trying to establish such a smooth, efficient work flow. By and large, a police agency has little control over the rate at which its most important work must be done. Reports of crimes, traffic accidents, and other police matters simply arrive at random and must be dealt with promptly.[1]

This is not true for all police work. Some of the tasks of a law enforcement agency, such as preventive patrol, can be scheduled in advance or performed as time is available from less predictable activities such as responding to calls for service. Nevertheless, the fact that a police agency's responsibility includes responding to emergencies that are essentially unpredictable creates a substantial burden on the administrator to arrange the agency's personnel so that at any time, someone will be available to deal with any crisis that occurs.[2]

In almost all American police agencies, the need to have personnel available to respond to emergencies is solved by dividing the agency's personnel into two main forces: the patrol force, whose members are expected to perform a broad range of tasks whenever they occur; and the various specialized units, whose members are expected to perform a more limited range of particular tasks at whatever rate is appropriate to the tasks themselves. The precise point at which tasks are divided between the generalists—that is, the patrol force—and the specialists may vary, with important consequences for the scheduling of work and the efficient use of the agency's human resources.

PRINCIPLES OF PATROL FORCE MANAGEMENT

Functions of the Patrol Force

Any number of specific functions and duties may be assigned to the patrol force, but this basic element of a police agency generally has four major functions.[3]

1. To answer complaints of crimes, investigate the complaint, and, if possible, identify and arrest the suspected offender.
2. To assist citizens, other police officers, and personnel of other emergency services (such as the fire department or emergency medical service) to protect lives and property during noncriminal crises.
3. To perform various services that may be only indirectly related to law enforcement, such as inspecting taxicabs, boarding houses, pawn shops, or other businesses and escorting funeral processions.
4. Observing the people and places in a specific territory for any suspicious activity or indications of a crime (preventive patrol).[4]

In addition to these basic functions, patrol officers also must perform various administrative tasks, such as making reports. A substantial part of a patrol officer's time may be taken up by duties such as transporting prisoners to the jail, presiding over the booking procedure, searching for missing persons, and appearing in court to give testimony.

Patrol officers also must be given a certain amount of time while on duty for meals and personal business (not to mention occasional rest breaks). Finally, patrol officers may be taken off the street to attend training sessions, to have their vehicles or other equipment repaired or maintained, or to perform miscellaneous errands for the agency such as providing chauffeur service for officials or visiting celebrities.

We refer to all the items in the preceding paragraph as *nonproductive tasks* because they take time away from the patrol force's principal functions. Clearly, the more time spent on nonproductive tasks, the less productive a patrol force will be. However, it is also clear that reducing the time spent on nonproductive tasks to zero is impossible. If nonproductive time can be kept to about 10 or 15 percent of the patrol force's total duty time, administrators may be well satisfied.

The four major functions listed previously are not necessarily equal in importance and certainly do not involve equal amounts of a patrol officer's time. Most police administrators (and probably most citizens) feel that the first two functions—responding to criminal complaints and assisting in noncriminal crises—are clearly more important than the latter two, and administrators generally seek to concentrate the patrol force's efforts on these two functions.

Administrators typically resist the kinds of tasks we have listed as the third major function (performing services that are only incidentally related to law enforcement).[5]

The significance of the fourth major function (preventive patrolling) has become a major point of controversy in law enforcement, largely because of experiments conducted in Kansas City in the early 1970s.

Preventive Patrolling

We have mentioned the Kansas City experiment earlier in this text, but this is an appropriate point at which to discuss it further. Briefly, the purpose of the experiment was to determine whether conventional patrol techniques are effective in reducing the incidence of crime.[6]

Several patrol districts in Kansas City were selected on the basis of having typical and similar characteristics such as population, geographic size, and historical crime rates. The districts were divided into three groups. In one group, the number of patrol units was increased and officers were instructed to devote as much time as possible to preventive patrolling (that is, cruising their beats and observing any indications of criminal activity).

In a second group of districts, officers were instructed to spend *no* time on preventive patrol but to concentrate entirely on responding as quickly as possible to all calls for service. In the third group of districts, no changes were made in the previously existing patrol techniques.

After several months, the number of criminal incidents reported, the number of cases cleared by arrest, the average time of police response to calls for service, and other statistics were compared for the three groups. Officials were surprised to discover that there were no significant differences among the three. Contrary to their expectations, increasing the patrol force's effort in preventive patrolling had no apparent impact on the number of reported crimes and did not reduce the need for law enforcement services. On the other hand, eliminating preventive patrols altogether also had no apparent effect on crime rates.

Many law enforcement authorities concluded that preventive patrol, despite its appeal to common sense, is in fact a waste of time and resources. However, other authorities have criticized the Kansas City experiment on several grounds and have insisted that the experiment did not provide a valid test of preventive patrols. Nearly two decades after the experiment was reported, very few American municipal police agencies have completely abandoned the practice of preventive patrolling.[7]

Perhaps the most practical assessment of the Kansas City experiment would be that the value of preventive patrolling cannot be assumed, but must be proved on the basis of each police agency's actual experience in its own community and indeed in each patrol district. The concept of preventive patrolling needs to be reexamined in light of the *possibility* that it has little or no value *under certain circumstances*.[8]

The principle of preventive patrolling can be traced back to the original "watchman" function of the earliest municipal police agencies. As we discussed in Chapter 1, it is unlikely that a prospective criminal will attempt to commit a crime if there is a high probability that the attempt will be observed by a police officer. Therefore, placing police officers where they are able to observe the most likely places for crimes to occur should discourage criminals.

Because it is impractical to station a police officer at every potential crime location, it is necessary to assign officers to patrol a territory that may contain a number of potential crime locations. As long as prospective criminals are not certain exactly where the patrolling officer will be at any given time, they are discouraged by the possibility that they will be observed.[9]

Placing patrol officers in automobiles increases their mobility and decreases the predictability of their movement. In theory, this should be even more discouraging to prospective criminals than the practice of foot patrols. The criminal should calculate that a police officer in a patrol car might appear at any time.

Unfortunately, what seems logical in principle may be invalid in practice. Under the best of circumstances, patrol officers cannot spend all of their time patrolling. In most cases, preventive patrolling does not occupy more than about 25 percent of a patrol officer's time. Three-fourths of the time, the officer is occupied by other duties, and criminals are therefore unlikely to be observed.

Criminals may take a variety of precautions to avoid detection, such as using a police-frequency radio receiver (commonly called a scanner) to keep track of the whereabouts of patrol units, or they may commit crimes out of public view, or they may act so quickly that they are able to complete a crime and leave the scene before there is any possibility that a patrol officer will appear. Criminals also have access to automobiles and other means of rapid mobility.

The crimes that are most likely to be observed by patrol officers are those that are not planned in a calculating manner and that occur in public view. It is therefore not surprising that the crimes most often detected on view by patrol officers are public drunkenness, traffic violations, disorderly conduct, and the like.[10]

The administrators of each agency must decide how much of their patrol officers' time should be devoted to patrol, considering the kinds of criminal offenses that are likely to be detected and prevented. If less time is spent on preventive patrols, more time can be devoted to responding to calls for service and assisting in noncriminal cases; or fewer officers will be needed in the patrol force; or patrol officers can be assigned to larger beats without affecting their ability to respond rapidly to calls for service. These decisions must be based on the administrators' best judgment of the needs of the community, public opinion concerning the proper role of the police, and similar factors.[11]

Twenty-Four-Hour Service

Usually the first demand of the public is to have police service available twenty-four hours a day. People feel more secure when they know that competent help is available at all times, no matter what their emergency might be. Indeed, many people unrealistically want a police officer to appear instantaneously whenever trouble occurs. If it takes an officer more than five minutes to respond to a call for help, there will be complaints that the police are not "doing their job."

The problem is especially acute for the very small police agency (those with fewer than ten commissioned officers). In order to provide twenty-four-hour service, it might be necessary to have only one patrol officer on the street at a time. This would mean three officers on daily shifts and one relief officer filling in on the regular officers' days off. But this arrangement is likely to mean that the officers on duty during the day have more work than they can handle efficiently, while the officer on duty in the middle of the night has little to do.

The usefulness of twenty-four-hour patrol service in rural areas is highly questionable. Few incidents requiring immediate police response are likely to occur between midnight and six o'clock in the morning. In fact, usually the heaviest demand for police service in *any* community occurs between about six o'clock in the evening and midnight; between six o'clock in the morning and six o'clock in the evening, the demand for police service varies considerably.

Therefore, it would be more sensible in most cases to assign patrol officers to daytime shifts, with an extra officer on duty during the evenings, and none at all in the middle of the night. The occasional need for police service late at night can be met by having officers on call, either at the police station or at their own homes.

In most communities, the telephone company can arrange to have the police department's telephone calls forwarded to the home of whichever officer is on call. An officer on call would not respond as quickly as an officer who is already on duty, but a somewhat greater response time late at night should be outweighed by the increased quality of service during the day.

In a slightly larger department (one with between ten and twenty-five officers), it is likely that the public will insist on having twenty-four-hour service. Again, however, this does not necessarily mean that the same number of officers must be available at all hours. A single patrol officer may be able to handle virtually anything that is likely to arise on the late-night watch, and there could be one or more officers on call for unusual crises.

One- and Two-Officer Patrol Units

One of the most intense areas of controversy in police administration is the question of whether one-officer patrols are preferable to two-officer patrols. Usually the question is stated in terms of efficiency versus safety.[12]

It is clear enough that two officers in separate vehicles will be able to patrol a larger area more effectively than the same two officers in one vehicle. Thus, one-officer patrols are presumably more efficient. However, many police officers believe fervently that two officers working together are much safer than two officers working alone.[13]

The advocates of one-officer patrols claim that two officers working together do not give each other as great a margin of safety as one might think. The reason is human psychology: When two people spend a great deal of time together, they are likely to devote more attention to their personal relationship than to their work. As a result, they become careless. When a dangerous incident arises, both officers are likely to be inattentive until it is too late.

Furthermore, when two officers work together they may come to rely on one another excessively. An officer may act rashly, precipitating a dangerous situation, in the belief that his or her partner will provide adequate backup. If both officers do this at the same time, each overrelying on the other, both may get into serious trouble.

The safety issue cannot be ignored. Statistics reported by the FBI consistently show that police officers are most likely to be injured or killed in either of two potentially dangerous situations: when responding to disturbance calls (especially family disturbances) and when making an arrest, including routine traffic stops.

Between 1980 and 1989, a total of 801 law enforcement officers were killed in the line of duty in the United States. Of these, about 20 percent were working out of two-officer patrol vehicles and 78 percent were working alone out of one-officer vehicles. At first glance, these statistics suggest that the one-officer vehicle is more dangerous. However, it must be remembered that the majority of police agencies in the United States use one-officer patrol exclusively, and virtually all police agencies have more one-officer patrol vehicles than two-officer vehicles. Furthermore, the figures for the same period also show that more than one-half of the officers killed had no backup assistance at the time of the incident. The one-officer vehicle is safe—provided officers respond to calls together and back one another up in dangerous situations.[14]

If one-officer vehicles are assigned to large, separate patrol districts, they may have difficulty in providing backup to one another. On the other hand, putting two officers in separate vehicles and assigning both to the same district reduces the efficiency of one-officer vehicles. A sensible compromise is to assign one-officer patrols to overlapping districts, so that most areas are covered by both officers and they are near enough to one another to provide backup (see Figure 14.1).

The use of one-officer vehicles also requires coordination of radio calls. When a dispatcher has a call for any police service that is potentially dangerous such as a disturbance complaint, crime in progress, or suspicious activity, the call should be assigned to the officer in whose district the incident occurs and to a backup officer. Similarly, when an officer is about to take any action that could be dangerous, the dispatcher should be notified and a backup should be assigned immediately. No officer should ever initiate any action that has the potential for danger without first notifying the dispatcher. The assignment of a backup then should be automatic, whether or not the officer asks for one.

These policies require a high degree of flexibility on the part of officers and communications personnel. Patrol districts should not be treated as though their boundaries were impassable walls; those boundaries exist primarily as an

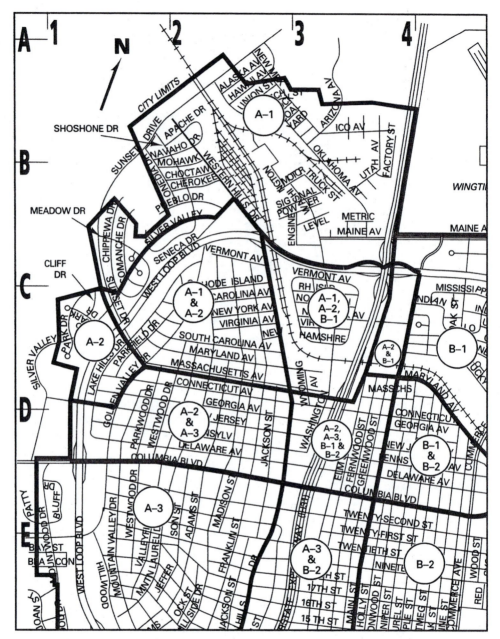

Figure 14.1. Overlapping patrol beats ensure that backup assistance is readily available to every patrol unit.

administrative convenience. Whenever an officer is involved in providing a service that is likely to be time-consuming (investigating a crime or a traffic accident), the rest of the officer's district should not be left vacant while officers in adjacent districts have nothing to do. It should be the responsibility of the dis-

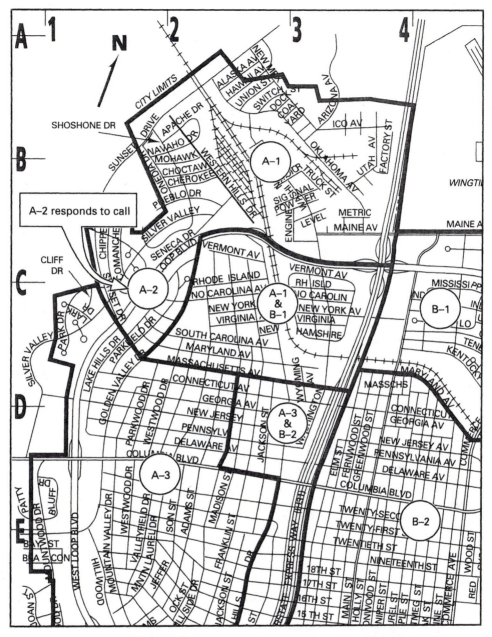

Figure 14.2. When patrol beats overlap (see Figure 14.1) it is easier to coordinate assignments so that all areas are covered when one unit (in this case, A-2) is temporarily unavailable.

patcher to see that all patrol areas are covered at all times, by temporarily rearranging district boundaries whenever necessary. Figure 14.2 illustrates how this can be done.[15]

COMMUNITY AND TEAM POLICING

Once upon a time, law enforcement was virtually a neighborhood enterprise. Police officers were recruited off the street in the communities they were to serve. Foot patrolmen were intimately acquainted with the shopkeepers, housewives, children, and the criminals on their beat. In the larger cities, the police department often was divided into several relatively small precincts, each of which had its own command structure and, for most purposes, operated as if it were an autonomous, independent police agency. Each precinct commander accepted personal responsibility for the protection of the people and property in his precinct, and each patrol officer assumed similar responsibilities for his beat.

In Chapter 1, we mentioned some of the factors that changed the picture of policing in America: the drive toward professionalism in law enforcement; the introduction of new technologies such as the automotive patrol and two-way radio; and the centralization of command in an effort to increase efficiency and eliminate corruption. By the early 1960s, the traditional community-centered personalized police service had all but disappeared in most urban areas.

But traditions die hard and often seem to return from the grave with new names. One consequence of the social upheavals of the late 1960s was a public demand for a return to community-centered policing. The public wanted police officers who were familiar with the community they served, who were accessible to citizens when they needed help, and who were subject to some degree of local control.

The police, for their part, desperately needed to reestablish a closer, more cooperative and less hostile relationship with the public. Somehow, it seemed as if the drive for professionalism had caused the police to lose contact with the people they were trying to serve.

Thus the concept of *community policing* was born in the early 1970s. In its most basic form, community policing simply means that the focus of a police agency's attention should be on providing a range of law enforcement services to the community, not just on "catching crooks."[16]

This basic idea is broad enough to cover many variations. Unfortunately, this very flexibility has created a good deal of confusion about exactly what constitutes community policing.[17] Some agencies have used the term to define what almost amounts to a revival of the precinct organization, except that the precincts are now called teams: a unit of the police agency is established and given sufficient resources to assume full responsibility for all law enforcement functions within a certain community.[18] At the other extreme, some agencies have adopted what they called community policing when in fact they have done nothing but instruct their patrol officers to be more aware of the people and events in their assigned territories.

Out of twenty-six police agencies studied in the late 1970s that had adopted one or another form of team policing, seventeen had found it successful enough to continue. These seventeen agencies ranged in size from thirty to about two thousand commissioned officers. According to the authors of the study, the successful team-policing systems had eight fundamental characteristics.[19]

1. Patrol units, or teams, were assigned to relatively small geographical areas for relatively long periods.

2. Interaction and cooperation among team members were encouraged.

3. Specific methods were adopted to facilitate coordination and communication between teams.

4. Communication between team members and the public, within each team's service area, was given a high priority.

5. Each team area was under the full-time control of one supervising officer (usually a lieutenant or captain).

6. As far as possible, each team was given authority to set its own operating policies and procedures.

7. Each team was responsible for the delivery of *all* routine police services.

8. Patrol and investigative functions were combined or integrated as much as possible.

Although this study concentrated on agencies that called their programs "team policing," the same set of characteristics could be applied to many of the police programs that were labelled "community policing." Discussion and evaluation of both types of program has been hampered by the confusing and diverse definitions that have been used.[20] From now on, we refer to both types of innovative police programs as community policing.

Despite these common characteristics, the exact method of implementing community policing varies considerably. However, all forms of community (or team) policing have the greatest impact on patrol deployment, policies, and practices.

Implementing a community policing program can be accomplished by establishing one of four types of patrol systems:

1. *Basic patrol teams.* Only the patrol force is affected. Instead of assigning patrol officers to conventional beats and districts, with one officer to each beat, the patrol force is divided into several teams, each of which is responsible for a "service area." Each service area is larger than a typical beat but often smaller than a typical district. The idea is that all members of a team are jointly responsible for all patrol activity in their service area.

2. *Patrol-investigative teams.* Teams are assigned to service areas much like those described earlier, but each team is responsible for both patrol services and most, if not all, investigations. Usually one or more investigators are assigned to each team, but patrol officers themselves are expected to perform most investigative functions.[21]

3. *Patrol-community service programs.* Patrol officers are not given expanded responsibility for criminal investigations, but they are expected to provide a variety of community services in addition to conventional patrolling such as conducting crime-prevention meet-

ings for residents of their service area and performing crime-prevention inspections of homes and businesses. Often one or two officers are assigned specifically to community-service activities. They may be placed in neighborhood "store-front" police stations, completely separate from the regular police force.

4. *Full-service teams.* Within each team service area, virtually all police services are provided by team members: conventional patrols, criminal investigations (except the most complex cases), community services, and so on.[22]

This diversity in organizational forms indicates that there may be many different goals and purposes in adopting a community policing structure. The most common objectives, according to the 1978 study,[23] were the following:

1. To improve the management of police services; to make police services more responsive to public demand; and to manage the police work load more efficiently.

2. To make policing a more personally satisfying occupation; to expand the role of the patrol officer, say, by giving responsibility for investigations and community services; and to enable patrol officers to work together more productively.

3. To improve the public's degree of satisfaction with its police service; to establish and maintain better relations between the community and the police; and to decrease the public's fear of crime.

As with all innovative police practices, of course, the ultimate goal of community policing is to reduce the actual incidence of crime through more effective and efficient law enforcement service.

Community policing should not be presented as the solution to every problem of law enforcement. Crime will not vanish as soon as the patrol force is reorganized into teams. Clear objectives, thorough advance planning, and careful implementation are necessary if the change is to be accomplished successfully.[24]

DETERMINATION OF PATROL FORCE REQUIREMENTS

Patrol Force Size

Given the fact that personnel resources are limited in every police agency—no police administrator ever has as many officers as might be desired—what proportion of the force should be assigned to patrol?

First, there is no magic number, and no rule of thumb that can provide guidance. In small agencies, it is not uncommon for 80 or 90 percent of the force to be devoted to patrol. In very large agencies, the proportion might be 50 percent or less.[25]

The single most important factor is the number and nature of the services that the patrol force is expected to provide. If patrol officers are required to make complete investigations of every criminal incident reported or discovered on their beats, plus respond to all noncriminal crises, plus devote a considerable amount of time to preventive patrolling, plus handle a variety of nonproductive tasks, then certainly a large number of patrol officers will be needed.

If the patrol officers' responsibility for criminal investigations is very limited (for instance, if they are expected merely to obtain a preliminary report of the incident and forward the report to detectives), and if the police department assumes only limited responsibility for other kinds of services, then a smaller proportion of the agency will be needed on patrol.

Geographical and population factors also influence the need for patrol officers. If population density is relatively high (for example, in the inner city), a single officer may be kept busy responding to calls for service within a small geographical area. If population density is low (for example, in rural areas), one officer may be able to handle all the calls that arise in a very large area. However, response time may be unacceptably large because of the long distances that an officer must travel to respond to a call.

These are not the only factors that affect the size of the patrol force. The basic efficiency of the agency and the productivity of the patrol officers themselves have an important influence. If administrative and operational procedures are designed to assist officers in carrying out their tasks quickly and effectively, and if the officers are competent, well trained, and highly motivated, fewer officers will be needed to handle a given quantity of work.

But the ruling factor, in practical terms, usually is the size of the agency's budget. Few police administrators are given a budget large enough to hire all the officers they would like to have. Consequently, the usual procedure is to take the agency's overall budget, subtract the nonpersonnel costs, and then subtract the personnel who must be assigned to nonpatrol duties. Whatever is left determines the number of patrol officers available. This base number may be decreased by shifting one or more patrol officers to other duties, or it may be increased by shifting nonpatrol officers to patrol—or by persuading the parent government to increase the agency's budget.

Decreasing the size of the patrol force is not always a bad idea. For example, in a small department it may be the standard practice for patrol officers to perform all of the tasks involved in booking their prisoners including fingerprinting, photographing, assigning a jail cell, and so on. This may be a time-consuming procedure. At some point, it is likely to be preferable to assign one officer as the full-time booking officer, thereby reducing the amount of time that the patrol officers must spend off the street. Even if this means there will be one less patrol officer on duty, the increased efficiency of the entire patrol force may outweigh the loss. However, if the agency has a booking officer whose duties are not sufficient to keep the officer occupied full-time, it might be preferable to shift the booking officer to patrol and require the patrol officers to do their own booking of prisoners, or to assign other duties to the booking officer.

Patrol Deployment: Shifts and Beats

Once the base number of patrol officers has been determined, the administrator must decide how the officers are to be deployed: in other words, what geographical area each officer is to patrol and for what period.

The simplest procedure is to take the total number of officers available and divide by four. One-fourth of the total roster is assigned to each of three daily shifts; the fourth group is the *relief shift* that fills in for the other three on their days off. Each shift is eight hours long.

Since at any given time, one-fourth of the total patrol force is on duty, the agency's total jurisdiction is divided into that number of patrol districts or beats. The beats usually are laid out to be approximately equal in area, and their boundaries are determined by whatever natural or artificial boundaries (major highways or thoroughfares, rivers, etc.) exist in the community (see Figure 14.3).

For example, if the agency has sixteen patrol officers, the community would be divided into four patrol beats, usually by simply quartering the town along the major north-south and east-west streets.

This procedure would not be entirely unsatisfactory if the need for police service were more or less constant and evenly distributed. However, that is clearly not the case. Although incidents occur at random, in the sense that they do not invariably occur at the same times and places, there are patterns in the distribution of incidents. Those patterns are not accidental; they are a product of the way people live.

Since most people are at home asleep during the late-night hours, relatively few crimes are committed and relatively few noncriminal crises occur between midnight and the early daylight hours.

During the morning and early afternoon, most people are either at work or engaged in their individual personal business at home. Criminal activity is comparatively light, although somewhat heavier than late at night, and there are more traffic accidents and noncriminal incidents that require police attention.

During the evening hours, many people are involved in family activities at home, and many others are engaged in social activities in public places. The increased social activity (whether at home or in public) creates more situations in which people come into conflict. Also, more people use alcohol during the evening hours. There are therefore more family disturbances, criminal assaults, serious traffic accidents, and other incidents requiring police involvement during the evening hours.

These patterns vary considerably from one community to the next, within a single community, and from one day to the next. Even changes in the weather affect the number of calls for police services![26] Police incidents are more frequent and more serious in areas that have denser populations, where socioeconomic status is lower, where different ethnic groups come into contact, and where young people have limited opportunities for productive recreation or employ-

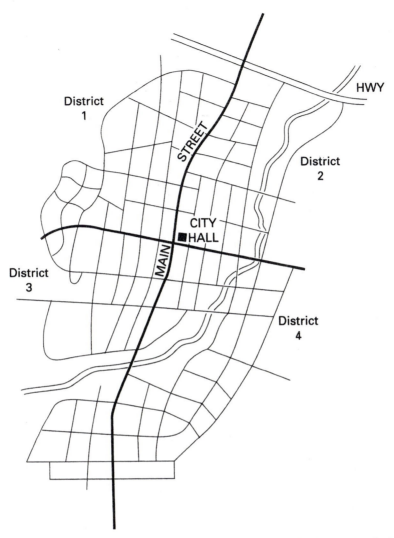

Figure 14.3. Conventional patrol beats. The area is simply divided into as many beats or districts as there are patrol officers, with the districts roughly equal in area regardless of the service workload.

ment. Police incidents are also more frequent and more serious on weekends and on some holidays.[27]

In short, what we might call the *natural workload* for the police is not the same from place to place nor from time to time. The efficient deployment of the patrol force must consider these variations in workload. Otherwise, some patrol officers will have too little work to do, while others will not be able to handle all of their work effectively.

Experienced and perceptive administrators may be able to predict the natural workload patterns in their communities with surprising accuracy. However, the most accurate method of determining the workload pattern is to examine the historical record of police activities in the community.[28] If accurate, complete records have not been kept in the past, considerable effort may have to be devoted to setting up the necessary records system. Usually it is not practical to attempt to recreate the natural workload pattern from haphazard records.

The records should show the number of police incidents that have occurred by location, by day of the week, and by the time of day. For each incident, it is necessary to know the number of officers whose services were required and the length of time each officer was involved.

This examination of records should result in three analyses: the geographic distribution of police incidents, the distribution of police incidents over time, and the total workload per patrol officer.[29]

The analysis of geographic distribution can be presented in the form of a map of the agency's jurisdiction, with pins or shading to show the number of recorded incidents in each area. The analysis can be made more thorough by using color codes or some similar device to indicate different categories of incidents, such as crimes (which may be further subdivided into felonies and misdemeanors, or some other categories), noncriminal emergencies, and services indirectly related to law enforcement. Nonproductive tasks should not be shown in terms of their geographic distribution, since they are to be eliminated as far as possible. An example of an incident distribution analysis for the hypothetical City Police Department is shown as Figure 14.4.

The distribution of incidents over time can be presented in the form of graphs. Again, the analysis is improved if different categories are displayed. An analysis for one day is shown as Figure 14.5, and a graph for one week is shown as Figure 14.6.

The average workload for a single patrol officer can be determined by taking the total number of incidents in a given period of time, such as one eight-hour shift, and dividing by the number of patrol officers available on that shift. This number is then multiplied by the average time required to handle each type of incident. The result is the average *service workload* per officer. Time spent on nonproductive tasks and time that is devoted to preventive patrolling must be added to the service workload to arrive at the total workload per officer.

For example, let us suppose that City Police Department has eight patrol officers on duty during the daytime shift on, let us say, Tuesday. According to Figure 14.5, the average number of police incidents during this time is 56, of which 8 are felony cases, 24 are misdemeanor cases, and the rest (24) are noncriminal incidents. The agency's records also show that City Police Department's patrol officers spend on the average 45 minutes on each felony case, 30 minutes on each misdemeanor case, and 20 minutes on each noncriminal incident. On a typical Tuesday, a day-shift officer would spend 45 minutes on one felony case, 90 minutes on three misdemeanor cases, and 60 minutes on three noncriminal incidents. This is a total of 195 minutes, or 3 hours and 15 minutes, in the officer's service workload. To this would be added (again, based on the agency's records) perhaps 2 hours on nonproductive tasks such as meal breaks and

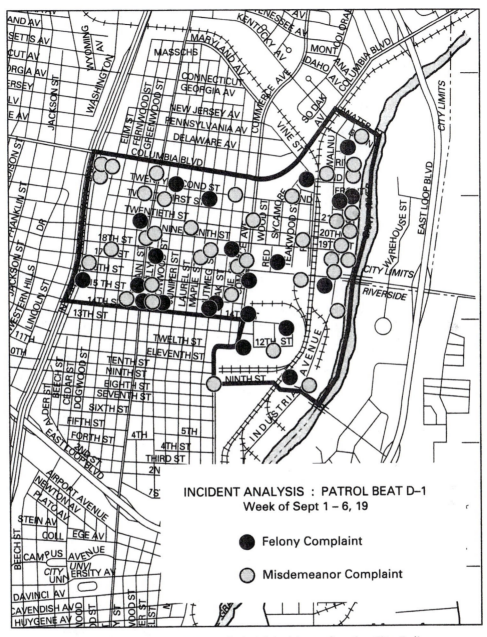

Figure 14.4. A geographic analysis of incidents for the City Police Department.

administrative duties. Out of an 8-hour shift, there are 2 hours and 15 minutes left for preventive patrolling (see Figure 14.7).

However, these figures are based on averages and do not take into account the geographic distribution of incidents. An officer assigned to an area that has a

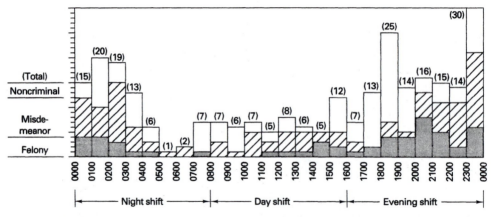

Figure 14.5. An incident analysis for an average day in the City Police Department.

relatively high number of felony incidents might have no time at all for preventive patrolling. Or, on a particular day, the number of noncriminal incidents might be either twice as great, or only half as great, as the average.

For a small agency, it may be possible to adjust patrol beats according to the geographical distribution of incidents by simply shifting the boundaries of the beats. Thus, an officer assigned to a busy area would have a smaller beat to patrol, and an officer assigned to a quiet area would have a larger beat. As long as none of the beats are so large that *response time* (the total time required for a call for service to be dispatched and for the assigned officer to reach the scene of the incident) is not excessive, shifting beat boundaries in this manner is a satisfactory technique. However, in a large agency the problem can become extremely complicated, and merely shifting beat boundaries to achieve roughly equal workloads may be unsatisfactory.

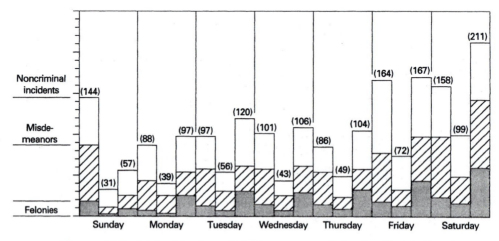

Figure 14.6. An incident analysis for one week in the City Police Department.

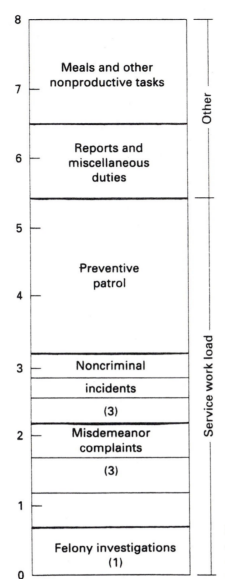

Figure 14.7. The service workload and total workload for one patrol unit in the City Police Department.

The uneven distribution of incidents over time creates a different set of problems, even for the small agency. If the day is simply divided into three equal periods, or shifts, and the same number of officers is assigned to each shift, there may be too many officers on duty at some times, and not enough at others.[30]

Ideally, there should always be a sufficient number of officers available at any time to handle any incident that might arise. In reality, this ideal is nearly impossible to achieve and would be very expensive if it could be achieved. A more realistic goal is to have on duty at any given time enough patrol officers to handle 80 percent of the predicted peak workload. This means that, during peak periods, there will not be enough officers available to handle every call for service immediately. Some calls will have to be set aside until an officer becomes available. In effect, the natural workload is spread out over time.

For a small agency, if the distribution of incidents over time is not too uneven, adopting a policy of handling calls for service according to a specific system of priorities may solve the problem.[31] At all times, at least one officer should be available immediately to handle first-priority calls or incidents (such as crimes in progress, traffic accidents with injuries, civil disturbances, and so forth). All other calls for service are handled in the order in which they are received, except that when there are more calls waiting to be handled than there are officers available, officers are dispatched according to the priority of the call.

Devising a system of priorities is a matter for administrative judgment, based on the community's needs and the kinds of incidents that occur most frequently. Usually, after the kinds of first-priority incidents we have just mentioned, second-priority incidents include felony complaints where the crime is no longer in progress, traffic accidents that do not involve injuries, assistance to other emergency services, and misdemeanor public disturbances. Third-priority incidents include all other misdemeanor complaints, minor traffic accidents and any other incidents that disrupt traffic (such as stalled vehicles or malfunctioning traffic signals), and some calls for services that are indirectly related to law enforcement. Any other kind of incident or call for service would be treated as fourth priority.

Assigning priorities to the different types of incidents is certainly better than merely handling all incidents, regardless of their nature, in the order in which they are received. The use of a system of priorities may be sufficient to smooth out the distribution of incidents so that it is not unreasonable to have the same number of officers on duty at all times of the day. In fact, most small- to medium-sized agencies have adopted this practice in one form or another, although the system of priorities often is not worked out specifically, but instead is left to the discretion of the radio dispatchers or their supervisor.

This approach may not be satisfactory in larger agencies. During peak workload periods, third-priority and even second-priority calls may stack up to such an extent that unacceptable delays occur. While there are no universal standards of acceptable response time, police administrators generally agree that all first-priority calls should be handled in the shortest possible time, preferably not more than 5 minutes (total response time). Second-priority calls should be handled within 10 to 15 minutes. Third-priority calls should be handled in not more than 20 to 30 minutes. Longer response times seriously degrade the quality of service that is being provided to the public and may pose an unacceptable threat to the safety of lives and property.[32]

If calls cannot be handled within an acceptable time during peak workload periods, the only solution is to increase the number of officers available. However, adding more patrol officers to a busy shift does not mean that more officers should also be added to a relatively quiet shift. Each shift should be considered separately and assigned the number of officers that is appropriate to the average workload for that part of the day.

In some communities, this may mean that there should be twice as many officers on duty during the day as at night and three times as many on duty during the evening as at night. Instead of dividing the total available force into three equal shifts, the total force is divided by six. One-sixth is assigned to the night shift (that is, the early morning hours); two-sixths (or one-third) is assigned to the day shift; and three-sixths (one-half) is assigned to the evening shift. Of course, it is not necessary to divide the total force by any even fraction. If there' are twenty-five patrol officers, three might be assigned to the night shift, eight to the day shift, and fourteen to the evening shift.

The starting and ending times for each shift also can be adjusted to accommodate variations in workload. For a variety of reasons, usually it is not practical to vary the lengths of the shifts so that, for example, the night shift only works seven hours while the evening shift works nine; in any three-shift rotation, each shift should be eight hours long. However, the shifts do not have to begin at midnight, 8:00 a.m., and 4:00 p.m. Depending on the distribution of incidents over an average day, in some communities it might be preferable to start the shifts at 2:00 a.m., 10:00 a.m., and 6:00 p.m.

Finally, in some communities the peak workload period may be longer than eight hours, or there may be two peak periods. For example, there might be one peak period during the evening rush hour and a second peak period (usually a somewhat smaller peak) between 10:00 p.m. and 2:00 a.m. when people are ending their evening's social activities. Because it is impractical to have the entire evening-shift patrol force on duty continuously from 4:00 p.m. to 2:00 a.m., the best solution is to have either staggered shifts or a fourth peak-period shift. In a staggered-shift system, all of the night shift might work from 2:00 a.m. to 10:00 a.m. and all of the day shift from 10:00 a.m. to 6:00 p.m. Half of the evening shift might work from 4:00 p.m. to 12 a.m. and the other half from 6:00 p.m. to 2:00 a.m. Thus, the twenty-four-hour schedule might look like the following:

SHIFTS

Period	Night	Day	Evening A	B	Total
12 a.m.–2:00 a.m.	0	0	0	7	7
2:00 a.m.–10:00 a.m.	3	0	0	0	3
10:00 a.m.–4:00 p.m.	0	8	0	0	8
4:00 p.m.–6:00 p.m.	0	8	7	0	15
6:00 p.m.–12 a.m.	0	0	7	7	14

Of course, once again it is not essential for the staggered shift to be divided equally in half, if it would be preferable to divide it unequally (so that the evening A shift might have nine officers and the evening B shift five, for example).

The use of a peak-period shift is similar to the use of staggered shifts. Instead of dividing one shift into two units, or subshifts, a separate fourth shift is assigned to the peak workload period, typically from 4:00 p.m. to 10:00 p.m., although that may vary.

One advantage in the use of either staggered shifts or peak-period shifts is that the shift assignments can be varied considerably from one part of the week to another, merely by changing the shift starting times. For example, the peak-period shift (or one subshift) might work from 4:00 p.m. to 10:00 p.m. on weekdays, but from 5:00 p.m. to 11:00 p.m. on Fridays and Saturdays, and from 3:00 p.m. to 9:00 p.m. on Sundays, if the distribution of incidents in the community requires these variations.

Relief Shifts

One common practice in many agencies, especially smaller ones, is the creation of a fourth relief shift whose members substitute for the regular-shift officers on their days off. Since the eight-hour work day and five-day work week are virtually universal in the United States, most people expect to have two consecutive days off each week.[33] Law enforcement services must be provided seven days a week, so there must be a substitute available to fill in two days a week for each patrol officer. In a three-shift system, if all of the shifts are equal in size, this can be done in the following manner:

RELIEF SHIFTS

Day of Week	12 a.m.–8:00 a.m.	8:00 a.m.–4:00 a.m.	4:00 p.m.–12 a.m.
Sunday	Relief	Shift 1	Shift 2
Monday	Relief	Shift 1	Shift 2
Tuesday	Shift 3	Relief	Shift 2
Wednesday	Shift 3	Relief	Shift 2
Thursday	Shift 3	Shift 1	Relief
Friday	Shift 3	Shift 1	Relief
Saturday	Shift 3	Shift 1	· Shift 2

This plan, although simple in concept, has a number of disadvantages. First and foremost, it means that the relief shift must work during a different time of the day every other day, which imposes a severe physical burden on the relief shift officers. Whenever the relief shift is on duty, there is no one from the regular shift present to follow up on uncompleted work from the previous day.

Notice also that the relief shift must work six consecutive days and has only one day off. In some agencies this is in fact the usual practice. The relief shift officers are either the newest recruits or officers who are out of favor with their supervisors, and therefore are considered to be not entitled to two days of rest each week.

An alternative solution is to have all officers on all shifts work six days consecutively, followed by two days off. Thus, in the schedule shown above, the relief shift would be on duty again the following Monday (on the night shift). This means that, for all officers, different days off are taken each week, a situation that can be very disruptive to their personal and family life.

Another alternative is to divide the relief shift into two squads and give each squad an additional day off during the week. Thus, for instance, on

Monday only half the relief shift would replace Shift 3, and on Tuesday only half the relief shift would substitute for Shift 1.

Clearly, none of these arrangements is very satisfactory. A better approach is to assign a sufficient number of extra personnel to each regular shift to provide the necessary substitutes. If there are six officers on a given shift, there should be two additional officers assigned to the shift for relief. In large agencies, where the patrol force is divided into precincts or districts, each precinct or district should have its own group of relief officers, so that they will be familiar with the territory in which they are working.[34]

Another problem results from the need to provide twenty-four-hour-a-day police service. Many people, including police officers, consider working in the evenings or at night less desirable than working during the day. In order to accommodate the preference for day-shift assignment, some police agencies have adopted the practice of rotating all shifts periodically.[35] For example, every six months the day-shift officers are reassigned to the evening shift, the evening-shift officers to the night shift, and the night-shift officers to the day shift. Some agencies even include the relief shift in the rotation. Thus, over some period (in this case, two years), all officers are required to work on every shift.

This arrangement, however, becomes very complicated when the shifts are not equal in size. Instead of rotating all shifts at the same time, it is necessary to reassign individual officers (since there are more officers on some shifts than on others) at approximately equal intervals. What usually happens, in fact, is that at least a few officers are rotated every week, and the intervals become very unequal. Being rotated to a desirable shift becomes a privilege, and being rotated to an undesirable shift becomes an informal type of punishment. Both privileges and punishments may be dependent on the personal favoritism of supervisors, a condition that is often harmful to morale.

If seniority is used as the basis for awarding privileges, the result is that the least-experienced officers are assigned to the undesirable shifts for longer periods of time. The evening and night shifts, although they may be less desirable, are also the times of the day that are often busiest and that impose the greatest demands on the experience and good judgment of officers. Loading these shifts with the least-experienced personnel may seriously reduce the agency's effectiveness and productivity.

There is no perfect solution to this problem. Fortunately, some officers actually prefer to work in the evenings or at night, leaving them free to pursue their personal business during the day. As far as possible, these preferences should be encouraged and accommodated.

It is not likely that the undesirable shifts can be completely filled with officers who prefer those working hours, and in any case preferences may change from time to time. Rotating all shifts periodically may be the worst solution, since it requires all shifts to be equal in size. Rotation on an individual basis is certainly better, provided the rotation policy does not permit shift assignments to be used unfairly to reward or to punish the individual officers on the basis of supervisors' favoritism, and provided the rotations do not leave the evening and night shifts filled with inexperienced personnel.[36]

COMPUTERIZED ALLOCATION SYSTEMS

So far we have discussed some of the factors that must be taken into consideration in the deployment of the patrol force. In particular, we have emphasized the need to allocate officers to different shifts and beats on the basis of equalizing their natural workload.

In a small- to medium-size agency (up to about one hundred patrol officers), the allocation techniques we have discussed can be carried out manually. At least once each year, preferably more often, the agency's records must be analyzed to determine the geographical and time distribution of the service workload in each patrol area. Beat boundaries can then be adjusted and, if necessary, the size of the shifts can be changed to keep workloads even.

In larger agencies, the allocation process becomes increasingly complicated. Performing it manually is excessively time-consuming and the results are likely to be less than optimal.[37] Equalizing the workload is not the only goal to be achieved by patrol force allocation. Other performance factors also should be taken into consideration. For example, improving response time may be more important than equalizing workloads. In some cases, *unequal* workloads may be desirable if they permit officers on some beats to devote more time to preventive patrolling.[38]

Furthermore, care must be exercised to ensure that the calculated workload accurately reflects the services being performed. For example, one patrol beat may have a very high number of reported crimes, while another beat might have a relatively low number. This could mean that the first beat should be reduced in size and the second one increased, to equalize the workload.

However, closer analysis might indicate that the crimes being reported on the first beat are relatively minor ones that are being given very little follow-up investigation, but the crimes reported on the second beat tend to be more serious and require more time-consuming work by both patrol officers and investigators.

The low number of complaints on the second beat also might reflect the reluctance of people in the area to report minor crimes, because they know that the officers on their beat are already overburdened. In fact, *both* beats may need to be reduced, or boundaries with adjacent beats rearranged, to give the patrol officers on the first beat adequate time to investigate the numerous crimes reported to them, and to give officers on the second beat time to respond to more calls.

Manual systems of patrol force allocation (including those that are based on mathematical formulas for equalizing the workload) simply cannot incorporate all of these performance factors, especially in an agency with more than about fifty patrol officers.[39]

Fortunately, there are computerized patrol allocation systems that offer several advantages over any manual system.[40] The computerized systems are widely available in the form of standard computer programs that can be run on most computers, including the smaller personal computers that are used by many police agencies for record keeping and other administrative purposes. Any agency that uses CAD certainly has the ability to use a computerized patrol allocation system.

The various computerized systems differ widely in their manner of operation and in the number of performance factors they are designed to take into account. However, they all work in a similar fashion. Records of an agency's operations for a given period are fed into the computer, which then analyzes the distribution of incidents by geography and time. In the more sophisticated systems, basic information about the agency's jurisdiction (a map of the community, information about population density, and so forth) also is fed into the computer. The computer then produces a set of patrol force assignments to meet either the criteria built into the program (such as equalizing work loads or reducing response time), or, in the more elaborate systems, according to whatever criteria the agency decides to use.

Some systems are designed specifically to be compatible with standard CAD programs. As the dispatching program records calls for service, these records are automatically accumulated by the allocation program. In essence, the allocation program performs a continuous analysis of the agency's operations. A new patrol force allocation can be called up at any time.

The development of computerized patrol allocation systems is one area in which substantial progress has been made in recent years, and the progress is continuing as more refined and sophisticated programs are being designed and tested.

AUTOMOTIVE AND NONAUTOMOTIVE PATROLS

As we pointed out in Chapter 6, the use of the automobile for police patrol is probably the single most important characteristic of modern law enforcement. The automobile with its two-way radio has become indispensable to all police services.

However, it does not necessarily follow that all patrol services require the use of an automobile. We briefly discussed some of the uses of different types of vehicles such as motorcycle and aircraft, and we mentioned the fact that foot patrols have reappeared in many urban police forces in recent years. Managers should decide which of many different patrol techniques is appropriate to each patrol assignment and should then allocate the most appropriate type of vehicles.

The standard patrol car will continue to be the most widely used vehicle for both general patrol and traffic patrol. An automobile designed or equipped for police service is able to transport from one to five persons (one or two patrol officers, prisoners, witnesses, and so forth) rapidly and comfortably over relatively long distances and is able to carry most of the equipment that a patrol officer might need in various circumstances. The patrol car also can be used in nearly all weather conditions. In areas where the natural workload is relatively low, a single patrol officer in an automobile can cover a beat of ten or more square miles and still reach the scene of most incidents within a reasonable response time. In rural areas, it is not unusual for a sheriff's deputy or highway patrol officer to cover, by automobile, an area of several hundred square miles.

The automobile is less attractive as a patrol vehicle under certain circumstances. For example, in urban areas where traffic is dense, a motorcycle may be preferable because of its greater maneuverability. Both two-wheel and three-wheel motorcycles are able to cut in and out of traffic lanes where an automobile would be blocked. A motorcycle is able to transport a single officer, with a limited amount of equipment, rapidly over shorter distances. Two-wheel motorcycles usually are capable of being operated at relatively high speeds, making them more effective as pursuit vehicles on urban expressways and city streets.

Motorcycles also have a number of disadvantages. By their nature, they are limited to carrying one person, or at most two if a sidecar is attached (a practice that is so unsafe that it cannot be recommended), and a minimal amount of equipment. Three-wheel motorcycles generally are incapable of being operated at high speeds, and all motorcycles are dangerous to operate because of their inherent instability and the complete lack of protection for the operator. Motorcycles are even more dangerous, and extremely uncomfortable, in poor weather.

Two-wheel motorcycles are probably best suited to traffic patrol because of their maneuverability and pursuit capability in urban areas, especially on expressways.

Three-wheel motorcycles are limited to low-speed uses such as urban street patrol (both traffic and general law enforcement) and are widely used for the even more limited purpose of parking law enforcement.

Wherever motorcycles are used for patrol they must be backed up by automotive patrols for transportation of prisoners and for the delivery of equipment that is too bulky to carry on a motorcycle. Motorcycle officers also must have either the use of an automobile or an alternative assignment for days when the weather prevents the use of cycles.

In very dense urban areas, bicycles should be considered as an appropriate patrol vehicle. A bicycle is even more maneuverable than a motorcycle, especially off the street, in alleys, parks, railroad yards, docks, and other areas where motor vehicles are too cumbersome. A bicycle, although it is obviously a very low-speed vehicle, does allow an officer to travel from one place to another more rapidly than on foot and in some cases permits an officer to reach an off-the-street location more quickly than an automobile.

Horses have recently made something of a comeback in police service, too. A number of large cities have found horse patrols to be extremely useful in crowd management and even in traffic control.[41]

Finally, as many urban police agencies have discovered, the foot patrol is far from being obsolete. In dense urban areas, especially in commercial districts that have a very high concentration of potential (or actual) crime sites, foot patrols are considerably more effective than any type of vehicular patrol. A foot patrol officer is able to develop and maintain a personal relationship with citizens, particularly shopkeepers, that would be very difficult for a motorized patrol officer to achieve. An officer patrolling on foot is able to answer citizens' questions and perform countless small, personal services that may not be closely related to law enforcement, but do help to uphold the agency's public image.

Just as the development of mobile two-way radios made the automobile practical for patrol use, so too the development of compact, efficient portable two-way radios has made the revival of the modern foot patrol possible. With appropriate radio equipment, a foot patrol officer is always accessible to the dispatcher and has rapid access to backup assistance.

Foot patrol officers should be provided with a vehicle of some sort to transport them to their beat, to carry extra equipment they might need while on duty, and occasionally to enable them to transport a prisoner, victim, witness, or some other person to the police station. All but the last of these functions could be served by a motorcycle rather than an automobile.

Combined motor-and-foot patrols are useful in many communities. An officer patrols most of a beat by automobile or motorcycle, but parks the vehicle to walk through a shopping mall or commercial district of a few blocks. This is clearly the most flexible way to deploy patrol officers in areas where neither the foot patrol nor the automotive patrol is entirely satisfactory by itself.

Foot patrols are rarely needed on a twenty-four-hour-a-day basis. In most cases, foot patrols are effective in commercial areas during business hours and sometimes only during the peak business hours. For most commercial districts, these hours would be between 10:00 a.m. and 6:00 p.m. In an entertainment district, of course, the peak hours occur during the evening.

Whatever the local conditions may dictate, it is not necessary for foot patrol shifts to be exactly the same as motorized patrol shifts. For example, the standard day shift might be from, say, 8:00 a.m. to 4:00 p.m., but foot patrol officers on the day shift might work from 10:00 a.m. to 6:00 p.m.

To summarize both this section and the chapter as a whole: The selection of patrol techniques, the choice of patrol vehicles, and the assignment of officers to shifts and beats should reflect an analysis and understanding of the needs of the community an agency is serving. Arbitrary methods of deploying personnel, such as dividing the community into beats of equal geographic size or dividing the patrol force into four equal shifts, are unlikely to provide the close match between the natural workload and the human resources that must handle that workload. Given the fact that all law enforcement agencies are faced with limited resources, administrators must use deployment strategies that make the best and most efficient use of their personnel.

REVIEW

1. Police administrators face a peculiar problem in trying to establish a smooth, efficient work flow because

   (a) Police officers vary in their competence and enthusiasm.

   (b) Most police work is extremely dangerous.

   (c) Tasks to be performed occur at unpredictable, random intervals.

   (d) All employees expect to work eight hours a day, five days a week.

   (e) There is never enough money to hire the number of people needed.

2. True or False: The crimes most likely to be detected on view by a patrol officer are those that are not planned in a calculating manner and that occur in a public place.

3. The major objectives of team policing are

    (a) To make police services more responsive to community needs.

    (b) To ensure that preventive patrol is effective.

    (c) To expand the role of patrol officers.

    (d) To improve the relationship between the police and the public.

    (e) To eliminate the need for trained criminal investigators and other specialists.

4. The natural workload of a patrol force includes

    (a) The number of police incidents that occur in a given community, distributed geographically and over time.

    (b) The total amount of useful work performed by each patrol officer.

    (c) The amount of work an officer is able to perform in a given shift without excessive effort.

    (d) The nonproductive tasks that are unavoidably assigned to each officer.

    (e) Only those incidents that require an immediate police response.

5. True or False: Despite several disadvantages, the modern automobile is the only practical, effective vehicle for use by a police patrol force.

NOTES

[1]Herman Goldstein, *Problem-Oriented Policing*. New York: McGraw-Hill, 1990, pp. 18–21.

[2]Ibid.

[3]See also Thomas F. Adams, *Police Field Operations*, 2nd ed. Englewood Cliffs, N.J.: Prentice Hall, 1985, pp. 2–7.

[4]Charles D. Hale, *Police Patrol, Operations, and Management*. New York: Wiley, 1981, pp. 16–30.

[5]O. W. Wilson, *Police Planning*, 2nd ed. Springfield, Ill.: Charles C Thomas, 1962, p. 80.

[6]Geoffrey P. Alpert and Roger C. Dunham, *Policing Urban America*, 2nd ed. Prospect Heights, Ill.: Waveland Press, 1992, p. 136.

[7]John J. Broderick, *Police in a Time of Change*. Prospect Heights, Ill.: Waveland Press, 1987, pp. 198–200.

[8]William H. Bieck, William Spelman, and Thomas J. Sweeney, "The Patrol Function," in William A. Geller, ed., *Local Government Police Management*, 3rd ed. Washington, D.C.: International City Management Association, 1991, p. 61.

[9]Ibid., p. 63.

[10]Robert Tojanowicz et al., *An Evaluation of the Neighborhood Foot Patrol Program in Flint, Michigan*. East Lansing, Mich.: Michigan State University, 1982, p. 30.

[11]Herman Goldstein, *Problem-Oriented Policing*, has produced a particularly provocative approach to law enforcement.

[12]Hale, *Police Patrol*, pp. 97–100.

[13]Stephen D. Mastrofski, "The Prospects of Change in Police Patrol," in *American Journal of Policing*, vol. 9, no. 3, 1990, p. 31.

[14]Victoria L. Major, "Law Enforcement Officers Killed, 1980–89," in *FBI Law Enforcement Bulletin*, vol. 60, no. 5, May 1991, pp. 2–5.

[15]See also Richard C. Larson, *Urban Police Patrol Analysis*. Cambridge, Mass.: Massachusetts Institute of Technology Press, 1972, pp. 66–69.

[16]Malcolm K. Sparrow, Mark H. Moore, and David M. Kennedy, *Beyond 911*. New York: Basic Books, 1990.

[17]Bieck, Spelman, and Sweeney, "The Patrol Function," in Geller, *Local Government Police Management*, p. 61.

[18]Hale, *Police Patrol*, pp. 111–13.

[19]Ellen J. Albright and Larry G. Siegel, *Team Policing: Recommended Approaches*. Washington, D.C.: U.S. Department of Justice, National Institute of Law Enforcement and Criminal Justice, 1979, pp. 5–7.

[20]Frank Schmalleger, *Criminal Justice Today*. Englewood Cliffs, N.J.: Prentice Hall, 1991, pp. 183–86.

[21]Thomas F. Hastings, "Criminal Investigation," in Bernard L. Garmire, ed., *Local Government Police Management*, 2nd ed. Washington, D.C.: International City Management Association, 1982, pp. 166–67.

[22]Albright and Siegel, *Team Policing*, pp. 8–12.

[23]Ibid., pp. 1–3.

[24]Jerome H. Skolnick and David H. Bayley, *The New Blue Line*. New York: Free Press, 1986, pp. 220–29.

[25]Larson, *Urban Police Patrol Analysis*, p. 5.

[26]James L. LeBeau and Robert H. Langworthy, "The Linkages Between Routine Activities, Weather, and Calls for Police Services," in *Journal of Police Science and Administration*, vol. 14, no. 2, June 1986, pp. 137–45.

[27]Edward P. Ammann and Jim Hey, "The Discretionary Patrol Unit," in *FBI Law Enforcement Bulletin*, vol. 58, no. 1, January 1989, pp. 18–22.

[28]Hale, *Police Patrol*, pp. 162–68.

[29]Wilson, *Police Planning*, pp. 92–106.

[30]Hale, *Police Patrol*, pp. 169–76.

[31]Charles D. Hale, "Patrol," in Bernard L. Garmire, ed., *Local Government Police Management*, 2nd ed. Washington, D.C.: International City Management Association, 1982, p. 120.

[32]Larson, *Urban Police Patrol Analysis*, pp. 15–20.

[33]However, this is not always the case. The Illinois Department of State Police has reported success in scheduling patrol officers for four 10-hour shifts per week. See Daniel T. Moore and J. Glen Morrow, "Evaluation of the Four/Ten Schedule in the Illinois Department of State Police Districts," in *Journal of Police Science and Administration*, vol. 15, no. 2, June 1987, pp. 105–9.

[34]Wilson, *Police Planning*, p. 106.

[35]Hale, *Police Patrol*, p. 177.

[36]Wilson, *Police Planning*, pp. 108–9.

[37]For example, see Larson, *Urban Police Patrol Analysis*, pp. 149–69.

[38]R. L. Sohn and R. D. Kennedy, *Patrol Force Allocation for Law Enforcement*. Washington, D.C.: U.S. Department of Justice, Law Enforcement Assistance Administration, 1978, p. 10.

[39]Ibid., pp. 5–6.

[40]See, for example, Jan M. Chaiken and Warren E. Walker, *Patrol Car Allocation Model*. Santa Monica, Calif.: Rand Corporation, 1985.

[41]Stephen E. Doeren, "Mounted Patrol Programs in Law Enforcement," in *Police Studies*, vol. 12, no. 1, Spring 1989, pp. 10–17.

TRAFFIC LAW ENFORCEMENT ADMINISTRATION

There are four primary traffic law enforcement functions: (1) to observe the general flow of traffic and, when necessary, remove or direct traffic around any obstructions that would otherwise interrupt the flow; (2) to observe individual motorists and pedestrians for violations of the law; (3) in the event of a traffic accident, to protect the persons and property involved from further harm, and to obtain a report of the incident (for purposes that we will discuss shortly); and (4) to educate motorists, and the public in general, about traffic safety.[1]

Many law enforcement agencies also have accepted some degree of responsibility for providing general assistance to motorists, such as providing directions, helping persons with a disabled vehicle to correct the problem or arrange for towing service, and so on. However, some administrators feel that these services are too far removed from the law enforcement function and too time-consuming for their officers to undertake. The possibility of lawsuits if an officer attempts to provide a nonenforcement service and something goes wrong also has discouraged many agencies from offering such services.

Police agencies in this country began to assume responsibility for traffic law enforcement after World War I, when automotive traffic grew rapidly with the introduction of relatively inexpensive, mass-produced vehicles.[2] From the beginning, traffic law enforcement has been a troublesome responsibility for the police.

Traffic laws are different from the criminal laws that are the principal concern of the police, in the sense that traffic laws involve the maintenance of public order and the protection of lives and property but not the maintenance of a social code of morality. Exceeding the speed limit or running a stop sign, *if no harm to another person results,* is not an offense against the community's moral code; it is only an offense against a rather arbitrary set of rules imposed to prevent traffic accidents. An armed robbery or burglary is an offense against the whole community *whether or not* the immediate victim is substantially harmed.

This is not to say that traffic laws are unimportant. On the contrary: Far more people are injured, and far more property is lost or damaged, as a result of traffic accidents than through armed robberies or burglaries. The national death toll from traffic accidents has been widely regarded as a tragedy and a disgrace for many years.[3]

One important consequence of the distinction between traffic law enforcement and general criminal law enforcement is that the former brings police officers into contact, in the role of authoritarian law enforcer, with many citizens who do not regard themselves as criminals. Anyone who has ever received a traffic ticket will recall the strong feelings of embarrassment and resentment that the encounter provoked.[4]

Police officers, too, often have strong feelings about these encounters. On the one hand, citing motorists for violating the law is their duty, and (at least in theory) it is an essential means of preventing the traffic accidents that kill or maim hundreds of thousands of citizens every year.

On the other hand, most traffic law violations are not the product of criminal intent, but simply negligence or carelessness. All drivers violate some or all of the traffic laws most of the time, and police officers (whose own driving habits are not always better than those of the general public) realize that only a small minority of the violators are detected and punished. It seems unfair to penalize a few people for actions that are not deliberately malicious, that usually go unpunished, and that in most cases result in no real harm to anyone.

Traffic law enforcement also exposes police officers to danger. A motorist stopped for a traffic violation may become so resentful that a violent confrontation develops, or the motorist may have committed some other, more serious offense and may be determined to avoid capture. In recent years, traffic stops have been among the most common situations in which police officers have been assaulted, sometimes fatally.[5]

In short, the police have had several reasons to wonder whether traffic law enforcement is an appropriate responsibility for them to assume and, if so, how to discharge it effectively without exposing officers unnecessarily to physical danger and to public resentment.

One possible solution is to separate the traffic law enforcement function from the general criminal enforcement function. This can be done to some extent by forming a traffic law enforcement unit within the agency and assigning to it some or all of the traffic law enforcement functions. Most medium-sized and larger police agencies have, in fact, adopted this strategy with varying degrees of success.[6]

TYPES OF TRAFFIC LAW ENFORCEMENT UNITS

For the sake of brevity, we refer to the four main traffic enforcement functions as (1) directing traffic, (2) detecting and citing violators, (3) investigating accidents, and (4) safety education.[7]

Small police agencies (those with twenty-five or fewer patrol officers) generally do not have sufficient personnel to form a separate traffic enforcement unit, and for that matter the natural workload for traffic matters is not large enough to warrant a completely separate unit. The first three functions are therefore assigned to the general patrol force. However, all but the smallest police agencies usually have at least one person, often with the rank of lieutenant or above, whose responsibilities include traffic safety education.[8]

Agencies with more than fifty patrol officers vary widely in the size and assigned duties of their traffic units. Some municipal police departments maintain only a small traffic safety education unit with perhaps five or ten officers. Their primary purpose is to collect information on traffic problems in the community (to advise traffic engineers and other local officials), to conduct safety education and promotional campaigns for the general public, and to assist general patrol officers in carrying out the other three traffic enforcement functions.

Other agencies go to the opposite extreme: their traffic enforcement unit consists of a complete patrol force, covering the agency's geographical jurisdiction (often, though not necessarily, on a twenty-four-hour-a-day basis). Traffic patrol officers perform virtually all accident investigations, all planned surveillances for traffic law violators (such as traffic radar stops), and any traffic-directing activities that may be necessary. General patrol officers are relieved of all responsibility for traffic enforcement, except that they are expected to stop and cite motorists for any flagrant violations they happen to observe, and they are expected to assist in any serious traffic accidents that occur on their beats.

Finally, many agencies adopt a middle ground. The traffic unit is principally concerned with safety education and the collection of traffic accident information for enforcement purposes. In addition, the unit has a small patrol force that conducts planned surveillances and that can be called on to direct traffic when necessary. The patrol component may or may not assume responsibility for accident investigations. Whatever functions are not assigned specifically to the traffic unit are left for the general patrol force.

Most police agencies do not assign specific responsibility for general assistance to motorists to either the traffic unit or the general patrol force. Instead, any officer who has an opportunity to provide a service to a motorist or any other citizen is encouraged by agency policy to do so.

The number of officers needed in a traffic law enforcement unit obviously depends on several factors, the first of which is the kind of responsibilities assigned to the unit. The size of the community, the public's demand for services, and the availability of competent personnel are other significant factors.

OPERATIONS OF THE TRAFFIC LAW ENFORCEMENT UNIT

Traffic Safety Education

The educational function is almost universal among both large and small police agencies, yet the specific methods used to carry out this function are quite diverse. Some agencies rely primarily on the mass media (newspapers, radio, television, and so on) to promote safe driving behavior. Others concentrate on training young people to become safe drivers; traffic safety officers visit schools, organize student safety patrols, promote safety poster contests, and, in some cases, conduct driver training courses.

Nearly all traffic law enforcement officials accept the principle that citing violators is in itself an "educational opportunity." Presumably, a motorist who is given a ticket for a traffic violation is forcefully reminded of the law's existence and is strongly encouraged to obey the law in the future. Some officers conscientiously deliver a brief safety lecture along with the ticket.[9] Unfortunately, the real effectiveness of this kind of education is open to question. People who believe that they have been unfairly punished for doing what almost everyone else does and that their actions have caused no harm to anyone are not likely to be very impressed with this form of education.

The fact that citing motorists for violations may be somewhat ineffective as an educational device does not mean that violators should not be cited. It simply means that the police, if they are seriously interested in educating motorists, must find other means of doing so.

Promotional campaigns, using the mass media, speeches to civic organizations, and so on, certainly have some value. Depending on the skill and ingenuity with which they are prepared and delivered, promotional campaigns can be useful in giving the public information about safe driving techniques and in reminding motorists about the technical details of traffic laws.

Traffic safety campaigns aimed at young people are more likely to be effective. If safe driving habits and attitudes can be taught to young people *before* they begin their driving careers, at a time when they are intellectually and emotionally receptive to being educated, the effects should be substantial and long-lasting. In recent years, there has been some controversy about the effectiveness of driver education courses and some statistical evidence that students who complete these courses are not necessarily safer drivers than students who have not taken the course at all. However, many traffic safety experts believe that the evidence is open to many different interpretations. It may be true that some driver education courses are relatively ineffective, but it has not been proved that all courses are ineffective, or that ineffective courses could not be made more effective (see Figure 15.1).

In most communities, formal driver education courses are the responsibility of the public schools, not of the police. If there are no driver education courses available, the police should exercise leadership in trying to get them established

Figure 15.1. Traffic safety presentations to school children help to form good safety attitudes. (Photo courtesy Texas Dept. of Transportation.)

and should work cooperatively with school authorities to ensure that the courses being offered are taught by competent teachers.

Similarly, the police should work with the schools to develop and carry out traffic safety programs at all grade levels. These programs may consist of classroom visits by traffic safety officers, the organization of student safety patrols, bicycle safety training, poster contests, and any other activities that will reinforce in students' minds the importance of safe driving.[10]

Traffic Accident Investigation

Whenever a motor vehicle collision or other mishap occurs on the streets, most people immediately notify the police. Actually, the laws in many states do not require the police to be notified unless someone is injured or property damage exceeds a minimum amount. Nevertheless, the police are called routinely for even the most minor "fender-benders" and "bumper-bashings."

The responsibilities of a police officer responding to an accident are: (1) to protect the persons and property at the scene from any further harm, (2) to restore the smooth flow of traffic as quickly as possible, and (3) to obtain a complete, accurate report of the accident.[11]

The importance of the first responsibility is obvious and requires no further comment here. The importance of restoring the flow of traffic is not so obvious until one considers that any obstruction to traffic increases the danger of further collisions. Motorists passing the scene of even a minor accident are notoriously inattentive to their own driving, as they try to see what has happened, and are therefore all too likely to bend their own fenders. Except for the most serious accidents that require an extensive investigation, the officer should not hesitate to move the vehicles involved in an accident out of the traffic lanes and to remove any debris from the roadway.

Traffic accident investigation techniques have grown increasingly sophisticated in recent years. There are now computer programs capable of recreating an accident, based on measurements taken at the scene, to demonstrate the movement of each vehicle involved and, thereby, show very accurately how the accident occurred.[12]

For most minor collisions, an elaborate investigation is not required. Statements should be taken from each driver, and the names, addresses, and telephone numbers of any witnesses at the scene should be recorded. The exact location of the vehicles and their paths of motion before the collision should be recorded.

In general, a police investigation of a traffic accident is *not* intended to determine who was at fault. Under the laws of most states, traffic accidents are primarily a civil matter. Fault, liability, and the payment of damages are issues to be settled between the parties, or, if they cannot agree, to be decided in a civil court. The record of the police investigation may be helpful in settling these issues, but the investigation is not for that purpose.

The primary purpose of the police investigation is to improve traffic safety for the public in general. This purpose is served by analyzing traffic accident records to determine patterns of incidents that may indicate a need for more intensive traffic law enforcement at certain locations, or the presence of physical hazards (such as an obstructed view at an intersection) that need to be corrected by engineering means. Thus, the investigating officer's efforts should be devoted to determining exactly *what* happened, not *who* caused the collision.[13]

Most police agencies permit, and indeed encourage, officers to issue citations at the scene of an accident if it appears that one or more of the drivers committed a violation, whether or not the violation directly contributed to the accident.

However, this practice has been called into question, and some administrators now discourage their officers from citing anyone at an accident scene unless the evidence of a violation is overwhelming (for example, if one of the drivers is intoxicated). The reasons are: (1) issuing a citation for a violation that was not witnessed by the officer, but is merely reported by another driver or witness, is legally doubtful (generally, citations for misdemeanor violations are not issued on the basis of hearsay but only on the basis of the officer's own observations); and (2) a citation may be interpreted incorrectly as evidence that the cited motorist caused the accident and therefore bears liability for the damages.

State laws and court precedents regarding liability in traffic accidents are extremely complicated, and the fact that one motorist involved in an accident was cited for a violation may confuse the issue more than clarify it.

Traffic Enforcement

In simplest terms, the enforcement of traffic laws means observing violators and arresting them; in that sense, traffic law enforcement is no different from any other type of law enforcement. As is true for all criminal law enforcement, the officer must follow a set of strictly defined procedures in order to ensure that a violator is convicted and punished.[14]

However, because most traffic law violations are relatively minor, state laws permit officers to issue citations rather than make physical arrests in most cases (see Figure 15.2). Legally, a *citation* is a notice to an offender that a charge is being filed and that the offender must appear in court to answer the charge. Most states require a citation to be signed by the offender, as evidence that he or she is aware that a charge has been filed. Otherwise, it would be possible for police officers to write and file a ticket without a motorist's knowledge.

Whether or not the violator's signature is required on the citation, the officer must obtain positive identification of the person being charged with a violation. This means that the offender must be stopped and asked to produce identification.

In some states, only a valid driver's license can be accepted as identification, and the failure to produce a license is in itself a violation. In other states, a police officer may accept other forms of identification in place of the driver's license, but the motorist must be able to prove to a judge's satisfaction that a valid license was in effect at the time of the offense, or an additional violation will be charged.

In most states, a police officer has some discretion in whether to issue a citation or perform a physical arrest. A physical arrest usually is required for certain serious violations such as driving while intoxicated, reckless driving, or leaving the scene of an accident.

Depending on state laws and agency policies, an officer may be able to make a physical arrest if, in the officer's judgment, the driver is incompetent to operate a motor vehicle, does not have a valid driver's license, poses a continuing danger to other persons, has previously been charged with some number of violations, or is likely to flee to escape the charges.

Persons who are passing through the community and who have an out-of-state license (or, sometimes, persons who reside in another county) are often arrested and taken immediately to court. However, a number of states have established an agreement whereby an out-of-state citizen may be given a citation and, if the citizen fails to appear in court or to pay the fine without contesting the charge, the police in the citizen's home state are authorized to act on behalf of the state in which the citation was issued.

Physical arrest for a traffic violation is, of course, a considerable inconvenience to a motorist. In some cases, especially at night when a judge is not available to settle the charges immediately, an arrest is a greater penalty than the fine that may be imposed. Policies regarding the use of arrest in place of issuing a citation should be worked out carefully and uniformly applied, to be sure that the officers' discretion is exercised consistently and fairly.

Some states' traffic statutes permit officers to issue written warnings for relatively minor violations. Where state law allows this practice, local agencies must adopt clear policies governing the use of warnings.

Generally, *warnings* are given only for certain kinds of infractions such as operating a vehicle whose safety equipment (a brake light, perhaps) is defective or for violations that involve some degree of judgment on the part of the motorist, such as entering an intersection after the signal has turned to yellow (when, in the officer's judgment, the driver could and should have stopped).

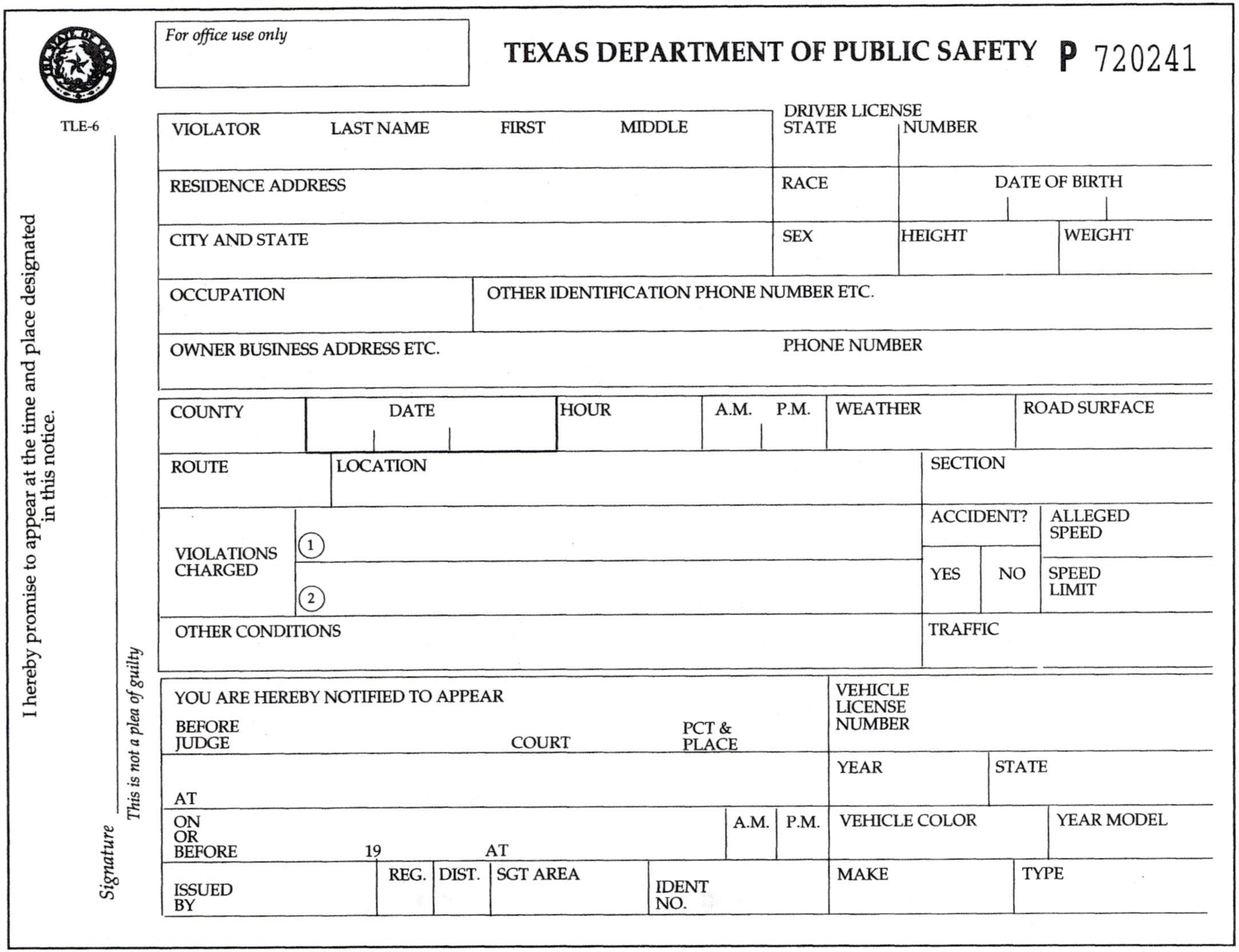

TEXAS DEPARTMENT OF PUBLIC SAFETY P 720241

For office use only

TLE-6

VIOLATOR   LAST NAME   FIRST   MIDDLE

DRIVER LICENSE
STATE   NUMBER

RESIDENCE ADDRESS

RACE   DATE OF BIRTH

CITY AND STATE

SEX   HEIGHT   WEIGHT

OCCUPATION   OTHER IDENTIFICATION PHONE NUMBER ETC.

OWNER BUSINESS ADDRESS ETC.   PHONE NUMBER

COUNTY   DATE   HOUR   A.M.   P.M.   WEATHER   ROAD SURFACE

ROUTE   LOCATION   SECTION

VIOLATIONS CHARGED ①   ACCIDENT?   ALLEGED SPEED

②   YES   NO   SPEED LIMIT

OTHER CONDITIONS   TRAFFIC

YOU ARE HEREBY NOTIFIED TO APPEAR

BEFORE JUDGE   COURT   PCT & PLACE

AT   VEHICLE LICENSE NUMBER

ON OR BEFORE   A.M.   P.M.   YEAR   STATE

19   AT

ISSUED BY   REG.   DIST.   SGT AREA   MAKE   VEHICLE COLOR   YEAR MODEL

IDENT NO.   TYPE

I hereby promise to appear at the time and place designated in this notice.

Signature

This is not a plea of guilty

Figure 15.2. A traffic citation form. (Courtesy Texas Dept. of Public Safety.)

Warnings also may be used for minor violations where there are extenuating circumstances, such as failing to make a complete stop at a stop sign when there is clearly no other traffic approaching the intersection.[15]

Not all police administrators favor the use of warnings. Some feel that if a violation occurs, an officer should issue a citation and let the motorist settle the matter with a judge. Certainly expanding an officer's discretion could lead to abuses and inconsistencies, especially if the agency's policies are not clear and officers are not properly trained and supervised.

On the other hand, the use of warnings puts more emphasis on the education of motorists than on merely enforcing the law. In many cases, a potentially unpleasant confrontation with a hostile motorist can be converted into a productive contact; the motorist is first relieved at not getting a ticket and second (presumably) grateful for the officer's attention and consideration.

If warnings are used, the form should be designed to resemble a citation; there should be appropriate space for the officer to indicate exactly what violation has been observed and for the motorist to sign the warning, as an acknowledgment that it was received. A copy of the warning should be filed at the police agency (see Figure 15.3).

If a motorist accumulates several warnings, especially if they occur in a brief period, a letter should be sent advising the motorist that no further warnings will be given. The motorist's name and license number then should be published on a "no-more-warnings" list that should be distributed periodically to all patrol units.

An even better procedure, if the agency has the proper equipment, is to maintain the warnings file in a computer. Whenever an officer intends to give a written warning, first the computer file should be checked and, if too many warnings have already been given, the motorist should be cited instead.

Consistent policies also must be developed concerning the surveillance techniques to be used by traffic enforcement officers. Most police agencies expect all officers to observe traffic whenever they are on patrol and to make a traffic stop and issue a citation (or make an arrest, if necessary) whenever a violation is observed.

However, such *casual surveillance* is not enough to form the basis for a thorough traffic enforcement program. Other surveillance techniques should be applied to specific locations where they are needed. Later, we will discuss the importance of, and procedures for, *selective enforcement*.

Surveillance techniques used in traffic enforcement may be categorized as *overt* or *covert* and as *stationary* or *mobile*. Overt surveillance means simply that the officer is readily visible to motorists; in fact, the officer's visible presence should be sufficient to discourage most violations. Covert surveillance means that the officer is not visible to potential violators, but is able to observe traffic well enough to detect and apprehend violators. A stationary surveillance may be either overt or covert, depending on where the officer is placed and whether that location is in view of passing motorists. Mobile surveillance is almost always overt; the major exception would be the use of an unmarked patrol car.[16]

HP-3 R-9-80

TEXAS DEPARTMENT OF PUBLIC SAFETY

| DATE | HOUR |
| | M. |

| DRIVER NAME | LAST | FIRST | MIDDLE |

| VEHICLE MAKE & TYPE | LICENSE NO. | STATE | D.L. NO. | STATE |

| COUNTY | LOCATION & ROUTE |

VIOLATION(S)

THIS IS A WARNING ONLY for an infraction of the traffic laws committed to a minor degree or with extenuating circumstances present. No penalty will be assessed and no further action on your part is necessary other than to comply with the traffic laws in the future. This does not become a part of your driving record.

SIGNATURE   X

This warning is given to you in an effort to secure your cooperation in better observance of the traffic laws thus helping to prevent traffic accidents. The Texas Department of Public Safety believes that good citizens will comply with traffic laws when reminded of their provisions and of the importance of strict compliance with them.

ISSUED BY

| SERVICE REGION/AREA | I.D. NO. |

MOORE BUSINESS FORMS, INC., AUSTIN, TX. M

Figure 15.3. A traffic warning form. (Courtesy Texas Dept. of Public Safety.)

Covert mobile surveillance is a somewhat controversial technique in some communities, where the public regards it as unfair. For that matter, some motorists regard *any* covert surveillance as unfair, including the placement of radar stops behind bushes, just beyond the crest of a hill or around a bend in the road.

Of course, from a law enforcement viewpoint, the perceived unfairness of covert surveillance is more than outweighed by the seriousness of traffic enforcement and the likely consequences of unsafe driving.

Aerial surveillance, in which either small airplanes or helicopters are used to observe and detect violations, is an extremely effective technique. An airborne officer can spot many violations and report them to a unit on the ground, which then makes the stop and issues a citation. Although airborne units are not actually hidden from view, the fact is that motorists are rarely aware of their presence, and therefore aerial surveillance has many of the advantages of covert surveillance.

The issue of fairness also has been raised with regard to aerial surveillance, and this technique has been prohibited in some jurisdictions. Aerial surveillance also is a relatively expensive technique, because of the operating costs of aircraft and the need to have one or more ground units available at all times in addition to the airborne unit.

Directing of Traffic

Finally, the traffic responsibilities of the police include controlling the flow of traffic, particularly around obstructions or in areas of extreme congestion. At one time, directing traffic was a major function of the police in almost all communities. The development of mechanical and electric traffic signals has reduced the need for this kind of activity, which is extremely expensive because of the personnel time required.

Modern traffic signal systems can be controlled by a computer that reacts to the actual flow of traffic more rapidly and more effectively than any system of direction by officers. Nevertheless, there are circumstances when officers must direct traffic themselves. Even the most sophisticated traffic signal system may break down.

With or without modern signals, traffic congestion may occur at the scene of an accident, near a community event such as a dance or football game, around construction areas, or any place where the volume of traffic exceeds the carrying capacity of the streets (such as downtown intersections during the rush hour).

The techniques of directing traffic are fairly simple and should be learned by all police officers in recruit school. From an administrator's point of view, traffic direction requires some policies concerning the circumstances under which officers will be used for this purpose: who is to have the responsibility of assigning officers to direct traffic, and in cases in which traffic must be directed over a large area, how the flow of traffic is to be coordinated.

General Assistance to Motorists

There are two major purposes to this aspect of the traffic law enforcement function: (1) to improve the safe and efficient movement of traffic by removing potential obstacles, and (2) to encourage positive contacts between officers and the public.

The kinds of assistance to be offered should be determined by agency policy, which in turn will be influenced by the administrators' perception of public demand and by the availability of resources (especially personnel).

Usually, police officers should freely offer directions to motorists who are unfamiliar with an area, since people wandering about are especially likely to cause a traffic hazard. Officers also should make some effort to assist motorists stranded by a disabled vehicle.

Depending on the circumstances (including the availability of other assistance), an officer might help an elderly motorist change a tire, or might push a disabled vehicle off the road onto a safer parking lot or shoulder, or might even perform simple on-the-spot repairs (such as reconnecting a loose battery wire or advising a motorist on how to correct a flooded carburetor). An officer certainly should be willing to call for a tow truck or other assistance and remain with the disabled vehicle until help arrives.

Some agencies go so far as to equip patrol vehicles with tow chains, emergency tire and fan-belt repair kits, and other special equipment to assist motorists. The practice of a few agencies—carrying gasoline in the trunk of a patrol car to help motorists who have run out of gas—should be discouraged because of the safety hazard to the patrol officers.

POLICE AND TRAFFIC ENGINEERING

In the past, many local law enforcement agencies have been given, or have assumed, responsibility for traffic engineering; that is, designing streets and highways, right-of-way devices (guard rails, pavement markings, signs, and signals), and traffic flow patterns.

Today, this function is more often assigned to other governmental agencies at the state and local level. Nevertheless, many police administrators still feel that they should have a role in traffic engineering, and quite a few medium-sized and larger law enforcement agencies have full-time traffic engineering staffs.[17]

Whether this practice is necessary should be decided as a matter of local governmental policy. In our view, the principal responsibility of the police with regard to traffic engineering should be limited to supplying information, based on analyses of the agency's traffic incident records, and offering advice or recommendations. It may be true that smaller communities often cannot afford to maintain a full-time professional traffic engineering department, but in that case the equally limited resources of the police department should not be stretched to make up the deficiency. The proper function of a law enforcement agency is law enforcement, not engineering.[18]

However, this is not to say that the police have no role in traffic engineering. A police agency's records of traffic accidents and patterns of violations should be regarded as a vital resource for traffic engineers. The agency's own traffic law enforcement staff should analyze these records periodically and prepare recommendations to the engineers responsible for street and highway design in the community. If the police agency's staff does not have the expertise or the time to do the analyses, at least the raw information should be provided to the highway engineers for their use. Quite often, a pattern of accidents or violations may reveal a defect in the street or highway design[19] (see Figure 15.4).

SPECIAL PROBLEMS IN TRAFFIC LAW ENFORCEMENT

The great majority of traffic violations are not very serious. Drivers exceed the legal speed limit, "slide" through stop signs, change lanes without signalling, and otherwise ignore the rules of the road, not out of malice but out of ignorance or carelessness. Enforcement action consists merely of stopping the violator, obtaining identification, explaining the nature of the violation and its potential consequences, and issuing a warning or citation. The motorist generally realizes that he or she is guilty and pays the fine without protest. Whether the motorist also changes his or her driving behavior is, as we said earlier, somewhat problematical.

Some violations, however, are considerably more serious, and some traffic situations are considerably more difficult for traffic law enforcement officers to handle.

Enforcement of Driving-While-Intoxicated Violations

Perhaps the most serious traffic violation is DWI, whether under the influence of alcohol or of drugs, including prescription medication. It is widely believed that about half of all fatal accidents involve one or more persons who are intoxicated.[20] In some communities, intensive campaigns to enforce the DWI laws have had impressive effects on the accident rate.

Making a case against an intoxicated driver is not always easy. Not all intoxicated drivers are affected by alcohol or drugs to such an extent that their behavior is clearly aberrant. Sometimes it is difficult even to detect the intoxicated state. Furthermore, by the time a motorist cited or arrested for DWI appears in court, usually enough time has elapsed that the person is sober; in the absence of any other evidence, the prosecutor has only the officer's word concerning the motorist's state of intoxication.

The first problem is to detect the level of intoxication (if any) when a motorist is stopped. Whenever a motorist is stopped for any unlawful or dangerous behavior, the officer must consider the possibility that the motorist is intoxicated; however, it is only a possibility.[21] People who are extremely fatigued, ill, suffering from unanticipated side effects of medication, or simply incompetent may behave in the same manner associated with intoxication. Conversely, a motorist may be coherent, able to converse intelligently, and apparently sober when, in fact, he or she is under the influence of alcohol or drugs.

When intoxication is suspected, an officer should require the motorist to leave the vehicle and move to a safe place (either a sidewalk or, on a rural road,

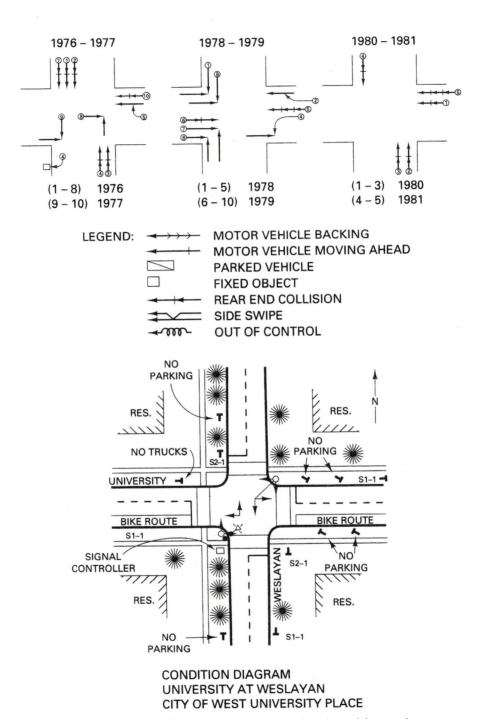

CONDITION DIAGRAM
UNIVERSITY AT WESLAYAN
CITY OF WEST UNIVERSITY PLACE

Figure 15.4. A traffic accident analysis, developed by engineers from accident report data. (Courtesy Texas Dept. of Transportation.)

well off the pavement). The officer then should conduct a *field sobriety test*. The test consists of giving the motorist a series of instructions for simple acts, such as closing one's eyes and touching the end of one's nose with the index finger. Several of these tests have been developed and used with varying degrees of success.[22]

In most states, a police officer is authorized to ask a driver to submit to a *breath analysis test*. This test consists of having the subject exhale into a device that performs a chemical analysis to detect the presence of alcohol in the breath. Some breath analysis machines also have been designed to detect certain drugs, such as marijuana, although the effectiveness and accuracy of these devices have not been thoroughly established.[23]

Both the field sobriety test and the breath analysis test are merely screening devices; that is, they neither prove nor disprove the fact that the subject is intoxicated. What matters is not the motorist's score on a field test, but the officer's overall impression of the motorist's behavior (before and after being stopped) and performance on the tests. If a motorist refuses to submit to either a field sobriety or breath analysis test, in most states the officer is empowered to arrest the motorist on suspicion of intoxication. In some states, refusal to submit to a field test may be grounds for suspension of the driver's license.

Whatever the outcome of the field tests, if the officer remains convinced that the motorist *may* be intoxicated, the motorist should be arrested and transported to the police station. At the police station, more thorough tests may be performed. A more elaborate and accurate breath analysis device may be used, or samples of the subject's urine or blood may be taken by qualified medical personnel. These types of tests are designed to detect the presence and degree of concentration of alcohol or drugs in the bloodstream.[24] Again, in some states a person suspected of DWI who refuses to submit to these tests may face license suspension, even if the person turns out not to have been intoxicated.[25]

Persons arrested on suspicion of DWI also should be observed at the police station, and if possible their behavior should be recorded on videotape. Most courts accept videotape recordings, when made under properly controlled circumstances, as evidence. A tape recording showing an intoxicated motorist's behavior often proves to be very persuasive in court.

Gaining convictions of intoxicated drivers is not as difficult as many people believe.[26] In recent years, a great deal of attention has been devoted to this serious problem. State laws concerning DWI have been tightened, and in many jurisdictions the courts and juries have begun to understand the serious nature of this common violation. If nothing else, the conventional attitude that drunk driving is socially acceptable has changed. Most people realize that drunk driving is not amusing, and that innocent people are entitled to travel on the streets and highways without being endangered by drunks.

Expressway Enforcement

Expressways, freeways, turnpikes, toll roads—whatever they are called—limited-access, high-speed highways generate several problems for traffic law enforcement officers.

In urban areas, expressways are often designed so that the roadway takes up all of the right-of-way. There is little or no space alongside the roadway for motorists to pull off, out of the line of traffic. There is no place for a traffic patrol officer to have a motorist pull out of traffic for an enforcement stop, without creating a serious hazard to other traffic. It may be necessary for the patrol officer to follow the violator to the nearest exit ramp and leave the expressway altogether. Because exit ramps may be spaced a mile or more apart, this is not an entirely satisfactory situation.

Expressways are designed for high-speed travel. Many expressways were designed for speeds of 65 or 70 miles per hour (mph). Consequently, motorists often find themselves exceeding the legal speed limit even when they do not intend to do so. Those who deliberately disregard the speed limit may be traveling at 80 mph or more. In light traffic and on rural stretches of expressway, speeds in excess of 100 mph are not uncommon. Stopping a violator who is traveling at such very high speeds is difficult and dangerous.[27]

We will discuss the special problems of pursuit driving shortly. With regard to expressway traffic enforcement, one practical solution is to use two or more patrol units for speed enforcement. The two units may be spaced about a mile apart on one side of the expressway. The first unit serves as a spotter and the second as an interceptor.

When the spotter observes a vehicle in violation of the speed limit, or other traffic laws, a description of the violator's vehicle is radioed to the interceptor. The interceptor should then move onto the highway *ahead* of the violator and, using flashing lights as appropriate, signal the violator to move off the road. If the interceptor waits until the violator has passed and then pulls onto the highway, a pursuit situation is created, in which case much of the advantage of the two-unit procedure is lost.

Another problem with expressway traffic enforcement is that, in some communities, jurisdiction over limited-access highways may be divided among several different law enforcement agencies. In some cases, a separate police agency is responsible for enforcement on a freeway, turnpike, or toll road; the local municipal police may not have enforcement authority. In other cases, expressways or freeways that are part of the interstate highway system may be under the authority of the state highway patrol. In yet other cases, the route of an expressway passes through a series of small suburban communities and in and out of the city limits of the central city. Thus, jurisdiction may be fragmented among several municipal police departments.

In all such situations, close cooperation among all the affected law enforcement agencies must be arranged at the highest administrative levels and maintained at the operating level. Most states have *"hot pursuit"* laws, under which a police officer in pursuit of a suspected offender (whether for a traffic violation or some other crime) may make a legal arrest beyond his or her proper jurisdiction. However, the precise definition of hot pursuit and the offenses to which it applies vary from state to state.[28]

Rather than relying on hot pursuit laws, police officials in adjacent communities, or those who share jurisdiction over limited-access highways, should establish procedures under which an officer in one agency may request and receive assistance from officers in another agency. For example, there should be no legal barrier to using officers from two different agencies as spotters and interceptors for expressway speed-limit enforcement, if mechanisms exist for the spotter (the officer who initially observes the violation) to provide evidence against an offender who is cited or arrested in the interceptor's jurisdiction.

Mandatory Safety Restraints and Devices

Federal laws prescribe the minimum safety equipment that must be present, and in working order, on all automobiles sold in the United States. Many states also have laws regarding minimum safety equipment such as lights, windshield wipers, and horns. A majority of the states require all licensed vehicles to be inspected once a year (in some states, at more frequent intervals) to detect any defective safety equipment. However, the inspection programs do not prevent an officer from enforcing the safety equipment laws if defective equipment is observed on the highway.

All states require motorists to place children in federally approved *child safety seats* or *restraints*. Unfortunately, there are major variations in the wording and intent of these laws, and they are difficult to enforce. In many cases, it is impractical for officers to enforce mandatory child safety restraint laws on the basis of their observations of passing traffic. Only the most flagrant violations, such as a motorist whose small child is standing on the front seat or sitting in the driver's lap, are readily observable. In other cases, an officer's only opportunity to observe a violation occurs when a motorist is stopped for some other reason (such as a speed-limit violation), and the officer finds a child in the vehicle without a proper restraint.

Despite the enforcement problems, child safety restraint laws are important. Traffic accidents are the leading cause of death among children under four years of age in the United States, and child safety restraints are a proven method of reducing this tragic statistic. All officers, and particularly traffic patrol officers, should be thoroughly familiar with the law and should be encouraged to enforce it.

At present, most states have mandatory restraint laws applying to adults. Again, enforcement is difficult; usually enforcement occurs only when motorists are stopped for other violations, or when an accident occurs and the driver or passenger was not properly restrained.

Unlicensed Drivers

Current law in every state forbids the operation of a motor vehicle on any public road without a valid operator's license or permit. The courts have found consistently that the holding of a driver's license is a privilege granted by the state, and that the state has the right to withdraw, suspend, or revoke that privilege for reasonable cause.

Unfortunately, it is not possible for law enforcement officers to determine whether a motorist is licensed or not merely by observing the motorist's behavior. Thus there are only two possible methods of enforcing the driver's license laws: checking licenses whenever a motorist is stopped for some other violation and conducting periodic license inspections at random roadblocks. The first method is standard procedure for all traffic enforcement. The second method, however, has been successfully challenged in some courts.

A few courts have ruled that random stops for license inspections or other purposes violate the constitutional rights of citizens (usually on the grounds that they are unreasonable searches). However, most state and federal courts have ruled just the opposite: that random stops are permissible *if* all motorists are stopped for the same purpose and the purpose is limited to checking licenses, vehicle safety equipment, detecting drunk drivers, and the like.[29]

Random stops and inspections for violations of other laws (such as possession of drugs or controlled substances) are less likely to be permitted by the courts, and in some states evidence of criminal offenses that may be found in the course of a driver's license inspection may be inadmissible. Again, however, court decisions on these questions have varied from state to state, and even the federal courts have been inconsistent. Patrol officers must be made aware of the most current applicable decisions, and agency policies must be established accordingly.

Commercial Vehicles

Trucks, vans, buses, taxicabs, and other motor vehicles used to transport people or goods for hire are subject to many laws that do not apply to private vehicles. Furthermore, commercial vehicles generally are on the roads more often, for longer periods, than private vehicles. The drivers of commercial vehicles may be under pressure from their employers to do their work as quickly as possible. As a result, commercial drivers sometimes are careless about observing traffic laws.

Traffic law enforcement officers may be reluctant to enforce the laws against commercial drivers, knowing that a driver could lose his or her livelihood as a result of being convicted for one or more violations. Officers may reason that most commercial drivers, because of their special training and extensive experience, are more competent than most other drivers, and so their violations should be tolerated to some extent.

The dangers in this attitude should be obvious. If commercial drivers are permitted to flout the laws with impunity, other drivers are also encouraged to disregard the rules. It may be true that many commercial drivers are more competent than the average driver, but that is certainly not true in all cases. Commercial vehicles, because they generally are larger and heavier than private vehicles, pose a greater safety hazard. In a collision, the extent of damage and the likelihood of injury are significantly higher when a heavy commercial vehicle is involved. Furthermore, some commercial vehicles are used to transport extremely hazardous materials that, in a collision, could cause enormous property damage and loss of life.[30]

If anything, traffic enforcement should be even more stringent for commercial drivers than for others. Commercial drivers should be expected to know and understand the laws, and there should be no excuse for their failure to comply.

Because their income depends on their holding a commercial operator's license, they should be more conscientious, not less, about observing the law. Traffic officers should not hesitate to cite a commercial driver for speeding, failing to yield the right of way, or other common offenses. Agency policy should demand strict enforcement.[31]

The special laws applying to commercial vehicles also should be strictly enforced. In particular, laws regarding a commercial vehicle's load should receive the attention of traffic officers.

It is difficult for traffic officers to determine by casual observation whether a truck is over the legal weight limit, but other loading violations are observable. A load that is too large for the vehicle or that is improperly secured can interfere with the driver's ability to control the vehicle safely, and an improperly secured load may scatter debris into the path of other motorists. These violations have direct, sometimes drastic, effects on the safety of all travelers and should not be tolerated.

Overweight vehicles can be detected if the vehicle is clearly unable to maintain a normal speed, if the tires and suspension are severely strained, or if the driver appears to have difficulty in performing normal maneuvers such as turning, braking, and shifting gears.

When an overweight vehicle is suspected, the vehicle should be stopped and the driver questioned. If the officer continues to believe that the vehicle may be overloaded, the driver should be required to proceed to the nearest scale for further inspection.

In some cases, unfortunately, scales may not be readily available. However, the police usually can arrange to use privately owned scales at warehouses, scrap recycling facilities, and other businesses. So-called portable scales, designed to be carried in the trunk of a patrol car, are not very satisfactory: They are heavy, bulky, and awkward to use. Generally they are useful, if at all, only by officers whose specific responsibility is checking vehicle weights.

Commercial drivers who operate in more than one state are now required to hold a federally issued license. They are no longer permitted to obtain separate licenses in several states. Eventually, all violations committed by federally-licensed drivers will be recorded in a central computer system. At present, violation records are maintained by the states.

Parking Enforcement

Many municipal police agencies regard parking violations as too trivial to deserve their attention. Parking meters and restrictions on on-street parking seem designed primarily to produce revenue and to protect the privileges of local businesses. Especially where a separate agency has been established to enforce the parking laws, the police may feel free to ignore parking violations.

However, parking violations do have important implications for traffic safety and for the efficient movement of traffic. If metered spaces are monopolized by a few people, other drivers will be forced to waste time and fuel searching for a parking space; as they grow increasingly frustrated, they also become inattentive to their driving. If cars are parked in restricted spaces, they may obstruct moving traffic or interfere with motorists' view of oncoming traffic at intersections. And if people are permitted to leave their cars in no-parking spaces and otherwise ignore the parking regulations, a general attitude of disregard for all traffic laws may be encouraged.[32]

Consequently, all police officers should be expected to observe parking violations and to take appropriate enforcement action. Even in communities where the primary responsibility for parking enforcement has been assigned to a specific police unit or another agency, general patrol officers and traffic officers should at least enforce the laws against more serious violations such as leaving an unattended vehicle in the street (double parking), parking in restricted spaces (near a fire hydrant or too close to an intersection), and parking in an unsafe manner (parking on the wrong side of the street, facing oncoming traffic).

SELECTIVE ENFORCEMENT

Police officials have realized for many years that it is impossible to enforce every violation of the traffic laws. Most violations go unobserved; they occur when no police officer is present.

Many other violations are so minor that taking action to enforce the law would create a greater hazard than tolerating the violation. For example, when a driver changes lanes without signalling in heavy traffic, unless the maneuver seriously and dangerously interferes with another driver, there is little to be gained by pursuing and stopping the offender.

Some violations are so common that it is impossible to stop and cite every offender. It is a common experience on limited-access highways that more than half of the drivers exceed the speed limit by anywhere from 5 to 15 miles per hour.[33]

The only possible course for the police is to enforce the laws selectively. *Selective enforcement* means that a specific effort is made to enforce particular laws, primarily on the basis of evidence that certain violations contribute to the frequency or seriousness of accidents. Enforcement also can be selective regarding the places where the laws are enforced.[34]

For example, an agency will analyze its records of traffic accidents and citations to determine what patterns exist. The analysis might show that an abnormally high number of accidents have occurred at a certain intersection, and that the cause of most accidents has been the failure of drivers to heed a stop sign or traffic signal.

This analysis would provide the basis for a selective enforcement effort. Officers would be assigned to observe traffic at that intersection and to concentrate on right-of-way violations.

Selective enforcement can be a powerful tool in achieving traffic safety goals. For this reason, the National Highway Traffic Safety Administration, an agency of the U.S. Department of Transportation, has established a special program to encourage police agencies to adopt selective enforcement projects. The *Selective Traffic Enforcement Program (STEP)* provides federal funds to support these projects. The funds are allocated to every state on the basis of a formula

TEXAS HIGHWAY PATROL
MAJOR ACCIDENT
INVESTIGATION

(REQUIRED ON ALL FATALS)

PLACE WHERE
ACCIDENT OCCURRED

COUNTY _____   CITY OR TOWN _____

SHOW ONLY IF INSIDE CITY LIMITS

IF ACCIDENT WAS OUTSIDE CITY LIMITS,
INDICATE DISTANCE FROM NEAREST TOWN _____ MILES ☐ ☐ ☐ ☐ OF _____

NORTH   S   E   W   CITY OR TOWN

ROAD ON WHICH
ACCIDENT OCCURRED _____   CONSTR. ☐ YES   SPEED

BLOCK NUMBER   STREET OR ROAD NAME   ROUTE NUMBER OR STREET CODE   ZONE   ☐ NO   LIMIT ___

INTERSECTING STREET
OR RR X'ING NUMBER

BLOCK NUMBER   STREET OR ROAD NAME   ROUTE NUMBER OR STREET CODE   CONSTR. ☐ YES   SPEED

NOT AT INTERSECTION _____ ☐ FT. ☐ ☐ ☐ ☐ OF _____   ZONE ☐ NO   LIMIT ___

☐ MI. N S E W   SHOW MILEPOST OR NEAREST INTERSECTING NUMBERED HIGHWAY.   GRID SQUARE

IF NONE, SHOW NEAREST INTERSECTING STREET OR REFERENCE POINT.

DATE OF   DAY OF   ☐ A.M. IF EXACTLY NOON
ACCIDENT _____ 19 _____   WEEK _____ HOUR _____   ☐ P.M. OR MIDNIGHT, SO STATE

TROOPER _____   AREA _____
   Investigator

Figure 15.5. There are more than 30 pages of forms in this manual, to be used in the investigation of serious traffic accidents. The information can be used for engineering, safety education, and legal purposes. (Courtesy Texas Dept. of Public Safety.)

PLACE WHERE
ACCIDENT OCCURRED

COUNTY _____   CITY OR TOWN _____   LOC. NO. _____

SHOW ONLY IF INSIDE CITY LIMITS

IF ACCIDENT WAS OUTSIDE CITY LIMITS.
INDICATE DISTANCE FROM NEAREST TOWN _____   MILES ☐ ☐ ☐ ☐   OF _____   DO NOT WRITE IN THIS SPACE

NORTH   S   E   W   CITY OR TOWN

DPS. NO. _____

ROAD ON WHICH
ACCIDENT OCCURRED ___ BLOCK NUMBER ___ STREET OR ROAD NAME ___ ROUTE NUMBER OR STREET CODE   CONSTR: ☐ YES ☐ NO   SPEED LIMIT _____   ZONE: LOC. _____

INTERSECTING STREET
OR RR X'ING NUMBER ___ BLOCK NUMBER ___ STREET OR ROAD NAME ___ ROUTE NUMBER OR STREET CODE   CONSTR: ☐ YES ☐ NO   SPEED LIMIT _____   ZONE: CODE ___ SEVERITY _____

NOT AT INTERSECTION _____ ☐ FT. ☐ ☐ ☐ ☐   OF _____   TYPE _____
    ☐ MI.   N   S   E   W   SHOW MILEPOST OR NEAREST INTERSECTING NUMBERED HIGHWAY.
    IF NONE, SHOW NEAREST INTERSECTING STREET OR REFERENCE POINT.   FAT. REC _____ DR. REC _____

DATE OF
ACCIDENT _____ 19_____   DAY OF WEEK _____ HOUR _____   ☐ A.M.   IF EXACTLY NOON ☐ P.M.   OR MIDNIGHT, SO STATE

UNIT NO. 1   – MOTOR VEHICLE   VEH. IDENT. NUMBER _____

YEAR MODEL _____   COLOR & MAKE _____   MODEL NAME _____   BODY STYLE _____   LICENSE PLATE _____

    YEAR   STATE   NUMBER

DRIVER'S NAME _____   PHONE NUMBER _____
    LAST   FIRST   MIDDLE   ADDRESS   CITY   STATE

DRIVER'S LICENSE _____ D.O.B. _____ RACE _____ SEX _____ OCCUPATION _____
    STATE   NUMBER   CLASS/TYPE   MONTH-DAY-YEAR

LESSEE OWNER _____   PEACE OFFICER OR FIRE FIGHTER ON EMERGENCY?
    NAME (ALWAYS SHOW LESSEE IF LEASED, OTHERWISE SHOW OWNER)   ADDRESS   CITY   STATE   NO   YES   IF YES, DESCRIBE IN NARRATIVE

LIABILITY INSURANCE   YES   NO _____ INSURANCE COMPANY NAME _____ POLICY NUMBER _____ VEHICLE DAMAGE RATING _____

UNIT NO. 2   – MOTOR VEHICLE ☐ TRAIN ☐ PEDALCYCLIST ☐ PEDESTRIAN ☐   VEH. IDENT. NUMBER _____
TOWED ☐ OTHER ☐

YEAR MODEL _____   COLOR & MAKE _____   MODEL NAME _____   BODY STYLE _____   LICENSE PLATE _____

    YEAR   STATE   NUMBER

DRIVER'S NAME _____   PHONE NUMBER _____
    LAST   FIRST   MIDDLE   ADDRESS   CITY   STATE

DRIVER'S LICENSE _____ D.O.B. _____ RACE _____ SEX _____ OCCUPATION _____
    STATE   NUMBER   CLASS/TYPE   MONTH-DAY-YEAR

LESSEE OWNER _____   PEACE OFFICER OR FIRE FIGHTER ON EMERGENCY?
    NAME (ALWAYS SHOW LESSEE IF LEASED, OTHERWISE SHOW OWNER)   ADDRESS   CITY   STATE   NO   YES   IF YES, DESCRIBE IN NARRATIVE

LIABILITY INSURANCE   YES   NO _____ INSURANCE COMPANY NAME _____ POLICY NUMBER _____ VEHICLE DAMAGE RATING _____

DAMAGE TO PROPERTY OTHER THAN VEHICLES $ _____ DAMAGE ESTIMATE

OBJECT _____ NAME AND ADDRESS OF OWNER _____ FEET FROM CURB _____

OBJECT _____ NAME AND ADDRESS OF OWNER _____ FEET FROM CURB _____ $ _____ DAMAGE ESTIMATE

LIGHT CONDITION ☐	WEATHER ☐	SURFACE CONDITION ☐	TYPE ROAD SURFACE ☐	DESCRIBE ROAD CONDITIONS (INVESTIGATOR'S OPINION)
1- DAYLIGHT 2-DAWN 3-DARK-NOT LIGHTED 4-DARK-LIGHTED 5-DUSK	1- CLEAR/CLOUDY 6-SMOKE 2-RAINING 7-SLEETING 3-SNOWING 8-HIGH WINDS 4-FOG 9-OTHER 5-BLOWING DUST	1- DRY 2-WET 3-MUDDY 4-SNOWY/ICY 5-OTHER	1-BLACKTOP 2-CONCRETE 3-GRAVEL 4-SHELL 5-DIRT 6-OTHER	_____ _____ _____

IN YOUR OPINION, DID THIS ACCIDENT RESULT IN AT LEAST $500.00 DAMAGE TO ANY ONE PERSON'S PROPERTY?   ☐ YES   ☐ NO

CHARGES FILED   CITATION NUMBER _____

NAME _____ CHARGE _____

NAME _____ CHARGE _____   CITATION NUMBER _____

TIME NOTIFIED OF ACCIDENT _____ DATE _____ HOUR _____ M   HOW _____   TIME ARRIVED AT SCENE OF ACCIDENT _____ DATE _____ HOUR _____ M

TYPED OR PRINTED NAME OF INVESTIGATOR _____ DATE REPORT MADE _____   IS REPORT COMPLETE ☐ YES ☐ NO

SIGNATURE OF INVESTIGATOR _____ ID NO. _____ DEPARTMENT _____ DIST./AREA _____

Figure 15.5. Continued.

CODE FOR TYPE SPECIMEN TAKEN FOR ALCOHOL/DRUG ANALYSIS	CODE FOR TYPE RESTRAINT USED	CODE FOR INJURY SEVERITY (Use only the most serious one in each space for injury.)
A–Breath B–Blood C–Other N–None R–Refused	A–Seat Belt & Shoulder Strap B–Seat Belt & No Shoulder Strap C–Child Restraint D–Air Bag Deployed E–Shoulder Strap Only N–None	K–Killed A–Incapacitating Injury – Severe injury which prevents continuation of normal activities. Includes broken or distorted limbs, internal Injuries, crushed chest, etc. B–Nonincapacitating Injury – Evident injury such as bruises, abrasions, minor lacerations which do not incapacitate. C–Possible Injury – Injury which is claimed, reported or indicated by behavior, but without visible wounds. Includes limping, momentary unconsciousness or complaint of pain. N–Not injured

UNIT NO. 1

DAMAGE RATING _____   VEHICLE REMOVED TO _____

BY _____

Item No.	OCCUPANT'S POSITION	COMPLETE ALL DATA ON ALL OCCUPANTS' NAMES, POSITIONS, RESTRAINTS USED, ETC.; HOWEVER, IT IS NOT NECESSARY TO SHOW ADDRESSES UNLESS KILLED OR INJURED.		TYPE SPECIMEN TAKEN	RESULT	TYPE RESTRAINT USED	AGE	SEX	INJURY CODE
		NAME (LAST NAME FIRST)	ADDRESS						
1	DRIVER	See Front							
2									
3									
4									
5									
6									

UNIT NO. 2 (Complete only if Unit No. 2 was a motor vehicle.)

DAMAGE RATING _____   VEHICLE REMOVED TO _____

BY _____

Item No.	OCCUPANT'S POSITION	COMPLETE ALL DATA ON ALL OCCUPANTS' NAMES, POSITIONS, RESTRAINTS USED, ETC.; HOWEVER, IT IS NOT NECESSARY TO SHOW ADDRESSES UNLESS KILLED OR INJURED.		TYPE SPECIMEN TAKEN	RESULT	TYPE RESTRAINT USED	AGE	SEX	INJURY CODE
		NAME (LAST NAME FIRST)	ADDRESS						
7	DRIVER	See Front							
8									
9									
10									
11									
12									

COMPLETE IF CASUALTIES NOT IN MOTOR VEHICLE

Item No.	PEDESTRIAN, PEDAL CYCLIST ETC.	CASUALTY NAME (LAST NAME FIRST)	CASUALTY ADDRESS	TYPE SPECIMEN TAKEN	RESULT	AGE	SEX	INJURY CODE
13								
14								

DISPOSITION OF KILLED AND INJURED — IF AMBULANCE USED, SHOW

ITEM NUMBERS	TAKEN TO	BY	TIME NOTIFIED	TIME ARRIVED AT SCENE	NO. ATTENDANTS INC. DRIVER

INVESTIGATOR'S NARRATIVE OPINION OF WHAT HAPPENED (ATTACH ADDITIONAL SHEETS IF NECESSARY)

DIAGRAM   ☐ ONE WAY   ☐ TWO WAY   ☐ DIVIDED

○ INDICATE NORTH

FACTORS AND CONDITIONS LISTED ARE THE INVESTIGATOR'S OPINION

TRAFFIC CONTROL	
0 - NO CONTROL OR INOPERATIVE	5 - TURN MARKS
1 - OFFICER OR FLAGMAN	6 - WARNING SIGN
2 - STOP AND GO SIGNAL	7 - RR GATES OR SIGNALS
3 - STOP SIGN	8 - YIELD SIGN
4 - FLASHING RED LIGHT	9 - CENTER STRIPE OR DIVIDER
10 - NO PASSING ZONE	11 - OTHER CONTROL

FACTORS/CONDITIONS CONTRIBUTING

	1	2	3
UNIT 1			
UNIT 2			

OTHER FACTORS/CONDITIONS MAY OR MAY NOT HAVE CONTRIBUTED

	1	2
UNIT 1		
UNIT 2		

1. Animal on Road – Domestic
2. Animal on Road – Wild
3. Backed Without Safety
4. Changed Lane When Unsafe
5. Defective or No Headlamps
6. Defective or No Stop Lamps
7. Defective or No Tail Lamps
8. Defective or No Turn Signal Lamps
9. Defective or No Trailer Brakes
10. Defective or No Vehicle Brakes
11. Defective Steering Mechanism
12. Defective or Slick Tires
13. Defective Trailer Hitch
14. Disabled in Traffic Lane
15. Disregarded Stop and Go Signal
16. Disregarded Stop Sign or Light
17. Disregarded Turn Marks at Intersection
18. Disregarded Warning Sign at Construction

19. Distraction in Vehicle
20. Driver Inattention
21. Drove Without Headlights
22. Failed to Control Speed
23. Failed to Drive in Single Lane
24. Failed to Give Half of Roadway
25. Failed to Heed Warning Sign
26. Failed to Pass to Left Safely
27. Failed to Pass to Right Safely
28. Failed to Signal or Gave Wrong Signal
29. Failed to Stop at Proper Place
30. Failed to Stop for School Bus
31. Failed to Stop for Train
32. Failed to Yield ROW – Emergency Vehicle
33. Failed to Yield ROW – Open Intersection
34. Failed to Yield ROW – Private Drive
35. Failed to Yield ROW – Stop Sign
36. Failed to Yield ROW – To Pedestrian

37. Failed to Yield ROW – Turning Left
38. Failed to Yield ROW – Turn on Red
39. Failed to Yield ROW – Yield Sign
40. Fatigued or Asleep
41. Faulty Evasive Action
42. Fire in Vehicle
43. Fleeing or Evading Police
44. Followed too Closely
45. Had Been Drinking
46. Handicapped Driver (Explain in Narrative)
47. ILL (Explain in Narrative)
48. Impaired Visibility (Explain in Narrative)
49. Improper Start From Parked Position
50. Load Not Secured
51. Opened Door into Traffic Lane
52. Oversize Vehicle or Load
53. Overtake and Pass Insufficient Clearance
54. Parked and Failed to Set Brakes
55. Parked in Traffic Lane

56. Parked Without Lights
57. Passed in No Passing Zone
58. Passed on Right Shoulder
59. Pedestrian Failed to yield ROW to Vehicle
60. Speeding - Unsafe (Under Limit)
61. Speeding - Over Limit
62. Taking Medication (Explain in Narrative)
63. Turned Improperly - Cut Corner on Left
64. Turned Improperly - Wide Right
65. Turned Improperly - Wrong Lane
66. Turned When Unsafe
67. Under Influence - Alcohol
68. Under Influence - Drug
69. Wrong Side - Approach or In Intersection
70. Wrong Side - Not passing
71. Wrong Way - One Way Road
72. Other Factor (Write in on Line Below)

Figure 15.5. Continued.

that considers population, number of registered motor vehicles, accident rates, and other factors. These funds, in turn, are parceled out by a state coordinating agency. Any law enforcement agency may apply for a STEP grant. In some states, the competition for STEP funding is intense; in other states, there are ample funds available.

Ordinarily, STEP grants are used to pay overtime for officers who conduct selective traffic enforcement projects in addition to their normal duties. Because STEP funds are limited, usually they are not available to purchase extra equipment or to hire additional personnel. Also, STEP grants are limited to one year, even though a STEP project may be planned for a two-year or longer period. The reason is that there is no assurance that federal funds will be available for more than one year at a time.

STEP projects are intended to be demonstration projects; the purpose is to show that selective enforcement, even using a limited number of officers on an overtime basis, can be effective in improving traffic safety. Once the effectiveness of the project has been demonstrated, it is hoped that the local government will be able to provide adequate funds for a permanent selective enforcement program.

In some cases, STEP projects have demonstrated not only that selective enforcement reduces accident rates, but that the revenue from fines paid by offenders exceeds the cost of the program. Although the production of revenue should not be the goal of any law enforcement activity, this fact is sometimes more persuasive to local political officials than the reduction of accident rates. When the savings from reduced accident incidence are also shown—savings from reduced property damage, personal injuries, and personnel time to attend to accident victims and perform investigations—the cost effectiveness of selective enforcement may be overwhelming.[35]

PURSUIT DRIVING

One of the major areas of controversy in traffic law enforcement, and in patrol administration generally, is the practice of pursuing fugitives at high speed. We have referred to this issue several times in this text.

High-speed pursuit is an exceedingly dangerous kind of police operation. It is dangerous not only for the police officer and the fugitive, but equally so for innocent citizens who happen to be in their path. Once a high-speed pursuit begins, it is unlikely that the fugitive will stop voluntarily. From the fugitive's point of view, there is nothing to be gained by stopping as long as the police are unsuccessful in the pursuit. All too often, a high-speed pursuit ends only when either the fugitive or the officer is involved in a collision, often a fatal one.[36]

If the likelihood of overtaking and stopping a fugitive vehicle is doubtful, the rationale for continuing a high-speed pursuit is absent. In other words, patrol officers should not continue to pursue a fugitive if it is apparent that the fugitive will merely increase his or her speed and that the patrol vehicle is incapable of overtaking the fugitive rapidly.[37]

Does this mean that police officers should merely look the other way and let the fugitive go? Of course not. It means that a different set of techniques must be used.[38]

When a fugitive vehicle is observed, the patrol officer should make one attempt to stop it. If the fugitive flees at high speed, the patrol officer should follow at a safe, reasonable speed, keep the fugitive in sight, and report the incident at once. In many cases, the dispatcher should be able to direct other patrol officers to intercept the fugitive.

Experienced patrol officers will see two major difficulties with this procedure. First, the fugitive, by making a series of rapid turns, may escape from the pursuing officer's observation. Second, it often takes more than one police vehicle to intercept a fugitive because one car cannot control the entire width of a street.

If the fugitive is unwilling to stop, which is usually the case, the intercepting officer may be tempted to block the roadway. A patrol car is not large enough to block an entire road and is not an appropriate tool for this purpose.

The first problem may not be as serious as it seems. Even if the fugitive vehicle escapes from sight, the pursuing officer at least knows the general vicinity and direction of travel before the pursuit began. Once the fugitive loses the pursuing officer, he or she will once again head for his or her original destination. Intercepting officers may still be able to head off the fugitive. If the pursuing officer is able to get the license number of the fleeing vehicle, an officer might be waiting on the fugitive's doorstep when he or she arrives home. This is by no means a perfect solution; the prospects of capturing the fugitive remain uncertain, but it is still better than chasing at high speeds until someone is killed.

The second problem is more difficult. A patrol car makes a poor blockade at best, and two or three cars may be needed to control the entire width of a street. Getting several patrol cars into the proper position ahead of a vehicle traveling at high speed requires an extremely well-coordinated operation. If the roadblock is successful, the fugitive may be unable to stop in time to prevent a violent collision, or may lose control of the vehicle while attempting to avoid the blockade.

All of these situations pose obvious dangers to the fugitive, to the police officers involved, and to any innocent citizens who happen to be in the vicinity. For these reasons, most police agencies prohibit their officers from using their vehicles as roadblocks.

Many agencies also prohibit a practice we will call "high-speed parading," in which several police vehicles join in a pursuit. There are instances of twenty and thirty police vehicles involved in a high-speed parade. Probably the most notorious example was depicted in Steven Spielberg's motion picture, *The Sugarland Express* (based on an actual case), in which more than a hundred police vehicles chased one fugitive for nearly two hundred miles! Again, the dangers are obvious, and there is simply no rational justification for this practice.

Agency policies typically state that no more than two police vehicles may be involved in a pursuit at any one time. Other officers who wish to assist should either follow at a safe, reduced speed, or take an alternate route in an effort to intercept the fugitive.

Airborne surveillance, using a helicopter or small airplane, is extremely useful in pursuit situations. An airborne officer can maintain constant surveillance of a fleeing fugitive and can direct intercepting officers no matter what path the fugitive takes. In many cases, the fugitive will be unaware of the airborne surveillance and will slow down as soon as the pursuing patrol car is out of sight.

This kind of air and ground operation requires a high degree of coordination among the airborne officers and patrol officers on the ground.

Even these admittedly imperfect techniques are preferable to high-speed pursuit. Police officers do not have the right to endanger themselves and innocent citizens by driving with reckless abandon. There is no rational justification for taking the lives of innocent bystanders even in the process of apprehending the most dangerous of criminals.[39]

The "chase-at-all-costs" attitude has been fostered not by sound police philosophy, but by the romantic notions and clever techniques of television and movie producers. Hollywood may take deserved credit for promoting excellent police practices and a positive image of law enforcement through many films and television series. However, Hollywood also must share the blame for the deaths of many police officers and innocent citizens.

The fact is that neither police vehicles nor any other ordinary automobiles will perform the way they appear to in television shows and movies. The cars used in Hollywood stunt work are carefully, ingeniously prepared—sometimes built from the ground up—for each separate stunt. Often what appears on the television or movie screen as one car going through a series of hair-raising flights, crashes, and careening turns was, in fact, several different cars, each built and prepared specifically for one stunt. The stunts are filmed separately and edited together into a continuous sequence. Quite often a single stunt destroys the stunt car; what appears in the next shot is actually a different car.

Since these facts are not common knowledge, many people (even police officers who should know better) sometimes try to drive their ordinary cars in the manner of Hollywood stunt cars, with disastrous results.

Police officers should be trained in high-performance driving not to encourage recklessness, but to ensure that they are competent to operate a vehicle under extreme conditions in an emergency.[40] Nevertheless, *even when an emergency exists*, a police officer must operate his or her vehicle in the safest manner possible, with full regard for the safety of other motorists and pedestrians. Continuing a high-speed pursuit, heedless of the danger involved, is every bit as inappropriate and senseless as shooting at a fugitive on a crowded sidewalk.[41]

REVIEW

1. The four main functions of traffic law enforcement are
   (a) Investigating accidents.
   (b) Directing traffic.

(c) Parking law enforcement.

(d) Detecting and citing violators.

(e) Assisting disabled motorists.

(f) Safety education.

(g) Routine patrol.

2. True or False: Nearly all traffic law enforcement officials accept the principle that citing violators is an educational opportunity.

3. Covert surveillance means that

(a) Officers are hidden from motorists' view.

(b) Officers use aircraft to observe traffic.

(c) Motorists are hidden from officers' view.

(d) Officers observe traffic from their own moving vehicles.

(e) All police officers should observe traffic and arrest or cite offenders.

4. A _____ can be used to screen a motorist who is suspected of driving while intoxicated.

(a) Blood sample analysis.

(b) Videotape recording.

(c) Written examination.

(d) Field sobriety test.

(e) Written warning.

5. A STEP project is intended to

(a) Recognize exceptional traffic officers for their efforts by granting them extra pay.

(b) Increase the revenue from traffic fines.

(c) Demonstrate the value of selective enforcement.

(d) Reduce traffic accidents by eliminating most common violations.

(e) Ensure that frequent violators are punished more severely than occasional, minor violators.

NOTES

[1]Noel C. Bufe and Larry N. Thompson, "Traffic Services," in William A. Geller, ed., *Local Government Police Management,* 3rd ed. Washington, D.C.: International City Management Association, 1991, pp. 162–68; Bruce A. Hand, Archible W. Sherman, Jr., and Michael E. Cavanagh, *Traffic Investigation and Control,* 2nd ed. Columbus, Ohio: Charles E. Merrill, 1980, p. 86.

[2]Hand, Sherman, and Cavanagh, *Traffic Investigation and Control,* pp. 1–2.

[3]James B. Jacobs, *Drunk Driving: An American Dilemma.* Chicago: University of Chicago Press, 1989, p. 17.

[4]Terry C. Cox and Mervin F. White, "Traffic Citations and Student Attitudes Toward the Police," in *Journal of Police Science and Administration,* vol. 16, no. 2, June 1988, pp. 105–21.

[5]Victoria L. Major, "Law Enforcement Officers Killed, 1980–89," in *FBI Law Enforcement Bulletin*, vol. 60, no. 5, May 1991, pp. 2–5.

[6]Bufe and Thompson, "Traffic Services," in Geller, *Local Government Police Management*, pp. 168–69.

[7]*Police Traffic Responsibilities* (Highway Safety Program Management Series, Volume II). Washington, D.C.: International Association of Chiefs of Police, 1976, pp. 25–28.

[8]For an example of a small agency that did establish a separate traffic unit, see Thomas Guthery, Henry Frawley, and James Orcutt, "A Traffic Enforcement Unit," in *Law and Order*, vol. 37, no. 8, August 1989, pp. 71–72.

[9]Hand, Sherman, and Cavanagh, *Traffic Investigation and Control*, pp. 122–24.

[10]Bufe and Thompson, "Traffic Services," in Geller, *Local Government Police Management*, p. 170.

[11]*Introduction to Police Traffic Services Management* (Highway Safety Program Management Series, Volume I). Washington, D.C.: International Association of Chiefs of Police, 1976, pp. 51–52.

[12]Two of the best sources of information on accident investigation procedures are: The Traffic Institute, Northwestern University, Evanston, Illinois; and The Texas Transportation Institute, Texas A&M University, College Station, Texas.

[13]Connie Fletcher, *Pure Cop*. New York: Villard, 1991, pp. 155–99.

[14]John Gales Sauls, "Traffic Stops: Police Powers under the Fourth Amendment," in *FBI Law Enforcement Bulletin*, vol. 58, no. 9, September 1989, pp. 26–31.

[15]*Introduction to Police Traffic Services Management*, pp. 161–64.

[16]Hand, Sherman, and Cavanagh, *Traffic Investigation and Control*, pp. 87–96.

[17]Bufe and Thompson, "Traffic Services," in Geller, *Local Government Police Management*, pp. 160–62.

[18]Hand, Sherman, and Cavanagh, *Traffic Investigation and Control*, p. 38.

[19]President's Task Force on Highway Safety, *Mobility without Mayhem*. Washington, D.C.: Government Printing Office, 1970.

[20]*Alcohol Enforcement Countermeasures* (Highway Safety Program Management Series, Volume IV). Washington, D.C.: International Association of Chiefs of Police, 1976, p. v. However, see also Jacobs, *Drunk Driving*, pp. 27–36, regarding questionable assumptions about the incidence of DWI fatalities.

[21]Jacobs, *Drunk Driving*, pp. 91–92.

[22]L. P. Watts, *Some Observations on Police-Administered Tests for Intoxication*. Chapel Hill, N.C.: University of North Carolina Press, 1966.

[23]*Testing the Drinking Driver*. Chicago: National Safety Council, 1970.

[24]Jacobs, *Drunk Driving*, p. 72.

[25]Ibid., pp. 96–98.

[26]Ibid., pp. 98–99.

[27]R. Dean Smith and David A. Espie, *Guidelines for Police Services on Controlled-Access Roadways*. Washington, D.C.: International Association of Chiefs of Police, 1968.

[28]E. F. Hennessy, T. Hamilton, K. B. Joscelyn, and J. S. Merritt, *A Study of the Problem of Hot Pursuit by the Police*. Washington, D.C.: U.S. Department of Transportation, National Highway Safety Bureau, 1970.

[29]James M. Pellicciotti, "The Law and Administration of Sobriety Checkpoints," in *Journal of Police Science and Administration*, vol. 16, no. 2, June 1988, pp. 84–89.

[30]Bufe and Thompson, "Traffic Services," in Geller, *Local Government Police Management*, pp. 176–77.

[31]Bedford M. Crittenden, *Truck Violation Survey*. Sacramento, Calif.: California Highway Patrol, 1965.

[32]Bufe and Thompson, "Traffic Services," in Geller, *Local Government Police Management*, p. 181.

[33]Ibid., p. 170.

[34]Hand, Sherman, and Cavanagh, *Traffic Investigation and Control*, p. 83.

[35]*Introduction to Police Traffic Services Management*, pp. 19–21.

[36]Geoffrey P. Alpert and Roger C. Dunham, *Policing Urban America*, 2nd ed. Prospect Heights, Ill.: Waveland Press, 1992, pp. 188–89.

[37]Geoffrey P. Alpert, "Questioning Police Pursuits in Urban Areas," in *Journal of Police Science and Administration*, vol. 15, 1987, pp. 298–306. Reprinted in Roger C. Dunham and Geoffrey P. Alpert, eds., *Critical Issues in Policing*. Prospect Heights, Ill.: Waveland Press, 1989.

[38]I. Gayle Shuman and Thomas D. Kennedy, "Police Pursuit Policies: What Is Missing?," in *American Journal of Policing*, vol. VIII, no. 2, 1989, pp. 23–28.

[39]Alpert, "Questioning Police Pursuits in Urban Areas," p. 305.

[40]Alpert and Dunham, *Policing Urban America*, pp. 190–93.

[41]Shuman and Kennedy, "Police Pursuit Policies," p. 22.

ADMINISTRATION OF CRIMINAL INVESTIGATION

The thorough, scientific investigation of crime was not a function of the earliest police departments in this country. As investigative techniques began to be developed in France and Britain, around the middle of the last century, a few American cities established completely separate patrol and investigation departments. After the Civil War, however, the investigative function was more often assigned to a detective bureau within the municipal police department, and that is the usual arrangement in nearly all American police agencies today.[1]

FUNCTIONS OF CRIMINAL INVESTIGATION UNITS

In simplest terms, the mission of a criminal investigation unit is to discover, for every crime that comes to the attention of the police, what crime was committed, by what means, and by what person or persons.[2]

This mission can be divided into a series of specific functions that must be performed, to some degree, for each crime that is investigated. First, a crime must become known to the police. Occasionally a police officer on patrol will observe a crime in progress or will happen upon a crime scene. However, it is relatively rare for crimes to become known to the police in this manner.

The police go to great lengths to become aware of certain types of crimes, and this becomes one of the principal functions of certain investigative units. Officers in organized crime units, vice units, drug abuse units, and sometimes juvenile crime units devote most of their time to discovering what crimes are being committed and by whom. The principal technique used for this function is undercover investigation, in which officers conceal their identity in an attempt to penetrate the criminal activity. Once the nature and extent of that criminal activity have been learned, the investigation can focus on gathering sufficient evidence to charge offenders with specific crimes.

The vast majority of crimes that come to the attention of the police are discovered through a report or complaint from a victim, witness, or other person who has knowledge of the crime. Usually the initial report is made by telephone to the police emergency switchboard. Occasionally a victim or other person with knowledge about a crime will visit the police station in person to file a report.

This initial report, however it is made, becomes the basis on which the police begin their investigation, and therefore it has great significance for the prospective success of the investigation.[3] From now on, we refer to the initial report of a crime (or possible crime) as a *complaint*, regardless of how or from whom it is received.

Before we continue this discussion, we should note that, according to numerous surveys, only a small fraction of all crimes ever become known to the police. Naturally, criminals take whatever steps they can to avoid the discovery of their offenses. Unfortunately, victims and witness, who presumably would have an interest in letting the police know that a crime has occurred, frequently fail to complain.

Some research indicates that for every crime that becomes known to the police, perhaps ten or twenty crimes go undiscovered.[4] The reasons for nonreporting of crime are difficult to pin down. Some people seem to distrust the police, while others are so traumatized by the criminal event that they are psychologically incapable of responding to it in a positive manner, and still others are fearful of retaliation by the criminal if they report the matter. Since the police cannot do anything about a crime until they know of its existence, nonreporting is a serious problem and police administrators must do whatever they can to encourage prompt reporting of all crimes.

Once a crime becomes known to the police, a series of actions may be taken. If the crime is fresh—either still in progress, or recently committed—usually a patrol officer is dispatched to assist the victim, protect the crime scene so that further investigation is possible, and make an initial or *preliminary investigation*. If the offender is still on the scene, of course, the patrol officer is expected to make an arrest. If the crime is not fresh (for example, when the victims return from vacation and discover that their home has been burglarized sometime during their absence), an urgent response is not necessary, but either a patrol officer or a detective is assigned to conduct the preliminary investigation.

The purpose of the preliminary investigation, no matter when it is done or by whom, is to determine the circumstances of the crime and to gather as much information as possible about how it was committed and, if known, the identity of the perpetrator.

The quality of the preliminary investigation often is the most significant factor in determining the prospects for a successful investigation. The information and evidence obtained in this step usually provide the basis for any further investigation of the crime.[5] According to some studies, the majority of crimes are solved only if the identity of the perpetrator is learned during the preliminary investigation (either because the victim knows who committed the crime, or physical evidence leading to the perpetrator is found at the scene).[6]

Even so, many criminal investigations must continue beyond the preliminary stage. If the identity of the perpetrator is known, it may still be necessary to obtain and develop evidence to prove that the suspect committed the crime.

If the perpetrator is not known, but there is sufficient information or physical evidence to give some hope that the criminal's identity can be discovered, or if the crime is sufficiently serious to warrant an all-out attempt to identify the perpetrator, further investigation is necessary. Whether the follow-up investigation is conducted by patrol officers or specialist investigators depends largely on the policies and organizational structure of the police agency.[7]

Finally, a successful investigation ends when the suspected perpetrator is not only identified but also located and arrested, and sufficient evidence is assembled to obtain a conviction in court. Again, both patrol officers and specialist investigators may have a role at this point, depending on agency policy and structure.[8]

Summarizing the investigative process, there are four distinct phases or stages.

1. Receiving the initial report or *complaint*.
2. Conducting a *preliminary investigation*.
3. Conducting a *follow-up investigation*, if necessary.
4. *Concluding* the investigation by arresting the accused perpetrator and assembling evidence for presentation in court.

TYPES OF INVESTIGATIVE UNITS

The relationship between the general patrol force and the criminal investigation element is one of the most important characteristics of a modern law enforcement agency. Briefly, three common patterns are found in American police departments.[9]

Specialized Investigation.

The general patrol force has very limited responsibility for criminal investigations. Patrol officers responding to a crime complaint are expected to make an arrest if a suspect is present at the scene. Otherwise, the patrol officer merely obtains a brief report from the victim or witnesses, often by filling out a one-page form, and forwards the report to the agency's investigators.

In some serious cases, or if it appears that substantial physical evidence is available at the scene, the patrol officer may notify investigators by radio or telephone, and protect the scene until an investigator or evidence technician arrives.

Virtually all felony cases and most serious misdemeanor cases (other than those in which the offender is present at the scene when patrol officers arrive) are the responsibility of investigators. The investigators are divided into a number of specialized organizational units according to categories of crimes, so that each investigator concentrates on a particular type of offense.[10]

Integrated Investigation.

The general patrol force conducts most preliminary investigations. The patrol officers responding to a crime complaint are expected to obtain a thorough report from the victim or witnesses and to collect any physical evidence that can be found at the scene. The report and evidence (if any) are turned over to specialist investigators for follow-up.

Evidence technicians and other technical specialists may be available to assist the patrol officers. Investigators may be divided into a number of organizational units, according to crime categories, or they may be consolidated into two or three relatively large and generalized units (such as "crimes against persons," "crimes against property," "vice," and so on). Investigators retain most of the responsibility for completing the investigation of most felonies and serious misdemeanors, although they are encouraged to involve the patrol force whenever possible (for example, in searching for suspects, seeking informants, and so on).

Patrol-Oriented Investigation.

The patrol force is responsible for completing most investigations. After obtaining the preliminary report, patrol officers notify the investigators of their findings, but the responsibility for most follow-up activities is left with the patrol unit. The investigators' function is primarily one of support for the patrol force: advising patrol officers on investigative techniques and procedures, assisting with records checks, and analyzing information from a number of sources. Investigators may assume full responsibility only for the most serious or complicated cases that would be too time-consuming for patrol officers to pursue to completion.[11]

Thus, the nature of the investigators' responsibilities and the organizational structure of the investigative element of a police agency depend mostly on the extent of the patrol force's responsibilities for investigation. If the patrol force performs very little investigation, the investigative unit must be relatively large, and usually is divided into some number of specialized subunits. If the investigative unit is kept small and its personnel serve primarily as technical advisors, the bulk of investigative work must be done by the patrol force.

Until fairly recently, the specialized pattern described above was by far the most common among medium-sized and larger American police agencies. Only small agencies, because of their limited human resources, expected patrol officers to be involved beyond the preliminary investigation. Most police authorities assumed that patrol officers were too inexperienced and limited in technical competence to be trusted with such important tasks.

Furthermore, as we discussed in Chapter 3, specialization was assumed to be valuable in itself. By specializing in specific types of investigations, detectives presumably would become more knowledgeable and skillful. Many police authorities believed that typical criminals tended to specialize in committing certain types of crimes, and therefore that specialized investigators should become familiar with the habits and personalities of the offenders with whom they would have to deal.[12]

However, these assumptions are not necessarily valid. Since the beginning of this century, American police agencies have worked hard to upgrade the capabilities of their patrol forces. Today, entry-level officers are, on the average, better educated and more highly motivated than their predecessors, and they receive better training in all aspects of police science. There is little reason to assume that a patrol officer is incapable of performing a sound, thorough investigation.

There are also several reasons to regard specialization as a mixed blessing. Although specializing may indeed permit an investigator to reach a high level of knowledge and skill in a certain type of investigation, it also tends to narrow the range of the investigator's competence.

It is simply not true that "typical" criminals specialize in specific types of crimes. Some do, but many others do not, and it is not unusual for one person to be responsible for several different types of property crime. Also, some criminal incidents are not so easy to categorize. If an incident involves both a burglary and an armed robbery, or if it is not clear which type of crime has occurred, who should be given the responsibility for investigating it: the burglary detective or the robbery detective?

Specialization also creates administrative problems in the management of the natural workload. At one time, there might be many more burglaries than armed robberies to be investigated, while at another time the reverse might be true. Thus, in the first instance the burglary unit would be overworked and the robbery unit might have too little work, whereas in the second instance the robbery unit would be overworked and the burglary unit idle.

For these and many other reasons, police administrators in recent years have tended to despecialize their investigative units. What we have called the integrated pattern has become more common: patrol officers are given more responsibility for the early stages of most investigations, while investigators are given responsibility for a broader range of the more serious and complicated cases. Some agencies, especially the smaller ones, have gone even further, adopting the patrol-oriented system.

As we pointed out in Chapter 14, both the integrated and patrol-oriented systems demand more of the patrol force's time. Both the integrated system and the patrol-oriented system, therefore, depend on having more patrol officers than the conventional specialized system. On the other hand, the integrated and patrol-oriented systems also require fewer detectives, fewer specialized units (each of which requires its own administrative mechanisms for supervision), and more flexible arrangement of workloads.

TECHNICAL SUPPORT FOR INVESTIGATION

No matter which organizational system is used, the fundamental techniques and procedures of criminal investigation are essentially the same. The basic steps we described earlier must be carried out either by patrol officers or by specialist investigators, or by both to some degree.

Most of these steps should be within the capabilities of any police officer. Every officer should be trained to complete at least the preliminary investigation of a crime including the completion of a basic crime report form and the interviewing of the victim, witnesses, or complainants. Similarly, every officer should be able to perform a legal arrest once a suspect has been identified.

However, the search for, acquisition of, and analysis of physical evidence involves an array of technical skills that may not be possessed by all officers. The laws that govern searching for evidence are complicated and subtle. The actual search for physical evidence may involve such varied skills as those of a photographer, a chemist, and a physicist. Much of the evidence that might be found at a crime scene must be examined and analyzed in a laboratory. These are all areas that seem to demand the experience, knowledge, and skills of specialists.

Investigative technical specialists can be considered in two general categories: crime-scene technicians and laboratory technicians.[13] The former might include photographic technicians, latent print acquisition technicians, and physical evidence technicians. Laboratory specialists might include chemists, physicists, print examination and identification specialists, firearms examination specialists (sometimes called ballistics technicians), and photo analysts.

All of these technical specialists may be called *criminalistics technicians*, or, more simply, *criminalists.* Another type of specialist, the polygraph examiner, also should be considered a criminalist, although the use of the polygraph (and other so-called lie-detection techniques) does not actually involve physical evidence.[14]

The number of technical specialists available to support criminal investigations depends primarily on the size of the police agency and secondarily on the extent of the training and skill possessed by other personnel. The following list suggests the distribution of personnel and the various ways that technical support for investigations may be provided in police agencies of varying sizes:

Small-sized agency (under 10 officers)	No criminalistics specialists; all officers are trained in basic evidence acquisition techniques; assistance is requested from county or state police in complicated cases; laboratory work is done in a regional cooperative lab, state police lab, or, in some cases, the FBI lab in Washington, D.C.
Small- to medium-sized agency (10–50 officers)	One or two officers (patrol officers or detectives) receive special training in such technical fields as police photography, print examination, and firearms examination; laboratory work is done elsewhere.

Medium-sized agency (50–100 officers)	One or two full-time evidence technicians may be available to assist in relatively complex cases; a small laboratory may be available for photography, print examination, and some routine evidence processing, with more complicated work done elsewhere.
Medium- to larger-sized agency (100–250 officers)	Two or more evidence technicians may be available, some with a high degree of special training in photography, latent prints, or firearms; a small laboratory with one to five full-time criminalistics technicians may be available for most routine work, with more complicated work done elsewhere.
Large-sized agency (more than 250 officers)	Approximately one evidence technician for every 10 patrol officers on duty at a given time; technicians specialize in various types of evidence collections; extensive laboratory facilities with several criminalistics technicians, capable of handling nearly all types of evidence examination procedures.

The extent to which technical support personnel are needed also depends partly on the capabilities of the patrol force and regular investigators and on the extent to which they are expected to carry out their own investigations. Obviously, if all patrol officers and detectives are thoroughly trained in such matters as crime scene photography, latent print acquisition, evidence collection, and so on, there is less need for full-time specialists. However, if patrol officers are expected to perform time-consuming evidence searches, more patrol officers will be needed to make sure that there is adequate coverage.

As noted in the preceding list, medium-sized and smaller-sized agencies generally cannot support a full-time criminalistics laboratory. There are several alternatives. Three or four small agencies may establish a cooperative regional laboratory. A smaller-sized agency may rely heavily on a laboratory operated by the state police agency (or, in a few cases, by the county sheriff's office). Or evidence that requires elaborate analysis may be sent to the FBI's laboratories in Washington.[15]

Another option is to use the resources of a nearby college or university for complex evidence analyses that require expertise in chemistry, physics, or other scientific disciplines. However, this practice introduces problems in maintaining the integrity of evidence, and care must be taken to ensure that non–law enforcement personnel, acting as technical consultants, understand the full legal significance of their participation.

Agencies that do have their own criminalistics laboratory usually do not place it under the administrative supervision of the criminal investigations unit. More often, the lab is an element of a technical services bureau or equivalent division, along with other support elements (communications, fleet services, and so on).

If the lab provides services for the patrol force, traffic law enforcement, and perhaps special investigative units, as well as the investigations unit, this organizational structure is appropriate. However, closer coordination can be achieved by placing the lab directly under the authority of the investigations unit if that unit accounts for nearly all of the lab's workload.

ORGANIZATION OF CRIMINAL INVESTIGATION UNITS

Traditionally, criminal investigation units have been divided into more and more specialized subunits, following the principle expressed by O. W. Wilson: "Specialization in detective assignments should be the invariable rule."[16]

Even a small agency with only four or five detectives would divide them into at least two subunits, a Crimes against Persons Unit and a Crimes against Property Unit. Larger agencies would further subdivide the detective bureau into as many small units as possible, each with a distinct and specific category of crimes to handle. There might be a Homicide Division, Theft Division, Burglary Division, Robbery Division, Auto Theft Division, Fraud Division, and so on and so on.

Each division, in a very large agency, would be further subdivided into subsections, sometimes called "details"; the Theft Division might have a Larceny Detail, Shoplifting Detail, Pickpocket Detail, and so on. Each detail might contain anywhere from one to a dozen detectives.

In an integrated structure, in which both the patrol force and the detective force are involved in most criminal investigations, such elaborate specialization is neither necessary nor justified. An investigations unit with a total of twenty officers or less may be divided into two, or at most three, sections: Crimes against Persons, Crimes against Property, and perhaps a separate unit for Organized Crimes. Larger agencies may find it useful to divide these sections into subsections according to broad categories of crimes. For example, the Crimes against Persons Section could be divided into a Homicide Unit, Sex Crimes Unit, and Assault Unit.

In a patrol-oriented investigations system, where patrol officers perform virtually all investigative functions and the detectives serve primarily as resource consultants to the patrol officers, there is little need to divide the detective force into more than two or three sections. Any further subdivisions would simply encourage the kind of specialization that patrol-oriented investigation is intended to overcome.

Organized Crime Investigative Units

The kinds of organizational structures we have just described are appropriate for investigative units that are concerned with, for lack of a better term, "common" crimes. However, some police administrators believe that a different structure is necessary for investigative units that are concerned with organized crimes.

Organized crime is not a precise term. Usually it includes the various categories of crimes that are called vice: prostitution, pornography (to the extent that it is still a criminal offense), gambling, some liquor law violations, and violations of narcotics and controlled-substances laws.[17]

The key elements that set these crimes apart from common crimes are: (1) criminals engage in these activities primarily for profit, sometimes as their principal source of income; and (2) the "victims" are willing participants, usually as consumers of illicit goods and services. Whether the consumers should be regarded as victims or as contributing offenders is a matter of some debate.[18] These are organized crimes in the sense that the perpetrators have established themselves in an illegal business, in which they engage over some period of time, and which requires some degree of cooperation and coordination among several participants.[19]

For example, a successful gambling operation requires one or more "banks," where money is collected, odds are established, and bets are paid off; usually several "books" where bets are recorded; and any number of runners who deal directly with the customers, accepting bets and payments, and, for the lucky few, making payoffs. The operation also may require one or more enforcers to collect money from losers. Well-established gambling operations often have several functionaries who are responsible for placing bribes with corrupt police officers and officials. Many gambling organizations also maintain profitable loan-sharking operations to accommodate heavy gamblers who are unable to pay their debts. Thus, a gambling organization might involve anywhere from a dozen to several hundred individuals, some on a full-time basis and many others on a part-time or occasional basis.[20]

Other vice operations are similarly organized. Drug trafficking may involve importers who bring huge quantities of illegal substances into the country, distributors who process and adulterate the drugs, and dealers who sell small quantities to individual users (many of whom also resell even smaller quantities to their friends).[21]

Vices are not the only crimes that lend themselves to organization. Stolen goods, whatever their nature, must be sold in order for the thief to realize an income from his or her work. Since the trade in stolen goods is itself illegal, some degree of organization is necessary to protect the traders from discovery.

Sometimes the trader, or "fence," organizes groups of thieves to steal specific kinds of goods for which a ready market exists. Auto theft rings, for example, have been known to accept "orders" for specific makes and models of cars—sometimes with optional accessories also specified—to be delivered to their customers. The order is then assigned to professional thieves who scout parking lots for the desired vehicle, steal it, modify it to meet the specifications if necessary, and deliver it. Auto parts, electronic consumer and industrial products, jewelry, clothing, medical supplies, and any number of other goods are traded by well-organized fencing operations.[22]

Since all of the participants in organized crime—the organizers, the street-level functionaries, and the consumers—are willing participants and go to great lengths to avoid discovery, law enforcement authorities cannot wait for a complaint before they investigate. Instead, active investigation is necessary just to discover what crimes are being committed and by whom.

The principal police technique in all organized crime investigations is undercover surveillance performed by an *undercover agent*. Police officers pose as prospective customers or as minor participants. They attempt to infiltrate the criminal organization in order to gather evidence about the kinds of crimes being committed and identities of the participants. Understandably, if an undercover investigator's true identity is revealed, the agent's usefulness is lost and, more important, his or her life may be in jeopardy. Therefore, extreme secrecy is essential to the undercover investigator's work.[23]

Organized criminals are well aware that an active police force will attempt to discover their activities. The criminals rely on two approaches to protect themselves. If possible, they use bribery, blackmail, and similar methods to corrupt the police and criminal justice officials.[24] If corruption fails, the criminals may attempt to infiltrate the police agency itself and obtain information about current investigations.

Usually organized criminals do not resort to overt violence, since that simply calls attention to their activities and gives the police more opportunity to arrest the key individuals. Violence is used to enforce discipline within some criminal organizations, however. Usually the organization hires a professional enforcer, not otherwise associated with the criminal organization, for this purpose.

Because of the need for secrecy to protect the undercover agents and because of the need to protect sensitive information and to prevent criminals from sabotaging the investigative effort, special measures must be taken to maintain security. Those measures may be reflected in the agency's organizational structure.

For example, in many agencies the organized crime investigative unit is completely separate from the criminal investigations unit. The organized crime investigators are placed under the direct supervision of, and are responsible only to, the agency's chief executive. In large agencies, the identities of the undercover investigators may be unknown even to other police officers. In small agencies, such extreme secrecy usually is not practical. Nevertheless, undercover agents are discouraged from frequenting the police station and associating with other officers, and their identities are protected as far as possible.

Reasonable measures to protect organized crime investigators are essential, both in the interest of effective law enforcement and in the interest of protecting the agents' lives. However, there is a danger that the undercover officers and the entire organized crime unit may evolve into a privileged elite of secret agents who, because of the secretive nature and great importance attached to their work, are permitted to operate in ways that would not be tolerated among the conventional force.

For example, the niceties of preparing detailed, accurate reports of investigative activities may be ignored. Agents may be permitted to use tactics whose propriety, and even legality, are doubtful. Constitutional guarantees of freedom from unreasonable searches and improper arrests may be overlooked. Abuses such as the excessive use of force, the entrapment of suspects (by inducing them to participate in crimes for the sole purpose of catching them red-handed), and the commission of criminal acts by agents to gain the confidence of suspects may be tolerated. And finally, undercover agents are unavoidably exposed to enormous opportunities for corruption: soliciting and accepting bribes, subverting the law itself, and becoming active participants in criminal enterprises.[25]

These dangers exist even in agencies that otherwise scrupulously avoid any taint of corruption or impropriety. Undercover investigation is, to be sure, difficult and dangerous. Even the most conscientious administrator may be tempted to make allowances when an overzealous or frustrated investigator goes beyond the bounds of proper police technique. The temptation is even greater when the impropriety is obscured within a secretive organizational structure. As long as no one else knows what is being done, it is easy for the administrator to look the other way or to rationalize that the ends justify the means.

Precisely because of these dangers, and the ultimate consequences that may result from them, it may be a serious mistake to remove the organized crime investigative unit from the built-in chain of accountability in the agency's conventional structure. If the abuses we have mentioned are tolerated, the likely consequences include the corruption of undercover officers and, eventually, other members of the force. If improper police actions become public knowledge, the agency will suffer a devastating loss of public confidence. These dangers must be weighed against the sensitive nature of the task and the need for security.

It also may be a serious mistake to overlook the need for close coordination between the organized crime unit and other elements of the agency. Undercover agents usually do not participate in arrests, since that would reveal their identity as police officers. Therefore, at some point the undercover agents must rely on regular investigators and patrol officers to complete their work. For example, the undercover agent might gather sufficient evidence to justify a raid on a crime site (such as a house of prostitution or a gambling bank), but the raid itself is carried out by patrol officers and other investigators.

Conversely, the activities of regular investigators and patrol officers may interfere with undercover investigations, if the regular officers are not aware of the undercover agents' activities. For example, a patrol officer who happens to observe suspicious activity might interrupt a drug deal that has been painstakingly arranged by an undercover agent through months of work.

In short, regular investigators and patrol officers must be aware of undercover investigations, at least to the extent that the regular officers will not accidentally interfere with the undercover agents. This awareness can only come through the exchange of information among the patrol force, regular detectives, and organized crime investigators.

Coordination and the free exchange of information are infinitely easier when the agency's organizational structure does not place unnecessary barriers between the various elements. In simple terms, this means that the organized crime investigative unit should not be separated from the rest of the agency's criminal investigations element. Organized crime investigation should be treated as a specialized element, but it also should be placed within the agency's organizational structure where proper coordination and supervision can be maintained.

Special Crimes Unit

Depending on the nature and needs of the community, many law enforcement agencies have established special investigative units to deal with crimes that demand unique investigative skills and techniques.

For example, the police in a state capital may have jurisdiction over any illegal acts performed by officials of the state government including legislators and other elected officials. In a community whose economy is dominated by a particular type of industry—perhaps mining, finance, high-technology manufacturing, or maritime trade—the police may be concerned with certain types of crime that are peculiar to that industry.

In recent years, many American law enforcement agencies have assumed new responsibilities for the investigation of illegal acts that once were considered private matters, subject only to civil laws. For example, business crimes such as embezzlement, industrial espionage and sabotage, consumer fraud, and other so-called white-collar crimes formerly were regarded as civil disputes between the parties involved. Today they are regarded as crimes that require police intervention. Similarly, offenses that were once considered purely family matters such as child abuse and incest are now very much the concern of the criminal justice system.

Each type of crime we have mentioned requires unusual investigative skills and techniques. That is one reason the police traditionally have been reluctant to involve themselves in these areas. Public pressure, in some cases very intense pressure, has brought about a change of attitudes in many communities, and today the police cannot ignore the demand for enforcement activity.

The particular skills and techniques needed for the successful investigation of special crimes depend on the nature of the crime being investigated.

For example, the investigation of political crime (that is, offenses such as interference with an election, bribery of a legislator or other public official, misapplication of the laws for private gain, or misuse of public property) requires extraordinary discretion. The mere fact of an investigation may have serious consequences for the person being investigated, which may be unfair if the person turns out to be innocent. Powerful officials, whether they are innocent or guilty, may use their power to retaliate against the police.

The investigation of political crime generally involves conventional investigative procedures including interviewing victims and witnesses, collecting documents, and so on. Occasionally, unconventional techniques may be required. In the Abscam series of cases in the late 1970s, the FBI and other law enforcement agencies used agents who posed as Arab sheiks and offered bribes to members of Congress and other officials. The offers were made in a series of specially arranged meetings, which were recorded on videotape. The spectacular results included the conviction of a few corrupt officials and the destruction of the political careers of several others.[26]

Business crime investigations also require a high degree of discretion and, in some instances, special techniques. More often, the major requirement is special knowledge of the business practices that are being criminally attacked.

Embezzlement, for example, often is perpetrated by accountants, bookkeepers, cashiers, and sometimes financial officers who are intimately familiar with the cash-handling and bookkeeping procedures of their company. Their crimes can be discovered and evidence acquired only by investigators who understand thoroughly how the financial records are kept. Because virtually all companies and governmental agencies now use computers to keep their financial records, knowledge and understanding of computer systems also may be required.

Few police officers are trained in bookkeeping and accounting, much less the esoteric practices of, say, the banking industry. In order to investigate such crimes, the police may require the assistance of experts. Once again, discretion is required. Expert consultants not only must be competent in their field of expertise but also must be aware of the full legal significance of their participation. Consultants who are asked to assist in an investigation must understand that their involvement is absolutely confidential, no matter what the results, except that they may be required to testify in court. If an investigation fails to produce enough evidence to bring charges against the offender, the police and their consultants could be sued for slander, invasion of privacy, or other civil offenses.

The widespread use of computers has opened up whole new fields for criminal activity, fields that are filled with landmines for investigators.[27] Computers can be used as tools to commit crimes, or the computers themselves and the information they contain can be the objects of criminal attack.[28]

Criminals using computers have gained entry to bank computers and have ordered them to transfer to their own accounts the fractional cents that arise from the calculation of interest. When depositors' interest is computed, often the amount to be credited to each account includes part of one cent. It is not possible to pay a fractional cent in cash, so the computer is programmed to set aside the fraction. But in order for the bank's books to balance, the fractional cents must be accounted for. The amount may accumulate into the hundreds of thousands of dollars in a large bank. A knowledgeable criminal may be able to penetrate the computer system and divert the fractional-cents account into his own account, withdraw the money periodically, and disappear before anyone realizes what has happened.[29]

This particular crime is not especially common, and most computerized accounting systems now have safeguards to prevent it. However, it is a good illustration of the complexity of many business crimes, and it is easy to see that detecting and investigating such a crime would be quite difficult.

The difficulties are even greater when the object of criminal activity is not money but information. State and federal laws are inconsistent on the basic question of whether information is property. If it is property, it belongs to the person or company who possesses it, and stealing it is a crime. If information is not property, it cannot be stolen; a person who takes it without authorization is not a thief.

Nevertheless, it is certainly true that information has value, sometimes an enormous value. The owners or managers of any business would gain a significant advantage if they knew the sales plans, product designs, and other information possessed by their competitors, suppliers, or customers. Some people are sufficiently dishonest to steal such information if they can, and the use of computers to produce and store information may create the opportunity to do so.

Many computers are connected to telephone and other communications systems for the exchange of information between terminals and processors, or between computer systems at different sites. It may be possible for an intruder to gain access to the communications system and thereby into the computers themselves, and then remove information without leaving any trace of the intrusion.[30]

Of course, information can be stolen without the involvement of computers at all. A thief may enter a business and steal or copy plans, business documents, and so forth, or may be able to obtain information from dishonest or naive employees.

Clearly, detecting and investigating these instances of espionage require an intimate understanding of the nature of the information, the way the information is created and stored whether or not computers are involved, and the methods used to steal it. Even when these facts can be ascertained, a criminal offense may or may not be involved, depending on the exact content of state and federal laws.[31] In some cases, industrial spies who have been found guilty of stealing information worth millions of dollars have been punished only for the misdemeanor offense of taking the pieces of paper or the computer tapes on which the information was recorded.

Consumer fraud is another area of law enforcement that has grown rapidly in recent years, but that poses special difficulties. A dissatisfied purchaser of goods or services may complain that the product was misrepresented (it could not do what the seller claimed), or that it was defective (it did not do what it was supposed to do, or it caused some harm when it was used). The seller, however, may claim either that the purchaser misunderstood what the product or service was supposed to do or misused the product and is therefore responsible for its failure and any harm that may have resulted.

Consumer protection laws are the result of legislation at the federal and state level and of numerous court decisions, mainly in civil courts, although the criminal courts also have dealt with this subject. Some consumer frauds may involve criminal violations; others remain purely civil matters; still others may involve both. Prosecuting a consumer fraud may require an understanding of such terms as "puffery," "implied representation," "merchantability," "contributory negligence," and other technical concepts. An accusation of consumer fraud might be merely the result of a misunderstanding between an honest merchant and a dissatisfied customer. In a few cases, the "victimized" customer is actually attempting to bilk the merchant by claiming damages that never occurred!

The most important investigative tool in consumer fraud cases is the investigator's open mind. Some cases are very clear-cut, as when an advertiser offers a product with certain claimed features, but actually delivers a cheap substitute. Far more often, the investigator must determine not only whether the product is in fact defective, but whether there was any criminal intent (such as deliberate misrepresentation) on the part of the merchant, or whether indeed the customer is the one with fraudulent intent. At the same time, a fraud investigator must be thoroughly familiar with the laws and court decisions that affect this kind of situation. What begins as a criminal investigation may wind up as a civil matter, or vice versa, or may be settled privately between the parties involved.[32]

Crimes within a family are among the most difficult of criminal matters for the police. Traditionally in our society, family relationships have been considered purely private and not subject to interference by outsiders.[33] But in recent years, growing evidence of violent abuse within the family has brought about a change in public attitudes. Many people now believe that society has a right and a need to protect the innocent from abuse wherever that abuse occurs, and that abusers cannot draw a veil of privacy over their crimes.[34] New laws have been enacted to spell out the kinds of abuse that are deemed criminal and to require the police to intervene.[35]

Again, the presence of criminal intent may be an issue. Most state laws regarding child abuse (whether the abuse is violent, sexual, or even emotional) explicitly disregard the question of intent. Even unintentional abuse, caused by ignorance, neglect, or the emotional problems of the abuser, is criminal and must be prosecuted. However, intent is very much an issue when the victim of abuse is an adult—either spouse or parent. Thus, the investigation of alleged abuses within a family must determine not only what happened but why it happened and whether there are extenuating circumstances.

In summary, the investigation of special crimes usually requires (1) experienced and sensitive investigators who are aware of the need for discretion and due regard for the delicate nature of each situation; (2) extensive knowledge of the legal and technical aspects of the alleged criminal acts; and (3) an open-minded attitude and ability to conduct the investigation without prejudging the parties involved. Great ingenuity may be needed to develop investigative procedures that fit the circumstances. Expert consultants may be called on to advise the investigator or to participate in discovering the nature and methods of the criminal activity.

We have discussed these matters in some detail because of the implications for the administrator. Special crimes demand special treatment. In particular, special crimes units require close administrative supervision. Personnel assigned to these units must have personal qualities that are different from (not necessarily superior to) those required in regular investigators or even organized-crime investigators. Extensive training may also be required, and frequent opportunities for continuing training and education are necessary.

For all of these reasons, the investigation of special crimes usually demands the establishment of separate units. As in the case of organized crime units, the special crimes units usually should be under the administrative authority of the detective bureau (or whatever the investigative element is called); the requirements of coordination and close supervision are the same.

ADMINISTRATION OF INVESTIGATIVE OPERATIONS

Administrators are particularly concerned with five aspects of the operations of all investigative elements.

1. The assignment of cases for investigation.
2. The methods used to evaluate the solvability of each case.
3. The coordination of investigative activities.
4. The collection and distribution of information.
5. The legal implications of investigative activities.

Case Assignment

Every incident that comes to the attention of the police may require investigation. Some cases are generated by the investigative units themselves; this is most often true for organized crime investigations. Other cases are usually the result of a complaint.

If the investigative unit is highly specialized, the nature of the alleged crime usually determines the unit to which the case is assigned. Thus, a complaint of a theft is assigned to the theft division. Assignment of cases within each specialized unit may be left to the discretion of the unit's supervisor or manager or may be based on some mechanical principle such as assignment by rotation.[36]

In a patrol-oriented organization, case assignment usually is based on geography. That is, each officer is responsible for investigating every case, regardless of the nature of the alleged offense, that occurs within his or her beat. This means that beats must be carefully designed to equalize the patrol officers' workload. Otherwise, one officer may have an excessive number of cases while other officers have little or nothing to do.

Case assignment is a major concern in an integrated organizational structure. The initial investigative steps, such as the preliminary investigation, may be assigned to patrol officers on the basis of geography, but further investigation usually is assigned to detectives. The investigators also may be allocated to geographic districts, with each detective responsible for every investigation within his or her district; this system is sometimes a feature of team policing structures. Or the responsibility for case assignment may be given to the chief of the detective bureau or other manager.

In the first system, the assignment procedure must be designed to ensure the reasonable equalization of workloads and cases must be given to investigators who have the appropriate training to perform successful follow-up investigations.

If a manager is responsible for making assignments, there must be some assurance that the assignments are made in a fair and reasonable manner. Favored investigators should not be given all of the easiest or most interesting cases, while others are given an overwhelming number of cases that are either dull and routine or unlikely to be successful.

Evaluation of Solvability

One of the unfortunate facts of life is that not all crimes can be solved, no matter how ingenious or persistent the detective may be. Furthermore, the effort to solve a given crime may be considered disproportionate to the crime itself.

Given the limited resources available to the police, it is only sensible to devote those resources to the more serious crimes that are more likely to be solved.[37]

In short, one of the important tasks of an investigator (whether detective or patrol officer) is to evaluate the solvability of each criminal case, and on the basis of that evaluation to determine how much effort should be spent in the investigation.[38]

There are several ways that these evaluations can be performed. Some rather complicated evaluation procedures have been published, in which numerical values are assigned to various factors. At the opposite extreme, investigators may decide to proceed or not on the basis of their intuition, hunches, whether they already have a backlog of unsolved cases, whether they are feeling pressure from their supervisors, competition among the detective force, and any number of other informal factors.

Case evaluation is too complex an issue for us to discuss at length here. The point we wish to make is that administrators have an interest in the evaluation procedures that determine how investigative resources are used. If poor procedures are used, investigators' time and effort may be wasted while the backlog of unsolved crimes grows ever larger.[39]

Case evaluation also involves important policy matters. For example, investigators may have a "rule of thumb" that burglaries in a certain neighborhood are so common, and so often are committed by residents of the area, that trying to investigate them thoroughly is a waste of time. But the same investigators may feel that any burglary committed in another neighborhood must be investigated thoroughly because the residents of that area are wealthy and influential, and are likely to make trouble if they feel the police are inattentive to their needs.

These evaluation procedures may never be written down, but may have developed informally. An administrator might not even realize that patrol officers and sergeants are making policy decisions about how much law enforcement service will be given to citizens in different parts of town. Obviously, the administrator may have some very different notions about the priorities that should be given to burglaries and other crimes.

Coordination of Investigative Activity

Criminal activity does not occur in a vacuum. Some criminals, especially those involved in organized crime, tend to specialize in certain types of crimes, but many others commit a variety of offenses, depending on what opportunities arise. Thus it is not unusual for an individual to be involved in burglaries, thefts, robberies, assaults, and other offenses.

If investigative cases are assigned to specialized units, it is easy to overlook the possibility that a suspect in one case may be involved in a number of other crimes. According to one study of criminal investigation practices, investigators rarely discuss their cases with one another unless there is some specific procedure that encourages or requires them to do so.[40]

In an integrated or patrol-oriented system, the need for coordination is more obvious. Investigators and patrol officers cannot do their work successfully unless they cooperate with one another. The means for achieving and maintaining cooperation must be provided, partly through the organizational structure itself and partly through administrative policies and regulations.

Some very basic policies should be adopted by every law enforcement agency. First, no case under investigation belongs to an individual officer, whether patrol officer or investigator. Every case must be assigned to someone, who serves as the principal investigator and who is responsible for bringing the case to conclusion. However, any other officer may contribute information that might influence an investigation, and every officer should have access to whatever information has been obtained.

A second basic policy should be that enforcement activities directed toward related criminal activities must be carried out in a coordinated manner. Earlier we pointed out the urgent need for coordination between conventional and undercover units. In fact, the efforts of all units should be directed to reinforce one another.

The coordination of activities among several elements of an organization is the responsibility of the managers. Certainly it is much easier for the managers in a small agency to know what all of their subordinates are doing. In a larger agency, a greater effort must be expended to ensure that each manager has the necessary information and acts on it. Operating-level personnel (patrol officers and detectives) must be encouraged to share information and to cooperate with one another. It is the responsibility of their supervisors and managers to see that they do so.[41]

Collection and Distribution of Information

Every police investigative unit is, in a sense, an information-processing machine. Investigators collect information from victims and witnesses, obtain information from physical evidence, compare information from different cases, and process information to arrive at an understanding of how a crime was committed and by whom.

The administrators' principal concern is that all of this information be collected and processed in an efficient manner. At a crime scene, the patrol officers, detectives, and evidence technicians must collect all of the information that is available. Once collected, the information must be fully and accurately recorded. Once recorded, the information must be reported fully and accurately to everyone who may need it, including investigators who are responsible for follow-up investigations and managers who are responsible for coordinating investigative activity.

In Chapter 8 we pointed out the importance of maintaining a single, central record of all crime-related information, and we discussed what kinds of records should be kept. We also discussed some of the ways that information-processing equipment such as computers could be used to make the investigative process more efficient and effective. Naturally, administrators must be deeply involved in planning such a system. Much thought and effort will go into designing the entire information-processing system, purchasing the necessary equipment, and training all personnel to use it properly. Beyond that, administrators must enact appropriate policies and procedures to be sure that the system is used as it was intended.

The investigative units need direct, immediate access to any computerized information system. Some police communications and information systems have been designed primarily to serve the needs of administrators and of the patrol forces; criminal investigators' specific needs have not gotten much attention. In our view, the best information system is one that gives every investigator immediate, direct access to every scrap of information possessed by the agency.

Legal Implications

It should not be necessary to warn administrators that every police activity has potential consequences and implications affecting the legal and constitutional rights of citizens. Unfortunately, experience suggests that too many administrators are either ignorant of these implications or overlook them in favor of what they regard as "efficient police practices."

A moment's reflection on history (see Chapter 1) will help one to understand why Americans traditionally have distrusted police authority and have gone to great lengths to limit the privileges of the police.

In the most sweeping terms, every citizen has the right to be left alone by the police. To put the matter another way, the police have no right or authority to act against the interests of any citizen unless there is compelling evidence that the citizen has committed, or knows something about, a particular crime.

Even victims and witnesses cannot be forced to give information, although many states have laws that place an obligation on citizens to cooperate with the police. When an officer begins an interview with a complainant, the officer must be aware that the complainant has a right to remain silent about anything other than the particular crime being investigated. If the officer wants to search the crime scene for evidence, either the consent of the property's owner must be given or a court order must be obtained. If an arrest is to be made, the specific rights of the suspect must be observed. Otherwise, not only will the suspect go free and the crime go unpunished, but the police themselves may wind up defending a civil suit.[42]

Police officers cannot reasonably be expected to know and understand every fine interpretation of every law that applies to common police procedures and techniques. Administrators have a duty to inform their officers of the most basic and significant laws that affect their work. This information should be conveyed to each new officer during training, and to all personnel through inservice training and the distribution of periodic bulletins.

However, there may remain large areas of legal uncertainty concerning the particular applications of the law to particular cases. In every police agency, someone should be responsible for keeping current with court decisions and new legislation in the areas of suspects' rights, the laws of search and seizure, the laws of arrest, and the liability of police officers for improper acts.

Ideally, an agency should have the benefit of a trained, experienced legal counsel. Either an in-house attorney should be employed by every agency or, if that is impractical, at least the agency should have access to an assistant city attorney, a member of the attorney general's staff, or (as a last resort) a private attorney. Whenever officers, especially investigators, are considering any action that goes beyond routine, conventional police procedures, the advice of a qualified attorney should be solicited.

REVIEW

1. The four stages of the investigative process are (list in the correct order)
    (a) The preliminary investigation.
    (b) Undercover surveillance.
    (c) Conclusion of the investigation.
    (d) Conviction of the perpetrator.
    (e) Receipt of the complaint.
    (f) The follow-up investigation.
    (g) Coordination with other units.

2. In an integrated investigative system
    (a) Investigators are divided into specialized units that are responsible for specific types of crimes.
    (b) Patrol officers perform nearly all investigations.
    (c) All investigators assist in each case that is investigated.
    (d) Investigators are responsible for most follow-up investigations.
    (e) Cases are assigned according to the ethnic background of the complainant.

3. A *criminalist* is
    (a) A person who engages in crime for monetary gain.
    (b) A technical specialist in the discovery, collection, and analysis of evidence.
    (c) Anyone who investigates a crime.
    (d) Anyone who reports a crime.
    (e) An expert in the psychology of criminals.

4. True or False: Because undercover agents must operate in great secrecy, it is necessary to allow them to use police techniques and procedures that might be questionable in other investigations.

5. Administrators are particularly concerned with the following aspects of the investigative operation:
    (a) The use of standardized reporting techniques.
    (b) The assignment of cases for investigation.
    (c) The legal implications of investigative activities.
    (d) The effects of crime on its victims.
    (e) The coordination of investigative activities.
    (f) The use of specialized investigative techniques.
    (g) The collection and distribution of information.
    (h) The methods used to screen cases for solvability.

NOTES

[1]James N. Gilbert, *Criminal Investigation*, 2nd ed. Columbus, Ohio: Charles E. Merrill, 1986, pp. 3–11.

[2]Ibid., pp. 44–48.

[3]Richard V. Ericson, *Making Crime*. Toronto: Butterworth and Company, 1981, pp. 68–70.

[4]Parviz Saney, *Crime and Culture in America*. Westport, Conn.: Greenwood Press, 1986, pp. 19–20; Robert Trojanowicz et al., *An Evaluation of the Neighborhood Foot Patrol Program in Flint, Michigan*. East Lansing, Mich.: Michigan State University, 1982, p. 28.

[5]Gilbert, *Criminal Investigation*, pp. 59–64.

[6]Peter W. Greenwood, Jan M. Chaiken, and Joan Petersilia, *The Criminal Investigation Process*. Lexington, Mass.: D. C. Heath, 1977, p. 121.

[7]Gilbert, *Criminal Investigation*, pp. 64–67.

[8]Ibid., pp. 67–68.

[9]John E. Eck and Gerald L. Williams, "Criminal Investigations," in William A. Geller, ed., *Local Government Police Management*, 3rd ed. Washington, D.C.: International City Management Association, 1991, p. 137.

[10]Thomas F. Hastings, "Criminal Investigation," in Bernard L. Garmire, ed., *Local Government Police Management*, 2nd ed. Washington, D.C.: The International City Management Association, 1982, pp. 166–67.

[11]Ibid., pp. 166–67.

[12]O. W. Wilson, *Police Planning*, 2nd ed. Springfield, Ill.: Charles C Thomas, 1967, p. 114.

[13]Jay Cameron Hall, *Inside the Crime Lab*. Englewood Cliffs, N.J.: Prentice Hall, 1974, p. 16.

[14]Richard H. Fox, "Criminalistics," in Garmire, *Local Government Police Management*, pp. 363–79.

[15]Richard Saferstein, *Criminalistics*, 3rd ed. Englewood Cliffs, N.J.: Prentice Hall, 1987, pp. 1–12.

[16]Wilson, *Police Planning*, p. 113.

[17]Michael D. Maltz, "Toward Defining Organized Crime," in Herbert E. Alexander and Gerald E. Caiden, eds., *The Politics and Economics of Organized Crime*. Lexington, Mass.: Lexington Books, 1985, pp. 21–35.

[18]Robert W. Ferguson, *The Nature of Vice Control in the Administration of Justice*. St. Paul, Minn.: West Publishing, 1974, pp. 3–8.

[19]Charles H. Rogovin, "Organized Crime," in Garmire, *Local Government Police Management*, pp. 181–96.

[20]Howard Abadinsky, *The Criminal Elite*. Westport, Conn.: Greenwood Press, 1983, pp. 95–96.

[21]Susan Harlan, "Undercover Agents in Hollis, N.H." in *Law and Order*, vol. 38, no. 6, June 1990, pp. 40–42.

[22]Darrell J. Steffensmeier, *The Fence*. Totowa, N.J.: Rowman and Littlefield, 1986.

[23]Ferguson, *Vice Control*, pp. 53–54.

[24]Peter A. Lupsha, "Networks versus Networking: Analysis of an Organized Crime Group," in Gordon P. Waldo, ed., *Career Criminals*. Beverly Hills, Calif.: Sage Publications, 1983, pp. 59–88.

[25]Gary T. Marx, *Undercover: Police Surveillance in America*. Berkeley: University of California Press, 1988.

[26]Frederick M. Kaiser, "Executive Investigations of U.S. Legislators," in *Police Studies*, vol. 13, no. 1, Spring 1990, pp. 14–25.

[27]Clyde M. Stites, "PCs—Personal Computers, or Partners in Crime?" in *Law and Order*, vol. 38, no. 9, September 1990, pp. 161–65.

[28]Donn B. Parker, *Fighting Computer Crime*. New York: Scribner's, 1980.

[29]Ibid., p. 107.

[30]Buck BloomBecker, *Spectacular Computer Crimes*. Homewood, Ill.: Dow Jones-Irwin, 1990.

[31]David Stamps, "CPU Crimes Weak," *MISWeek*, September 28, 1983, p. 12.

[32]Thomas F. Adams, *Police Field Operations*, 2nd ed. Englewood Cliffs, N.J.: Prentice Hall, 1985, pp. 208–14.

[33]Susan E. Martin and Edwin E. Hamilton, "Police Handling of Child Abuse Cases," in *American Journal of Policing*, vol. 9, no. 2, 1990, pp. 1–24.

[34]Sylvia D. Stalnaker and Daniel J. Bell, "Police Referrals of Family Violence: A Texas Family Code Process Assessment," in *Police Studies*, vol. 13, no. 2, Summer 1990, pp. 67–73.

[35]Susan R. Paisner, "Domestic Violence: Breaking the Cycle," in *Police Chief*, vol. 58, no. 2, February 1991, pp. 35–38.

[36]Ericson, *Making Crimes*, pp. 70–72.

[37]Eck and Williams, "Criminal Investigations," in Geller, *Local Government Police Management*, pp. 137–38.

[38]Gilbert, *Criminal Investigation*, p. 63.

[39]Hastings, "Criminal Investigation," in Garmire, *Local Government Police Management*, pp. 164–65.

[40]Greenwood, Chaiken, and Petersilia, *The Criminal Investigation Process*, p. 15.

[41]Hastings, "Criminal Investigation," in Garmire, *Local Government Police Management*, pp. 175–76.

[42]J. Shane Creamer, *The Law of Arrest, Search, and Seizure*, 2nd ed. Philadelphia: W. B. Saunders, 1975.

MANAGEMENT OF AUXILIARY AND SUPPORT UNITS

The patrol, traffic, and criminal investigation units are the backbone of the police service in any law enforcement agency. Their functions include all of the major aspects of the police mission. Many smaller agencies consist entirely of those three elements, plus a fourth very small administrative element.

Larger agencies may have an array of specialized units to provide police services for which the public, through its elected officials, has expressed some demand, or for which the administrator perceives a need. Larger agencies also require more extensive technical and administrative support, if only to provide for the routine "housekeeping" tasks that are necessary to keep the agency in operation. In this chapter, we consider some of the different special operational units and technical support elements that are found in many police agencies.

SPECIAL OPERATIONAL UNITS

Before the turn of the century, many progressive police administrators recognized that neither the watchman concept of crime prevention nor the trooper concept of maintaining public order by suppressing crime was sufficient to ensure the safety and well-being of the public. Law enforcement agencies, in order to fulfill their fundamental mission, would have to take action to identify and, as far as possible, correct those conditions in society that contribute to lawlessness.

This thinking led to the establishment of a range of services that, in retrospect, seem rather peculiar. Some police departments set up soup kitchens, transient shelters, employment offices, and other social welfare units, in a well-meaning but ultimately futile attempt to attack crime by getting at its presumed sources: poverty, unemployment, and social deprivation in general. Most of these welfare-oriented services fell by the wayside as later police reforms concentrated on professionalizing the law enforcement functions.[1]

One part of the social welfare program remained within many police agencies, however, and continues to this day: the juvenile crime prevention unit. Partly because of the unique structure of the American juvenile justice system, and partly because of the philosophy that diverting juveniles from crime is a crucial means of preventing future adult crime, most police agencies continue to treat juvenile crimes in an entirely different manner from adult crimes.

Juvenile Crime Units

A typical juvenile crime unit is designed not to identify and arrest offenders, but to identify young people who are (or might be) involved in criminal activities so that they can be rehabilitated and diverted into law-abiding citizenship.[2]

In every state, laws specify at what age a person is regarded as an adult for various purposes. All states presently allow citizens to vote at age 18. Most states permit young people to obtain motor vehicle licenses at age 16, sometimes with the qualification that they must have passed a driver education course. In some states, young people may marry without their parents' consent at age 14. In others, the age of consent is 16 or even 18. In most states, young people may enter into binding contracts, including contracts for debts and for the purchase or sale of automobiles, at age 18. In a few states, the minimum age is 16.

Similarly, the age at which a person who commits an offense is treated as an adult varies from state to state. In most states, an offender is an adult at age 16; in some states, at age 17 or 18. Several states leave the decision to a judge or jury. In Texas, for example, anyone younger than the age of 18 who is accused of a criminal offense must be certified as an adult before he or she can be tried. Offenders as young as 14 years old have been certified as adults.

Persons who have, or may have, committed criminal offenses but who are not legally adults are diverted into the juvenile justice system. The accused offender is not tried and cannot be convicted. However, the offender may be the subject of a hearing to determine what kind of rehabilitative treatment is appropriate.[3]

Treatment may include assignment to an institution (variously called reform school or rehabilitation center, among other terms) or release into the custody of his or her parents or some other responsible adult. Some imaginative judges have "sentenced" young offenders to various sorts of community-service work such as cleaning up trash along roadways or assisting in the emergency room of a hospital. Other judges simply impose fines, which are most often paid by the offenders' parents.[4]

The role of the police in the juvenile justice system is, in a word, ambiguous. On the one hand, the police have a general duty to enforce the law and prevent crime. On the other hand, when the criminals are juveniles, the usual crime-suppression tactics are not available.

Most police authorities agree that, in principle, children cannot be held responsible for their antisocial behavior, even when that behavior includes violent crimes. Second, it is more important to rehabilitate juvenile offenders than to punish them. Third, the formalities of the adult criminal justice system are not entirely appropriate for application to juveniles.[5]

Nevertheless, the fact remains that juveniles commit crimes, that rehabilitation is very far from foolproof, and that diversion from the adult criminal justice system often seems to mean wandering in a quagmire of good intentions, confused motives, and uncertain results.[6]

All of which means, in practical terms, that the functions and activities of the juvenile crime unit are considerably different from those of the patrol and criminal investigation units. Officers assigned to the juvenile unit must be thoroughly familiar with the laws relating to juvenile offenses and with the workings of the juvenile justice system.[7]

We should also note here that smaller law enforcement agencies often have a "juvenile unit" that consists of just one officer, usually a regular patrol officer who has received some special training in this field.

Juvenile offenders come to the attention of the juvenile crime unit in any of several ways. Some juveniles are taken into custody by regular police officers in the course of investigating a crime; others are reported by victims or witnesses. Most juvenile crime units also take active measures to identify juvenile offenders. Officers from the juvenile unit attempt to maintain contact with youth gangs (which are not necessarily involved in criminal behavior) and organizations that cater to young people such as recreation centers. Whenever possible, juvenile crime officers attempt to identify offenders before they are arrested, preferably even before they become involved in serious crimes.

The first step in dealing with a juvenile offender (or prospective offender) is counseling with both the child and his or her parents. The object of counseling the young offender is twofold: (1) to discover whether there is some reason, such as emotional illness, for the antisocial behavior; and (2) to persuade the young person to devote his or her energies to productive, lawful activities. Counseling with the parents also has a dual purpose: to determine whether the child's problem is symptomatic of more serious social or psychological problems in the entire family, and to persuade the parents to accept responsibility for their child's behavior.

On the whole, the success rate for these counseling efforts is not very high. By the time young people come to the attention of the police, it is not likely that they will be persuaded to mend their ways. Still, many juvenile crime officers are very successful in their counseling. Even if the success rate is not very high, *any* success is better than none.[8]

If counseling fails, or if in the officer's judgment it is hopeless to begin with, the next step is to turn the young offender over to the juvenile justice system. Usually this means that a social worker will assume responsibility for the child's custody, although in many cases the child will continue to live at home, and a hearing will be held before a judge or juvenile magistrate.[9]

At this point, the only role of the police is to provide information concerning the young offender's record.

This brief outline of the police role in the juvenile justice system should not be taken as anything more than an introduction. The field of juvenile justice is exceedingly complex, and any thorough discussion of it would require much more time and space than we can devote to it.

In many states, juveniles are subject to any number of noncriminal laws that do not apply to adults. Juveniles may be arrested, taken from their parents' custody, and sentenced to institutions for noncriminal offenses, such as being truant from school, running away from home, or violating a local curfew ordinance. The custody of children where the parents are divorced or legally separated also involves the juvenile justice system in many states.[10]

Crime-Prevention Units

Juvenile crime units and crime-prevention units are often one and the same. Many police administrators adhere to the idea that the only really effective method of preventing adult crime is to divert young people from criminal activity.[11] Therefore, the only element of the police agency whose specific mission includes crime prevention is the juvenile unit, except to the extent that patrol officers are supposed to engage in preventive patrolling.

In recent years, however, a different philosophy of prevention has developed in American law enforcement. Police administrators and law enforcement theorists have concluded that most citizens can protect themselves from crime if they are taught how to do so. Educating the public to prevent crime therefore becomes the main responsibility of the crime-prevention unit.

The crime-prevention measures usually advocated by the police include such elementary matters as securing one's home or business by installing better locks on doors and windows. An extensive literature has developed on the subject of residential and commercial security. In fact, a whole industry has grown up to make and sell door locks, "burglar bars," and alarm systems. Some crime-prevention units have become little more than salespeople for the vendors of commercial security systems.

Crime-prevention units also may be concerned with teaching people how to deal with violent attack. Some police departments go so far as to conduct workshops and courses on self-defense techniques or the use of self-protection devices. Special workshops are often conducted for particular segments of the public such as women and the elderly.

Most of these efforts are certainly appropriate for a police agency. Whether a particular workshop enables a particular citizen to avoid a particular crime may be beside the point. If nothing else, the involvement of the police in these crime-prevention activities brings them into positive relationships with the community. These activities also promote citizen involvement in law enforcement, to the benefit of both the police and the community.

In some instances, crime-prevention officers may become a bit overenthusiastic and promise more than they can deliver. No security system, no matter how elaborate, will protect a home or business from an experienced and determined intruder. Some of the commercial systems offered to the public are notoriously unreliable. By the same token, all the self-defense and self-protection lessons in the world will do little good when a citizen is confronted by a determined, armed attacked. If false expectations are raised and citizens are given false confidence, the results could be unfortunate.

More effective methods of crime prevention certainly need to be developed. We may reasonably hope that as they are developed, more police agencies will share them with the public through educational campaigns, workshops, and other methods. In the last analysis, for reasons that we have discussed earlier, the police themselves can never hope to be entirely successful in preventing crime. The more responsibility for crime prevention is assumed by individual citizens, the more successful the police will be and the less crime ridden our communities will become.

Community Relations Units

Modern American police agencies are not likely to be involved in the kinds of social welfare schemes that were common before the turn of the century. Nevertheless, it is still true that crime is a social problem, not just a legal problem, and the source of all crime is the community in which people live. For those reasons, the police continue to have a legitimate concern for social conditions in the community they serve.

Although they do not operate soup kitchens, transient shelters, and employment offices, many police agencies do offer some services to assist the people in their community to improve their social welfare.

At the same time, the police urgently need to be aware of social conditions in their community. They may be able to do little about slum housing and chronic unemployment, but they need to know which slum landlords fail to provide their tenants with heat in the winter and which unemployed laborers are reaching the breaking point of desperation. This kind of knowledge can be gained only by establishing and maintaining a close, continual contact with the community.

Both social services and what we might call community intelligence functions are sometimes assigned to a *community relations unit*. In fact, all kinds of non–law enforcement services and functions may be assigned to the community relations unit. One danger in establishing this organizational element is that it may become a grab bag of assorted, miscellaneous operations.[12]

Some community relations units operate "store-front police stations" in low-income neighborhoods and other areas where it is especially important to maintain good relations with the community.[13] The function of a store-front operation is primarily to assist citizens in their dealings with the police and other public agencies, and generally to build a favorable image for the police.

Store-front community relations officers have one mandate: to help people in any way they can without compromising the primary police mission. They might help a young mother fill out an application for food stamps or counsel a young man on how to apply for a job. They might act as interpreters for a person who is qualified for free medical assistance but is unfamiliar with the bureaucratic procedures of a public clinic or hospital. They certainly would offer advice on crime prevention and self-protection and might assist a citizen in filing a complaint against a noisy neighbor.

Community relations units often sponsor youth groups such as Boys Clubs or Police Athletic Leagues, and adult volunteer organizations such as Citizens Band emergency monitoring clubs, "police buddy" organizations, or neighborhood crime watches. In some instances, the community relations unit is responsible for administering the agency's police cadet program.

Finally, the community relations unit may be combined with the crime-prevention unit. Both operations involve establishing and maintaining a positive relationship between the police and the community. The specific objective of any of these activities may have relatively little to do with criminal law enforcement, but the ultimate goal is to keep open the lines of communication between the police and the community.[14]

Tactical Units

Not all special operational units are concerned with services that are only indirectly related to law enforcement. Some police departments have established special units to concentrate on specific types of criminal activities. These units, often called tactical squads or tactical patrols, are similar in concept to the selective traffic enforcement operations we discussed in Chapter 15.

Tactical units are most likely to be effective in larger agencies serving metropolitan areas. If the agency is able to identify a particular pattern of criminal activity, the tactical unit can be assigned to take whatever action is appropriate.[15] For example, if there has been a series of armed robberies of convenience stores in one part of town, tactical units might be assigned to patrol that area intensively.

The specific actions taken by a tactical unit should be based on the type of crime. In some cases, intensive patrol of an area may be appropriate, using standard, marked patrol vehicles. In other cases, patrolling in unmarked vehicles might be more effective. In yet other cases, tactical officers might stake out businesses that are likely crime targets. The choice of tactics depends on the amount of information that is available about the type of crime being fought.[16]

The value of a tactical unit depends largely on the fact that criminals, individually and collectively, often commit offenses in a series or pattern. If a person is successful in holding up a liquor store, he or she is encouraged to hold up a second liquor store, then a third and a fourth, and to continue until he or she is caught or moves on to some other crime.

Furthermore, if one person is successful in holding up several liquor stores, other people may try their hand at it.

To be sure, not all crimes occur in series, and not all criminal activity exhibits a clear pattern. Many crimes occur because of a unique opportunity or because of the personal relationships between the offender and victim, and these crimes are less likely to be repeated. Some criminals are either not clever enough or too clever to follow any particular pattern of activity.

Nevertheless, when a clear pattern emerges from the daily flood of crime, the police have an opportunity to choose a specific tactic to disrupt the pattern.

Often the only tactic necessary is to inform patrol officers or investigators and encourage them to concentrate their efforts on a specific enforcement goal. However, patrol officers and investigators may already have a heavy workload, or the organizational structure may not permit the kind of flexibility needed to assign officers to intensive patrolling or other specific tactics. In such cases, the formation of a tactical unit may be effective.

Members of the tactical unit, which may vary in size from three or four officers to fifteen or twenty, usually are drawn from the general patrol force and typically are among the more experienced and successful patrol officers.

The assignment of a tactical unit to a particular task may come from the head of the patrol force, the head of the investigations element, or more rarely the head of a crime analysis or intelligence unit. The head of the tactical unit should be an experienced patrol commander with an excellent record for leadership.

For each tactical assignment, the commander and other unit heads should determine exactly what tactics are to be employed, where, by whom, and for what period. These instructions then are given to the tactical officers. The instructions should specify whether marked or unmarked vehicles are to be used, whether officers are to be uniformed or in plainclothes, what actions are to be taken when criminal activity is observed, whether an individual officer should make an immediate arrest or simply observe the suspect and call for backup assistance, and anything else concerning the mission.

As with selective traffic enforcement, a tactical unit can be an extremely effective weapon against certain kinds of crime. There is a large body of evidence to show that a majority of crimes are committed by a minority of criminals.[17] These are the people who commit essentially the same crime repetitively. The very fact that they are able to repeat their crime indicates that routine patrol and investigative techniques are unsuccessful.

Tactical units, because they are not limited to a single patrol area nor to the time-consuming, after-the-fact approach of conventional investigation, intensify the enforcement effort and focus it on those criminals who are otherwise all too likely to escape detection.

Special Situations Units

Some police agencies, especially in the major metropolitan areas, have established another kind of specialized operational unit, known variously as SWAT teams, special weapons units, and special situations units. Whatever they are called, these units are organized to provide rapid police response to certain kinds of critical events, usually of an emergency nature.

A special situations unit may be organized on a permanent basis, or may be established on an on-call basis. In the latter case, members of the special situations unit are regular patrol officers and investigators who have received advanced training in special-situations techniques. Ordinarily they are assigned to their usual patrol or investigative duties. However, when a special situation arises, the unit is activated and members report to a prearranged point for their special assignment.

What constitutes a special situation? That is the first and perhaps most important question to be answered *before* a special situations unit is established. Generally, these units are called up for some or all of the following kinds of incidents:

1. A hostage-taking, barricaded suspect, or other situation in which a criminal has taken refuge in some location that is not readily accessible, or has taken one or more hostages.
2. Sniper incidents, where the criminal is in an inaccessible location and is firing a weapon at police officers or innocent bystanders.
3. VIP security, where a public official or other celebrity requires special protection, usually because the celebrity is a controversial person, or because the celebrity's safety has been threatened.

Special situations units also may be used as the primary force, supplemented by regular officers, in suppressing riots and other major disturbances, parades or demonstrations that have a potential for violence, or any other incident that, because of its magnitude and potential for violence, is beyond the capabilities of the regular patrol force.

Members of a special situations unit usually are trained in a variety of advanced skills and techniques, including the use of both lethal and nonlethal weapons, hand-to-hand combat, and techniques for gaining access to inaccessible locations, such as rappelling. They also should be trained in crowd-control and riot suppression techniques and in the psychology of hostage situations.

SUPPORT SERVICES

Every police agency, large or small, requires a certain number of housekeeping and supportive services that do not directly involve law enforcement, but are nevertheless necessary for the law enforcement officers to do their jobs. In addition, a police agency usually requires certain technical support services to be performed in order to carry out the primary law enforcement functions.

These functions may be assigned to a single element, often called the technical services unit or auxiliary support unit. A partial list of these auxiliary support functions would include the following:

1. Building maintenance and custodial service.
2. Vehicle maintenance.
3. Secretarial, clerical, and office management.
4. Financial management (bookkeeping, accounting).
5. Personnel management (recruitment and selection, management of the promotion system, recordkeeping, administration of contracts with unions or unionlike police organizations, administration of benefit programs, etc.).
6. Purchasing and property management.

7. Communications and information services.

8. Public information, relations with news media.

9. Legal counsel concerning both law enforcement and non–law enforcement matters.

10. Personnel training services.

11. Criminalistics (evidence collection and analysis).

12. Crime analysis, intelligence, and planning.

13. Jail operations.

Some of the support services listed here are often provided by the parent government of small and medium-sized agencies. Larger agencies, however, generally prefer to be responsible for their own support services, thereby giving the administrators better control over their own resources.

In the past, many medium-sized and larger police agencies assigned police officers to these auxiliary functions. There were two reasons for this practice. The police, like an army, preferred to be entirely self-reliant. By entrusting even the most menial housekeeping tasks only to fellow officers, police administrators kept civilians at bay and maintained a "police mystique," according to which all police work was necessarily secretive and dangerous.[18]

A second reason was that most lower-level police officers were so poorly paid that it was simply cheaper to use patrol officers, and even sergeants, as secretaries and mechanics rather than hiring civilians for these functions and paying them competitive wages.

Modern police practice has changed considerably. Today, civilians—that is, personnel who are not commissioned law enforcement officers—are found throughout most police agencies, performing most of the auxiliary support services listed above.

There is a straightforward, practical reason for this change: Today's police officers, even at the entry level, cost too much to be assigned to menial tasks. In almost all police agencies, even the smallest ones, an entering patrol officer's salary is several thousand dollars higher than a typical clerk's or mechanic's salary. In addition, a patrol officer represents a considerable investment in training that should not be wasted on tasks more suitable to noncommissioned personnel.[19]

Aside from these practical considerations, there also have been changes in the attitudes of most police administrators. Maintaining the police mystique is no longer seen as necessary, or even desirable. Modern administrators are more interested in building bridges between the police and the public—not in stocking the moat with alligators. By bringing noncommissioned personnel into the agency at all levels, the administrator helps to break down the fortress mentality that once permeated police thinking.[20]

Noncommissioned personnel also bring to their jobs a broader range of skills and talents than are likely to be found among police personnel. People with special ability and experience in, say, computer programming can make a substantial contribution to a police agency.[21]

However, noncommissioned support personnel usually require some training in law enforcement and police policies if they are to apply their particular skills successfully. For example, a secretary should know enough about basic police procedure to decide whether "filed," "filled," or "field" is the correct word in this sentence:

Officer Jones submitted a _____ interview report on this subject yesterday.

The basic training needs of noncommissioned auxiliary support personnel may be satisfied by providing a fairly brief orientation course through the agency's own academy or by encouraging new noncommissioned personnel to enroll in an introductory course on the criminal justice system at a local community college. In addition, all new noncommissioned personnel should be given a planned on-the-job training program under the supervision of a commissioned police officer, usually the officer who will supervise the new employee.

There are some disadvantages in using noncommissioned personnel in auxiliary support roles. Personnel turnover is generally higher than it is among police officers. Because noncommissioned employees have not gone through the rigorous screening and selection process applied to police recruits, they may have, on the average, lower qualifications in terms of educational levels, work experience, and personal backgrounds. They may be somewhat less resistant to the opportunities for corruption that are unavoidably present in law enforcement.

Noncommissioned personnel also have no legal authority as law enforcement officers, and therefore it may be necessary to protect some sensitive records and police activities from them.

Finally, commissioned officers may resist accepting direction from noncommissioned employees, which means that the latter are restricted to subordinate positions.

Nevertheless, despite the disadvantages, the practical benefits of using noncommissioned personnel in most auxiliary support roles are so great that this practice is likely to grow. Upgrading the qualifications and abilities of police officers is one of the highest priorities of modern police administrators, but one of the side effects is that highly qualified, well-trained professional officers are simply too valuable to waste on routine housekeeping tasks.

Some of the support services listed above demand the special training and qualifications of commissioned law enforcement officers. At the very least, an experienced senior officer should manage such technical elements as communications, personnel training, crime analysis, and the jail. In some states, all jail personnel must be either commissioned law enforcement officers or certified correctional officers.

Building Maintenance and Custodial Services

If this function is not served by the parent government or, in the case of rented facilities, the owner of the building, the police agency must employ a custodial staff and some maintenance personnel. The skills required include those of an electrician, plumber, and carpenter. Sometimes it is possible to find individuals who have two or more of those skills and who are able to handle the agency's work load. The building maintenance and custodial staff should be under the supervision of the police administrator responsible for administrative services. In a small agency, this is likely to be the agency's chief executive.

Vehicle Maintenance

We have previously discussed police fleet maintenance (see Chapter 6). Usually it is preferable to use noncommissioned personnel for this function. They should be under the supervision of the official responsible for administrative or technical services.

Secretarial, Clerical, and Office Management Services

Most police agencies have a secretarial and clerical pool whose members provide basic office services to all personnel as needed. Larger agencies may have a few secretaries assigned to individual administrators or to units such as records and identification. Pool employees should be supervised by the chief of administrative services. Secretaries or clerks assigned to specific officials or units should be supervised by them.

Financial Management Services

This is one function that is almost always provided by the parent government. Nevertheless, every police agency except perhaps the very smallest ones should have at least a bookkeeper to maintain, on a day-to-day basis, accurate accounts of expenditures. Larger agencies may require a separate accounting staff to handle revenues from various sources, payroll accounting, and purchases and other disbursements. These personnel should be under the supervision of the chief of administrative services.

Personnel Management

Again, this function often is provided by the parent government and sometimes by a completely separate Civil Service or merit system agency. Still, every police agency should have at least a personnel clerk, and preferably a staff of experienced personnel administrators, to plan and carry out recruitment and selection campaigns, to maintain records of promotions and other personnel actions, to supervise the administration of union contracts, and to ensure that personnel benefits such as health insurance and pension plans are properly administered. The personnel management function should be supervised by the chief of administrative services or by the agency's chief executive.

Purchasing and Property Management

We discussed this function at considerable length in Chapter 10, so there is no need to repeat it here. Purchasing and property management services should be supervised by the chief of administrative services. The management of evidence, found property, and other personal property that does not belong to the agency should be at least supervised by a commissioned officer.

Communications and Information Services

Again, these functions were described in Chapters 8 and 9. Because of their technical nature and vital importance to effective police operations, communications and records services must be under the direct management of experienced police administrators.

Many administrators feel that only commissioned officers should act as radio dispatchers, partly because of the sensitive information that must be handled and partly because the dispatcher actually serves as the de facto commander of most routine patrol operations. However, many agencies have had great success in using noncommissioned personnel as dispatchers, provided that an experienced commissioned officer is constantly present to supervise the dispatching operation and assume command in emergencies.

Public Information

Police activities and criminal incidents are of constant interest to the general public, and the news media demand access to information about these matters. Every police agency should have one designated senior official who is authorized to act as the official spokesperson for the department.

In addition, except in very small agencies, the public information officer should make a positive effort to inform the community, through the news media, of significant activities and events. Larger agencies, especially those in major metropolitan areas and those whose jurisdiction is statewide, may require a staff of public information specialists: writers, audiovisual materials producers, and others. The latter do not need to be commissioned officers, but should be under the direct supervision of a commissioned officer.[22]

Legal Counsel

Every police agency should have access to the services of the city or county attorney or the state attorney general for legal counsel. However, those individuals are not always readily available when an urgent question arises. Arrangements can be made to employ a private attorney on a retainer basis; a regular fee is paid, entitling the client to a certain amount of service. This may be a satisfactory solution for many small and medium-sized agencies. Larger agencies usually should have at least one qualified attorney on the agency's staff. Very large agencies may require several attorneys, along with their own support personnel. Usually the department's attorney or chief legal counsel is considered an administrative officer and is responsible directly to the chief executive.

Personnel Training

Both preservice and inservice training of all commissioned and noncommissioned personnel should be provided by a permanent training staff. In smaller agencies, this function may be combined with the personnel management function. Training staff members should include experienced police officers. However, special expertise in training and education is also needed and may be provided by noncommissioned personnel. In some agencies, the training staff is directed not by a senior police officer but by a qualified educator.[23] Whatever the composition of the staff may be, they should be supervised by the chief of administrative services.

Criminalistics Services

Evidence technicians and laboratory personnel do not need to be commissioned officers. Conversely, however, there is no reason that commissioned officers with special interests in criminalistics cannot be assigned to that function. As we noted in Chapter 15, agencies that are too small to maintain a full-time criminalistics staff may rely instead on a regional laboratory or some other cooperative arrangement. Criminalistics should be under the command of the chief of technical services or the head of the agency's criminal investigations unit.

Crime Analysis, Intelligence, or Planning Services

Earlier we discussed the use of tactical units to attack specific types of crimes, on the basis of analyses of criminal activity.[24] Those analyses may be performed by a technical staff, sometimes called the *crime analysis* unit; other terms for the same function are *intelligence unit* and *planning unit*. In some agencies, the intelligence unit is concerned primarily or exclusively with information about organized crime, a field that is considerably narrower than that of a crime analysis unit.[25]

In some agencies, the planning unit is responsible for a broader range of functions than a crime analysis unit. Planners may not only collect and analyze information about criminal activities, but also develop plans for long-range programs, special projects, personnel development, and budgeting.

Regardless of the administrative structure adopted, the crime analysis, intelligence, or planning unit should be commanded by an experienced police officer. Other members of the staff do not need to be commissioned officers. Often it is preferable to employ technical specialists such as demographers (persons who study population trends), computer operators, statisticians, and so on. Depending on its specific functions, the unit may be supervised by the chief of the criminal investigations element, by the head of the organized crimes unit, by the official responsible for the administration of technical services, or by the chief executive of the agency.

Jail Operations

Most municipal police agencies are required to operate some kind of detention facility to hold persons who have been arrested and are awaiting a hearing before a magistrate. Prisoners awaiting trial who have not been released on bail may be held either by the municipal agency or, more often, transferred to a county or state facility. Usually prisoners in a city detention facility are held only for a short period, as little as one or two hours and not more than a week.

Prisoners in county and state facilities are more often held for longer periods, either awaiting trial or serving a sentence. A county jail or state prison is not merely a detention facility, it is a correctional institution whose responsibilities include the rehabilitation of inmates.[26]

The requirements of both short-term detention facilities and long-term correctional institutions are spelled out in state laws and, increasingly, federal laws and court decisions. In the most general terms, the operators of a jail or prison have two major duties: to prevent prisoners from leaving and to protect the prisoners' health and welfare while they are being detained. The specific means of discharging both responsibilities are far too complex to discuss here.

As noted earlier, most states require that custodial personnel in jails and prisons be either commissioned law enforcement officers or certified correctional officers. Other jail staff members (cooks, rehabilitation specialists, librarians, and so forth) usually are not commissioned officers. Any detention facility operated by a municipal police department must be under the immediate management of a senior police official, often the chief executive.

This brief overview of support services is intended only to suggest the kinds of services that are required by most police agencies and the kinds of administrative and management structures that are likely to be successful. However, each agency has its own needs, and police administrators have their own preferences. The organization of support services is one area in which there are few universal rules. Administrators should apply their creativity and ingenuity to provide for necessary services in the most effective and efficient manner possible.

REVIEW

1. The primary purpose of a juvenile crime unit is

   (a) To obtain sufficient evidence against juvenile offenders that they will be convicted for their crimes.

   (b) To remove juvenile offenders from the influence of their parents, peers, and others who are responsible for their antisocial attitudes.

   (c) To identify the victims of juvenile crimes.

   (d) To identify juvenile offenders so that they can be rehabilitated.

   (e) To prevent adults from committing crimes against juveniles.

2. Community relations units generally are

   (a) Ineffective in promoting a positive image of the police.

(b) Designed to deliver social welfare services to the poor and oppressed minorities.

(c) Intended partly to provide the police with information about social conditions in the community.

(d) Established to provide the news media with accurate information from the police viewpoint.

(e) Organized only as a response to pressure from community political leaders.

3. Noncommissioned personnel are often employed in police agencies because

(a) Commissioned officers are too valuable to use for routine housekeeping chores.

(b) Noncommissioned personnel often have special skills or talents that are not available among the commissioned officers.

(c) Noncommissioned personnel usually are paid less than commissioned personnel.

(d) Personnel turnover is often lower among noncommissioned staff.

(e) Noncommissioned personnel are usually better educated and more highly motivated than regular police officers.

4. A retainer may be used

(a) To guarantee that highly paid personnel will not leave the agency unexpectedly.

(b) To secure a minimum level of services from a private attorney.

(c) To ensure that a city or county attorney will be available at all times to answer questions.

(d) To protect a police agency from any legal problems that might arise.

(e) To reduce the need for commissioned officers to be assigned to non–law enforcement duties.

5. The major responsibilities of jail administrators are

(a) To prevent prisoners from leaving.

(b) To operate the jail as economically as possible.

(c) To assist police officers in making legal arrests.

(d) To protect the health and welfare of prisoners.

(e) To deliver prisoners promptly to the courts for trial and sentencing.

NOTES

[1]William J. Bopp, "Police History and Social Consciousness," in William J. Bopp, ed., *Police Administration: Selected Readings*. Boston: Holbrook Press, 1975, p. 77.

[2]William S. Davidson II et al., *Alternative Treatments for Troubled Youth*. New York: Plenum Press, 1990, p. 12.

[3]Larry Miller and Michael Braswell, *Human Relations and Police Work*, 2nd ed. Prospect Heights, Ill.: Waveland Press, 1988, p. 49.

[4]*Task Force Report: Corrections*. Washington, D.C.: President's Commission on Law Enforcement and the Administration of Justice, 1967, pp. 23–24.

[5]Miller and Braswell, *Police Work*, p. 51.

[6]John D. Wooldredge, "Age at First Court Intervention and the Likelihood of Recidivism Among Less Serious Juvenile Offenders," in *Journal of Criminal Justice*, vol. 19, no. 6, 1991, pp. 515–24.

[7]Miller and Braswell, *Police Work*, p. 49.

[8]Davidson et al., *Alternative Treatments*, pp. 34–37.

[9]*Reports of the National Juvenile Justice Assessment Centers: The Shadows of Distress.* Washington, D.C.: U.S. Department of Justice, Law Enforcement Assistance Administration, 1980, pp. 34–43.

[10]For a more thorough discussion of the juvenile justice system, see Ira M. Schwartz, *(In) Justice for Juveniles*. Lexington, Mass.: Lexington Books, 1989.

[11]Charles D. Hale, *Fundamentals of Police Administration*. Boston: Holbrook Press, 1977, p. 142.

[12]Arthur Niederhoffer and Alexander B. Smith, *New Directions of Police-Community Relations*. San Francisco: Rinehart Press, 1974, pp. 2–6.

[13]Gary Enos, "Cities Take Diverse Paths to Community Policing," in *City and State*, vol. 9, no. 14, July 27, 1992, pp. 16–17.

[14]Niederhoffer and Smith, *New Directions*.

[15]Jerry L. Carlin and Colin L. Moodie, "A Comparison of Some Patrol Methods," in Bopp, *Police Administration*, pp. 251–52.

[16]Edward P. Ammann and Jim Hey, "The Discretionary Patrol Unit," in *FBI Law Enforcement Bulletin*, vol. 58, no. 1, January 1989, pp. 18–22.

[17]Alfred Blumstein, Jacqueline Cohen, Jeffrey A. Roth, and Christy A. Visher, eds., *Criminal Careers and "Career Criminals."* Washington, D.C.: National Academy Press, 1986.

[18]William J. Bopp, "The Traditional Approach to Police Organization," in Bopp, *Police Administration*, pp. 79–80.

[19]Robert D. Pursley, "Traditional Police Organization: A Portent of Failure?" in Bopp, *Police Administration*, pp. 83–85.

[20]Stephen D. Mastrofski, "The Prospects of Change in Police Patrol," in *American Journal of Policing*, vol. 9, no. 3, 1990, pp. 21–22.

[21]"Tempe's 'VIPs'," in *FBI Law Enforcement Bulletin*, vol. 16, no. 7, July 1992, pp. 4–5, describes the recruitment, screening, assignment, training, and supervision of noncommissioned volunteers in the Tempe, Arizona, police department.

[22]Gerald W. Garner, *The Police Meet the Press*. Springfield, Ill.: Charles C Thomas, 1984.

[23]Bruce L. Berg, "Who Should Teach Police," in *American Journal of Policing*, vol. 9, no. 2, 1990, pp. 79–100.

[24]Richard H. Snibbe, "A Concept for Police in Crime Prevention," in Bopp, *Police Administration*, pp. 204–10.

[25]Robert W. Ferguson, *The Nature of Vice Control in the Administration of Justice*. St. Paul, Minn.: West Publishing, 1974, pp. 16–20.

[26]*A Handbook on Jail Administration*. Washington, D.C.: National Sheriffs' Association, 1974, pp. 9–10.

POLICE AND THE COMMUNITY

Throughout this book we have been concerned principally with the internal operations of police agencies, particularly with the duties of administrators to ensure that their agencies operate effectively and efficiently. We have not taken the time to examine the relationship between a police agency and the community it serves, except when that relationship has an immediate bearing on the workings of the agency.

The tasks that have occupied our attention are virtually universal. All police administrators must have an understanding of fundamental principles of management. They also must provide their personnel with the facilities and equipment necessary to do their job. All police administrators must recruit, select, train, and motivate their employees, and must organize them into effective operating units to carry out their responsibilities.

What we have not emphasized, until now, is that *how* police administrators perform their duties must take into account the communities they are serving including the needs of those communities for law enforcement services.[1]

The first point we must make is that *no law enforcement agency, no matter how small, exists to serve only a single community*. To see why this is true, and how it affects policing, we should begin by defining "community."

WHAT IS A COMMUNITY?

In the most general sense, a *community* is *any group of people who have something in common*.[2]

Communities are often defined by geography: All of the people who live in a certain place have at least that much in common. When people speak of "the community" without further explanation, they usually mean a city or some part of it, such as a neighborhood.

People who live near one another are likely to have a number of characteristics in common. They are more likely than not to be of the same ethnic background and economic status, to have the same general range of educational attainment, and to have similar kinds of employment, religious and political values, and so forth.[3]

Social scientists use the concept of *class* to summarize a number of characteristics of a population into three broad categories: upper, middle, and lower. "Working class" is sometimes used as an intermediate category, overlapping the top of the lower class and the bottom of the middle class. Among the factors that contribute to a person's class status are income, educational attainment, and occupation. Sociologists have found that people of the same class tend to be similar in their values and attitudes. However, these are very broad generalizations that are useful only in giving clues to the social behavior of large groups of people.

Some communities are much more *homogeneous* (all of one kind) than others. Usually older neighborhoods tend to be more homogeneous than newer ones, but that is not always true. Older neighborhoods sometimes go through transitions during which one group of people is moving in while another group is moving out. For a while at least the neighborhood may be highly *heterogeneous* (a mixture of dissimilar parts). There are also stable communities that remain heterogeneous more or less permanently.

It is important to understand that all communities are *atomistic*; they can be divided into smaller and smaller parts, down to the level of individuals, and those parts can be recombined in an infinite variety of ways. Even the most homogeneous neighborhood is still made up of individual human beings who, though they may be similar in many respects, do not necessarily have identical interests, values, concerns, and needs.

Since we have defined community as any group of people having something in common, it is possible to speak meaningfully of a community based on similar occupation ("the business community," "the farming community," "the banking community," "the merchant community," and even "the police community" and "the criminal community"), ethnicity ("the African-American community," "the Hispanic community," and "the Asian community"); national origin ("the Portuguese-American community" and "the Vietnamese community"); religion ("the Hasidic community," "the Amish community," and "the Methodist community"); or any other characteristic one can imagine. Furthermore, the characteristics can be combined; one can speak of the "African-American Catholic community" or the "Irish shopkeeper community" or the "Hispanic female police officer community."

From what we have said so far, several conclusions can be drawn:

- *Most people are members of several communities, not just one.*
- Any geographical area contains a number of different communities, even though it may be homogeneous in some respects.

- The police, if they are to take into account community needs and interests, must deal with multiple communities defined not only by geography but by such characteristics as occupation, ethnicity, economic status, and perhaps others.[4]

Every law enforcement agency has a specific *jurisdiction*, the geographic area within which it has been given authority to enforce the laws. When police administrators give some thought to the needs and interests of their community, there is a natural tendency to think only in terms of the dominant group in their jurisdiction as if it were the only community that matters. From what we have just said, the dangers in this approach should be obvious.

Ironically, one of the reasons that police administrators have not always recognized the varying needs and interests of the communities they serve is the desire to make policing more professional.[5] The police reform movement, as we discussed in Chapter 1, began more than a century ago with several specific aims, one of which was to remove law enforcement from the corrupting influence of local politicians and other narrow interests. The reformers believed that policing should be based on scientifically derived, universal principles of crime prevention and suppression, and that these principles should be applied consistently without regard to local peculiarities.

One major difficulty with this concept of professionalism was the utter lack of scientifically derived principles of crime prevention and suppression. In their place were opinions, prejudices, hunches, and untested theories. To this day, many of the traditional practices of law enforcement agencies are based not on facts and evidence, but on someone's "commonsense" ideas of what ought to work. Motorized random crime-prevention patrol is one of those ideas.[6]

Another "commonsense" idea that has become enshrined as conventional wisdom is that the sole duty of the police is to enforce the law. The professional law enforcement officer is said to have neither the authority nor the responsibility to maintain order in the community unless a specific law has been violated.[7]

These ideas about police professionalism have created barriers that have prevented police administrators from recognizing, much less serving, the interests and needs of their communities.

In all fairness, we must say that not all police administrators have been oblivious to these problems. As early as the 1950s, many law enforcement agencies, especially in the major metropolitan areas, established police-community relations units whose specific mission was to improve the connections between the police and the general population. Unfortunately, this mission typically was conceived in terms of promoting a favorable image of the police, not in terms of discovering the needs of the community and finding ways to serve those needs.[8]

Since the widespread urban disorders of the 1960s, there has been a greater willingness among police administrators to reconsider the issue of police-community relations. Several times in earlier chapters we have mentioned the popularity of the term "community policing" and how it influences the management of police resources and operations. We have also pointed out that community policing is a very ambiguous term, applied indiscriminately to all sorts of variations on police tactics and organization. It has seemed to us that community policing, like the team policing concept that it has supplanted, is in danger of becoming merely a faddish slogan with little real meaning.

Conversely, the search for ways to make policing more responsive to community needs and interests is neither a fad nor a slogan; it is nothing less than a redefining of what professional law enforcement means.[9]

In the next section of this chapter, we discuss some of the characteristics of different kinds of communities and how they affect the principal mission of the police. Later, we explore some of the ways police departments have tried to identify and to serve the needs of their communities.

Minority Communities

In many cases, the relationship between the police and the minority population of their communities has been marked by hostility, mutual suspicion, and occasional outbursts of violence. This has been true not only for the past few decades, but throughout the history of American law enforcement. It is difficult to imagine how this would not be true.

Remember that the earliest municipal police agencies in many southern cities were organized in the early nineteenth century partly for the specific purpose of tracking down and bringing back runaway slaves. In the northern cities, the relationship between the police and minority groups also was often tense. Police officers were often recruited from newly arrived immigrant groups who had no past experience in dealing with people of different cultural and ethnic backgrounds.

One might reasonably suppose that the movement toward police professionalism would reduce police-minority tensions. After all, if the professional police officer treated everyone equally in accordance with universal principles, and concentrated entirely on enforcing laws that apply to everyone equally, the minority communities could hardly complain that they were being discriminated against.

Unfortunately, there is ample evidence that the police *have* discriminated against minorities.[10] One reason is that the ideal of professional law enforcement has not been realized, especially at the "street level" where patrol officers interact directly with citizens. Another reason is that the universalist ideals themselves fail to recognize the special character, and the special needs, of minority communities.

We defined *minority* in Chapter 11 generally as any identifiable group that makes up less than half of the total population. In a more limited sense, African-Americans, Hispanics, Native Americans, and Asians are identified as "minorities" in certain laws, mostly at the federal level. For our purposes here, we will use the general definition. Thus, in one city, white Anglo-Saxon Protestants may be the dominant group and African-Americans may be a minority. In another city, exactly the opposite may be the case.

The police are by definition representatives of the dominant group in their community because that is the group with the power to make laws and allocate resources to enforce the laws.[11] The relationship between the police and the minority community depends partly on the history of relations between the dominant and minority groups and partly on present circumstances.

We will use the experience of African-Americans as an example, because the relationship between the police and the African-American community has been studied more extensively than other police-minority relationships. However, much of what we will say would apply equally, allowing for different historical details, to Hispanics, to Asians in some parts of the country, and in certain regions to other minority groups.

After the Civil War and the abolition of slavery, African-Americans were subjected to discriminatory laws and customs not only in the South but in virtually every part of the country. Southern states adopted the doctrine of "separate but equal," meaning that African-Americans and whites were allotted separate schools, medical facilities, sections of buses, public rest rooms, jail cells, and every sort of public facility whether publicly or privately owned. By any standard one wishes to apply, the facilities were *not* equal.

Furthermore, African-Americans were denied the right to vote, usually on some pretext such as a literacy test as a prerequisite for voter registration, and were systematically excluded from all but the most menial employment.

Discrimination in northern and western parts of the country depended on custom rather than law. For example, African-Americans were not prohibited from attending white schools, but the school boundaries were carefully drawn to ensure that children in African-American neighborhoods were kept out of schools that served children from white neighborhoods. Similarly, there were no laws prohibiting African-Americans from living in white neighborhoods, but most banks would not give mortgages to African-Americans who wanted to buy a home in a white neighborhood.

It is a matter of common experience among African-Americans that some of these extralegal customs still exist, despite more than half a century of effort on the part of African-Americans themselves and sympathetic whites to eliminate discrimination. Anti–African-American laws in the South came under attack beginning in the early 1950s and have been abolished. Anti–African-American attitudes and customs have proven much harder to eliminate.

Historically, then, African-Americans as individuals and as a group have had good reason to be suspicious and resentful of whites. However, it has been more than thirty years since the Civil Rights Movement culminated in the passage of the federal Civil Rights Act, Voting Rights Act, Equal Employment Opportunity Act, and other landmark federal laws. More than a dozen other federal laws have been enacted to prohibit discriminatory practices, and many states have passed their own antidiscrimination laws. Surely, then, present circumstances are very different.

Measured objectively, it is certainly true that there is far less *legal* discrimination against African-Americans than there was thirty, or even twenty, years ago. Nevertheless, the effects of past discrimination continue to affect the minority community, and subtler forms of discrimination have not been eliminated.

For example, because of past discrimination, in most cities African-Americans have been relegated to specific neighborhoods. There probably is not a single city in the United States with a population of more than 5000 in which the white and African-American population is evenly dispersed. As a result, African-American neighborhoods tend to be heterogeneous in social class: upper-, middle-, and lower-class African-Americans live in the same neighborhoods, often next door to one another. White neighborhoods tend to be much more homogeneous in social class.[12]

To some extent, the class heterogeneity of African-American neighborhoods has been a positive factor. The wealthier, better educated, and more community-minded upper- and middle-class African-Americans have provided a degree of community leadership that otherwise would be difficult to find. However, ethnic segregation of housing has limited the opportunities of even the most successful African-Americans, who have been forced to offer their goods and services mostly to each other, and has prevented African-Americans and whites of comparable social and economic status from associating.

The suburbanization of America also has affected African-American communities to their detriment. Since World War II, most of the huge cities of the Northeast and Midwest have gradually lost population. People in the cities who could afford to do so have moved to the suburbs or to the less densely built cities of the southwestern and western states, some of which are hardly cities at all but more like collections of suburbs.

Before the 1960s, upper- and middle-class African-Americans were prevented from joining the suburbanization movement by the laws and customs we have mentioned. After the mid-1960s, changes in the law and in society's tolerance for discrimination have permitted many African-Americans to leave the segregated neighborhoods and move to white-dominant areas.

These changes are, for the most part, beneficial to the individuals who can take advantage of them. Upper- and middle-class African-Americans, especially those in the generation born since 1960, have seen a virtual explosion of opportunities. However, the African-American communities in many cases have been devastated by the loss of social leadership. The void has been filled by an influx of poorer, lower-class African-Americans and other minorities, moving into the central cities from rural areas and from other countries (primarily Latin America and Southeast Asia).[13]

As a result, leadership in the African-American communities of most cities has been split between those few upper- and middle-class African-Americans who, for whatever personal reasons, have chosen to remain, and a slowly emerging lower-class leadership corps. The established leaders tend to be elderly and relatively conservative. The emergent leaders tend to be young, inexperienced in community organization, and relatively radical. The two groups of leaders differ greatly in their perceptions of their communities' needs, in their ideas about the actions that should be taken to satisfy those needs, and in their willingness to negotiate compromises.[14]

The present circumstances of most African-American communities, then, are characterized by a growing degree of lower-class homogeneity. In this respect, African-American communities are increasingly similar to other minority-dominant communities and to lower-class white-dominant communities, but with some residual effects of historical anti–African-American discrimination.

How do these characteristics affect the needs of African-American communities for law enforcement services? (Again, the reader is cautioned to remember that we are using sweeping generalizations that may not apply, or may need to be substantially modified, in particular situations.)

Alcoholism and drug abuse may be more common, and certainly are much more visible, in lower-class communities than in middle- and upper-class areas. Mental retardation, mental illnesses, and physical diseases ranging from lice and impetigo to syphilis and tuberculosis are also more common in lower-class communities than elsewhere.

Some studies have indicated that lower-class young males tend to be concerned primarily with immediate gratification, without much regard for the long-term consequences of their actions or the effects on others.[15]

If these generalizations are accurate, it is not difficult to predict that many types of crime would be common in lower- class communities. People who are impulsive, indifferent to the effect of their behavior on others are likely to be involved in brawls, spouse abuse, aggravated assaults, and homicides, and statistics on reported crime indicate that this prediction is all too accurate.[16]

Common sense might suggest that property crime would be less common in lower-class neighborhoods. It seems reasonable to suppose that lower-class individuals bent on committing property crimes to augment their income would target middle- and upper-class neighborhoods, where their crimes are more likely to be worth the risk. However, careful analysis of reported crimes and of arrest records indicates that property crimes are overwhelmingly more common in lower-class neighborhoods than in upper- and middle-class areas, and that the perpetrators and their victims are, more often than not, neighbors.[17]

The commonsense predictions fail because most property crimes are *not* carefully planned in advance. They are impulsive and opportunistic.

Perhaps the single most important characteristic of crime in lower-class communities, according to sociologists and criminologists, is that the vast majority of all crimes are committed by young males, between ages 17 and 30. In recent years, the incidence of crimes committed by young females has risen sharply, but still represents a negligible percentage of all crimes.

Furthermore, lower-class young males are far more likely than upper- and middle-class young males to be arrested for such crimes as burglary, auto theft, armed robbery, aggravated assault, and homicide.[18] Upper- and middle-class young males are as likely as their lower-class counterparts to be involved in such crimes as shoplifting, larceny, vandalism, violations of liquor laws, drug abuse, disorderly conduct, and similar crimes, usually misdemeanors and usually occurring out of public view. Some social scientists are convinced that upper- and middle-class males also commit as many serious property crimes as their lower-class counterparts, but are less likely to be accused of such crimes.

Finally, young males of all classes, but especially lower-class young males and, increasingly, young females place great emphasis on their association with their peers. They have a natural inclination to form and join gangs, and to conform their behavior to the gang's expectations. It is hardly any wonder that gangs of young hoodlums have been regarded as a social problem at least since the days of the ancient Greeks![19]

We would conclude, therefore, that the primary law enforcement need in the African-American community, and for that matter in most lower-class minority communities, is for the control of young males. To the extent that teenagers and young adults are kept occupied with productive, or at least harmless, activities under the supervision of responsible adults, the incidence of virtually all types of crimes should be reduced.

Naturally, the people in an African-American community are just as interested as anyone else in avoiding serious crime. What is less obvious, perhaps, is the need in an African-American community for the exercise of police authority to control noncriminal social disorder.[20]

Precisely because of the characteristics of a lower-class neighborhood and of the individuals in such a neighborhood, maintaining social order is a difficult problem. In an upper- or middle-class community, social order is maintained primarily through informal and interpersonal controls. People in such neighborhoods tend to be considerate of their neighbors, if only to avoid incurring their neighbors' disapproval. In fact, intervention by the police or other authority figures is rarely needed and might be resented.[21]

In a lower-class neighborhood, informal social controls are either absent altogether or very ineffective. The problem is made worse by the tendency in lower-class neighborhoods to "live in the streets." People who live in overcrowded, uncomfortable housing are understandably inclined to spend as much time as possible in public places, where they can pass the time socializing. But this very social activity can, and often does, lead to conflicts, resulting in loud arguments, fights, retaliatory attacks on property (especially automobiles), and other forms of disorderly conduct. Drunks fall asleep on the sidewalk; prostitutes solicit prospective customers at the curb; drug dealers meet their customers at designated street corners. At the boundaries where lower-class neighborhoods meet working- and middle-class neighborhoods, panhandlers molest pedestrians.[22]

Some of these examples of disorderly conduct involve clear violations of the law. Others are marginally illegal, or not illegal at all. Some of the clearly illegal activities are not only tolerated but promoted by members of the community. Other activities, though not illegal, are considered disruptive or annoying.[23]

A lower-class African-American community, or any other minority community, is entitled to the same quality and quantity of law enforcement service as, say, an upper-class white community. But what standards of quality should apply? And how does one measure the quantity of service to which the community is entitled? These are the kinds of questions that a police administrator must address if there is to be a serious effort to meet the specific needs of each community.

Before we suggest some ways that those needs can be addressed, let us consider some other kinds of communities and what their needs might be.

Dominant Community

In most American police jurisdictions (whether a city, county, or state), whites of European origin—commonly though innacurately called "white Anglo-Saxon

Protestants"—are the dominant social and ethnic group. They constitute a majority, sometimes a slim one, of the population, and historically have held control of the social and political institutions.

However, at least since the Civil War there have been a few communities in which African-Americans have been dominant. In the past, these communities were small and scattered, either suburbs on the fringes of northern metropolitan areas or rural villages in the South. More recently, African-Americans and in some cases Hispanics have become the majority population of several of the largest cities in the United States.

Regardless of ethnicity or national origin, however, it is probably true that upper- and middle-class communities remain dominant. Even though the lower-class population comprises most of a city's or county's citizens, the people who own and manage important businesses such as banks and who are politically and socially active are the middle- and upper-class. Therefore, we will consider the law enforcement interests and needs of these dominant communities.

By and large, middle- and upper-class citizens want to be *left alone*. They do not want government agents, such as police officers, poking into their private affairs or interfering with their social lives. Police patrol cars cruising their neighborhood are likely to make them nervous and irritated, not reassured.

Middle- and upper-class citizens, perhaps influenced to a large extent by television and movie portrayals of crime tend to be more fearful of crime than actual crime statistics would justify, and tend to equate criminality with minorities.[24]

Of course, when a crime does occur, or whenever a citizen feels that there is an emergency of any sort, the police are expected to respond instantaneously. In fact, the most common complaint by citizens against their police department is not police brutality but slow response.[25] Ironically, studies have shown that the "lag time" between the occurrence of a crime, or its discovery by the victim, and the victim's notification to the police—usually by telephone—is almost always *greater* than the time required for a patrol officer to respond.[26]

In a very broad sense, the demands of the dominant community for police service are not that different from the needs of the minority community. In both communities, people want to be protected from criminals, but they also want the police to leave them alone. The important differences between the dominant and the minority communities are in such details as (1) what people regard as important crime problems, (2) who people fear as likely to commit crimes, and (3) their expectations as to what the police will do for them. Some other communities may have a very different set of priorities.

Business Community

The business community includes everything from the proprietor of the neighborhood barber shop to the chief executive of a worldwide corporation whose headquarters is located in a downtown skyscraper. Clearly, at these extremes, the needs for law enforcement services are not quite the same.

Nevertheless, there may be some useful generalizations to be made. All members of the business community are likely to be concerned about two kinds of problems: property crimes and behavior (whether or not criminal) that disrupts or interferes with commerce.

Burglary, armed robbery, theft (especially shoplifting), and vandalism are the main property-crime concerns of merchants and the owners or managers of small industrial facilities. Many small-business owners forego the considerable expense of insuring their property against criminal losses, in the hope that they will somehow be immune from attack, or at least that the police will be successful in catching the criminal and recovering their property. Statistics indicate that these hopes are, to say the least, overly optimistic. The clearance rates for commercial property crimes are among the lowest for all serious crimes.[27]

For merchants and other small-business owners in commercial districts, noncriminal disorderly conduct is nearly as big a problem as property crime. Panhandlers, "street characters," drunks, and other disorderly people discourage customers from visiting their stores, and generally create a climate that is unattractive to the consuming public.

Violent, assaultive crime is relatively rare in most commercial districts. Two exceptions do concern business owners: armed robberies of such opportunistic targets as liquor stores, convenience stores, and check-cashing businesses, and street muggings (robbery by threat or assault), both of which sometimes result in incidental violence. Otherwise, violent crimes on business premises are usually the result of personal relationships gone sour, such as an estranged spouse seeking revenge, a disgruntled employee venting a grudge, and so forth. Again, such incidents are rare, and by their nature virtually impossible to predict or prevent.

Criminal Community

It may seem odd at first even to consider the law enforcement needs of criminals. Obviously, most criminals prefer not to be identified as such, and therefore their primary need with regard to the police is to avoid them!

Sometimes the best way to gain perspective on a complicated subject is to look at it from a different angle. That is what we will try to do here, to consider what a criminal needs from the police to be successful, that is, to escape apprehension or punishment for criminal acts.

First, let us define *criminal* as *any person who commits a crime. Crime*, in turn, is defined very broadly as *any violation of the law that is specified as such by a legislature.*

Before anything useful can be said about criminals, one must recognize that different kinds of criminals commit different kinds of crimes. Crimes can be categorized in any number of different ways, such as the "index crime" categories used by the FBI, or the categories used by many police departments in subdividing their detective bureaus. For our purposes, we define four categories of crimes: private crimes, public crimes, criminal enterprises, and regulatory crimes.

Private crimes arise out of relationships between or among individuals. These relationships exist before, and sometimes after, the crime and are usually ongoing. The criminal and the victim are friends or family members, or have some other relationship. Many property crimes can be private crimes, such as a teenager stealing money from his mother's purse, or an employee embezzling from the company's payroll account. However, it is likely that the majority of private crimes involve some degree of assault including spouse abuse, child abuse, sexual assault, aggravated assault, and murder.

Most private crimes occur out of public view. Probably most are never reported to the police, unless the violence escalates to the point of homicide or other people (family members or neighbors, for example) intervene.

For a private criminal to be successful, he or she must keep the crimes out of public view and discourage anyone from intervening. The criminal's primary strategy is to intimidate his or her victims into keeping silent, usually by threatening even more violence or worse consequences. Ordinarily, unless someone reports something to the police or some other governmental agency such as a child protective agency or public health agency, the police will remain unaware that crimes are taking place.

Eventually, most private criminals are brought to the attention of the police either because a friend, family member, or neighbor intervenes, or because the violence of their crimes has brought them to public notice, or because a victim has overcome his or her fears sufficiently to complain to someone. In the case of lower-class private criminals, the crimes themselves sometimes spill over into the public arena and neighbors call the police if only to put a stop to the annoying disturbance.

Once the police become aware of a private crime, or are called to the scene of a domestic disturbance, the criminal's best hope is that the police will not consider the matter serious or perhaps even criminal. If the police can be convinced that "nothing really happened," that it was merely a private squabble that "got a little out of hand," and that the criminal is remorseful, they are likely to go away. Obvious evidence of physical injury can be explained away as the victim's own clumsiness or some accident unrelated to the argument.

In many domestic disturbances it is not very clear, without intensive questioning of all parties, who is the victim and who is the attacker. If the police can be satisfied with a promise to keep the noise down, or to "see a counselor first thing in the morning," they are unlikely to pursue the matter any further. The criminal may even make a reasonable effort to reform—for a few days.

Public crimes, in contrast, involve criminals and victims who are strangers, or at most slight acquaintances. Probably most property crimes, especially armed robbery, burglary, and auto theft, are public crimes. When violence occurs, it is often an incidental consequence of a property crime: a mugging or armed robbery that gets out of hand, or a homicide committed to cover up a burglary. Public crimes are likely to be reported to the police, usually soon after they occur or, in such cases as burglary and auto theft, soon after they are discovered, *only if the victim believes, accurately or not, that there is a reasonable chance of recovering the lost property.*

Traditional police practices, since they have proven to be remarkably ineffective in preventing public crime, are greatly favored by public criminals. If police officers patrol aimlessly in marked cars, it is fairly easy for the criminal to avoid them. Better yet, if patrol officers spend virtually all of their time responding to complaints and emergencies, they have no time to carry out any crime-prevention tactics.

When a crime is reported and a patrol officer shows up, it is in the criminal's best interests for the officer to merely take a perfunctory report, overlook any physical evidence that may be present, and send the report through interoffice mail to the detectives. The detectives will get the report a day or two later, perhaps look at it sometime later that week, and eventually call the victim to verify that the information in the original report was accurate. By that time, the chances of the criminal's being apprehended are virtually nonexistent.

Public criminals particularly do not want patrol officers to spend a significant amount of time in targeted crime prevention patrols, on foot or in vehicles. Most public crimes are essentially crimes of opportunity: The criminal sees a likely target or goes out looking for one, looks around to see whether there is much risk of being apprehended, and either goes ahead or abandons the attempt. Patrol officers who are thoroughly familiar with a neighborhood are too likely to know which targets are most appealing.

Public criminals know very well that their chances of being caught are slim in any case. Even if a victim or witness identifies them, there is a good chance of winning a reduced charge and probated sentence in exchange for helping the judicial system operate with its traditional, comfortable inefficiency. And if by some freakish chance a conviction results in an actual prison sentence, there is a better than 80 percent likelihood that less than half the sentence will actually have to be served. In short, even given an extraordinarily active and effective police agency, the public criminal faces only a modest risk of being caught, and an even smaller risk of facing any major consequences.

Criminal enterprises are essentially what we discussed as organized crime in Chapter 16: businesses engaged in illegal operations for the purpose of providing criminals with an income. Drug trafficking, prostitution, illegal gambling, and trafficking in stolen goods (fencing) are common examples of criminal enterprises. In addition, there are organized burglary and auto theft rings that operate as criminal enterprises, and many criminal enterprises have subsidiaries that provide important services, such as money laundering, legal representation, investment of crime-derived profits, and so forth.

As we noted in Chapter 16, criminals engaged in criminal enterprises normally avoid violent crimes except as an effective means of enforcing discipline within their own organization, and occasionally as a way of discouraging—or eliminating—competitors.

The essential feature of criminal enterprises is that the criminals act in a rational, pragmatic way to protect themselves and their businesses. Unlike private criminals, who may be driven by passion, emotional illness, or mental impairment, and unlike public criminals, who often act on spur-of-the-moment impulses without regard to the consequences, enterprise criminals carefully calculate risks and make every effort to avoid or reduce them.

From the criminal's point of view, then, the best law enforcement service is to be ignored. If the police are kept sufficiently busy with private and public crimes, answering emergency calls, and so forth, they may overlook the fact that criminal enterprises are flourishing. Enterprise criminals ordinarily try to keep a low profile in any case. If the police can be convinced that they do not exist, the criminal enterprise may be virtually invisible.

If invisibility does not work, the second-best law enforcement service is corruptibility. To corrupt a police force, the enterprise criminal needs police executives and elected officials who are more interested in politics than in law enforcement; poor or absent leadership at the management level; and a poorly organized, ineffective, demoralized police force at the street level. Patrol officers who are frustrated by their own ineffectiveness, who have no confidence in their superiors, and who believe that they have no support from the public are easily corrupted by modest bribes.

As corruption spreads through the operating-level ranks, it also seeps upward into lower-level management, and eventually, given sufficient time and the right environment, it reaches the executive level. Meanwhile, careful cultivation of officials outside of the police department—in the courts, prosecutors' offices, and elsewhere—can help to guarantee that the criminal enterprise will be able to continue generating its handsome profits.

Of course, corrupting an entire police department and associated agencies is expensive and imposes some risks of its own. Nevertheless, it is a strategy that has worked many times for enterprise criminals, and no doubt it is still working today.

Unfortunately, enterprise criminals are sometimes faced with a police agency that refuses to be corrupted, and that is not fooled by the "cloak of invisibility." The criminal's best hope now is that the agency will adopt the *mastermind strategy*. This is the police tactic, developed originally by the FBI and later adopted by most metropolitan police agencies, of ignoring street-level criminals while attempting to infiltrate the criminal operation and gather evidence sufficient to identify, arrest, and convict the organization's mastermind, the criminal executive who is presumably indispensable to the illegal business.

As long as the police concentrate their efforts on one elusive and extremely well-protected individual, the enterprise will continue to thrive. Even if the police succeed in identifying and removing the mastermind, there usually are trusted and experienced lieutenants ready to take over control of the operation. Every business executive knows that part of the responsibility of leadership is grooming one's successor.

Regulatory crimes are violations of laws that concern social and political processes. The most obvious example is the traffic laws. As we pointed out in Chapter 15, traffic laws are somewhat arbitrary in nature, and are not derived directly from a moral code as are laws against stealing or killing. The same is true for tax laws, immigration and naturalization laws, zoning and building codes, and some business laws. Violations of these laws do not involve a particular victim, although violations may *result* in harm, such as loss of property or personal injury, and the people who commit these crimes usually do not regard themselves as criminals. Nevertheless, that is what they are.

Regulatory criminals, like all criminals, prefer to be overlooked and ignored by the police. They are best served by a police agency that is too busy with serious crime, including public, private, and enterprise crimes, to bother with regulatory crimes. If the prosecutors and courts also can be persuaded that regulatory crimes are trivial and inconsequential, all the better.

We have been somewhat facetious in our description of the law enforcement needs of the criminal community. Our purpose is not to poke fun at the expense of the police but rather to consider the subject of policing from the viewpoint of the criminal. If it is the goal of the police to oppose the interests of the criminal, then in some sense the police ought to be concerned with doing exactly the opposite of what the criminal would like them to do.

We also should say something about the legitimate needs of criminals as citizens. This is often overlooked, but people who commit crimes are, for the most part, not very different from people who do not commit crimes.

Private criminals come from every social class, ethnic group, and part of town. They are not infrequently pillars of the community, active church members, and leaders of important institutions and social organizations. Public criminals are predominantly members of the lower class and are almost overwhelmingly young and male. Nevertheless, they too are citizens, and ordinarily spend most of their time in noncriminal pursuits.

Regulatory criminals are virtually everyone. It is a rare soul who does not sometimes drive faster than the speed limit, park in a restricted space, or "forget" to report some income on their tax return. Perhaps it is a bit unfair to label such people criminals. Nevertheless, traffic accidents kill approximately twice as many people each year as homicides.

If it is true that drunk drivers are responsible for about half of vehicular deaths, then it is reasonable to ask why murderers are given life sentences or capital punishment while drunk drivers are given modest fines and short, usually probated, sentences.

Enterprise criminals are another matter. At the upper levels of criminal enterprises, there indeed are masterminds, criminal business executives who are willing to use any means at their disposal to spread corruption. At the street level—the drug "retailers," gambling runners, street prostitutes, and so forth— are usually lower-class, young people, not much different from other public criminals. They are likely to regard crime not in moral terms, but in practical terms, as their best chance to rise above their lower-class origins. In fact, their calculation may be correct, provided they have the peculiar talents and abilities necessary to their chosen profession, and provided they avoid having their careers disrupted by the police.

RESPONSE TO COMMUNITY NEEDS

Let us now consider how the police might respond to the different needs of each of the communities we have described.

We begin by asserting a general principle, perhaps the one universal principle of law enforcement that has been discovered: *Police officers must be thoroughly familiar with each of the communities they serve.*[28] They must know as much or more about each community as they do about the details of criminal law and procedure. Their knowledge must be both general (the sociological characteristics of a typical lower-class neighborhood, the etiology of crime) and particular (who in the community are the likely victims of private crimes, which young people are most likely to spend their free time in criminal pursuits).

Furthermore, this knowledge cannot be gained entirely by studying statistics such as criminal incidence patterns. Police officers must devote a large part of their working time to collecting information by direct contact with citizens and with each other.

We suggest that these basic principles offer a powerful argument for the use of foot patrols wherever possible. Foot patrols were abandoned not because they were ineffective, but because new technologies—the automobile and the two-way mobile radio—seemed to offer greater efficiency. Now further technological change has made foot patrols once again practical: high-powered portable radios ("walkie-talkies") ensure that a foot patrol officer is never inaccessible or isolated.

Foot patrol should be used extensively in densely populated urban neighborhoods, in commercial districts, and anywhere else that people tend to congregate. In most cases, foot patrol should be combined with motorized patrols, either by requiring a patrol officer to spend part of his or her working time out of the vehicle, or by assigning both foot and motor officers to the same beat or overlapping beats.

Effective foot patrol requires more than an officer strutting about in a uniform weighed down with hardware. The officer must *talk to people*. This does not mean aggressive use of field interrogation. It means engaging in social conversation, in the course of which the officer can elicit a vast amount of information about the neighborhood and the people in it. Much of the information will have no direct law enforcement significance, but it will tell the officer, and other officers with whom the information is shared, what kinds of people live in the area, what their interests and concerns are, and how likely they are to be either perpetrators or victims of crime.

Patrol officers must be trained to recognize, collect, organize, and use information, just as they must be trained to use weapons, to perform legal searches and arrests, and to carry out their other duties. They must learn which information is likely to be valuable to officers assigned to other beats or to their own beats on different shifts, and there must be procedures and technical means to ensure that this information is shared.

Officers also must be prepared to give information to the community. Most citizens realize that the police cannot be expected to prevent every crime, that individuals must accept some responsibility for protecting themselves. Police officers should be resource experts on crime prevention. They should be prepared, and indeed eager, to share their expertise by advising citizens on the crime prevention steps they can take.

In upper- and middle-class communities, the police should take advantage of existing community organizations, both to gather and to share information. Police officers should routinely attend Parent-Teacher Association (PTA) meetings, church social events (regardless of their own church membership), civic club functions, and any other social gathering that is not closed to them.

Community involvement should not be relegated to speeches made by community relations officers or testimonials for the chief. It should be part of the duty of every patrol officer, particularly in low-population-density neighborhoods where foot patrols are inappropriate. Again, it is not necessary for these contacts to be formalized. Much can be accomplished by a patrol officer chatting with PTA members after a meeting. "Hi, I'm Sam Jones, I'm the cop who patrols this neighborhood," almost invariably will initiate a productive conversation.

Even in lower-class neighborhoods, there are almost always some existing community organizations: churches, parent-teacher organizations, social clubs, and so forth. The problem is that too few citizens are involved in these organizations. Indeed, an officer who attends the meetings of all of them will soon realize that the same people show up at one meeting after another.

Some police departments have experimented with developing their own community organizations, often centered around a neighborhood or storefront police station, or using the recreational facilities of apartment buildings and public housing complexes. In some cases, police-initiated community organization has been successful; in other cases, it has fallen flat. It is certainly worth trying, if there are no existing community organizations or if those that exist are ineffectual.[29]

Beat officers in lower-class communities should be particularly interested in encouraging young people to occupy themselves in productive, or at least harmless, activities. It may not be necessary for the police to organize an entire youth football league with full uniforms and championship trophies. Getting a property owner to clear off a vacant lot and let young people use it for informal flag-football games may be sufficient. Donated books might be used to establish a study library at the neighborhood station, with comfortable chairs and tables, and perhaps a couple of typewriters or, better yet, personal computers that students can use to do their homework.

Whether these devices will work depends on the particular character of a particular neighborhood. Our point is that it should be the responsibility of the beat officer to find out whether they will work, or, if not, what will work.

Clearly, in order for this kind of information-gathering to occur, the organization of the patrol force must allow ample time for it. There must be a sufficient number of patrol officers so that beats can be kept small, and officers must be assigned to a beat for a long enough period of time for informal contacts and community involvement to develop. Changing assignments every three months will not satisfy this need.

There also must be a rapid police response to genuine emergencies and crimes in progress or newly discovered. The public's demand for the "instant cop" is not entirely unreasonable.

However, it is not clear that this need has to be met only by beat patrol officers. It may be that each precinct or area should have a separate rapid-response team composed of officers housed at a precinct headquarters or area substation, whose sole responsibility is handling 911 calls. Beat officers certainly should be informed of such calls, and should be expected to respond to them along with the rapid-response officers when possible.

Beat officers should be required to follow up on all reported private crimes and most reported public crimes, whether or not the officer responded to the initial complaint. The purpose of the follow-up, which usually should involve a personal visit, is to provide whatever assistance is possible to the victim.

This could include, for example, helping the victim file an insurance claim, or a claim with a victim compensation program. It might involve advising a victim on sources of counseling, or how to report stolen credit cards and how to get a replacement driver's license. Even in upper- and middle-class neighborhoods, people who have been victimized are often too upset to think clearly about such bureaucratic procedures. *If the police cannot catch the crook, at least they can do something to help the victim.*

Follow-up calls are absolutely essential in the case of private crimes. The initial complaint may involve a relatively trivial offense, but the beat officer must not dismiss it as inconsequential. On the contrary, the beat officer should be held responsible for immediate, frequent, and persistent follow-up visits, even if the victim does not request them and does not seem to appreciate them. After one or two initial visits, most of the calls can be brief and may be accomplished by telephone rather than personal visits. The purpose of the calls is twofold: to give the officer an opportunity to determine whether there is continuing abuse and to let the victim know that help is immediately available at all times.

Diligent follow-up of a routine domestic disturbance call may well prevent far worse offenses including murder. This may be the single most important, and potentially effective, form of crime prevention available to the police.

Public crimes, especially property crimes, require follow-up only if there is some reasonable expectation of identifying and apprehending the perpetrator. When an officer responds to the initial complaint, whether it is the beat officer or a rapid-response officer, an initial determination of solvability should be made.

If the victim or a witness can identify the perpetrator, if there is physical evidence at the scene that can be used to identify the perpetrator, or if the crime seems to fit a pattern of related incidents, the crime may be solvable. If none of these factors is present, the crime probably is not solvable.

Unsolvable crimes should not be dismissed. The responding officer still must make a thorough investigation and must attempt to locate and collect whatever evidence exists. However, the officer also should explain to the victim that little or no follow-up investigation will be possible, and that the likelihood of recovering the lost property is very slim.

The responding officer also should determine, at least initially, whether the follow-up investigation of a solvable crime should be handled by the beat officer, a rapid-response officer, or a specialist investigator. This decision will depend on the nature of the crime, the nature of the evidence that points to a particular perpetrator, and the organization and policies of the department.

We are inclined to believe that most follow-up investigations are likely to be successful if they are carried out by whoever performs the initial investigation.[30] But if that person is a rapid-response officer whose day is spent mainly in handling 911 calls, he or she may not have the opportunity to perform an adequate investigation. In that case, perhaps the beat officer or a specialist investigator should have that responsibility.

What we have called enterprise crimes in this chapter are, as we said, essentially what we called organized crimes in Chapter 16, where we discussed some of the techniques used to suppress them. As our earlier discussion indicates, we are skeptical of the effectiveness of the "mastermind" strategy that has been the basis of most organized-crime investigative tactics in the past.

Enterprise criminals want, above all, freedom to sell their illicit goods and services. In many respects, enterprise criminals are not very different from legitimate business operators: an orderly, dependable market is essential to their success.

For illegal businesses, the market is the place where sales are made to consumers. If this market is disrupted, the criminal enterprise is stopped just as effectively as removing the mastermind. In fact, this strategy may be more effective, since it is much easier to replace one executive than an entire marketfull of customers.

Disrupting the market means adopting a policy of absolute intolerance of crime-related activity on the street, such as drug sales or soliciting for prostitution. It is not necessary for the street-level criminals to be punished severely. Lenient punishment is sufficient and probably appropriate.

What is important is that the "retail" trade becomes impossible because the street sellers, the prostitutes, or whatever, are constantly being arrested and taken off the street. Bond fees mount up, customers are chased away, and the upper-level criminals are stuck with goods they cannot sell, or their income from illegal services is cut off.

If at the same time the street-level criminals are being offered attractive opportunities to abandon their careers, such as treatment for drug or alcohol addiction, or job training, and if it becomes well known that the police do not tolerate street-level crime, soon the enterprise criminals will be looking for some other city to corrupt.

Finally, just to round out this discussion, the tactics that can be used effectively against regulatory crimes, especially traffic violations, are well known and widely practiced. This may be the one aspect of law enforcement that most American police agencies carry out reasonably well. There is ample evidence of the effectiveness of the selective enforcement techniques we described in Chapter 15. Similar techniques can be applied to other types of regulatory crime, to the extent that the police are responsible for enforcement.

POLICE AND SPECIAL COMMUNITIES

For most of this chapter, our concern has been the relationship between the police agency and the various segments of the public, which we have defined as communities. There are also some particular segments of the community with which the police have a special and important relationship: their parent government, the union or unionlike association that represents the interests of police officers, and the news media.

Police and the Parent Government

The first and most important characteristic of every American community is that its government is organized along the lines of a representative democracy. This means that the citizens choose individuals to represent their interests.

Elected representatives, in turn, appoint other individuals to manage the day-to-day affairs of government. Among the officials who are usually appointed are the chief executives of law enforcement agencies. An important exception is that most sheriffs, and a few state police executives, are elected directly by the citizens.

Police executives, whether appointed or elected, are subject to the authority and supervision of elected officials in all three branches of government. The head of the executive branch (mayor, county commissioner, or governor) is responsible for seeing that the laws are properly enforced. The legislators (city council or aldermen, county supervisors, and state legislators) enact the laws that are to be enforced, enact laws that define how enforcement agencies are to be organized and operated, and provide the public funds that are required by law enforcement agencies. The judicial branch (the judges and, to a lesser extent, prosecutors) determines the manner in which law enforcement authority is to be exercised.

It is hardly any wonder that police administrators are sometimes resentful of the politicians who have authority over law enforcement. The chief executive of the parent government, the legislators, and the judicial officers often have very different ideas about how the law enforcement mission should be carried out and often make different, even conflicting, demands on the police. In many municipal governments, yet another element, a city manager appointed by the city council, has direct authority over the police department.

Ever since the police reform movement began in the late nineteenth century, police administrators have tried to remove themselves from political influence, to give themselves as great a degree of freedom as possible to carry out their duties as they see fit.

Removing the police from the political system is impossible. Without the police, the political system would have no means of enforcing its decisions. Without the political system, the police simply would not exist.

What is possible, and indeed necessary, is to ensure that the political system does not influence the police in ways that are contrary to the interests of the public. No sensible person wants to go back to the kind of political domination of the police that existed a hundred years ago and that still lingers, in a few places, to this day.[31]

- The police are subject to the authority of the executive branch of government because that branch has the legal responsibility of carrying out the decisions made by the legislative branch.
- The police are subject to the authority of the legislative branch of government because that branch has the legal responsibility of making decisions. Among those decisions are the enactment of laws and the provision of public funds for public purposes.

- The police are subject to the authority of the judicial branch of government because that branch has the legal responsibility of ensuring that the laws are enforced fairly and of protecting individual citizens from the improper exercise of political authority.

Any attempt by the police to remove themselves from the authority of any of the three branches of government would be a direct repudiation of the principles of representative democracy. The police are not, and cannot be, a law unto themselves.

Elected officials often bring to office their own perceptions of the way government should operate. They also may have obligations to their political supporters. Conflicts between the ideologies and political interests of different officials may catch the police in a web of controversy, intrigue, and political maneuvering. One faction may demand stronger enforcement while another faction demands more economical government. The first faction may insist on assigning more police to vice enforcement, and the second faction may cause the agency's budget to be cut.

Within reason, these conflicts are legitimate. After all, the basic idea behind representative democracy is that citizens have different ideas about what kind of government they want, and those differences should be resolved through the democratic process. However, these noble principles are not much comfort to the police administrator caught in the cross fire.

The most dangerous course for a police administrator to take is to follow the dictates of whichever faction appears to have a momentary advantage. The second most dangerous course is to ignore the political battle altogether.

A police administrator's first duty, and first loyalty, must be to the public, not to the politicians. Currying the favor of any one political faction is likely to arouse the hostility of all others. Instead, the administrator must rely above all on the fundamental principles of criminal justice, expressed in the federal and state constitutions, that define and limit police authority.

Second, the administrator must gain and keep the confidence of the public. Politicians, no matter what their ideology, are less likely to attack an agency that enjoys a general and deserved reputation for efficiency, effectiveness, and fairness.

Police practices themselves are sometimes the subject of political controversy. When the budget for police service or the techniques of law enforcement are under attack, the best defense is the truth—preferably truth for which there is ample evidence. Facts about the incidence of crime, about the organization and operating policies of the agency, and about the agency's achievements are far more powerful than appeals to political interests or ideologies.

An administrator must be capable of explaining precisely why the agency needs the resources contained in its budget. If a budget cannot be defended, it probably does not deserve to be funded. An administrator also must be capable of giving clear, reasonable explanations of the agency's organization and operations. Otherwise, political leaders and the public as a whole are justified in losing confidence in the police.

The details of managing a police agency—buying supplies, recruiting and training personnel, assigning patrol beats, reviewing investigations—are not the crucial tasks for the police executive. As much as possible, those duties should be delegated to others. The most important duty of the executive is to represent the agency before the public and its elected representatives. If the executive fails in that duty, all the rest will count for very little.

Police Unions and Unionlike Organizations

Strictly speaking, the members of a police officers' union or unionlike organization are the personnel of the police agency and are therefore internal to it. However, we will treat them as if they were an external community because (1) a union or unionlike organization is an entity in itself, with interests and resources that usually are completely separate from those of the police department; and (2) the relationship between police administrators and the police officers' organizations is much like that of police administrators and other segments of the community.

We have used the cumbersome phrase, "unions and unionlike organizations," because historically the police and other public employees have been discouraged (if not actually prohibited) from joining the labor union movement.[32]

American police officers began to show interest in unionism partly as a reaction to the police reform movement of the 1870s, 1880s, and 1890s. Many of the reforms seemed, from the perspective of the rank and file, to undermine or destroy their privileges, such as gaining promotions by political favoritism. Beginning around 1897, low-ranking officers in several northeastern cities began to promote the idea of police unions.[33]

The first recorded effort of American police officers to form or join a union occurred in Cincinnati in 1918. Several hundred members of the Cincinnati police force joined the local Central Labor Council and went on strike to protest the arbitrary firing of four fellow officers. The strike lasted only a few days; the officers were persuaded to return to work, and their labor union activity soon dissolved.[34]

A year later, Boston police officers formed a union affiliated with the American Federation of Labor and attempted to negotiate collectively for better working conditions, which were, at the time, deplorable. Nineteen officers were immediately suspended for their participation in the union. A few weeks later, about three-fourths of the police force went out on strike. Looting and rioting followed. After four days, the striking officers were fired and replaced. Departmental working conditions and salaries were later improved, but the police officers' union was destroyed.[35]

Other police unions were formed before World War I and generally were met with the same intransigent response as in Boston. The public believed, and political leaders insisted, that public employees—especially police officers—had no right to form unions or hold strikes. During the 1920s and 1930s, few police officers dared to risk their livelihoods by openly joining unions.

However, there was an alternative route for police officers. They formed nonunion associations, such as the Fraternal Order of Police (FOP) and the

Patrolmen's Benevolent Association (PBA). The FOP was founded in 1915 to work for the welfare of police officers. PBA's, each independent and autonomous, were organized as early as 1892, partly as social organizations and partly to promote better working conditions and benefits.[36]

After World War II, established labor unions again attempted to organize police unions, this time with more success. Public attitudes about labor unions had relaxed, and union leaders were less inclined to use threats and inflammatory rhetoric to promote their cause. Some police unions renounced the right to strike. By the early 1960s, federal and some state laws prohibiting public employees from forming unions were overturned. At the end of the 1960s, some twenty states had specific provisions for collective bargaining by public employees, and similar laws have been enacted in many states since then.

There remains a good deal of public concern about the possibility that the police will go out on strike, leaving a community defenseless against criminals. The Boston strike is almost always cited as the cautionary example. However, studies of more recent police strikes indicate that looting and rioting do not immediately break out when the police walk off the job.[37]

Usually a police strike is the desperate act of officers who have found no other way to overcome the refusal of political leaders to consider their needs. Usually, too, there is ample time for police administrators to develop a contingency plan which, once the strike occurs, will provide at least a minimal level of police service.

It is not true that criminals "go on holiday" when the police strike. However, it does appear to be true that during a police strike only the most serious crimes are reported. Consequently, the official crime rates often go down drastically. In short, the fears of the public are somewhat exaggerated, if not entirely groundless.

It is important to realize that the police administrator can do little to prevent police officers from joining unions. Police officers have the same right to associate as any other citizens. Even if unions are prohibited, they will form fraternal orders or benevolent associations that eventually act like unions. In fact, many PBAs have affiliated with the AFL-CIO and other labor organizations.

It is also important to understand that the police administrator cannot, single-handedly, prevent officers from going on strike. Once a strike occurs, there is very little that the administrator can do about it, other than to put into effect a contingency plan to keep the community from being defenseless.

A police administrator is merely another public employee. He or she usually does not have the authority to either grant or refuse the officers' demands. That authority is held by the parent government and is exercised by the elected political officials. They may delegate authority to negotiate with the union, but ultimately the administrator can neither give nor withhold what the union wants.

Most police administrators do not even have the authority to negotiate with employee organizations. Probably this is for the best. Negotiations between employees and employers tend to involve controversial issues, and both sides typically approach negotiations as adversaries. Resentment, frustration, and hostility are not uncommon. The police administrator who takes an active part in union negotiations must step out of the role of leader and into the role of opponent. Whatever the outcome of the negotiations, it may be difficult for the administrator to take up the role of leader again. There may be a residue of bitterness that will poison the relationship between the administrator and officers for years to come.[38]

The political officials (including, in many cases, the city manager) have the authority to grant whatever benefits the union is demanding, and they are the ones who should conduct negotiations, or at least they should assume the responsibility for negotiations conducted by their immediate subordinates.

What the police executive can do, and certainly should do, is to give the negotiators factual information about the agency's policies, resources, and concerns. As negotiations progress, the negotiators should consult with the police executive, who should explain the possible implications or consequences of the various proposals being offered by both sides, and should suggest alternatives or compromises to any proposals or demands that the executive feels would be detrimental to the agency.

If negotiations fail and the union calls a strike or other job action, such as encouraging officers to call in sick, the administrator is well advised to let the political officials resolve the matter. The police administrator's duty is to see that essential police services are delivered, no matter what. This may require using managers and other ranking officers as temporary substitutes for field officers. In extreme cases, it may be necessary to request assistance from the state police or National Guard. Otherwise, the administrator should refrain from making public statements or taking any action that will further inflame the situation.

Fortunately, police strikes are rare, usually they last only a few days, and the consequences are not as disastrous as the public fears. In most cases, they are also futile. The striking officers do not get what they wanted, and sometimes a temporary victory leads to retaliation by the political officials that is more costly than the union leaders anticipated.

Police and the News Media

The news media occupy a special place in American society. Even before the right of a free press was guaranteed in the First Amendment to the Constitution, a tradition of journalistic freedom and independence had begun to evolve. In its modern embodiment, this tradition is expressed not only by print journalists but by broadcast journalists as well.

One of the elements of that tradition is the right of journalists to investigate, question, and criticize the actions of government. Any police officer or administrator who thinks that law enforcement should be immune from criticism is sadly mistaken. Journalists not only have a right to question the police but, according to their traditions, a duty to do so.

This does not mean that the police and the news media are necessarily enemies. It does mean that they are not necessarily friends. The best relationship between the police and the news media is one based on mutual respect, cooperation when that is appropriate, and understanding of one another's needs.

The police do need the news media. The public has a great interest in the activities of the police, not only because some people find police activities exciting and glamorous, but because citizens naturally want to know how well they are being protected from crime. By providing full, accurate information about police activity to the news media, who are then responsible for conveying the information to the public, the police can reassure the community that criminal behavior is being dealt with effectively.

The police often need the active assistance of the public. Very few crimes occur without someone besides the perpetrator and the victim having some knowledge about them. Citizens may observe a crime without realizing the significance of what they have seen, or may be reluctant to get involved. Appeals to the public for information and assistance, whether concerning a specific crime or criminal activity in general, are often remarkably effective. The most efficient way to make such an appeal is through the news media.

The techniques available to the police for developing and maintaining a satisfactory relationship with the news media are too numerous to describe here. Public information is a discipline in itself; valuable texts on the subject are widely available.[39]

However, we are concerned about the attitudes of police administrators toward the news media. Those attitudes seem to range from a naive belief that the news media exist to promote law enforcement, to an attitude of suspicion and hostility.

The news media do not exist to promote law enforcement or any other governmental function. They exist to inform the public about anything, including the activities of government, that is of interest to the public. How, and how well, they discharge that function is not the concern of the police.

Nor do the news media exist to discredit the police. Contrary to the belief of some police administrators, most journalists take seriously their duty to report information fully and accurately. Whether that information is favorable or unfavorable to the police, their duty is to report it. In the long run, newspapers or broadcasters do not profit by waging war on the police or any other governmental agency.

Not all journalists are conscientious and responsible, any more than are all police officers. It sometimes happens that a newspaper editor or a broadcast news director, for whatever reasons, will campaign against the police. When that occurs, the police administrator may feel defensive and frustrated; there are few weapons available to counterattack the news media.

However, even in small communities, it is rare for any one news medium to have an absolute monopoly. There may be only one newspaper, but there usually is at least a radio station, if not a television station, serving the same community. Even in rural counties, there are often two or more newspapers that circulate throughout the county and usually radio and television stations in nearby communities.

In other words, if one news medium treats the police unfairly, the administrator should seek other ways of getting full, accurate information to the public. Any effort to retaliate against the unfair medium is likely to be ineffective and probably counterproductive.

Withholding information from the news media is also ineffective and counterproductive. Competent journalists know that they can get the information they want from any of several sources. If a police administrator refuses to provide the information they need, for whatever reasons, they will find another source.

Sometimes this puts the police in an awkward situation. Detailed information about a criminal investigation should not be made public, since this may hinder the investigation and is likely to interfere with the prosecution of the offender. In such cases, the administrator has a legal duty to withhold information that journalists may feel strongly about. The only recourse for the administrator is to release as much general information as possible and explain candidly why additional information cannot be made public. Again, most journalists will act responsibly. They realize that it is not in their own interests to interfere with the criminal justice system.

Police administrators should try to learn as much as they can about the workings of the news media in their own community: what deadlines are imposed on reporters, which reporters and editors are conscientious and responsible in their work, which media are most successful in reaching different segments of the population. This knowledge will give the administrator the considerable advantage of being able to cooperate with the most competent and reliable members of the media and, through them, to ensure that full and accurate information reaches the public.[40]

REVIEW

1. A community is
   (a) A group of people living in a certain geographical area.
   (b) A place for young people to engage in productive activities.
   (c) The minority population of a city or county.
   (d) Any group of people who have something in common.
   (e) The number of people served by a law enforcement agency.

2. True or False: Most people are members of several communities.

3. A *heterogeneous* population is one that
   (a) Consists mostly of minorities.
   (b) Consists of people of various classes or ethnic groups.
   (c) Consists mostly of a single class or ethnic group.
   (d) Contains a population that is disposed toward criminal behavior.
   (e) Has suffered various types of discrimination in the past.

4. Which of the following statements, if any, are true?
   (a) Private crimes are more likely than public crimes to occur among people who have an ongoing relationship with one another.
   (b) Traditional police practices are largely ineffective against public crimes.

(c) Enterprise criminals generally commit opportunistic crimes with minimal planning or preparation.

(d) Regulatory crimes are violations of arbitrary laws designed to promote social order or to carry out governmental processes.

(e) People who engage in criminal behavior have thereby forfeited any right to the protection of the law.

5. True or False: Political considerations should not be allowed to influence the policies or operations of a police agency, because those considerations are typically based on the selfish interests of a few individuals.

NOTES

[1]Herman Goldstein, *Problem-Oriented Policing*. New York: McGraw-Hill, 1990, pp. 32–49.

[2]Robert C. Trojanowicz and Samuel L. Dixon, *Criminal Justice and the Community*. Englewood Cliffs, N.J.: Prentice Hall, 1974, pp. 6–8.

[3]Ibid., pp. 8–9.

[4]Malcolm K. Sparrow, Mark H. Moore, and David M. Kennedy, *Beyond 911*. New York: Basic Books, 1990.

[5]Jerome H. Skolnick and David H. Bayley, *The New Blue Line*. New York: The Free Press, 1986, pp. 19–20, 211–12.

[6]James Q. Wilson, *Thinking about Crime*, rev. ed. New York: Basic Books, 1983; see especially Chap. 5, pp. 75–89.

[7]David H. Bayley, "The Limits of Police Reform," in Bayley, ed., *Police and Society*. Beverly Hills, Calif.: Sage Publications, 1977, pp. 223–25.

[8]Trojanowicz and Dixon, *Criminal Justice*, pp. 49–57.

[9]Skolnick and Bayley, *The New Blue Line*, pp. 212–14.

[10]Stephen Leinen, *Black Police, White Society*. New York: New York University Press, 1984, pp. 163–66; Mary Jeanette Hageman, *Police-Community Relations*. Beverly Hills, Calif.: Sage Publications, 1985, pp. 89–94.

[11]John L. Cooper, *The Police and the Ghetto*. Port Washington, N.Y.: Kennikat Press, 1980, p. 33; Cooper, *You Can Hear Them Knocking*. Port Washington: Kennikat, 1981, pp. 28–40.

[12]Wilson, *Thinking about Crime*, p. 35.

[13]Ibid., pp. 38–40.

[14]Cooper, *The Police and the Ghetto*, pp. 9–12.

[15]Hageman, *Police-Community Relations*, pp. 66–72; Parviz Saney, *Crime and Culture in America*. Westport, Conn.: Greenwood Press, 1986.

[16]Wilson, *Thinking about Crime*, pp. 44–45; Saney, *Crime and Culture*, p. 11.

[17]Wilson, *Thinking about Crime*, p. 47.

[18]Ibid., pp. 92, 97.

[19]A thorough collection of information about criminals and criminal behavior is the two-volume report by Alfred Blumstein, Jacqueline Cohen, Jeffrey A. Roth, and Christy A. Visher, eds., *Criminal Careers and "Career Criminals."* Washington, D.C.: National Academy Press, 1986.

[20]Sparrow, Moore, and Kennedy, *Beyond 911*, p. 82.

[21]Cooper, *You Can Hear Them Knocking*, pp. 4–6.

[22]Pamela Irving Jackson, *Minority Group Threat, Crime, and Policing.* New York: Praeger, 1989.

[23]Stuart A. Scheingold, *The Politics of Street Crime.* Philadelphia: Temple University Press, 1991, pp. 1–28.

[24]Jackson, *Minority Group Threat*, pp. 25–32.

[25]Skolnick and Bayley, *The New Blue Line*, pp. 216–17.

[26]Wilson, *Thinking about Crime*, p. 71.

[27]See the most recent issue of *Crime in the United States*, published annually by the FBI. Washington, D.C.: Government Printing Office.

[28]Sparrow, Moore, and Kennedy, *Beyond 911*, pp. 1–15.

[29]Arthur Niederhoffer and Alexander B. Smith, *New Directions in Police-Community Relations.* San Francisco: Rinehart Press, 1974; Cooper, *You Can Hear Them Knocking*, pp. 7–12.

[30]The Royal Canadian Mounted Police have reported a study of student response to on-campus police officers in Richmond, B.C., and found that there was a strong, positive correlation between the students' trust toward an officer (as indicated partly by the number of "voluntary disclosures" of crimes and other problems) and the students' perceptions of an officer's *effectiveness*. In particular, students largely ignored on-campus officers who "patrolled" endlessly without actually doing anything, and who merely forwarded incident reports to some central headquarters rather than investigating or taking other appropriate action on their own. Although this one study is not conclusive, it suggests some fruitful areas for research that would apply to both vehicle-based and foot patrol officers. See Stephen A. Grant, "Students Respond to 'Campus Cops,'" in *School Safety*, Winter 1993, pp. 15–19.

[31]Skolnick and Bayley, *The New Blue Line*, pp. 180–85.

[32]Allen Z. Gammadge and Stanley L. Sachs, *Police Unions.* Springfield, Ill.: Charles C Thomas, 1972, pp. 76–84.

[33]Hervey A. Juris and Peter Ferrillo, *Police Unionism.* Lexington, Mass.: Lexington Books, 1973, pp. 13–15.

[34]John H. Burpo, *The Police Labor Movement.* Springfield, Ill.: Charles C Thomas, 1971, p. 3.

[35]Juris and Ferrillo, *Police Unionism*, pp. 16–17.

[36]Ibid., pp. 13–15.

[37]*Police Strikes: Causes and Prevention.* Washington, D.C.: International Association of Chiefs of Police, 1979.

[38]Juris and Ferrillo, *Police Unionism*, p. 183.

[39]For example, Gerald W. Garner, *The Police Meet the Press.* Springfield, Ill.: Charles C Thomas, 1984.

[40]Patricia A. Kelly, *Police and the Media.* Springfield, Ill.: Charles C Thomas, 1987.

Appendix A
ANSWERS TO REVIEW QUIZZES

Chapter 1

1. (a); 2. False; 3. (b); 4. (b); 5. (d).

Chapter 2

1. (a) iv, (b) iii, (c) ii, (d) i, (e) iii, (f) ii; 2. (c); 3. False; 4. (a); 5. (c).

Chapter 3

1. (c); 2. True; 3. (e); 4. (d); 5. (b).

Chapter 4

1. (c); 2. Experiment(ation); 3. (a); 4. Strategic; 5. (b).

Chapter 5

1. (d); 2. (b); 3. (a); 4. (c); 5. (b), (e), and (f).

Chapter 6

1. False; 2. (c); 3. (c); 4. False; 5. (b).

Chapter 7

1. True; 2. (a); 3. True; 4. (a) B, (b) A, (c) A, (d) C, (e) C, (f) B, (g) C, (h) D; 5. False.

Chapter 8

1. True; 2. (a); 3. (d); 4. (d); 5. (a), (c), (e), and (f). (In #5, answer [g], CAD, is incorrect because it is not a general application program.)

Chapter 9

1. (b), (c), (e), and (f); 2. (a); 3. (c); 4. False; 5. (b).

Chapter 10

1. (e), (a), (c), and (b); 2. (c); 3. (d); 4. (b); 5. (e).

Chapter 11

1. (d); 2. True; 3. (b); 4. False; 5. (e).

Chapter 12

1. (f); 2. False; 3. (d); 4. (a); 5. (b). (In #5, answer [a] is incorrect because a career development program has little, if any, effect on an agency's immediate personnel needs.)

Chapter 13

1. (a); 2. (e), (b), and (d); 3. False; 4. (a); 5. (b).

Chapter 14

1. (c); 2. True; 3. (a), (c), (d); 4. (a); 5. False.

Chapter 15

1. (a), (b), (d), and (f); 2. True; 3. (a); 4. (d); 5. (c).

Chapter 16

1. (e), (a), (f), (c); 2. (b); 3. (b); 4. False; 5. (b), (c), (e), (g), and (h).

Chapter 17

1. (d); 2. (c); 3. (a), (b), and (c); 4. (b); 5. (a) and (d).

Chapter 18

1. (d); 2. True; 3. (b); 4. (a), (b), and (d); 5. False.

APPENDIX B
GLOSSARY

Academy: A training and educational institution operated by a law enforcement agency, or cooperatively by several law enforcement agencies.

Access channels: In a cable television system, one or more channels reserved for use by the general public, educational institutions, or governmental agencies.

Account: In bookkeeping and budgeting, the heading or category under which all entries of the same type are recorded.

Accountability: In a hierarchical organization, the principle that each subordinate must report to his or her superior and account for his or her actions.

Accreditation: A procedure in which agencies such as police departments are judged according to an established set of standards and, if an agency is found to meet the standards, is awarded the status of "accredited."

Administration: In any organization, the element that is responsible for making and enforcing policies, allocating resources, and coordinating activities to fulfill the organization's primary goals or purposes.

Adoption: In the budget process, the point at which a budget proposal is approved for implementation.

Aerial surveillance: The observation of people or things from the air, as from an airplane or helicopter.

Affirmative action plan: A plan, usually but not always submitted for approval to the *Equal Employment Opportunity Commission*, stating an organization's goals for achieving equal employment opportunities for minorities and women, and how those goals are to be accomplished.

Amortized cost: The cost of a capital expense over the useful life of the item; see *capital budget*.

Analog signal: In communications, a signal that represents an actual sound or visual image (as opposed to a *digital signal*).

Application program: In a computer system, a program that is used to accomplish a particular type of useful work.

Area: In patrol organization, a portion of a district or precinct to which one or more patrol officers are assigned.

Armored shield: A type of protective device consisting of a metal or high-impact plastic shield that can be carried by an officer to protect his or her head and body.

Arrest report: The official record of the arrest of an accused offender, including the person's name and other identifying information, the nature of the alleged offense, and a report of the arrest itself.

Artificial intelligence (AI): In computer technology, any of various techniques used in designing software that simulates the intellectual behavior of humans.

Assessment-Center Method: A method of evaluating applicants for police jobs, by administering a series of tests, each of which is separately assessed by two or more evaluators; often but not necessarily involving a testing site chosen and prepared specifically for this purpose.

Atomistic: The quality of being divisible into smaller and smaller parts.

Attrition: Loss resulting from normal causes; for example, the loss of personnel due to retirement, voluntary termination, death, and so on.

Audit: A careful review of financial or other records to determine their accuracy and completeness.

Authority: In any organization, the right of an individual to make certain decisions and to require subordinates to perform certain tasks.

Auto-loader: A device used to reload a revolver's cylinder with several bullets at once.

AutoBid: A computer-assisted system for developing specifications for police vehicles and selecting the vehicles that most closely meet the desired specifications.

Automatic location display: A computer-based system that shows on a screen the address from which a telephone call is being made (such as to a 911 Center).

Automatic Print Identification: A computer-based system for the analysis and comparison of finger and other skin ridge prints.

Automatic Vehicle Location (AVL): A system using various arrangements of radio receivers and transmitters, usually with computer assistance, to provide dispatchers with the current location of patrol and other police vehicles.

Background investigation: An investigation into the personal history, characteristics, and associations of a candidate for police employment; usually one of the final steps in the selection process.

Backup officer: An officer assigned to assist another patrol officer, or to be ready to come to another officer's aid if needed.

"Bad-apple" theory: The theory, generally discredited, that virtually all police officers are honest, and that any dishonest or corrupt activity within a police agency is attributable to a few "bad apples."

Bailiff: Originally, a local English official appointed by the nobility whose principal responsibility was to maintain order and settle local disputes; in modern usage, a law enforcement officer assigned to the courts.

Ballistics technician: A specialist in the forensic examination of firearms and related physical evidence.

Base station: In a radio communications system, a radio station in a fixed location used to control communications with and among mobile radio units.

BASIC: A computer "language," or set of basic instructions used to develop computer software; an acronym for "Beginner's All-purpose Symbolic Instruction Code."

Baton: A wood or metal instrument, generally rod-shaped, used by an officer to prod an unruly individual or to ward off a blow; a nonlethal impact weapon, when used properly.

Beat: A geographical area to which a police officer is assigned for general patrol or other purposes.

Bid: In the purchasing process, a quotation from a vendor stating the price for which merchandise or services will be sold to the agency in accordance with the specifications given to the vendor by the agency.

Bill of Rights: The first ten amendments to the Constitution of the United States, defining the essential civil rights of all citizens.

Binary digit: The basic unit of information in a computer or other information and communications system, consisting of a single "on" or "off" signal.

Body armor: A device consisting of a vest or jacket made of, or filled with, shock- and impact-absorbing material, designed to protect the wearer from the impact of bullets or other projectiles; sometimes called a "bulletproof vest."

Breath analysis test: A device using an instrument to measure the presence of alcohol in a person's breath, which is presumed to indicate the presence of alcohol in the person's bloodstream and, therefore, the extent of intoxication; used as a screening device in suspected driving-while-intoxicated cases.

Budget: A plan for the receipt and expenditure of funds for specified purposes over a specified period.

Budget cycle: The series of events that include the development, approval, execution, and evaluation of a budget.

Bureau: In a police agency, one of the major organizational divisions, usually headed by a senior officer (deputy chief, major, or captain).

Bureaucracy: Any organization structured into a hierarchy consisting of specialized elements, each of which is responsible for a particular part of the organization's overall tasks; the theory of organization developed by Max Weber, a German sociologist.

Byte: In computer technology, a group of *bits* that represent a single character, number, or word.

Cable television: A system by which television and other electronic signals are distributed through cables (and other devices) to a number of locations; commonly a commercial system for distributing entertainment and informational programs to subscribers.

Cadet: A prospective police officer during training.

Calling number display: In a telephone system, a special feature that displays the telephone number from which a call is being made, so that if necessary the call can be traced or its source verified.

Camcorder: An electronic device in which a video camera and a video tape recorder are combined, permitting video recordings to be made inexpensively and conveniently.

Capital budget: The budget for the purchase of items that have an expected useful life of more than one year.

Career counselling: Generally, advice and guidance given to employees on ways they can advance their careers; in a *career development program*, a systematic procedure whereby each supervisor advises each of his or her subordinates on their current job performance and methods of attaining individual career goals.

Career development system: A systematic, planned procedure for coordinating the career aspirations of employees with the agency's human resource needs, through career counselling, career planning, job enrichment, and other methods.

Case assignment: The method by which reports of crimes are assigned for investigation to specific investigators.

Case evaluation: The procedures used to determine whether a criminal case, where the identity of the suspect is unknown, is likely to be solvable, and therefore what expenditure of effort is justified. Also sometimes called *case screening*.

Case report: The complete record of the initial report of a crime and of all police activities taken to identify and apprehend the perpetrator.

Casual surveillance: Incidental observation of the general public by a police officer while carrying out other duties (such as patrol).

Cellular telephone: A communications system in which individual telephones are connected by radio to a network of low-powered transmitter-receivers, each of which serves a relatively small geographic area or "cell"; the commercial system uses computers to control the radio connections, passing the signal from one transmitter-receiver to another as the mobile telephone travels from "cell" to "cell," and connecting the mobile units to the regular ("landline") telephone system.

Central processing unit (CPU):   In computer technology, the group of circuits that process information; the basic component of a computer.

Certification: Official recognition, such as by a state agency, that an individual has attained a specified level of proficiency by completing a required training course or by similar means.

Chain of command: In a hierarchical organization, the system of relationships between superiors and subordinates through which authority and accountability are connected.

Chemical Mace: A nonlethal chemical weapon consisting of an aerosol spray of a chemical similar to tear gas.

Chief of police: The chief executive officer of a municipal police department.

Citation: A written notice that an individual will be officially charged with a minor offense (such as a traffic violation) and must appear in court to answer the charge; used in place of a physical arrest.

Civil Service Commission: An agency, usually established by state law but often organized at the local level, responsible for ensuring that the best available qualified personnel are employed in public agencies; in some cases, also responsible for administering the parent government's personnel policies, grievance procedures, termination proceedings, and so forth.

Civil Service Reform Act: Also known as the *Pendleton Act*, the federal law passed in 1883 that established a merit selection and promotion system for most federal employees.

Class (social): The group of persons within society whose economic status, educational status, and other characteristics are approximately the same.

Clock rate: In a computer system, the number of operations performed by the *central processing unit* per second or other period of time, expressed as a frequency (usually in megahertz).

Command vehicle: A police vehicle, usually unmarked or unobtrusively marked, used by plainclothes officers and administrators.

Commander: The individual officer, regardless of rank, who is responsible for carrying out a specific operation or who has authority over a portion of the agency.

Commodities: Items used routinely or continuously, usually purchased in large quantities or on a contract basis for a period of several months to a year.

Communication: The transfer of information over time (by recording) and space (by transmission).

Communications medium: A device or system used for communication; particularly, that part of the system over which information is transmitted.

Communications system: A device or system used for communication, consisting of a transmitter or encoding device, the transmission medium, a receiver or decoding device, and a provision for feedback.

Community: Any group of people who have some characteristic in common (such as, but not only, residing in a given area).

Community police station: A police facility designed to house only those operational units that provide services directly to the public, usually located in a residential or commercial area away from the agency's central headquarters.

Community policing: Any of several organizational schemes designed to promote a closer relationship between the police and the community.

Community relations unit: An element of a police agency that is specifically responsible for promoting a positive relationship between the police and the community, and often is given responsibility for developing community-oriented police services.

Community service officers: Commissioned or, more often, noncommissioned personnel who provide non–law enforcement services such as crime-prevention counselling, crisis intervention, juvenile guidance, and so on.

Complaint: A report from a citizen that a crime has been or may have been committed; generally, any request for police assistance.

Computer-assisted dispatching (CAD):   A system whereby dispatchers are assisted by a computer in assigning cases to patrol officers.

Computerized patrol allocation system: A system whereby patrol officers are assigned to beats and shifts through the computerized analysis of incident records.

Constable: In medieval England, the person employed to care for the stables who, when the noble's guards were away from the village, acted as a watchman and caretaker; in modern usage, an elected or appointed county official with limited law enforcement responsibility.

Contingency plans: Plans that are designed to be implemented only if and when certain events (such as disasters) occur.

Contingency theory: The theory that an effective leader is one who is able to use whatever style or method of leadership is most appropriate to a given group or situation.

Coordination: In any organization, the act of providing for the performance of several tasks or a series of tasks in proper order and relationship to one another.

Corporal: In some police agencies, a rank intermediate between patrol officer and sergeant.

Corporate model: The general model of hierarchical organization in which authority and responsibility are established and tasks are assigned according to the principle of bureaucracy.

Cost-benefit assessment: A comparison of several proposed plans of action, all with the same or similar objectives, on the basis of the benefits that each is likely to obtain and the costs of each.

Covert surveillance: Observation of persons or things from a location that is hidden from public view.

Crime: The violation of any law or local ordinance that is so designated by a legislature.

Crime analysis unit: The element of a police agency that studies information about criminal incidents to detect patterns or trends of criminal activity that may be used to predict the need for specific police techniques.

Crime prevention: Any of various methods or techniques intended to discourage criminal activity.

Crime prevention unit: The element of a police agency that is specifically responsible for developing and implementing crime prevention programs.

Crime scene technician: A specialist in the detection, recovery, and recording of physical evidence at a crime scene.

Criminal: (1) A person who has committed a crime; (2) Having to do with crime or with persons who commit crimes.

Criminal enterprise: An illegal business operated for profit, more or less on a continuing basis.

Criminal investigation: The use of particular techniques and procedures to identify and apprehend the perpetrators of crimes and to gather evidence to demonstrate the perpetrators' guilt in court.

Criminalist: A person such as a crime scene technician who practices criminalistics.

Criminalistics: The gathering, processing, analysis, and evaluation of physical evidence.

Criminology: The scientific study of crime, including the nature and sources of criminal behavior.

Cross-training: The training of employees who are ordinarily assigned to different tasks, so that each is capable of performing the others' work when necessary or convenient.

Current duties: That portion of an administrator's responsibilities that are carried out currently, as distinguished from prospective and retrospective duties.

Current wants and warrants file: The file or record containing reports of individuals, and usually vehicles, that are wanted by the police or for whom arrest warrants have been issued.

Custodianship: The legal obligation to hold and protect certain property, such as, in the case of a police agency, property whose ownership is uncertain.

Daily bulletin: A police agency's official report of the previous day's activities and information about criminal activity.

Daisywheel printer: A type of computer printer that uses as the printing element a round wheel, divided into several slender petals, each of which contains one or more printing characters.

Data: Items of information, as in a computer system.

Database program: An application program designed to receive, store, index, sort, and retrieve a large volume of information.

"Dead-reckoning" system: An automatic vehicle location system in which a computer in each vehicle continuously records the vehicle's motions and calculates its movements from a known location, then reports the location (usually by radio signal) periodically.

Debt service: In budgeting, a provision to repay a portion of the agency's indebtedness; the amount of debt to be repaid during a given budget period.

Decoding: In a communications system, the process by which information is recovered from a signal.

Delegation: In any organization, the act of assigning a specific portion of one's authority to a subordinate, to enable the subordinate to carry out specified tasks.

Demography: The study of the characteristics of a given population, such as rates of births and deaths or rates of migration.

Depreciation: The loss of value of an item over a period of time due to its use and normal wear and tear; a provision in a budget for the eventual replacement of an item with an expected useful life of more than one year.

Deputy chief: A senior administrator in a police agency, subordinate only to the chief of police; usually the commander of a major division or bureau.

Desktop publishing program: An application program designed to facilitate the development of printed matter such as books, pamphlets, newsletters, posters, and so on.

Detective: The functional title of a police officer whose principal duty is the investigation of reported crimes.

Digital pager: An electronic device that receives digital radio signals and decodes them to provide a limited amount of information, such as an indicator that the person carrying the pager has a message waiting.

Digital signal: In a communications system, information encoded in digital form for accurate and rapid transmission, as opposed to an *analog signal*.

Digital voice signaling: In a radio system, the use of digital signals to encode voice messages, thereby reducing the frequency bandwidth required for transmission and, incidentally, making the signals difficult for anyone to intercept and understand.

Direct costs: In planning and budgeting, the costs directly attributable to a particular program or activity.

Disbursement: Any expenditure of funds for a particular purpose.

Disciplinary hearing: A formal procedure in which several appointed officers consider evidence regarding an accusation against a police officer, determine whether the officer is guilty of the alleged offense, and, usually, propose appropriate punishment.

Discipline: Adherence by subordinates to the policies and rules of an organization; more generally, obedience to a teacher or leader.

Discretion: The act of making decisions or engaging in activities on one's own initiative and, implicitly, assuming responsibility for the consequences.

Discrimination: Acting toward an individual or group on the basis of prejudice; acting toward an individual or group in a manner substantially different from actions toward other individuals or groups, and usually contrary to the interests of the person or group discriminated against.

Dispatch card: The form on which a telephone operator records essential information about a reported incident or complaint, used by the dispatcher to assign an officer to investigate, and on which the dispatcher records the actions taken.

Dispatcher: A person (either commissioned or noncommissioned) whose principal duty is to assign officers to investigate reports of criminal incidents or requests for police services, usually by transmitting radio messages to patrol officers.

District: A geographical division of the territory within a police agency's jurisdiction.

Diversion: In the criminal justice system, and particularly the juvenile justice system, any procedure whereby accused offenders are directed into counseling or other rehabilitative programs rather than being prosecuted.

Division: An organizational element of a police agency, usually part of a bureau, containing several subdivisions.

Domestic disturbance: An incident involving a violent or potentially violent confrontation among persons who have a previously established relationship, such as family members or friends.

Dominance: In the social sciences, the condition in which certain individuals exert influence over the other members of a group by persuasion, force, or otherwise.

Dominant community: The members of a community who tend to have the greatest influence over political and social policy; often but not necessarily the ethnic majority in a community.

Dot-matrix printer: A computer printer in which the printing element consists of an arrangement of small metal pins that are caused by electromagnets to project from a solid head, so that when the head impacts an inked ribbon, a pattern of dots forming a letter or character is transferred to paper.

Dual jurisdiction: An arrangement in which two or more police agencies share responsibility for certain law enforcement duties within a single geographical area.

Due notice: In law, information given to the general public or to certain affected individuals, warning them that specified acts are illegal, or advising them of the law's requirements.

Efficiency: The accomplishment of the greatest amount of work for the smallest expenditure of resources.

Electronic mail: A computer-based system for transmitting and storing messages.

Eligibility list: A list of employees, such as police officers, who are qualified for appointment to a given job, or for promotion.

Encoding: In a communications system, the process by which information is converted into a form, such as a signal, that can be transmitted.

Evaluation: The process of determining the value of someone or something; in a personnel system, the specific procedures used to determine whether an employee is performing in a satisfactory manner.

Evidence: In law enforcement, anything that bears or conveys information about a criminal act, such as physical evidence (an object that pertains to a crime) or testimonial evidence (the reports or statements of persons who have knowledge about a crime).

Evidence technician: A person whose assigned job is to discover, collect, and analyze physical evidence.

Execution (of a budget): The general process of receiving and disbursing funds in accordance with an approved budget.

Executive: In administration, the highest position in a hierarchy; the person who has authority over and responsibility for all elements of the organization; in government, the branch that is given the duty of enforcing the laws established by the legislature.

Expert system: In computer technology, a type of software that is intended to imitate the thought processes of a human expert; see also *artificial intelligence.*

Expressway: A roadway designed for high-speed automotive travel, with limited access and few if any intersecting roads.

External audit: An audit performed by accountants who are not employees of the company or agency being audited.

External memory: In a computer system, any memory device that is not included in the central processing unit's random-access memory (RAM); the device, such as a magnetic disk drive, may be contained in the same cabinet as the CPU.

Facsimile machine: A communications device in which printed matter is encoded into an electronic signal, transmitted through the telephone system, and decoded by a receiver back into printed matter.

Fax: An abbreviation for *facsimile machine.*

Federal Bureau of Investigation (FBI): The primary law enforcement agency of the United States government, a division of the Department of Justice.

Feedback: In a communications system, the means by which the receiver of a message indicates to the sender that the message was accurately received.

Fence: A person who, in a criminal enterprise, traffics in stolen merchandise.

Fiber-optic: A technology in which information signals are encoded as digital patterns of light and transmitted through transparent glass or plastic fibers.

Field interview report: The official report made by a police officer of a contact with a person who is not a victim, witness, or suspect in a particular crime, but whose behavior indicated to the officer that the person might be involved in or have some knowledge of criminal activity.

Field sobriety test: Any of several methods used by police officers to determine whether a driver suspected of driving while intoxicated is, in fact, intoxicated.

Field training: That portion of a training program that takes place outside of an academy or other formal classroom setting, and that includes actual experience in police activities by the trainee under the immediate supervision of an experienced officer.

Field training officer: An experienced police officer assigned to supervise an inexperienced officer during the latter's field training.

File management program: A computer program designed to store, index, sort, and retrieve a large volume of information; see also *database program.*

Firearms examination specialist: An evidence technician who specializes in the examination of firearms, projectiles, and explosive devices; see also *ballistics technician.*

Fixed-wing aircraft: An aircraft that maintains flight by the rapid passage of air over a nonmoving airfoil; an airplane, as opposed to a *helicopter*.

Floppy disk: In a computer system, a data storage system that uses removable magnetic disks as the storage medium.

Flow-through requirements: In budgeting, any legal requirements or rules that are attached to funds from an outside source, especially when the ultimate source is two or more levels removed; for example, rules governing the use of funds provided by the federal government through a state agency to a local police department.

Follow-up investigation: In criminal investigation, the police activities undertaken after the initial, or preliminary, investigation is complete, in an effort to identify or apprehend the perpetrator of a crime.

Forecasting: In planning, the systematic development of predictions about future conditions.

Fraternal Order of Police (FOP):   An organization of police officers, established as a mutual assistance organization, that in some circumstances has acted as a labor union.

Frequency bandwidth: A particular portion of the electromagnetic spectrum that is designated for or being used for a specific purpose, such as a radio communications channel or group of channels; a range of electromagnetic frequencies, usually expressed in Hertz (cycles per second) or in wavelength (meters).

Full duplex: In a communications systems such as a radiotelephone system, an arrangement whereby it is possible for all operators to speak and to hear one another simultaneously.

Function analysis: In planning, the procedure by which an organization's overall or ultimate goals or purposes are divided into a number of specific functions that are then distributed among various elements of the organization.

Functional assignment matrix: A graphic representation of the various functions of an organization and their assignment to specific organizational elements.

Funding: The provision of money from one or more sources with which a budget can be carried out.

Gas grenade: A nonlethal chemical weapon; an explosive device that, when it explodes, releases a cloud of tear gas.

Grid organizational development: A theory developed by social scientists Robert R. Blake and Jane S. Mouton to describe certain aspects of organizational behavior.

Grievance: An expression of dissatisfaction by an employee or group of employees.

Half-duplex system: In a communications system such as a radiotelephone system, an arrangement whereby mobile units transmit on one frequency or channel and receive on another, while a base station or repeater receives on the mobile-transmit channel and transmits on the mobile-receive channel; the result is that the operators cannot speak and hear one another simultaneously.

Handcuffs: Any of several types of police equipment used to restrain a prisoner's hands.

Hard disk: In a computer system, a data storage device in which a magnetic disk is housed permanently in a disk drive.

Hawthorne effect: According to the human relations theory developed by Elton Mayo, the temporary increase in productivity that often occurs when there is any change in the work environment, due to the increased attention and interest of the workers.

Heterogeneous: A mixture of diverse parts, as a community made up of various ethnic or social groups.

High-speed parading: The practice in which many police vehicles engage in the pursuit of a fugitive.

Homogeneous: A composition in which all of the parts are essentially the same, as a community whose members are of the same ethnic or social group.

"Hot pursuit" law: A law, usually enacted by a state, that specifies the circumstances under which a police officer may continue to pursue a fugitive beyond the borders of the officer's usual jurisdiction.

Human relations theory of management: The theory, originally developed by Elton Mayo and extended by others, that the social relationships among workers are an essential factor in determining the productivity of an organization.

Implementation: In budgeting and planning, the process of carrying out the budget or plan; see also *execution*.

Incident distribution analysis: In planning of patrol assignments, a procedure for examining records of criminal incidents to determine their distribution over time and their geographical distribution, to determine the natural workload of the officers.

Incident report: The original report of a crime or other request for police service; see also *case report*.

Incremental method: In budgeting, a presentation of proposed expenditures in which only the changes (increases or decreases) from the previous period's budget are explained.

Independent police station: A police facility that contains all elements of the agency, including administrative, technical, support, and operational units.

Indexing: The procedure used to record the location of specific items of information in a file.

Indirect costs: In budgeting, costs that are generally or proportionally shared among all organizational units, such as administrative and housekeeping expenses.

Information: The content and product of human minds; any signal or record of a change in the state, condition, or location of a physical object.

Information system: A system for acquiring, recording, indexing, storing, sorting, and retrieving information.

Ink-jet printer: A type of computer printer in which ink is sprayed from a set of tiny nozzles onto paper in a pattern to form letters and other characters.

Inspector: A police rank used by some agencies, usually superior to the rank of captain; also, a functional title used by some agencies, a commissioned officer (generally with the rank of lieutenant or higher) whose assigned duty is to detect, investigate, and report any instances of improper behavior on the part of other agency personnel; in British usage, a detective.

Institutional channels: In a cable television system, communications channels that may be used privately, usually for a fee, by governmental or commercial organizations, and that are not available to regular subscribers; such channels might be used to transmit video, voice, data, or other signals.

Insubordinate personality type: A type of personality characterized by an emotional or psychological inability to accept a reasonable degree of discipline and an inability to assume responsibility for one's own behavior.

Integrated circuit: A combination of numerous interconnected circuit elements in a single device, usually relatively small.

Integrated investigation system: The organization of a police agency such that criminal incidents are initially investigated by patrol officers, who are authorized to continue the investigation to its conclusion unless it is unduly complicated or time-consuming, in which case criminal investigators either assist the patrol officer or assume responsibility for further activity.

Intelligence unit: An element of a police agency that is responsible for analyzing criminal incident records to determine patterns or trends of criminal activity; in some agencies, the intelligence unit is limited to analyzing information about organized crimes.

Internal investigations unit: An element of a police agency that is responsible for detecting, investigating, and reporting incidents or alleged incidents of improper behavior by other members of the agency; see also *inspector*.

International Association of Chiefs of Police (IACP):   A professional organization of police officers and persons interested in the advancement of law enforcement as a profession.

Inventory: Any record of the ownership and location of items of property.

Investigation: The use of rational methods to obtain and analyze information, such as information about a reported crime.

Investigator: A police officer who is assigned to perform an investigation.

Invoice: A claim for payment submitted by a vendor after goods or services have been delivered.

Item order: The purchase of a single item or group of related items that are not bought on a continuing basis, as opposed to a *commodity*.

Job enrichment: Any measures taken to make employees' work more satisfying and personally rewarding.

Jurisdiction: The geographical area in which a law enforcement agency has legal authority; the types of incidents or circumstances for which a law enforcement agency has authority to provide police services.

Juvenile crime unit: An element of a police agency that is specifically assigned the task of identifying actual or potential juvenile offenders and bringing them into the juvenile justice system.

Juvenile justice system: A system of laws, rules, and institutions such as special courts and correctional facilities intended to discourage juveniles from becoming involved in criminal activity, and to rehabilitate those who are already committing criminal offenses.

Kansas City experiment: A research study conducted by the Kansas City, Missouri, police department in the early 1970s, intended to determine what level of police patrol would be most effective in preventing crime, but which seemed to show that preventive patrol had little or no effect on criminal activity.

Knapp Commission: A committee of respected citizens and experts on law enforcement, chaired by William Knapp, organized to investigate corruption in the New York City police department in the early 1970s, and one of whose conclusions was that the *"bad-apple" theory* is invalid.

Language: In computer technology, a set of instructions that can be combined in various ways into a program to operate the central processing unit; see also *BASIC*.

Laser printer: A type of computer printer that uses a low-powered laser beam, controlled by the computer, to create a static charge on a metal drum, which then attracts and holds powdered granules of graphite that are transferred to blank paper and fused by heat, thereby forming a pattern of letters and other characters.

Lateral entry: The transfer of a person from one police agency to a similar rank and position in another police agency.

Law enforcement: The practice of an agency of government, under the authority of law, to investigate reports of criminal offenses, identify and apprehend alleged offenders, and deliver the alleged offender with corroborating evidence to a court; and to provide other services intended to promote and maintain social order.

Law Enforcement Assistance Administration: A federal agency, established under the Omnibus Safe Streets and Crime Control Act of 1968, but gradually reduced and finally eliminated in the late 1970s; the first major effort by the federal government to assist state and local law enforcement agencies and to influence police practices.

Leadership: The practice or technique of inducing others, usually by persuasion, to attain certain goals or purposes; also, those persons who collectively exercise leadership, or are expected to do so, in an organization.

Leadership style: The general manner in which leadership is practiced.

Legislature: The branch of government that has the duty of establishing laws and providing funds and other resources for their execution or enforcement.

Lieutenant: In most police agencies, a rank intermediate between sergeant and captain; usually the rank of a supervisor of an operating element of an agency.

Line-item budget: A budget showing only the objects of expenditure: that is, the goods and services to be bought during the budget period, without explanation.

Local area network (LAN):   In computer technology, an arrangement whereby several individual computers or computer terminals are connected into a system that permits the sharing of programs and data.

Lowest qualified bidder: In purchasing, the vendor who submits the lowest price quotation for goods or services that fully meet the stated specifications, and who is known or believed to be capable of providing satisfactory goods or services.

Mainframe computer system: A large computer system consisting of one or more high-speed central processing units, various kinds of memory devices, and devices for the entry and retrieval of information; usually the central processors, memory devices, switching systems, and input-output units are separate devices.

Maintenance contract: In purchasing, a contract with a vendor to provide any needed repairs or maintenance on an item of equipment.

Maintenance of social order: Any police practices, activities, or techniques that are designed to promote social harmony and to discourage offensive behavior that is not necessarily illegal.

Management: In any organization, the hierarchical level that is responsible for supervising the operating elements, coordinating diverse operations, communicating orders and policies from the administrative level to the operating level, and reporting the results of operations to the administrators; the middle sections of the chain of command.

Management by objectives: The organizational theory that the most effective technique of managing an organization is to establish clear, unambiguous goals or objectives, usually with the participation of operating personnel.

Management information system: An information and communications system designed primarily to provide managers and administrators with information about the performance of the organization.

Management-for-quality theory: The theory, originally developed by W. E. Deming and extended by others, that the ultimate quality of goods and services is largely determined by the management practices in an organization.

Mandatory child safety restraint: Any of various devices, such as automobile child safety seats, designed to protect an infant or small child in the event of a collision, the use of which is required by law.

Marshal: An appointed law enforcement officer, especially in a village or unincorporated town; a law enforcement officer assigned to the federal court system.

Mass media: The commercial enterprises that provide entertainment and information services to the general public, such as newspapers, magazines, radio, television, sound and video recording, and motion pictures.

Mastermind strategy: The police practice of attempting to disrupt a criminal organization by identifying and arresting the key individual who controls and directs the criminal enterprise.

Mechanistic organization: In organizational theory, an organization that is structured according to presumably rational principles without regard to the nature of the organization's purposes or the character of its members.

Megahertz (MHz):   A unit of frequency: millions of cycles per second.

Merit system: A set of procedures and methods designed to ensure that the best-qualified persons are appointed to positions in public agencies; see also *Civil Service Commission*.

Message: In an information or communications system, a unit of information, especially during the process of transmission.

Microcomputer: A small computer system in which the central processing unit, a limited amount of memory, and (usually) additional memory devices and devices for the entry and retrieval of data are housed in a single cabinet, or several small cabinets designed to be connected directly together; specifically, a computer system whose central processing unit is a single, large-scale integrated circuit.

Millions of instructions per second: A measure of the rate at which a central processing unit (and by extension a computer system) performs operations; see also *clock rate*.

Minicomputer: A medium-sized computer system consisting of a central processing unit, one or more external memory devices, and one or more devices for the entry and retrieval of data, usually contained in several separate cabinets; a computer system intermediate in size between a mainframe system and a microcomputer system.

Minority: A member of an ethnic group that historically has been the victim of discrimination; in federal law, specifically members of African-American heritage, persons of Hispanic national origin, persons of Asian national or ethnic origin, persons of Native American and Alaskan ethnic groups, and certain other specified populations; more generally, any population that is identifiable by ethnic group, national origin, language, religion, or other social characteristic, and that is not dominant in the community.

Mobile data terminal (MDT):   A computer device for the entry and retrieval of data, designed to be used in an automobile or similar vehicle, connected by radio communications with a central computer system.

Mobile radio: A communications system in which radiotelephone transmitter-receivers, carried in vehicles or handheld, are connected by radio communications with one or more *base station* radios, sometimes with the aid of a *repeater*.

Mobile surveillance: The police technique of observing the public for suspicious or possibly criminal behavior, and for violations of traffic laws, while in a moving vehicle such as a patrol car.

Modem: A communications device that converts one type of electronic signal, such as a computer's digital signal, into another, such as a telephone signal, and vice versa; an abbreviation for "modulator-demodulator"; a device that enables computers to exchange data over the telephone system.

Modus operandi: The Latin phrase that is translated as "method of operation"; the typical behavior pattern of a particular criminal, as shown by the characteristics of the criminal's activities.

Monitor: In computer technology, a device used to retrieve information from the computer by displaying it on a television-like screen; sometimes also an attached keyboard used to enter data into the computer.

Multiagency facility: A building or group of buildings that contain elements of several law enforcement or other governmental agencies.

Multiple-assessment procedure: The procedure used in evaluating a prospective police officer by requiring the candidate to complete a series of tests, the results of all of which are used to determine the candidate's qualifications.

Multiple-hurdle procedure: The procedure used in evaluating a prospective police officer by requiring the candidate to complete a series of tests in a predetermined order, where the failure to pass any one test disqualifies the candidate from completing the series.

National Crime Information Center: An information and communications system, established and operated by the federal government, for the collection, storage, and exchange of information about criminal activity among local, state, and federal law enforcement agencies.

National Law Enforcement Telecommunications System (NLETS):   A teletype-based communications system, established and operated by the federal government, for the exchange of information among local, state, and federal law enforcement agencies.

Natural workload: In a police agency, the average amount of work required to be performed by each officer (or each officer in a particular unit, such as patrol) during a given period of time, as a result of reports of crimes and other matters requiring police services.

Needs assessment: In planning, the procedure by which the need for a particular type of police activity is established.

Neighborhood police station: A police facility designed to house various operational elements, such as patrol and community relations units, in or near a residential community.

Night watch: A band of men employed and organized to keep watch, usually in a commercial area, to discourage criminal activity.

911 system: A telephone system in which the directory number, "9-1-1," is reserved for emergency calls; usually based at a single location, the "911 Center," at which all such calls are received and routed to the appropriate agency.

Nonlethal impact projectile weapon: Any of several devices intended to subdue or restrain a prisoner without causing permanent injury or death.

Nonproductive tasks: In the analysis of patrol workloads, the number of tasks and the time required to perform them that do not contribute directly to the purposes of law enforcement.

Noxious nonlethal weapon: Any of several types of nonlethal devices intended to subdue or restrain belligerent persons by creating an unpleasant environment, such as by noise, strong odor, bright lights, and so forth.

Object-of-expenditure budget: A budget in which only the items to be purchased are listed and explained; see also *line-item budget*.

Objectives: The major purposes of an agency or organization, stated in terms of the ultimate goal or goals to be achieved, preferably in quantified terms.

Operating budget: A budget containing all expenditures for the operation of an agency during a particular period, excluding the capital budget but including such capital-related items as depreciation or debt service.

Operating personnel: The members of an organization or agency who are assigned to carry out the specific tasks that are necessary to achieve the organization's major goals or purposes; the lowest level of a bureaucratic hierarchy.

Operating system: In computer technology, the program that contains all of the instructions for the internal operations of the computer.

Operational definition: The statement of a goal or objective in quantified terms that permit evaluation of the extent to which the goal or objective is achieved.

Operational plans: Those plans that concern the routine, day-to-day work of an organization and that are intended to ensure the accomplishment of necessary tasks.

Optical disk: A data storage device in which information is encoded on a metal or plastic disk that can be "read" by a strong beam of light, usually a low-powered laser.

Oral interview: The procedure in the selection of candidates for appointment to police jobs, or in the selection of candidates for promotion, in which each candidate is interviewed by a committee of officers.

Organic system: In organizational theory, an organizational structure that develops naturally out of the character of the organization's members and the tasks they are expected to perform.

Organization: The arrangement of separate elements, such as personnel or groups of employees, into a structure through which authority and responsibility may be distributed.

Organized crime: The criminal activity of persons who engage in crime as a major source of personal income and who cooperate and coordinate illegal activity with one another for the production, sale, and delivery of illegal goods and services.

Overt surveillance: The observation of the public by a police officer, in plain view, to detect any suspicious or criminal activity, including traffic violations.

Packing density: The characteristic of any information recording system or medium that determines the quantity of information that can be stored in a given amount of space.

Parallel processing: In computer technology, the use of two or more central processing units, operating simultaneously on different parts of the same set of data.

Patrol car: An automobile or other vehicle, usually clearly marked as a police vehicle, used by patrol officers for transportation and as a mobile platform for preventive patrol.

Patrol deployment: The procedures used to determine the number of patrol officers required during a given period.

Patrol force: The operational element of a police agency assigned to general law enforcement duties including preventive patrol and immediate response to complaints and requests for police service.

Patrol officer: A commissioned police officer assigned to the patrol force; the lowest rank of commissioned police officer in most agencies.

Patrol wagon: A van or similar vehicle designed to transport prisoners from the point of arrest to a jail.

Patrol-oriented investigation: The organizational system in which patrol officers conduct virtually all criminal investigations from initial response through conclusion, with detectives available to provide technical assistance and advice when needed.

Patrolmen's Benevolent Association (PBA): A mutual-assistance organization for police officers, often acting in a manner similar to a labor union.

Peak-period shift: A patrol shift assigned to duty during the time of day when, according to historical patterns, criminal activity and requests for police services are likely to be most numerous.

Peer rating: A personnel performance evaluation system in which employees evaluate one another's work performance.

Pendleton Act: The federal *Civil Service Commission Act*.

Performance budget: A budget in which the quantity of work to be produced is shown for each proposed expenditure of funds.

Performance evaluation: A systematic procedure for determining the quality and quantity of work performed by an employee or group of employees.

Performance test: A procedure used to evaluate applicants for employment or promotion by requiring them to perform actual or simulated activities similar to those that they will be expected to perform if hired or promoted; generally, any test that measures the actual performance of a person rather than the person's personal qualities or characteristics.

Personal computer: A *microcomputer* system.

Personal identification files: Records maintained by a police agency on individuals including known and suspected criminals, victims, witnesses, and other persons who may be involved in some criminal activity.

Planning: The act or process of choosing goals to be attained in the future, methods of attaining those goals, the allocation of resources to carry out those methods, and methods of determining the extent or degree to which the goals are attained.

Planning criteria: The considerations that must be taken into account during a planning process, particularly constraints that limit the types of activities that can be carried out.

Planning unit: An element of a police agency that is specifically assigned to develop the agency's plans; often, though not necessarily, the same as the *crime-analysis unit* or the *intelligence unit*.

Planning-programming-budgeting system (PPBS): A procedure for developing an organization's plans and budget in a coordinated manner.

Plotter: A computer printing device, used primarily to print charts and graphic designs, in which one or more pens are moved across a sheet of paper.

Police academy: See *academy*.

Police aide: A person, either a paid employee or an unpaid volunteer, who is not a commissioned police officer but who performs various routine tasks to assist and support the work of the police agency.

Police blotter: An official record, kept in the form of a notebook or log, of activities at a police station or in a police operational unit.

Police cadet: See *cadet*.

Police commission: (1) A body of elected or appointed officials who are legally authorized to establish policies for, and generally to supervise the operations of, a police agency; (2) the legal authorization given to a person who meets established criteria to act as a police officer.

Police professionalism: The set of ideas, developed by various authorities during the late nineteenth century and since expanded, defining the general quality of police service that an American law enforcement agency is supposed to provide and the manner in which law enforcement officers are supposed to conduct their activities.

Police service information system: An information system designed to collect, record, store, sort, and retrieve information concerning crimes and police activities.

Policy: A statement of the work that is to be done and the manner in which it is to be done, issued by a person in authority to his or her subordinates.

Polygraph examiner: A technician who specializes in the use of the polygraph machine and other so-called lie detection techniques.

POSDCORB: An acronym, invented by early management theorist Luther Gulick, that is supposed to define all of the duties of an administrator or manager; the acronym stands for planning, organizing, staffing, directing, coordinating, reporting, and budgeting.

Precinct: A geographical area within a larger jurisdiction such as a city or county; a portion of a city defined as the area of administration for a section of a police agency.

Predictive validity: In reference to a test or system of evaluation, the extent to which the results of the test accurately indicate the future performance of the person or thing being tested.

Preliminary investigation: The initial stage of a criminal investigation, in which one or more officers respond to the complaint, interview the victim or witnesses, collect any physical evidence available at the crime scene, and attempt to determine the identity of the suspected perpetrator.

Preparation (of a budget): The stage in the budget cycle during which a proposed budget for a future period is being developed.

Preservice training: Any training given to a person before he or she is employed or, after the person is employed, before he or she is assigned to specific duties.

Preventive maintenance: Periodic examination and testing of equipment, such as vehicles, to determine the extent of wear and the need for repairs or replacement.

Preventive patrol: Various police techniques in which officers, usually members of the patrol force, observe the public for any suspicious or criminal activity.

Price quotation: See *bid*.

Private crime: A criminal act involving two or more persons who have a previously established, ongoing relationship; usually private crimes occur out of public view and, especially in the case of violent crimes, often are a result of spontaneous acts rather than premeditation.

Program: (1) In planning and administration, a specific activity or group of activities, to which personnel and other resources have been allocated, that is intended to continue for an indefinite period of time and to accomplish significant goals of the organization; (2) in computer technology, a list of instructions that controls a computer's operations.

Program budget: A budget in which proposed expenditures are arranged by program or organizational element, with an explanation of the activities that the expenditures will support.

Programmatic plans: In planning, those plans that define an organization's activities directed toward the achievement of specified goals or purposes; both *programs* and *projects*.

Project: In planning and administration, an activity or group of activities intended to be carried out within a definite, limited period of time; see also *program*.

Prospective duties: Those duties of an administrator or manager that concern preparations for the future, such as planning.

Public crime: Criminal acts involving persons who usually have no previously established or ongoing relationship, usually taking place in a public place or involving the intrusion of the criminal into a private place (such as burglary), and usually but not necessarily premeditated.

Public service announcement: An announcement broadcast by a radio or television station without charge to the originator, usually a nonprofit or governmental organization.

Purchase order: A contract between a purchaser and a vendor, authorizing the vendor to deliver a specified product or service at an agreed-on price and committing the purchaser to pay for the item upon delivery.

Purchase requisition: A request from an employee or unit of an agency to the agency's purchasing agent, asking that a specified item or service be purchased, and indicating the budget account from which the purchase may be made.

Purchasing agent: The person who is authorized to negotiate and issue purchase orders and otherwise control the purchasing process.

Pursuit driving: The practice of operating a police vehicle at high speed in pursuit of a fugitive.

Queuing: In a communications system, the means by which incoming calls are held in the sequence in which they are received until they are answered.

Radio channel: A particular band of radio frequencies designated for a particular type of communications.

Radiotelephones: A communications system in which voice messages are transmitted by radio.

Random access memory (RAM):   In computer technology, the portion of a computer system in which data are stored temporarily and retrieved in whatever order is required by the computer program.

Rank: In a police agency, a position in the organization's hierarchy, indicating a person's general degree of authority and, usually, implying the person's level of salary and other privileges and benefits.

Rapid-response team: An element of a police agency whose primary responsibility is to respond as quickly as possible to all reports of crimes and requests for police services.

Read-only memory: In computer technology, the portion of a computer system in which essential operating instructions are stored permanently and retrieved by the computer as needed.

Receipt: A record that purchased goods or services have been delivered in proper order, signifying that the agreed payment should be made.

Recruiting: The activity of seeking qualified applicants for employment.

Regulatory crime: Crimes that involve violations of laws or regulations whose primary intent is to maintain social order or to carry out the administrative functions of government.

Rehabilitation: Any procedure by which a person who has previously been involved in criminal activity is restored to the status of respectable citizen.

Reliability (of a test):   The extent or degree to which a test or evaluation procedure produces essentially the same results each time it is applied to the same or similar subjects; the ability of a test to reproduce comparable results.

Relief shift: In a police agency, a group of officers who have no regular duty assignment, but who substitute for other officers when the latter are absent.

Repeater: In a radio communications system, a device that receives a radio signal on one channel, amplifies the signal, and retransmits it on another channel, thus increasing the range over which the signal can be received.

Reprimand: A notice to an employee of unsatisfactory performance, issued either verbally or in writing; a written reprimand may or may not be placed in the employee's permanent record.

Request for proposals (RFP):   In purchasing, a document issued by a prospective purchaser, soliciting proposals from prospective vendors, usually for items (such as a communications system) that are relatively complex or expensive.

Requisition: See *Purchase requisition.*

Research: An activity designed to collect or evaluate information, in preparation for deciding what future activities may be needed or desired, or how such activities may be carried out.

Resolution: The characteristic of a video monitor or other display device that determines the fineness of detail that can be perceived.

Response time: The lapse of time between the receipt of a complaint or request for police service and the arrival of a police officer at the scene of the incident.

Responsibility: In administration, the duty of a person to perform certain acts as a necessary consequence of the authority assigned or delegated to that person.

Retrospective duties: The duties of an administrator or manager with regard to the past activities of the organization, such as evaluating past operations, auditing a previous budget period, etc.

Review (budget): The evaluation of receipts and expenditures during the previous budget period and comparison to the original budget for that period, to identify areas where significant discrepancies may have occurred.

Role-playing simulation: An exercise in which participants are assigned to portray certain roles or characters, and are then directed to interact as if they were involved in a defined situation; a method of training individuals to be aware of and sensitive to the emotional reactions of people under certain circumstances.

Rookie: A person such as a police officer who has no previous experience in the position to which he or she has been assigned.

Rotary aircraft: An aircraft that maintains flight by the rapid motion of two or more airfoils; a helicopter.

Sample audit: An examination of a portion of an agency's accounting records to evaluate their accuracy and completeness; see *audit*.

Scanner: (1) In radio communications, a receiving device that searches through all of the frequencies or channels in a particular bandwidth, stopping briefly when an intelligible signal is detected; (2) in computer technology, a device that converts printed matter into a digital signal that can be stored and manipulated in a computer.

Scientific management: The various theories, developed in the late nineteenth century and considerably extended since then, concerning the rational organization of work.

Secret Service: A federal law enforcement agency, part of the U.S. Treasury Department, whose original purpose was to suppress counterfeiting after the Civil War; later the agency was given responsibility for the protection of the president and his family; currently responsible for the protection of the president and vice-president, their families, candidates for those offices, and former presidents and vice-presidents.

Section: In a police agency, an organizational element, usually at the operational level; a subunit of a division or bureau.

Sector: In a police agency, a portion of the agency's geographic jurisdiction, usually smaller than an *area* but larger than a *beat*.

Security zones: The arrangement of a police facility into several distinct areas, each having security provisions that are appropriate to the functions carried out in each area.

Segregation: The practice of separating people, such as by race, ethnic group, social class, or other group characteristic.

Selection (of personnel): The techniques or procedures by which individual applicants are evaluated to determine their relative qualifications for employment or promotion.

Selective enforcement: The use of specific techniques and strategies to concentrate enforcement activity on particular violations, on areas where criminal activities (or traffic law violations) are believed to be frequent, or on specific suspected criminals.

Selective Traffic Enforcement Program (STEP): A program sponsored by the National Highway Traffic Safety Administration through which funds are made available to state and local police agencies to support selective traffic law enforcement demonstration projects.

Self-assessment: A stage in the *accreditation* process during which an agency compares its policies and activities with the published standards of the National Commission on Accreditation of Law Enforcement Agencies.

Sergeans: In medieval England, a watchman assigned to guard a town's or castle's gates; the original form of the military title, *sergeant.*

Sergeant: In many police agencies, a rank intermediate between patrol officer and lieutenant; the lowest rank at which an officer has supervisory authority over subordinates.

Service workload: The total volume of work required to be performed by a given officer or group of officers during a given period, including all police services and related duties, but excluding nonproductive activities.

Services: In purchasing, labor performed for an agency by someone who is not regularly employed by the agency, as opposed to merchandise.

Sheriff: The principal law enforcement official of county government in most states; usually elected by the public, and in more populous counties serves as chief executive of the county police agency.

Shift: A period during which an individual or group of employees are assigned to work.

Shire reeve: In medieval England, the chief official of a town or village, appointed by the local noble to keep the peace, adjudicate local disputes, and administer the noble's laws and regulations; the origin of the modern term, *sheriff.*

Simplex system: In a radio communications system, the use of a single frequency or channel for both transmitting and receiving messages, which therefore cannot be done simultaneously.

Simulation: Something that represents and appears in a form similar to something else; in personnel training, an activity that is intended to represent an actual police task or procedure, used to give trainees practical experience or to evaluate their response to realistic situations.

Situational test: A procedure used to evaluate applicants for employment or promotion, using a *simulation* to observe and assess the candidates' performance.

Socialization: The process by which an individual learns to exhibit the culture of a group in which the individual is a new member.

Software: In computer technology, the lists of instructions, or *programs*, that control a computer's operations.

Source list: In purchasing, a list of possible vendors for goods and services that may be required.

Span of control: The number of subordinate employees or organizational elements that are subject to the control and authority of a particular individual.

Special Crimes Unit: An element of a police agency that is assigned responsibility for the detection and investigation of particular types of offenses that are relatively uncommon or that require special expertise and police tactics.

Special fund budget: A budget account containing funds that may be used only for certain defined purposes.

Special Operational Unit: See *Special Crimes Unit.*

Special-Situations Unit: A police unit whose members are specially trained and equipped to deal with situations that involve high risks of violence, such as a barricaded sniper, hostage-taking, etc.

Special Weapons and Tactics (SWAT) Unit: See *Special-Situations Unit.*

Special Weapons Unit: See *Special-Situations Unit.*

Specialization: The principle that each employee should be assigned only to specific tasks, which the employee should learn (through training and experience) to perform in a highly effective and efficient manner; therefore, the principle that all of the work of an organization should be divided into as many specific tasks or types of tasks as possible.

Specialized building network: Several buildings designed to serve the total needs of an agency for various kinds of physical facilities, by designing, constructing, or locating each building to serve one or more specific purposes.

Specialized investigation: An organizational system in which most crimes are investigated by detectives who specialize in specific types or categories of offenses.

Specification: In purchasing, a statement of the type of goods or services to be purchased and the criteria that are to be satisfied.

Spoils system: A political arrangement whereby the party or faction that wins an election has the right to appoint its members or supporters to virtually all positions in public employment, and otherwise to take advantage of all opportunities for individual profit that are present in the government.

Squad: An element of a police agency consisting of a group of officers, usually at the operational level, who are jointly assigned to a particular territory or duty.

Squad car: Originally, a vehicle used to transport a squad of officers to the scene of a disturbance or crime; a predecessor to the patrol car.

Staggered shifts: In patrol force allocation, the procedure in which the personnel of one regular shift are divided into two or more subshifts to accommodate variations in the service workload during part of the day.

Standardization: The principle that all personnel assigned to similar duties should have essentially the same uniform and equipment.

Stationary surveillance: The police technique in which one or more officers remain at a fixed location to observe a person or place for suspicious or criminal activity.

Statistical quality control: The practices used in some industries to evaluate the quality of the goods being produced by testing a randomly selected sample and applying statistical methods to determine the likely distribution of faulty products.

Strategic plan: A general plan defining a series of methods or activities that are intended to attain a major goal or purpose, usually over a relatively long period of time; a summary of an agency's individual programs and projects, showing how the totality of them will contribute to the agency's ultimate purposes.

Stun gun: Various types of nonlethal weapons, including certain types of impact weapons and electrical weapons.

Subprogram: In computer technology, a portion of a *program* that accomplishes a particular task or carries out a particular operation.

Suburbanization: The social process in which residents of a metropolitan area tend to relocate to peripheral towns, or suburbs; this process, which is believed to have begun in the United States after World War II and continues at present, is said to have major sociological and political implications.

Supervisor: Any person whose assigned duties include observing, evaluating, and when necessary correcting the performance or behavior of subordinate employees; more generally, the first level of organization above the operational level.

Surveillance: Any technique or practice used to observe the general public or specific individuals in an effort to detect criminal behavior.

Systematic corruption: In any organization, especially a police agency, the presence of illegal and unethical behavior throughout the organization, involving most if not all personnel.

Systems approach: In organizational theory, the principle that all elements of an organization or process should be arranged to promote maximum efficiency in accomplishing the organization's purposes.

Tactical unit: An element of a police agency, usually part of the patrol force, that is assigned to perform specific tasks or to use particular techniques to prevent or detect specific criminal activities.

Taser: (Registered trademark) A type of electrical nonlethal weapon.

Team: In a police agency, a group of officers, often including officers of various ranks and functional specialties, who are assigned to work jointly on certain cases.

Team policing: A system of police organization in which a team of officers are assigned joint responsibility for all law enforcement services in a given area.

Tear gas: Any of several chemical nonlethal weapons.

Technological convergence: The process in which, due to advances in technology, certain kinds of devices become increasingly similar; for example, at the present time, various kinds of communications systems increasingly rely on the same technology used to design and build computers, while computers increasingly adopt the techniques of communications systems to expand their capabilities.

Technology Assessment Program (TAP): A program administered by the National Institute of Criminal Justice, part of the federal Department of Justice, that tests various kinds of police equipment for their safety, effectiveness, and other criteria.

Telegraphic call box: A communications device, now obsolete, that enabled foot patrol officers to send and receive messages between the central headquarters (or precinct station) and telegraphic instruments distributed at various locations.

Telephone call box: The successor to the *telegraphic call box*, and also now obsolete; a communications system in which telephone instruments were distributed at various locations for use by foot patrol officers to communicate with the central headquarters or precinct station.

Telephone log: A record of telephone calls made and received, usually listing the parties to each call, the time of day each call was made, and the general subjects discussed.

Teletype: Any of several communications systems in which a typewriter-like device is used to generate coded telegraph signals that are transmitted by wire or radio signal to receiving units, where the messages are displayed on a printer or television-like screen.

Terminal: In computer technology, any of several types of devices used to enter data into, or retrieve data from, a computer; in communications technology, any device for sending and receiving messages.

Theory X: In the behavioral management theory of Douglas McGregor, the belief among managers that most people do not want to work and will not accept responsibility for their performance unless they are forced to do so.

Theory Y: In the behavioral management theory of Douglas McGregor, the belief that most workers are eager to work productively and to accept responsibility for their performance, if they are given meaningful tasks and a fair opportunity to challenge themselves.

Traffic law enforcement: The element of a police agency that is responsible for enforcing the highway traffic laws, and related duties such as traffic safety education and assistance to traffic engineers.

Trait-rating system: A method of evaluating employees by judging them on a list of "traits" or personal characteristics, rather than by the quantity or quality of their work; see also *peer rating* and *work-product measurement*.

Transaction: In purchasing, the actual exchange of goods or services for an agreed-on price.

Transfer payment: In budget execution and accounting, the relocation of funds from one account to another, a transaction in which funds are not actually expended.

Transmission: In a communications system, the movement of a message from one place or time to another.

Triangulation: In an *automatic vehicle location* system, the technique of using the relative strength and direction of two or more radio signals to determine the location of the source of the signals.

Ultra-high-frequency (UHF) band: The portion of the radio frequency spectrum from 300 to 3000 MHz.

Undercover agent: A police officer assigned to investigate known or suspected criminal activity by posing as a prospective participant in the activity, in order to gain the confidence of criminals and thereby gather information.

Undercover investigation: The police technique of using one or more *undercover agents* to investigate suspected criminal activity, particularly *organized crime.*

Undercover vehicle: Any vehicle used by police officers during an undercover investigation, not marked as a police vehicle and usually intended to resemble the sorts of vehicles that are ordinarily used by the kind of person being portrayed by an undercover agent.

Underrepresentation: Regarding personnel, the condition in which a particular population segment, such as an ethnic minority, is found among the employees of an agency in a smaller proportion than the group's representation among the general population of the community.

Uniform Crime Reporting System (UCRS): A system for recording, collecting, and analyzing information about reported incidents of crime, voluntarily reported by state and local law enforcement agencies to the Federal Bureau of Investigation.

Uninterruptible power supply (UPS): In a communications or information system, the use of batteries, generators, and other devices to ensure that electrical power is continuously available to vital equipment.

Union: An organization that represents its members as employees of a business or governmental agency, and negotiates employment contracts on behalf of its members.

Unionlike organization: An association of employees that is not legally organized as a labor union but that carries out many of the same functions, such as negotiating employment contracts on behalf of its members.

Unit: An element of a police agency; generally the smallest organizational element other than individual officers.

Unity of command: The principle that each subordinate in an organization should be accountable to only one superior.

Very high frequency (VHF) band: The portion of the radio frequency spectrum from 30 to 300 MHz.

Video surveillance system: An information communications system using video (television) cameras to observe activity at various locations, with the cameras usually connected to a central monitoring or recording site.

Videotape recording: Any of various techniques and devices, and the use thereof, for recording visual images and associated sounds on magnetic recording tape.

Video teleconferencing: A communications system in which personnel at various locations are able to see one another, through the use of video cameras and monitors at each location, as well as converse through the use of telephone or radio communications.

Visual arrest records: The use of motion pictures or, more often, videotape recording to record the appearance and condition of persons during or immediately after they have been arrested, as a means of documenting the actions of both the suspected offender and the arresting police officer.

Walkie-talkies: Two-way radiotelephone devices that are designed to be hand-held.

Ward boss: A political functionary, either paid or a volunteer, who represents a political party or faction in a geographic area (ward), and often acts as an intermediary between the citizens and the municipal government.

Warning: An official notice, sometimes in writing, that a person has violated the law (such as a traffic law), but that no complaint will be filed; intended to encourage the offender to observe the law more carefully in the future.

Watch: In police organization, either an element of the organization (usually at the operational level), or a period to which officers are assigned (see *shift*).

Watchman: See *night watch*.

Word-processing program: A computer application program that is designed to facilitate the preparation and editing of written material, such as letters, memoranda, and so forth.

Work-product measurement: A technique for evaluating the performance of employees by measuring the quantity (or, more rarely, the quality) of their work.

Working class: In social science, a population segment, variously defined as equivalent to the lowest social class, or overlapping the lowest and middle classes; presumably, the class of people whose livelihood is based on working for wages, as opposed to the professional and entrepreneurial classes whose livelihood is autonomous.

Workstation: See *terminal* (in a computer system).

Xerographic copier: A device that copies printed matter by a process originally developed by the Haloid Corporation, later named the Xerox Corporation, in which powdered graphite is deposited electrostatically on a metal drum in the same pattern as the letters and characters on the original, then transferred and heat-fused onto blank paper, thereby creating a facsimile of the original.

Zero-base budget method: A budget in which all proposed expenditures are presented as if each type of activity had never been performed before, usually showing several variations in the amount of activity and the benefits that could be obtained at various levels of expenditure.

Appendix C
SELECTED BIBLIOGRAPHY

The following list includes those sources that we found to be most useful and that we believe to be most widely available. Many additional sources are listed at the end of each chapter in the text.

Adams, Thomas F. *Police Field Operations*. Englewood Cliffs, N.J.: Prentice Hall, 1985.

Alpert, Geoffrey P., and Roger C. Dunham. *Policing Urban America*, 2nd ed. Prospect Heights, Ill.: Waveland Press, 1992.

Ammons, David N. *Municipal Productivity*. New York: Praeger, 1984.

Auten, James H. *Law Enforcement Driving*. Springfield, Ill.: Charles C Thomas, 1989.

Barker, Thomas, and David L. Carter. *Police Deviance*. Cincinnati: Pilgrimage Press, 1986.

Bennis, Warren G., and Burt Nanus. *Leaders: The Strategies for Taking Charge*. New York: Harper and Row, 1985.

Blumstein, Alfred, Jacqueline Cohen, Jeffrey A. Roth, and Christy A. Vishers, eds. *Criminal Careers and "Career Criminals"* (2 vols.). Washington, D.C.: National Academy Press, 1986.

Bonifacio, Philip. *The Psychological Effects of Police Work*. New York: Plenum Press, 1991.

Bouza, Anthony V. *The Police Mystique*. New York: Plenum Press, 1990.

Brown, Michael K. *Working the Street*. New York: Russell Sage Foundation, 1988.

Bryman, Alan. *Leadership and Organizations*. Boston: Routledge and Kegan Paul, 1986.

Chaiken, Jan M., and Warren E. Walker. *Patrol Car Allocation Model*. Santa Monica, Calif.: RAND Corporation, 1985.

Cooper, John L. *The Police and the Ghetto*. Port Washington, N.Y.: Kennikat Press, 1980.

——*You Can Hear Them Knocking*. Port Washington, N.Y.: Kennikat Press, 1981.

Davidson, William S., et al. *Alternative Treatments for Troubled Youth*. New York: Plenum Press, 1990.

Davis, Keith. *Human Behavior at Work* (6th ed.). New York: McGraw-Hill, 1981.

Delattre, Edwin J. *Character and Cops*. Washington, D.C.: American Enterprise Institute, 1989.

Dunham, Roger C., and Geoffrey P. Alpert, eds. *Critical Issues in Policing*. Prospect Heights, Ill.: Waveland Press, 1989.

Farmer, David, ed. *Purchasing Management Handbook*. Brookfield, Vt.: Gower Publishing Company, 1985.

Flippo, Edwin B., and Gary M. Munsinger. *Management* (5th ed.). Boston: Allyn and Bacon, 1982.

Fyfe, James F., ed. *Police Management Today*. Washington, D.C.: International City Management Association, 1989.

Gilbert, James N. *Criminal Investigation* (2nd ed.). Columbus, Ohio: Charles E. Merrill, 1986.

Goldstein, Herman. *Problem-Oriented Policing*. New York: McGraw-Hill, 1990.

Graham, Cole B., Jr., and Steven W. Hays. *Managing the Public Organization*. Washington, D.C.: Congressional Quarterly Press, 1986.

Greene, Jack R., and Stephen D. Mastrofski, eds. *Community Policing: Rhetoric or Reality*. New York: Praeger, 1988.

Gunderson, D. F., and Robert Hopper. *Communication and Law Enforcement*. New York: Harper and Row, 1984.

Hageman, Mary Jeanette. *Police-Community Relations*. Beverly Hills, Calif.: Sage Publications, 1985.

Hale, Charles D. *Police Patrol, Operations, and Management*. New York: John Wiley and Sons, 1981.

Hand, Bruce A., Archible W. Sherman, Jr., and Michael E. Cavanagh. *Traffic Investigation and Control* (2nd ed.). Columbus, Ohio: Charles E. Merrill, 1980.

Holden, Richard N. *Modern Police Management*. Englewood Cliffs, N.J.: Prentice Hall, 1986.

Iannone, N. F. *Supervision of Police Personnel* (4th ed.). Englewood Cliffs, N.J.: Prentice Hall, 1987.

Jackson, Pamela Irving. *Minority Group Threat, Crime, and Policing*. New York: Praeger, 1989.

Jacobs, James B. *Drunk Driving: An American Dilemma*. Chicago: University of Chicago Press, 1989.

Juran, Joseph M. *Juran on Planning for Quality*. New York: Free Press, 1988.

Kelly, Patricia A. *Police and the Media*. Springfield, Ill.: Charles C Thomas, 1987.

Kenney, Dennis Jay, ed. *Police and Policing*. New York: Praeger, 1989.

Knapp Commission (William Knapp, chair). *Report on Police Corruption*. New York: George Braziller, 1973.

Kotter, John P. *The Leadership Factor*. New York: Free Press, 1988.

Kuykendall, Jack L., and Peter C. Unsinger. *Community Police Administration*. Chicago: Nelson-Hall, 1975.

Leinen, Stephen. *Black Police, White Society*. New York: New York University Press, 1984.

Local Government Police Management. Each volume in this series comprises a compendium of information on police administration, with separate chapters written by a multitude of contributors, each of whom is a recognized authority in his or her specialty. Because each "edition" is completely new, prior editions continue to be valuable. The 2nd edition, published in 1982, was edited by Bernard L. Garmire; the 3rd edition, published in 1991, was edited by William A. Geller. An earlier volume, *Municipal Police Administration*, was edited by George Eastman and published in 1974. All are published by the International City Management Association in Washington, D.C.

Lynch, Ronald G. *The Police Manager* (3rd ed.). New York: Random House, 1986.

Marx, Gary T. *Undercover: Police Surveillance in America*. Berkeley, Calif.: University of California Press, 1988.

Miller, Larry, and Michael Braswell. *Human Relations and Police Work*. Prospect Heights, Ill.: Waveland Press, 1988.

Missonellie, Joseph, and James S. D'Angelo. *Television and Law Enforcement*. Springfield, Ill.: Charles C Thomas, 1984.

Mitchell, Terence R., and James R. Larson, Jr. *People in Organizations* (3rd ed.). New York: McGraw-Hill, 1987.

Morgan, James E., Jr. *Administrative and Supervisory Management* (2nd ed.). Englewood Cliffs, N.J.: Prentice Hall, 1982.

Newman, William H., E. Kirby Warren, and Jerome E. Schnee. *The Process of Management* (5th ed.). Englewood Cliffs, N.J.: Prentice Hall, 1982.

Niederhoffer, Arthur, and Alexander B. Smith. *New Directions in Police-Community Relations*. San Francisco: Rinehart Press, 1974.

Potts, Lee W. *Responsible Police Administration*. University, Ala.: University of Alabama Press, 1983.

Ratledge, Edward C., and Joan E. Jacoby. *Handbook on Artificial Intelligence and Expert Systems in Law Enforcement*. New York: Greenwood Press, 1989.

Richardson, James F. *Urban Police in the United States*. Port Washington, N.Y.: Kennikat Press, 1974.

Rieder, Robert J. *Law Enforcement Information Systems*. Springfield, Ill.: Charles C Thomas, 1972.

Saferstein, Richard. *Criminalistics* (3rd ed.). Englewood Cliffs, N.J.: Prentice Hall, 1987.

Saney, Parviz. *Crime and Culture in America*. Westport, Conn.: Greenwood Press, 1986.

Saunders, Charles B., Jr. *Upgrading the American Police*. Washington, D.C.: The Brookings Institution, 1970.

Scheingold, Stuart A. *The Politics of Street Crime*. Philadelphia: Temple University Press, 1991.

Schmalleger, Frank. *Criminal Justice Today*. Englewood Cliffs, N.J.: Prentice Hall, 1991.

Schwartz, Ira M. *(In)Justice for Juveniles*. Lexington, Mass.: Lexington Books, 1989.

Scotti, Anthony. *Police Driving Techniques*. Englewood Cliffs, N.J.: Prentice Hall, 1988.

Skolnick, Jerome H., and David H. Bayley. *The New Blue Line*. New York: Free Press, 1986.

Sparrow, Malcolm K., Mark H. Moore, and David M. Kennedy. *Beyond 911*. New York: Basic Books, 1990.

Spielberger, Charles D. *Police Selection and Evaluation*. Washington, D.C.: Hemisphere, 1979.

Swank, Calvin J., and James A. Conser, eds. *The Police Personnel System*. New York: John Wiley and Sons, 1983.

Thibault, Edward A., Lawrence M. Lynch, and R. Bruce McBride. *Proactive Police Management*. Englewood Cliffs, N.J.: Prentice Hall, 1985.

Trojanowicz, Robert C., and Samuel L. Dixon. *Criminal Justice and the Community*. Englewood Cliffs, N.J.: Prentice Hall, 1974.

Trojanowicz, Robert C., et al. *An Evaluation of the Neighborhood Foot Patrol Program in Flint, Michigan*. East Lansing, Mich.: Michigan State University Press, 1982.

Waldron, Joseph A., Carol S. Sutton, and Terry F. Buss. *Computers in Criminal Justice*. Cincinnati: Pilgrimage Press, 1983.

Walton, Mary. *Deming Management at Work*. New York: Putnam, 1990.

Whisenand, Paul M., and George E. Rush. *Supervising Police Personnel: Back to the Basics*. Englewood Cliffs, N.J.: Prentice Hall, 1988.

Wilson, James Q. *Bureaucracy*. New York: Basic Books, 1989.

In addition to the sources listed previously, we found numerous articles in *Journal of Police Science and Administration*, *FBI Law Enforcement Bulletin*, *Police Studies*, and *Law and Order* to be useful.

INDEX

U

Ultra-high frequency (UHF) band, 233, 236–37
Undercover:
  agent, 408–9
  investigation, 400
  vehicle, 159
Underrepresentation, 274
Uniform Crime Reporting System (UCRS), 193, 204
Uninterruptible power supply (UPS), 234
Union, 458
Union-like organization, 458
Unit, 65
Unity of command, 51–53
Unlicensed drivers, 386–87

V

Vehicle file, 201
Vehicle maintenance services, 431
Very high frequency (VHF) band, 233, 236–37
Video, 206

surveillance system, 241
  tape recording, 241–42
  teleconferencing, 242
Visual arrest records, 241

W

"Walk-in" police station, 150
Walkie-talkies, 234
Ward bosses, 7–8
Warning, 376, 378, 379
Watch, 65
Watchmen, 5
Women, recruiting, 276–78
Word processing program, 215–16
Work:
  flow process chart, 31
  product measurement, 304
Working class, 438
Workstation, 212–13

Z

Zero-base budget method, 123–27